Informatics in
Radiation Oncology

IMAGING IN MEDICAL DIAGNOSIS AND THERAPY

William R. Hendee, Series Editor

Forthcoming titles in the series

Informatics in Radiation Oncology

Edited by
George Starkschall
R. Alfredo C. Siochi

CRC Press
Taylor & Francis Group
Boca Raton London New York

CRC Press is an imprint of the
Taylor & Francis Group, an **informa** business

A TAYLOR & FRANCIS BOOK

CRC Press
Taylor & Francis Group
6000 Broken Sound Parkway NW, Suite 300
Boca Raton, FL 33487-2742

First issued in paperback 2020

© 2014 by Taylor & Francis Group, LLC
CRC Press is an imprint of Taylor & Francis Group, an Informa business

No claim to original U.S. Government works

Version Date: 20130715

ISBN 13 : 978-0-367-57633-2 (pbk)
ISBN 13 : 978-1-4398-2582-2 (hbk)

Library of Congress Cataloging-in-Publication Data

Informatics in radiation oncology / edited by George Starkschall, R. Alfredo C. Siochi.
 p. ; cm. -- (Imaging in medical diagnosis and therapy)
 Includes bibliographical references and index.
 ISBN 978-1-4398-2582-2 (hardcover : alk. paper)
 I. Starkschall, George, editor of compilation. II. Siochi, R. Alfredo C. (Ramon Alfredo Carvalho), editor of compilation. III. Series: Imaging in medical diagnosis and therapy.
 [DNLM: 1. Medical Informatics. 2. Neoplasms--radiotherapy. QZ 26.5]

RC271.R3
616.99'40642--dc23
                                  2013027796

Visit the Taylor & Francis Web site at
http://www.taylorandfrancis.com

and the CRC Press Web site at
http://www.crcpress.com

Contents

SECTION I Introduction

SECTION II Information and the Radiation Oncology Process

Series Preface

Advances in the science and technology of medical imaging and radiation therapy are more profound and rapid than ever before, since their inception over a century ago. Further, the disciplines are increasingly cross-linked as imaging methods become more widely used to plan, guide, monitor, and assess treatments in radiation therapy. Today, the technologies of medical imaging and radiation therapy are so complex and so computer-driven that it is difficult for the persons (physicians and technologists) responsible for their clinical use to know exactly what is happening at the point of care, when a patient is being examined or treated. The persons best equipped to understand the technologies and their applications are medical physicists, and these individuals are assuming greater responsibilities in the clinical arena to ensure that what is intended for the patient is actually delivered in a safe and effective manner.

The growing responsibilities of medical physicists in the clinical arenas of medical imaging and radiation therapy are not without their challenges, however. Most medical physicists are knowledgeable in either radiation therapy or medical imaging and are experts in one or a small number of areas within their discipline. They sustain their expertise in these areas by reading scientific articles and attending scientific talks at meetings. In contrast, their responsibilities increasingly extend beyond their specific areas of expertise. To meet these responsibilities, medical physicists periodically must refresh their knowledge of advances in medical imaging or radiation therapy, and they must be prepared to function at the intersection of these two fields. How to accomplish these objectives is a challenge.

At the 2007 annual meeting of the American Association of Physicists in Medicine in Minneapolis, this challenge was the topic of conversation during a lunch hosted by Taylor & Francis Publishers and involving a group of senior medical physicists (Arthur L. Boyer, Joseph O. Deasy, C.-M. Charlie Ma, Todd A. Pawlicki, Ervin B. Podgorsak, Elke Reitzel, Anthony B. Wolbarst, and Ellen D. Yorke). The conclusion of this discussion was that a book series should be launched under the Taylor & Francis banner, with each volume in the series addressing a rapidly advancing area of medical imaging or radiation therapy of importance to medical physicists. The aim would be for each volume to provide medical physicists with the information needed to understand technologies driving a rapid advance and their applications to safe and effective delivery of patient care.

Each volume in the series is edited by one or more individuals with recognized expertise in the technological area encompassed by the book. The editors are responsible for selecting the authors of individual chapters and ensuring that the chapters are comprehensive and intelligible to someone without such expertise. The enthusiasm of volume editors and chapter authors has been gratifying and reinforces the conclusion of the Minneapolis luncheon that this series of books addresses a major need of medical physicists.

Imaging in Medical Diagnosis and Therapy would not have been possible without the encouragement and support of the series manager, Luna Han of Taylor & Francis Publishers. The editors and authors, and most of all I, are indebted to her steady guidance of the entire project.

William R. Hendee, Series Editor
Rochester, MN

Preface

The vast array of patient information acquired during the diagnosis and treatment of disease has created a large reservoir of data. Laboratory results, medical images, medication lists, allergies, and patient histories are among the many sets of data that have moved from paper to digital media, making the information easier to find and access. The move to electronic medical records (EMRs) has also made it more convenient to store large amounts of data, including previously uncaptured data, further opening the floodgates for these deepening pools of data. Government mandates to transition to EMRs will hasten the increase in the amount of data, and a growing number of clinical trials will prompt the collection of even more data.

This explosion of data has become very obvious in radiation oncology, where electronic databases are used to automate and verify the correct delivery of radiation treatments. These "record and verify" systems have been around much longer than EMRs and have lived side-by-side with paper charts for almost two decades. This long history has provided a wealth of experience that has led to the development of the modern radiation oncology treatment management system (TMS). Part EMR, part medical delivery device, the TMS stores not only patient demographics but also every single movement and combination of treatment machine settings for radiation treatments. With the increased complexity in treatments, beginning with intensity-modulated radiation therapy (IMRT) and continuing to such delivery techniques as volume-modulated arc therapy (VMAT), TMS has become essential. It is impossible to manually program contemporary dynamic delivery techniques, because they may have on the order of 100,000 parameters. Moreover, the treatment is only one part in the entire chain of the radiotherapy process. Imaging, simulation, planning, immobilization, localization, and image guidance contribute to the data stored in radiotherapy databases and picture archiving systems. With approximately 5000 treatments per accelerator per year, with potentially half of them receiving daily 3D imaging for patient setup, it is no wonder that data storage systems are now measured in multiples of terabytes.

With all these data at our fingertips, we are poised to move into the next phase of the transition into a paperless environment. For the most part, however, we have merely replaced the media that captures the data, mimicking paper forms in a digital environment. Efforts to tame this flood of data and harness its

power are enabling us to discover knowledge that could make our treatments more effective and our processes more efficient.

We will see improvements in healthcare as we process the data in new ways, automate systems for safety and efficiency, mine the data, and improve data access (e.g., through the Web). These improvements also demand new methodologies for security, privacy, and compliance with laws such as the Health Insurance Portability and Accountability Act (HIPAA). The field of radiation oncology can benefit from experts in informatics, as they apply their skills to our knowledge domain. Informatics in radiation oncology is, in many ways, still in its infancy. Many of the authors of this book have learned informatics on the job, but we need informaticists to become familiar with radiation oncology. This book is the first of its kind and fills this gap, with the hope of strengthening the collaborations between the domain knowledge experts and informaticists, enabling the full use of all the data that are being gathered in our data-rich environment. Similarly, this book will help researchers and physicists in radiation oncology deepen their knowledge of the information technology and informatics principles being applied to our field.

In the Introduction section, the first three chapters (1–3) describe the basics of informatics in general, while the next three chapters explore their connection to radiation oncology: where and how the data are stored (4), their relationships and connections to other parts of the hospital (6), and the terminology that facilitates the application of informatics to our field (5). The second section examines the process of healthcare delivery in radiation oncology: how information flows from one step to the next (7), the logical and physical connections among the various steps (10), and the resources we need and how to find them (11). Within that process, one can see our dependence on imaging and the benefits of working with radiology informatics (8). Drawing on their experience would be helpful since the image-guided radiotherapy era has brought with it the challenges of managing images in radiotherapy (9). A brief foray into the emerging field of radiogenomics (12) concludes the section.

The remaining sections dive deeper into the processes within radiotherapy, showing how we can benefit from applying informatics principles to them. Section III looks at how we share the results of our work, discussing teaching (13), clinical trials (14), information sharing and data federation (15), and tools that have been developed to promote informatics research in radiation oncology (16). Section IV examines imaging, describing the implications for open access clinical imaging archives in radiotherapy (17), the techniques for maximizing information from combinations of various types of images (18), and the roles of images in planning treatments (19). Informatics can also be applied to the improvement of our processes, as seen in Section V where we evaluate the quality and consistency of our treatment plans (20), design human interfaces to improve the safety and efficiency of our delivery systems (21), use images to guide the proper positioning of patients (22), and make use of tools for patient assessment (23). The last section shows us how we can ensure that the work we do achieves our goals. We have learned more about the effectiveness of our treatments by modeling outcomes (24) and improved the reliability of our processes through quality control informatics (25). Last but not least, we need to know that our informatics tools provide the knowledge we seek by performing quality assurance on the informatics environment (26).

It is our hope that the work presented in this book will inspire others to move the field forward, foster stronger collaborations between informatics experts and radiation oncology professionals, find new ways of applying informatics to radiation oncology,

and determine how best to serve our patients, partnering with them in their fight against cancer.

The authors are grateful to Bill Hendee for giving us the opportunity to work on this book. R. Alfredo C. Siochi also wishes to thank John Buatti, John Bayouth, and the Department of Radiation Oncology at the University of Iowa for building a strong clinical and academic environment that provides him the time and resources to work on projects like this. He is thankful for his wife Ann and his children Jeremiah and Katherine, who patiently waited for him to come home to a late dinner on many nights; for his mother, who taught him the value of hard work; for his brothers and sister who fostered a love of computing; and for his father, who was still developing databases and writing software in his seventies, long after retirement — he inspires him to keep learning new things. George Starkschall wishes to thank his colleagues in both the Department of Radiation Physics and the Department of Imaging Physics at the University of Texas MD Anderson Cancer Center for providing him with the intellectual environment to work on this book. He also wishes to thank Bruce Curran for some of the initial discussions that led to the development of the book. Both authors wish to thank our editor, Luna Han, for helping us take this book to publication, making our lives easier in the process.

Editors

George Starkschall, PhD, FAAPM, FACR, is a research professor in the Department of Imaging Physics and a distinguished senior lecturer in the Graduate School of Biomedical Sciences at The University of Texas MD Anderson Cancer Center. He earned his PhD degree in chemical physics from Harvard University and had a post-doctoral fellowship in the James Franck Institute at The University of Chicago. He is certified by the American Board of Radiology in therapeutic radiological physics and is a fellow of both the American Association of Physicists in Medicine and the American College of Radiology. He was formerly the editor-in-chief of the *Journal of Applied Clinical Medical Physics* and is presently the executive secretary of the Commission on Accreditation of Medical Physics Educational Programs. He is the recipient of numerous awards and honors, most recently the Edith H. Quimby Lifetime Achievement Award of the American Association of Physicists in Medicine.

R. Alfredo Siochi, MS, PhD, DABR, earned his PhD degree in physics from Virginia Tech in 1990 and his MS degree in radiological physics from the University of Cincinnati in 1995. He holds over 20 patents in radiotherapy and was selected as a Siemens Inventor of the Year in 2000. He has more than 40 publications in the field of medical physics, with a majority of them covering algorithm development in the subfields of medical image analysis, treatment delivery optimization, and informatics. He has lectured internationally, with over 60 presentations. He developed 17 (and counting) in-house software applications in clinical use. He is the director of Medical Physics Education and IT Operations in the Radiation Oncology Department at the University of Iowa. He is also the chair of the AAPM Working Group on IT and a co-chair of the Radiation Safety Stakeholders Initiative where he designed and manages the group's IT infrastructure. His research interests are in the areas of 4D radiation therapy, improved QA in paperless environments, and process improvement informatics.

Contributors

Bryan Allen
Department of Radiation Oncology
University of Iowa Hospitals and
    Clinics
Iowa City, Iowa

Aditya P. Apte
Department of Medical Physics
Memorial Sloan-Kettering Cancer
    Center
New York, New York

Peter Balter
University of Texas MD Anderson
    Cancer Center
Houston, Texas

John E. Bayouth
Department of Radiation Oncology
University of Iowa Hospitals and Clinics
Iowa City, Iowa

Collin D. Brack
The University of Texas Medical Branch
Galveston, Texas

Scott Brame
Radialogica LLC
Clayton, Missouri

Steven Conners
Department of Radiation Oncology
Mayo Clinic
Rochester, Minnesota

Joseph O. Deasy
Department of Medical Physics
Memorial Sloan-Kettering Cancer
    Center
New York, New York

Issam El Naqa
Medical Physics Unit
Department of Oncology
McGill University
Quebec City, Canada

Robert J. Esterhay
Department of Health Management and
    Systems Sciences
School of Public Health and Information
    Sciences
University of Louisville
Louisville, Kentucky

Timothy H. Fox
Radiation Oncology
Emory University School of Medicine
Atlanta, Georgia

John Freymann
Clinical Research Directorate
Clinical Monitoring Research Program
SAIC-Frederick, Inc., NCI-Frederick
Frederick, Maryland

Joseph M. Herman
Department of Radiation Oncology and
    Molecular Radiation Sciences
The Johns Hopkins University
Baltimore, Maryland

Michael G. Herman
Division of Medical Physics
Department of Radiation Oncology
Mayo Clinic
Rochester, Minnesota

Yu-Chi Hu
Department of Medical Physics
Memorial Sloan-Kettering Cancer Center
New York, New York

C. Carl Jaffe
Boston University School of Medicine
Boston, Massachusetts

George C. Kagadis
Department of Medical Physics
School of Medicine
University of Patras
Rion, Greece

Ivan L. Kessel
Department of Radiation Oncology
University of Texas Medical Branch
Galveston, Texas

Peter Kijewski
Department of Medical Physics and
Department of Radiology
Memorial Sloan-Kettering Cancer Center
New York, New York

Justin S. Kirby
Clinical Research Directorate
Clinical Monitoring Research Program
SAIC-Frederick, Inc., NCI-Frederick
Frederick, Maryland

Steve G. Langer
Department of Radiology
Mayo Clinic
Rochester, Minnesota

Frank Lohr
Department of Radiation Oncology
University Medical Center Mannheim
University of Heidelberg
Heidelberg, Germany

Gig S. Mageras
Department of Medical Physics
Memorial Sloan-Kettering Cancer Center
New York, New York

Charles Mayo
Department of Radiation Oncology
Mayo Clinic
Rochester, Minnesota

Stephen McNamara
Department of Medical Physics
Memorial Sloan-Kettering Cancer Center
New York, New York

Todd McNutt
Department of Radiation Oncology and
    Molecular Radiation Sciences
The Johns Hopkins University
Baltimore, Maryland

Daniel L. McShan
Department of Radiation Oncology
University of Michigan
Ann Arbor, Michigan

Alexis Andrew Miller
Radiation Oncology
Illawarra Cancer Care Centre
and
Centre for Oncology Informatics
University of Wollongong
Wollongong, New South Wales, Australia

Michael D. Mills
Department of Radiation Oncology
School of Medicine
University of Louisville
Louisville, Kentucky

Kevin L. Moore
Department of Radiation Medicine and
    Applied Sciences
University of California, San Diego
La Jolla, California

Sasa Mutic
Department of Radiation Oncology
Mallinckrodt Institute of Radiology
Siteman Cancer Center
Washington University School of Medicine
St. Louis, Missouri

Paul G. Nagy
The Johns Hopkins University
Baltimore, Maryland

Hai Pham
Department of Medical Physics
Memorial Sloan-Kettering Cancer Center
New York, New York

Joann I. Prisciandaro
University of Michigan
Department of Radiation Oncology
Division of Radiation Physics
Ann Arbor, Michigan

Fred Röhner
Department of Radiation Oncology
University Medical Center Freiburg
University of Freiburg
Freiburg, Germany

Yusuf N. Saghar
Department of Biomedical Informatics
Center for Comprehensive Informatics
Emory University
Atlanta, Georgia

Joel H. Saltz
Department of Biomedical Informatics
Center for Comprehensive Informatics
Emory University
Atlanta, Georgia

Luis Fong de los Santos
Division of Medical Physics
Department of Radiation Oncology
Mayo Clinic
Rochester, Minnesota

Frank Schneider
Department of Radiation Oncology
University Medical Center Mannheim
University of Heidelberg
Heidelberg, Germany

Eduard Schreibmann
Radiation Oncology
Emory University School of Medicine
Atlanta, Georgia

Ashish Sharma
Department of Biomedical Informatics
Center for Comprehensive Informatics
Emory University
Atlanta, Georgia

R. Alfredo Siochi
Department of Radiation Oncology
University of Iowa Hospitals and Clinics
Iowa City, Iowa

George Starkschall
University of Texas MD Anderson
    Cancer Center
Houston, Texas

Volker Steil
Department of Radiation Oncology
University Medical Center Mannheim
University of Heidelberg
Heidelberg, Germany

Stuart Swerdloff
OIS Particle Therapy
Elekta AB
Stockholm, Sweden

Mike Tilkin
American College of Radiology
Reston, Virginia

Gerald Weisser
Institute of Clinical Radiology and
    Nuclear Medicine
University Medical Center Mannheim
University of Heidelberg
Heidelberg, Germany

Frederik Wenz
Department of Radiation Oncology
University Medical Center Mannheim
University of Heidelberg
Heidelberg, Germany

Junyi Xia
Department of Radiation Oncology
University of Iowa Hospitals and Clinics
Iowa City, Iowa

Ying Xiao
Department of Radiation Oncology
Jefferson Medical College
Philadelphia, Pennsylvania

Jian-Ping Xiong
Department of Medical Physics
Memorial Sloan-Kettering Cancer Center
New York, New York

Deshan Yang
Department of Radiation Oncology
Washington University in St. Louis
St. Louis, Missouri

Introduction

1

What Is Biomedical Informatics?

Paul G. Nagy
The Johns Hopkins University

The field of biomedical informatics broadly encompasses the use of information in medical decision-making. Applied informaticians are involved with the specific uses of healthcare information technology (IT) to transform clinical practice. Researchers in informatics work to connect diverse areas of biomedical research to enhance knowledge and enable discovery (for example, in the evaluation of new drugs and treatments for disease). Shortliffe and Cimino (2006) assigned a formal definition to the field as the "discipline concerned with the study and application of computer science, information science, informatics, cognitive science and human–computer interaction in the practice of biological research, biomedical science, medicine, and healthcare." Figure 1.1 illustrates informatics as being at the intersection of technology, people, and information. The field focuses on the ways in which individuals can use evolving technologies to understand information in pursuing cognitive tasks (Friedman 2009; Hersh 2009). A common misunderstanding is that informatics as a field includes anything and everything with which computer support is associated (Friedman et al. 2001). As a result, current informatics experts work to make it clear that activities such as desktop support, systems administration, and basic computer tasks are not included in the purview of informatics (Bernstam et al. 2009).

1.1 History of Biomedical Informatics

Biomedical informatics as a field had its origins in the 1950s and came into focus with a series of articles by Ledley and Lusted published in the journal *Science* in 1959. These authors described the potential for the practical use of a systems approach in computer applications to provide medical diagnoses (Ledley 1959; Ledley & Lusted 1959a, 1959b). The growth of the biomedical informatics

field in the United States since that time has been advanced by several national initiatives and regulations. The Joint Commission of Accredited Hospitals was formed in 1951, and the Social Security Act of 1965 created the Medicare and Medicaid programs (Barton 2008). Both initiatives led healthcare organizations to invest in information systems to comply with reimbursement and quality requirements (Collen 1995; Sackett & Erdley 2002). In 1960, Ledley published a National Research Council report on the use of computers in biology and medicine (Ledley 1960). That report provided a comprehensive discourse on how medical scientists should be trained in computer science to assist in biomedical research. This report contributed to efforts by the National Institutes of Health (NIH) to support the creation of programs to develop academic biomedical computer centers and the first graduate programs in biomedical informatics (Ledley 1990).

The expansion in knowledge about biomedical informatics has paralleled the exponential growth in the IT market. In 1987, the term "medical informatics" was added as a Medical Subject Heading (MeSH) in the National Library of Medicine's (NLM) controlled vocabulary. During the two decades since then, the number of articles under this term has grown at an average of twelve percent per year, exceeding 10,000 publications in 2007 (DeShazo et al. 2009).

1.2 Components of Biomedical Informatics Training

To understand the range of activities in which an informatician may be involved, it is helpful to look at the training and curricula now offered under the heading of informatics. Significant efforts have been made over the past ten years to clarify and standardize

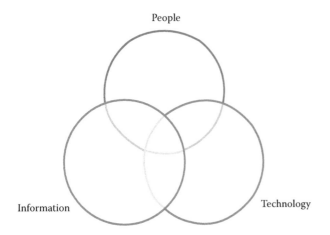

FIGURE 1.1 Informatics represents the intersection of people, information, and technology.

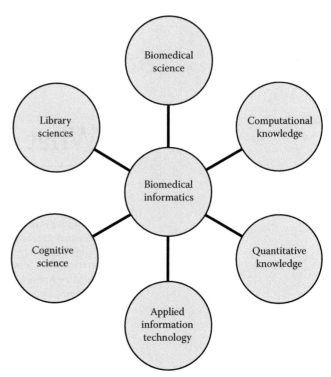

FIGURE 1.2 Training areas in informatics.

the core domains of knowledge required for training. In 1999, the NIH released a report identifying the need to elucidate and formalize the current and potential roles of biomedical informatics researchers—the Biomedical Information Science and Technology Initiative (Working Group on Biomedical Computing 2009). In 2004, using the NIH report as a model, the American College of Medical Informatics identified the specific application domains needed to build a combined training program in biology and computer science (Friedman et al. 2004). Although domains and data standards differ greatly, basic researchers in bioinformatics share common approaches and tools with clinically oriented informaticians (Johnson & Friedman 2007). Areas that typically are associated with informatics training include biomedical sciences, quantitative knowledge, computational knowledge, cognitive studies, library sciences, and applied information technologies (Figure 1.2). Each of these areas is reviewed briefly in the remainder of this section.

1.2.1 Biomedical Sciences

An informatician must have expert domain knowledge of the area in which he or she works to collaborate effectively with physicians and clinical researchers. Many informaticians take classes alongside medical students in biology and physiology and then go on to specialize in molecular biology or genetics for bioinformatics research or in public health or a medical discipline (or both) for more applied applications. To successfully build ontologies of knowledge, one must understand the relationships between entities and the nuance of differentiation. This requires considerable domain knowledge and experience in the field.

1.2.2 Quantitative Knowledge

Foundations in mathematics and statistics are needed to conduct effective research projects. Bayesian statistical methods are often

central to evidence-based medicine and clinical trial research. Numerical methods, simulations, and modeling are basic tools used by researchers in informatics and require a strong background in mathematics. Informatics researchers should also have expertise in quasi-experimental study design to minimize the effects of confounding factors (Harris et al. 2006).

1.2.3 Computational Knowledge

Informaticians should be fluent in at least one programming language (Berman 2007). Software development can include knowledge areas, such as natural language processing, neural networks, clustering analysis, genetic algorithms, Monte Carlo simulations, and many others (Altman 1998). Along with programming skills, an informatician should be adept in data manipulation. Data mining skills should include extraction, transformation, and loading from a variety of clinical and research data sources. Software development skills should include modern information architecture principles of service-oriented architectures.

1.2.4 Cognitive Studies in the Social and Organizational Uses of Information and Technology

Disciplines relevant to the field of informatics include human–computer interaction, information visualization, and organizational psychology. Human–computer interaction studies look at the many ways in which humans can interface with information systems and include everything from ergonomics to contextual inquiry, cognitive modeling, and behavioral theory (Sears & Jacko 2007). Usability analysis is a subdiscipline of human–computer interaction that works to make interfaces familiar and simple. Cognitive studies also focus on ways in which information can be easily disseminated from computer systems to users (Nielsen & Loranger 2006). Information visualization is a modern interdisciplinary field that has emerged from statistics, graphic design, and cognitive and behavioral sciences to focus on the best ways to represent information (Tufte 1990). Informaticians can use organizational psychology principles to effect change in an organization and understand the behavioral implications of information systems that are put into a work environment (Dunnette 1976).

1.2.5 Library Sciences

The library sciences study the classification, representation, archiving, and retrieval of information. In 1964, the NLM "went live" with the Medical Literature Analysis and Retrieval Systems, which catalogued and produced the *Index Medicus* (Schoolman and Lindberg 1988). By the late 1960s, more than 16,000 queries about more than 125,000 citations were being made annually of the Medical Literature Analysis and Retrieval Systems system. MeSH, a structured and controlled vocabulary, was created to help classify the citations. Today, PubMed, the public Web system from NLM with access to MEDLINE, has MeSH catalogued more than 19 million citations from more than 5000 journals and, on a typical day, handles more than 2 million electronic queries (Herskovic et al. 2007). One of the major challenges on which informaticists focus is knowledge representation. The Unified Medical Language Syntax project is an NLM initiative to cluster common terms into a common concept model and to define semantic relationships between those concepts to facilitate knowledge discovery and decision support.

1.2.6 Applied Information Technologies

IT is at the heart of the tools used by informaticians. Practical knowledge of IT systems is critical when bioinformaticians must run several hundred node processing farms to analyze genetic microarray data. Clinical informaticians must also have a strong practical knowledge of IT systems to design mission-critical, fault-tolerant IT architectures to meet the high availability demands of clinical information systems for the delivery of continuous care. A practical knowledge base of IT should include aspects of networking, server, storage, processing, and display technologies. Informaticians work with data encoding methodologies, such as Extensible Markup Language, as well as databases requiring proficiency with Standard Query Language. A large body of management science literature is available on methods for professionally delivering and supporting IT systems. The Information Technology Infrastructure Library (Behr et al. 2005) is a set of systems management practices that helps organizations deliver consistent service through activities such as configuration management, availability monitoring, change management, problem management, and capacity management.

1.3 Specialty Areas of Biomedical Informatics

The field of biomedical informatics has grown over the past fifty years to permeate every component of medicine, from basic research to the clinical environment. Specialized areas of informatics have grown to meet the unique needs of separate disciplines. Several example areas derived from the work of Shortliffe et al. (2006) are shown in Figure 1.3 and are described briefly in the paragraphs that follow. This list is not exclusive but is illustrative of the major "silos" of informatics expertise that have evolved to meet unique domain requirements.

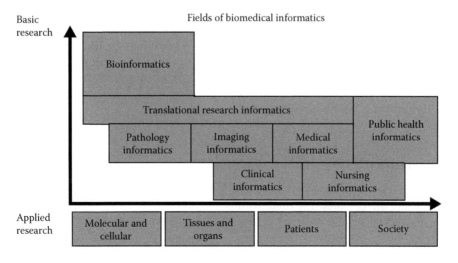

FIGURE 1.3 Fields within biomedical informatics.

1.3.1 Bioinformatics

Bioinformatics supports medical researchers in cellular and molecular biology with a focus on studying genetic expression patterns in human diseases (Butte 2008). High-throughput molecular measurement technology, such as that used in gene expression microarrays, has created an information explosion, with more than 105 billion nucleotides and 106 million sequences archived at the NIH GenBank since its inception in 1982 (Genetic Sequence Data Bank 2009). Leveraging the power of genetic information through bioinformatics is powering advancements into personalized medicine.

1.3.2 Clinical Translational Research Informatics

Clinical translational research informaticians help facilitate the work of investigators conducting clinical trials and clinical research. These informaticians play a crucial role in achieving the NIH's goal of accelerating research and innovations from the laboratory bench to the clinical bedside. They often assist researchers in managing requirements from institutional review boards to maintain human research protections by implementing "honest broker" systems. These systems deidentify protected health information from clinical records and assist in compliance with confidentiality regulations such as those of the Health Insurance Portability and Accountability Act (Office for Civil Rights, Department of Health and Human Services 2002). A major goal in translational informatics is the ability to aggregate data from isolated clinical trials into common data standards and develop reliable methods of comparability and exchange (Fridsma et al. 2008; Prokosch & Ganslandt 2009).

1.3.3 Pathology Informatics

The area of pathology informatics focuses on anatomic and cellular imaging as well as laboratory information systems in the delivery of clinical care (Sinard 2005). More than five billion laboratory tests are conducted within the United States each year (Berman 2007). The Systemized Nomenclature of Medicine—Clinical Terms is a comprehensive multilingual list of more than 311,000 clinical terms and concepts. Systemized Nomenclature of Medicine was developed by the College of American Pathologists and is currently supported by the International Health Terminology Standards Development Organisation (http://www.ihtsdo.org/snomed-ct/snomed-ct0/).

1.3.4 Nursing Informatics

Nurses trained in informatics help to address cultural barriers between the clinical world and the IT world. The commercial development of software for clinical environments has suffered from a lack of understanding of the complexity of routine workflow and medical decision-making. To paraphrase Wears and Berg (2005), software is normally designed as a linear, rational, and solitary set of functions, but the clinical world is often an interruptive, collaborative, and reactive environment. Nurse informaticians understand the use of information in decision-making in a clinical environment. They are typically extremely effective at helping to tailor clinical information systems to the needs of a variety of participants in the clinical workflow environment.

1.3.5 Public Health Informatics

Public health informaticians are involved in studying telehealth initiatives to leverage IT and medical home monitoring systems outside of the clinical setting. As computer literacy rises among patient populations, chronic disease management strategies have been shown to benefit by connecting virtual communities of patients. Patient-oriented medical record systems have empowered some patients to take control of their own health. Public health informaticians often use large sets of administrative claims data to study the effectiveness of treatment on healthcare costs and recurrence of disease, accidents, or other adverse events. Informaticians work on clinical data standards to allow medical records to be portable between medical institutions. Regional health information organizations have been piloted to allow cross-institutional exchange of electronic medical records.

1.3.6 Imaging Informatics

Imaging informaticians study the use of medical imaging in decision-making and delivery of care. Imaging informaticians have a specialized knowledge of the biomedical engineering and physics surrounding the image acquisition process as well as specialization in computer algorithms designed for image processing and machine vision–based (as opposed to textual syntactic) segmentation algorithms. The principal standard in medical imaging is the Digital Imaging and Communication in Medicine (DICOM) standard. The DICOM standard, developed by the National Electrical Manufacturers Association, defines the binary format for all medical imaging modalities along with methods for communicating those images within a picture archiving and communication system. The "home" for imaging informatics in the United States is the Society for Imaging Informatics in Medicine, which provides guidelines for fellowships in imaging informatics.

1.3.7 Clinical Informatics

Clinical informaticians are trained to lead large institutional IT programs such as computerized physician order entry and electronic medical record projects. Hospitals often create specialized administrative oversight (chief medical information officer and chief information officer) to direct these activities. Clinical informaticians are also closely involved with IT departments within a hospital or medical system. The American Medical Informatics Association created a certification program for clinical informaticians in 2009. The predominant standard used for the communication of medical information through clinical information systems in hospitals is Health Level 7, a text-based transactional messaging standard for broadcasting orders, reports, and patient demographic information.

1.3.8 Medical Informatics

Medical informatics focuses on the study of the use of information in the diagnostic process. Medical informaticians created the field of decision support in medicine, providing strategic tools with which clinicians can determine probabilistic pathologies based on indications. Medical informaticians are strong proponents in the evidence-based medicine movement and work to enable physicians to identify and disseminate best practices throughout the medical community. The American Medical Informatics Association, formed in 1979, is home to many of the specialties of medical informatics, including clinical, public health, and clinical translational research informatics.

1.3.9 Oncology Informatics

Oncology informatics lies at the crossroads of several major informatics disciplines. Clinical trial informatics is used extensively in cancer clinical trials to identify and evaluate new treatment techniques. Imaging informatics is applied in the analysis of tumors, and image processing and registration techniques can be used to detect changes in tumor morphology over time. Pathology informatics involves both imaging and cataloguing the cellular behavior of cancer. Oncology informatics is a perfect example of how researchers need to bridge several domains of informatics, with their corresponding knowledge structures, to be truly effective (Figure 1.4). The U.S. National Cancer Institute is addressing this challenge by creating the Cancer Biomedical Informatics Grid, with the goal of building an enterprise architecture for the exchange of cancer research data from these disparate domains (Oster et al. 2007).

1.4 The Medical Physicist and Informatics

In radiation oncology departments, the scientist with the training most complementary to informatics is the therapeutic medical physicist (Siochi et al. 2009). Informaticians in oncology are still scarce, and the informatics responsibilities often fall to the physicist, who already has experience as a liaison between physicians and IT support. Medical physicists should play an important role in the selection and implementation of clinical information systems in radiation oncology (Nikiforidis et al. 2006). A physicist is educated to apply physics to the medical environment, and an informatician is trained in the application of computer science. Both are scientists who must have expert knowledge of biological systems and the healthcare environment to be successful, and many of these knowledge areas are complementary. Many medical physicists have considerable computer programming skills gained during graduate work in conducting simulations or image analysis. Many physicists have contributed to the development of the DICOM standards. Physicists have also played prominent roles in the Society for Imaging Informatics in Medicine and on the IT committees of major radiology and radiation oncology professional organizations. It is not uncommon for physicists to oversee or participate with operations IT professionals in a department or practice.

Physicists with an interest in developing informatics skills should understand that although a physics education provides many of the prerequisites, additional areas of expertise must be acquired to become a successful informatician. Physicists without an awareness of these areas can encounter several kinds of pitfalls in implementing or developing tools in the clinical environment. Common pitfalls include underestimating the support cost and tools required to maintain a system in a clinical environment. Prototype research applications developed by graduate students are often unsupportable after the student's graduation. Having a foundation in systems management principles can be valuable, especially when the

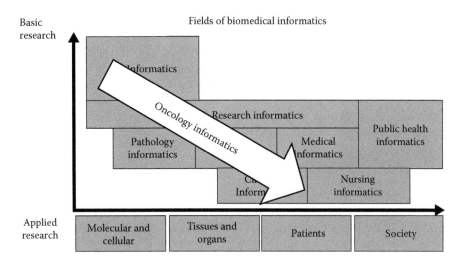

FIGURE 1.4 The field of oncology informatics bridges multiple domains of informatics.

physicist is involved with operations IT personnel or commercial IT vendors. Another potential pitfall is lack of awareness of common data elements in a data set that might prevent its integration with other information sources based on data standards. As computer scientist Andrew Tanenbaum has said, "The nice thing about standards is that you have so many to choose from" (Tanenbaum 1988). Finding the appropriate information model is a critical step in the development of any information system.

Acknowledgment

The author thanks Dr. Nancy Knight from the University of Maryland for her expert assistance in preparing this manuscript.

References

Altman, R. (1998). A curriculum for bioinformatics: The time is ripe (editorial). *Bioinformatics 14*, 549–550.

Barton, H. I., Health informatics, in *Encyclopedia of Library and Information Science*, 2nd edition, Drake, M., Ed., Taylor & Francis, Boca Raton, Florida, 2008, 1179.

Berman, J., *Biomedical Informatics*, Jones and Bartlett, Boston, Massachusetts, 2007.

Behr, K. et al., *The Visible Ops Handbook: Implementing ITIL in 4 Practical and Auditable Steps*, Information Technology Process Institute, Eugene, Oregon, 2005.

Bernstam, E. V. et al. (2009). Synergies and distinctions between computational disciplines in biomedical research: Perspective from the Clinical and Translational Science Award programs. *Academic Medicine, 84*, 964–970.

Butte, A. J. (2008). Translational bioinformatics: Coming of age. *Journal of the American Medical Informatics Association, 15*, 709–714.

Collen, M. F., *A History of Medical Informatics in the United States 1950–1990*, American Medical Informatics Association, Indianapolis, Indiana, 1995.

DeShazo, J. P. et al. (2009). Publication trends in the medical informatics literature: 20 years of "Medical Informatics" in MeSH. *BMC Medical Informatics and Decision Making, 9*, 7.

Dunnette, M., Ed., *Handbook of Industrial and Organizational Psychology*, Rand McNally and Co., Chicago, Illinois, 1976.

Fridsma, D. B. et al. (2008). The BRIDG project: A technical report. *Journal of the American Medical Informatics Association, 15*, 130–137.

Friedman, C. P. (2009). A "fundamental theorem" of biomedical informatics. *Journal of the American Medical Informatics Association, 16*, 169–170.

Friedman, C. P. et al. (2001). Toward a new culture of biomedical informatics. Report of the ACMI Symposium. *Journal of the American Medical Informatics Association, 8*, 519–526.

Friedman, C. P. et al. (2004). Training the next generation of informaticians: The impact of the "BISTI" and Bioinformatics—A report from the American College of Medical Informatics. *Journal of the American Medical Informatics Association, 11*, 167–172.

Genetic Sequence Data Bank (2009). NCBI-GenBank Flat File Release 173.0 Distribution Release Notes. August 15 2009. Available at ftp://ftp.ncbi.nih.gov/genbank/gbrel.txt. Accessed September 22, 2009.

Harris, A. D. et al. (2006). The use of interpretation of quasi-experimental studies in medical informatics. *Journal of the American Medical Informatics Association, 13*, 16–23.

Herskovic, J. R. et al. (2007). A day in the life of PubMed: Analysis of a typical day's query log. *Journal of the American Medical Informatics Association, 14*, 212–220.

Hersh, W. (2009). A stimulus to define informatics and health information technology. *BMC Medical Informatics and Decision Making, 9*, 24.

Johnson, S. & Friedman, R. (2007). Bridging the gap between biological and clinical informatics in a graduate training program. *Journal of Biomedical Informatics, 40*, 59–66.

Ledley, R. (1959). Digital electronic computers in biomedical science. *Science, 130*, 1225–1234.

Ledley, R., *Report on the Use of Computers in Biology and Medicine*, National Academy of Sciences National Research Council, Washington, District of Columbia, 1960.

Ledley, R. S., Medical informatics: A personal view of sowing the seeds, in *A History of Medical Informatics*, Blum, B. I. & Duncan, K., Eds., Association of Computing Machinery (ACM) Press History Series, ACM Press, New York, 1990, 84–110.

Ledley, R. & Lusted, L. (1959a). Probability, logic and medical diagnosis. *Science, 130*, 892–930.

Ledley, R. & Lusted, L. (1959b). Reasoning foundations of medical diagnosis. *Science, 130*, 9–21.

Nielsen, J. & Loranger, H., *Prioritizing Web Usability*, New Riders Press, Berkley, CA, 2006, xvi.

Nikiforidis, G. C. et al. (2006). Point/counterpoint. It is important that medical physicists be involved in the development and implementation of integrated hospital information systems. *Medical Physics, 33*, 4455–4458.

Office for Civil Rights, Department of Health and Human Services (2002). Standards for Privacy of Individually Identifiable Health Information. Final Rule. *Federal Register, 67*, 53181–53273.

Oster, S. et al. (2007). caGRID: A grid enterprise architecture for cancer research. *Archive of AMIA Annual Symposium Proceedings, 2007*, 573–577.

Prokosch, H.-U. & Ganslandt, T. (2009). Perspectives for medical informatics. Reusing the electronic medical record for clinical research. *Methods of Information in Medicine, 48*, 38–44.

Sackett, K. M. & Erdley, W. S., The history of health care informatics, in *Health Care Informatics: An Interdisciplinary Approach*, Englebardt, S. P. & Nelson, R., Eds., Mosby, St. Louis, Missouri, 2002, 453–477.

Schoolman, H. M. & Lindberg, A. B. (1988). The information age in concept and practice at the National Library of Medicine. *American Academy of Political and Social Science, 495*, 117–126.

Sears, A. & Jacko, J., *The Human Computer Interaction Handbook: Fundamentals, Evolving Technologies and Emerging Applications*, CRC Press, New York, 2007, 2–15.

Shortliffe, E. H. & Cimino, J. J., Eds., *Biomedical Informatics: Computer Applications in Health Care and Biomedicine*, 3rd edition, Springer, New York, 2006.

Sinard, J., *Practical Pathology Informatics: Demystifying Informatics for the Practicing Anatomic Pathologist*, Springer, New York, 2005, 7.

Siochi, R. A. et al. (2009). The Chief Information Technology Officer in a radiation oncology department should be a medical physicist. *Medical Physics, 36*, 3863–3865.

Tanenbaum, A. S., *Computer Networks,* 2nd edition, Prentice-Hall Software Series, Upper Saddle River, New Jersey, 1988, 254.

Tufte, E., *Envisioning Information*, Graphics Press, Cheshire, Connecticut, 1990.

Wears, R. L. & Berg, M. (2005). Computer technology and clinical work: Still waiting for Godot. *Journal of the American Medical Association, 293*, 1261–1263.

Working Group on Biomedical Computing, *Biomedical Information Science and Technology Initiative*, National Institutes of Health, Bethesda, Maryland, 2009. Available at http://www.nih.gov/about/director/060399.htm. Accessed September 22, 2009.

2

Ontology for Radiation Oncology

Daniel L. McShan
University of Michigan

2.1 Introduction

An "ontology" is a technology that is used to represent knowledge and information within a specified domain or area of interest. An ontology for a knowledge database is through the declaration of "concepts," "properties," "relationships," "axioms," and "instances." For radiation oncology domain, an ontology would elucidate concepts such as "Case," "PTV," "Treatment Plan," "Treatment Beam," and "Radiation Dose." These "concepts" are further defined and limited by the addition or association of "properties," such as "CreationDate" or "Dose Units" and through "relationships" between "concepts." For example, the concept "Treatment Plan" can be assigned a property "hasTreatmentBeam," which represents a relationship to the concept "RadiationBeam." In addition, "axioms" or rules can further constrain concepts.

Figure 2.1 shows screenshot from an ontology editor (Protégé) used to define an example ontology for radiation ontology.

Finally, these objects can be "instantiated" to describe a particular instance of an object. For example, an instance of a case with a single treatment plan consisting of parallel opposed beams would be modeled by creating an instance of a "Case" linked to an instance of "hasTreatmentPlan" property to an instance of "Treatment Plan" identifying specific properties such as "Name," "Description," etc., as well as relationship links to two instances of "TreatmentBeam" objects along with their properties.

Figure 2.2 illustrates this concept and their connectivity.

The ontology declarations together form a knowledge model for the specified domain. This ontological description of information is intended to both provide a consensus (human) understanding of concepts (semantics) and have that knowledge interpreted by machine (computer). The latter then allows queries and logical reasoning about the domain model.

For the bioinformatics and medical informatics field, the use of ontologies can provide significant advantages to the way we collect, organize, and characterize research and clinical findings. For clinical operations, a semantic understanding of data can facilitate interoperability among systems within an institution. Furthermore, logic rules and reasoning services based on ontological descriptions can be used to assist in clinical decision support.

For research purposes, a key advantage to using an ontology is the ability to share and coalesce findings among researchers at multiple institutions providing not only a tabulated listing of specific data findings but also a knowledge representation about that data using consensus concept nomenclature and the defined relationships between those concepts and their instantiated data instances.

In radiation oncology, the scope of knowledge to be described is quite diverse and extensive, ranging from detailed medical diagnosis information about disease (cancer) extent to details about planning and optimizing of a radiation treatment strategy and to the information about management of radiation treatment delivery. To date, a complete ontological model for radiation oncology does not exist. The size and complexity of the needed knowledge model make this a daunting task. Fortunately, with ontology modeling, it is always possible to reduce the domain scope and extent of knowledge described to be small enough to begin demonstrating and proving the value of ontologies for use in radiation oncology. Furthermore, it is possible to utilize (import) ontologies developed specifically for related areas such as modeling for medical imaging, anatomical modeling, medical domain modeling.

This chapter includes a background on "ontology" for knowledge representation and a summary of available tooling for working with ontology models. This will be followed by an overview of ontologies developed for biomedical related areas and how those efforts could be used for radiotherapy applications.

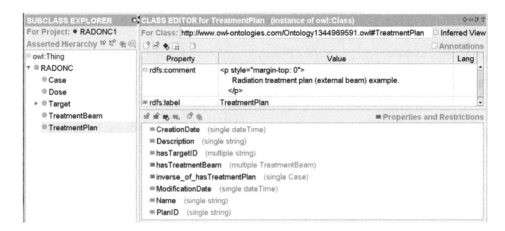

FIGURE 2.1   Example radiation oncology ontology editor showing properties assigned to the concept (class) "TreatmentPlan."

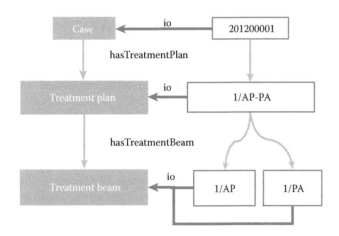

FIGURE 2.2   Simple ontology for radiation oncology. Concepts on the left are instantiated on the right. io = instance of.

2.2 Background

2.2.1 "Ontology" Definition

The word "ontology" originates from classic Greek where it was used to describe "the metaphysics dealing with the nature of being."[1] This area of metaphysics dealt with questions regarding things that exist and how such things are related in a hierarchy according to similarities and differences.

Ontology as used today in the field of information science similarly pertains to the identification of concepts within a domain of interest, along with the relationships between those concepts, and the attributes with which the concepts and relationships can be modified. As mentioned earlier, the usage is intended to be not only for human consumption (to foster a common understanding) but to also make them consumable by a computer to perform reasoning about and within the domain of interest.

The concepts of knowledge representation and information retrieval were the basis of much of the early work in artificial intelligence research in the 1970s and 1980s and that work formed the basis for the ontology developments for the information

sciences today. Tom Gruber in 1993 described "ontology" as a "consensus conceptualization of the real world."[2] Today, ontological descriptions are used in a wide range of domains. Much of the present day ontology work is being directed toward developing "the semantic Web," which promises to provide us (and our computers) with a wealth of data on a variety of topics.[3]

2.2.2 Representations and Knowledge-Based Systems

Smith (1982) states the knowledge-representation hypothesis as "A knowledge-representation which allows external observers to know the proposition that account for the knowledge that a computational process exhibits, and that the knowledge-representation plays an essential, casual role in engendering the computational behavior that manifests that knowledge."[4]

There are two kinds of computerized knowledge representations: procedural and declarative. Procedural representations are encapsulated in program code and generally within only specialized problem-solving context. Computerized radiation therapy planning systems incorporate knowledge in ways that assist a treatment planner to simulate radiation source placement, to calculate dose, to display dose distributions, and to analysis those results. Although these programs are highly efficient at doing those tasks, they are limited in terms of knowledge-based guidance, decision support, and ad hoc queries, and it is difficult to know what procedures and algorithms are used.

On the contrary, declarative knowledge as used in ontologies and illustrated in the simple example shown in Figures 2.1 and 2.2 uses a representation that is stored as propositions. Figure 2.3 shows an excerpt of the example ontology on which those figures were based.

These declarations must be interpreted by software for use. Depending on the scope and depth of these propositions, this form of knowledge representation can be used for a variety of purposes such as retrieving information about the model itself (which can be quite complicated for medical data) or retrieving the instance data (e.g., list of beams for a given case). However,

```
- <rdf:RDF xml:base="http://www.owl-ontologies.com/Ontology1344969591.owl">
    <owl:Ontology rdf:about=""/>
  + <owl:Class></owl:Class>
  + <owl:Class rdf:about="#PTV"></owl:Class>
  + <owl:Class rdf:ID="Dose"></owl:Class>
  + <owl:Class rdf:ID="TreatmentBeam"></owl:Class>
  + <owl:Class rdf:ID="GTV"></owl:Class>
  + <owl:Class rdf:ID="TreatmentPlan"></owl:Class>
  + <owl:Class rdf:ID="Case"></owl:Class>
  + <owl:Class rdf:about="#RADONC"></owl:Class>
  + <owl:Class rdf:ID="CTV"></owl:Class>
  + <owl:Class rdf:about="#Target"></owl:Class>
  + <owl:Class></owl:Class>
  + <owl:Class></owl:Class>
  + <owl:ObjectProperty rdf:ID="inverse_of_hasTreatmentBeam"></owl:ObjectProperty>
  + <owl:ObjectProperty rdf:ID="inverse_of_hasTreatmentPlan"></owl:ObjectProperty>
  + <owl:ObjectProperty rdf:about="#hasTreatmentPlan"></owl:ObjectProperty>
  + <owl:ObjectProperty rdf:about="#hasTreatmentBeam"></owl:ObjectProperty>
  + <owl:DatatypeProperty rdf:ID="PlanID"></owl:DatatypeProperty>
  + <owl:DatatypeProperty rdf:ID="Modality"></owl:DatatypeProperty>
  + <owl:DatatypeProperty rdf:ID="Description"></owl:DatatypeProperty>
  + <owl:DatatypeProperty rdf:ID="hasTargetID"></owl:DatatypeProperty>
  + <owl:DatatypeProperty rdf:ID="CreationDate"></owl:DatatypeProperty>
  + <owl:DatatypeProperty rdf:ID="Name"></owl:DatatypeProperty>
  + <owl:DatatypeProperty rdf:ID="ModificationDate"></owl:DatatypeProperty>
  + <owl:DatatypeProperty rdf:ID="PatientID"></owl:DatatypeProperty>
  + <owl:FunctionalProperty rdf:ID="usesBeamID"></owl:FunctionalProperty>
  + <owl:FunctionalProperty rdf:ID="TargetID"></owl:FunctionalProperty>
  + <owl:FunctionalProperty rdf:ID="Energy"></owl:FunctionalProperty>
  + <owl:FunctionalProperty rdf:ID="usesMachineID"></owl:FunctionalProperty>
  + <TreatmentPlan rdf:about="http://www.owl-ontologies.com/RADONC_Class4"></TreatmentPlan>
  </rdf:RDF>
```

FIGURE 2.3 Ontology declaration file (OWL/RDF file).

because the knowledge needs to be interpreted, it is relatively inefficient (slow) for complex problem solving.

There are important properties of any knowledge-representation system that claim to contain a full knowledge representation that can be used for logic reasoning:[5]

- Soundness: Knowing that, if a query is asked of the system, the result will be logically consistent with all other propositions in the knowledge base (i.e., you will get the right answer).
- Completeness: Knowing that, if it is possible to conclude a given proposition from a set of axioms, the system indeed will be able to make that conclusion (i.e., you will get all the answers).
- Decidability: Knowing that, if a query is asked of the system, the system will indeed return a result in a reasonable period of time (i.e., that you will get "an" answer).

In any implementation of knowledge representation, there is a fundamental tradeoff between how expressive to make the representation (i.e., being able to say everything you would like about the world being modeled) and how tractable it is (i.e., knowing that an inference is decidable and computationally efficient). Strategies that are used to manage complexity in a knowledge base are often to limit what is declared and to accept that the knowledge will be incomplete.

Knowledge-based systems require a model of the domain (e.g., radiation oncology) and tasks to perform (e.g., choosing an appropriate template treatment plan based on diagnosis, location, and other factors). A representation of that model allows one to make distinctions and inferences that are appropriate for the task. Representations can differ with respect to expressiveness and computational complexity enabling answering of certain queries. One of the advantages of a knowledge-based system is the ability to apply logic to reason about facts within the representation. Solutions by logic depend on construction of syntactically correct sentences (well-formed formulas).[6] First-order logic has been a major tool used for reasoning using knowledge representations. The adjective "first order" refers to the formulas that are dependent on single "predicates" or properties of object and not on derived predicates that depend on other predicates.

Like knowledge representation, reasoners need to be "sound" (only deriving correct results) and "complete" (able to derive any logically valid implications). First-order logic relies on a formal language for specifying formulas, but it is the semantic meanings of predicates and functions that supply meaning to the formulas and solutions within the domain for which the reasoning is done.

There are, however, a number of problems with first-order logic. Inferences may be undecidable (unable to achieve a yes or no answer). Representing knowledge as a list of propositions does not clarify relations and the structure among propositions (logical completeness does not guarantee cognitive tractability). Furthermore, the role that any logical sentence plays in reasoning cannot necessarily be determined simply by inspection. A solution to these issues that bring greater clarity is through the use of graph theory.[5]

Graph theory uses the notion of a set of vertices (or nodes) that are linked by a set of edges (or links).[7] One can further specify if edges are ordered (directed) or unordered (undirected). For knowledge representations, concepts (nodes) can be associated using relationships (links) and as such used to describe parts of a knowledge representation. A simple example, for radiation oncology, is shown in Figure 2.2. A more complex oncology is shown in Figure 2.4. (This was designed to be the basis of a new radiation treatment planning system framework.) As with relational databases, there are often restrictions on relations in terms of the number of objects that are allowed, that is, the "arity" of relations that range from unary (one to one), for example, date of birth, binary (true or false, male or female), or n-ary describing multiple related objects (list of treatment beams).

There are a number of difficulties when building ontologies. Defining relationships is generally arbitrary and often meanings are not well represented, distinctions between classes and instances are not obvious, and there are few unifying assumptions in different semantic-network representation. One obviously wants well-defined semantic relationships between general concepts, specific individuals, and values of attributes. Also, one wants clear and understandable structures with "reasonable" expressivity and tractability.

2.2.3 Frames

A popular construction technique for early ontologies consisted of developing "frames" introduced by Minsky in 1974 to theorize how one might represent the mental task of vision understanding.[8] In essence, a frame represents a remembered scene or situation that includes details about content and placement. In a more general sense, the theory can be used to form a knowledge representation that could be used for natural-language understanding and for understanding human problem-solving.

A "frame" is a data structure to which several kinds of information are attached describing roles and properties similar to that shown in Figure 2.1. Properties can also have specified facets that prescribe data type, cardinality, and default values. The information in a frame could include how to use the frame, what

one can expect to happen next, and what to do if these expectations are not confirmed. A frame can be thought of as a network of nodes and relations. Some of the nodes are fixed and represent fixed characteristics about a particular situation. Other nodes represent variable data that pertain to a specific instance of data. Nodes can themselves be subframes forming a subnetwork. A frame represents an encapsulated concept specifying a meaningful chunk of knowledge. A particular instance of the concept is "instantiated" by defining the specific data for the attached variable nodes. A frame can have one or more sets of individuals. For the scene understanding task, the individuals of a particular scene could represent different views or different time states. For nonvisual kinds of frames, the differences between the frames of a system can represent actions, cause-effect relations, or changes in conceptual viewpoint.[5,8]

Collections of related frames are generally linked together into "frame systems." These are themselves networks of frames with relationships that support hierarchy with inheritance.

The effects of important actions are mirrored by transformations among the frames of a system. These are used to make certain kinds of calculations economical, to represent changes of emphasis and attention, and to account for the effectiveness of "imagery."[8]

Much of the phenomenological power of the theory of frames hinges on the inclusion of expectations and other kinds of presumptions. A frame may contain a great many unneeded details to solve specific problems but have many uses in representing general information, most likely case classifications, techniques for bypassing "logic," and ways to make useful generalizations. A frame's properties are normally defined with "default" assignments. The default assignments are attached loosely to an instance, so that they can be easily displaced by new items that better fit a particular situation.

Some of the reasoning power of an ontology is inherent in its structure. Properties of classes are inherited by subclasses (or inherited classes). Properties for the inherited class can be modified from that inherited from its parent. A class can be a subclass of two different parents and thus inherit properties from both parents. This inheritance supports reasoning through generalization that we define for the parent nodes but can lead to difficulties because there are often exceptions. It is painful to have to declare many exceptions for a "general" concept. Furthermore, this leads to the need to perform nonmonotonic reasoning, where we make assumption for a new individual of a particular parent class; but when an exception occurs due to some particular property assignment, then the reasoning needs to retract some part of the assumptions as well as any downstream assumptions.[5]

An example of a frame-based ontology is shown in Figure 2.5. This ontology closely mimics the ontology developed for the "Athena/EON" project, which was developed to provide medical guideline decision support.[9–12] (That project has been subsequently morphed into the Sage project.[13–15]) In this example, however, the ontology has been modified to define guidelines for radiotherapy protocols. Shown in Figure 2.5 are listings of some

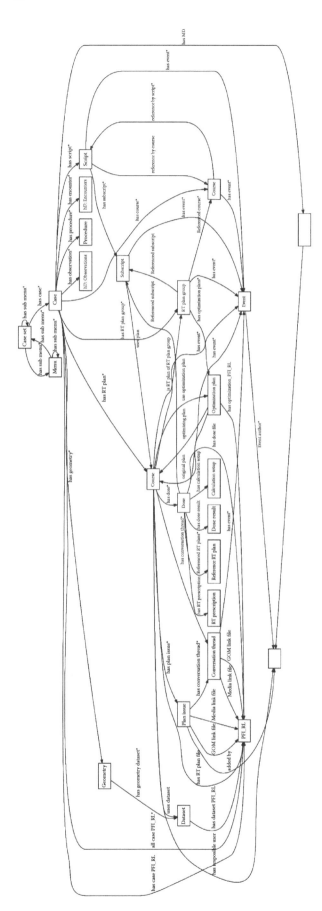

FIGURE 2.4 Ontology graph example (UMRO).

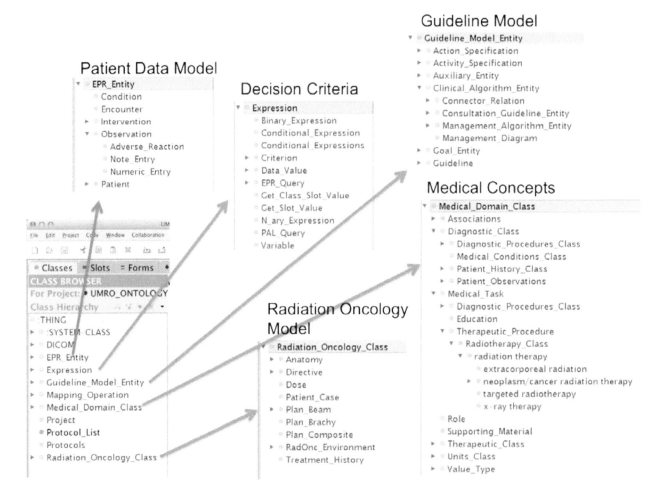

FIGURE 2.5 Guideline concept classes.

of the ontology concepts defined. These are grouped into the separated listing in Figure 2.5. There are medical domain concepts that define general concepts for disease identification and coding classification, descriptions of diagnostic tests and coding for results, and therapeutic classification terminology, methods, and coding. The patient model includes patient-specific findings and procedures that are planned or completed. Guideline concepts include protocol guideline-specific pathways that will be used to enforce a specified workflow. The decision criteria concepts are used to establish a vocabulary and algorithm support used in describing the protocol guideline implementation and support. The final list of concepts is specific to radiation oncology and includes more detail than found in the more general medical concepts.

Figure 2.6 shows the frame-based definition of protocol guideline ontology using Protégé (version 3.4.8), which is a free open-source knowledge editor and acquisition system (http://protege.stanford.edu). The larger under-laying window shows the main protocol listing screen for which various "clinical_trials" (list includes protocols defined at or used by the University of Michigan Department of Radiation Oncology) are listed by site. Selecting (or adding a new protocol) brings up a form

(frame) for displaying the properties for a particular clinical trial instance. Selecting the brain protocol MMMC 2003-83 is shown. The properties associated with each protocol includes the items: label, title, description, references, and scheme. Selecting the scheme brings up the top-laying window in Figure 2.6. The schema frame contains a label, comments, and the actual schema, which is presented graphically identifying various guideline (workflow) "action" steps, which include (a) scenarios (starting conditions represented as circles), (b) decision or choice steps (light blue triangles), (c) action or subguideline steps (square and rounded square), and (d) synchronization steps for waiting for multiple pathways to complete. The connecting lines describe the flow of care. A subguideline step can point to other defined schema, which can provide a more detailed guideline (e.g., the baseline study shown in Figure 2.6) for that step.

The knowledge modeling is only part of a fully functioning clinical decision support system. Figure 2.7 provides a schematic of the ontology-based decision support (guideline support) system.

The ontology knowledge model is read by a "Decision Support" program that gathers updated patient information from a variety of sources (treatment planning database, treatment history database, and hospital clinical database). The program then

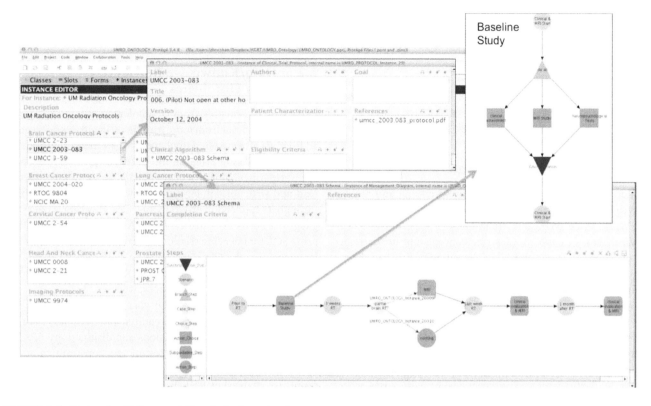

FIGURE 2.6 Protocol guideline acquisition for brain protocol.

checks the patient's progress based on the appropriate protocol guideline. The output of this program would then provide recommendations as to next steps in the patient management as dictated by the protocol and provide alerts to any health-related changes to the patient.

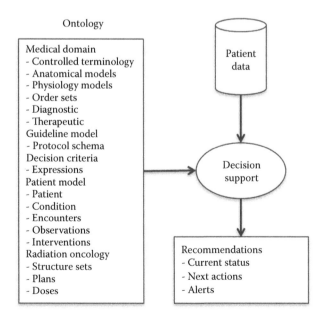

FIGURE 2.7 Schematic of ontology-based decision support system.

It should be noted that this example is similar to defining workflow. There are commercial workflow management systems that may be better equipped to robustly handle the various workflow aspects; however, an ontology provides a much more robust environment in terms of annotating and defining relationships to related concepts and reasoning services not directly part of the normal workflow tracking. Other examples of using an ontology to provide the knowledge source for driving workflow procedures have been implemented and reported.[16]

2.2.3.1 Frame Advantages

One of the advantages of frames is that classes and instances are organized as a "flat" knowledge base; in general, frame-based knowledge systems are easy to understand and can be reasoned about using first-order logic.[5] Frame systems have been in use for several decades and the tooling required to work with frames has matured. Frame-based systems do support some useful embellishments such as the creation of "metaclasses," which are basically template classes that specify properties that will be common to different derived classes. Also, frame ontologies support the use of axioms that allows additional logical sentences to be added to the ontology that do not directly fit into the language of frames.

2.2.3.2 Frame Disadvantages

On the contrary, there are problems with frames. With frames, one cannot easily represent negative factors. We can say that A is related to B, but we cannot say that A is NOT related to B. Further, we cannot specify disjunctions (if B and C are disjoint, then A can be related to B or C but not both). Quantification (saying "all" or "some") is also not part of the frame language. The semantics of relations are often not well defined. Classes may not be defined at useful levels of abstraction. Concept definitions that are overloaded (derived) are not always useful and hierarchical relationships are not uniformly taxonomic.[5]

2.2.4 Web Ontology Language

A more recent representation used for specifying an ontology is Web Ontology Language (OWL), which has been adopted by the World Wide Web Consortium (W3C) (http://www.w3.org). OWL makes it possible to describe concepts but uses a richer set of operators (e.g., "and," "or," "negation"). It is based on a different logical model that makes it possible for concepts to be defined as well as described. Complex concepts can be built up in definitions out of simpler concepts. The logical model allows the use of a reasoner that can check whether or not all of the statements and definitions in the ontology are mutually consistent and can also be used to determine which concepts fit under which definitions. A reasoner can help maintain the hierarchy correctly, particularly useful when dealing with cases where classes can have more than one parent.

The W3C OWL standard[17] specifies three species or sublanguages of OWL: "OWL-Lite," "OWL-DL," and "OWL-Full." A defining feature of each sublanguage is its expressiveness. OWL-Lite is the least expressive sublanguage. OWL-Full is the most expressive sublanguage. The expressiveness of OWL-DL falls between that of OWL-Lite and OWL-Full. OWL-DL ensures that description logic is always valid for the ontology. Description logic uses a closed world model that fully describes its concepts and logical dependencies.

The OWL language is typically stored and exchanged using a Resource Description Framework (RDF) format,[18] which is built on top of the Extensible Markup Language (XML) standard. The RDF implements a triple data store similar to a "noun"-"verb"-"object" sentence structure. For an ontology, the "noun" is a concept or class, the "verb" represents a relationship, and the "object" can be another concept object or a simple data object. In RDF, the concepts are represented using universal resource identifier. RDF files are designed for use on the Web, allowing an ontology to exist on the Web as a retrievable object.

Alternatively, an ontology can be stored in a relational database. Oracle Database 11g Semantic Technology[19] provides scalable and secure storage for RDF data (triplets). It should be noted that, although the ontology can be stored in a traditional relational database, the data are not organized as would be done for a typical relational database. A typical relational structure would contain many tables with rows of related data fields. There, of course, is linkage between tables defined by a database schema, which could be thought of as knowledge model. However, that schema is not easily changed. The ontology implementation to a relational database uses a simple schema that does not depend on the actual data model being implemented.

2.2.5 Data/Knowledge Retrieval from Ontologies

Retrieving information from an ontology is not much more difficult than retrieval from more traditional databases but requires additional software often implemented as a toolkit library along with an application program interface. For Java, the Jena package (http://incubator.apache.org/jena/) is commonly used to read and write an ontology and allows direct access to objects through its application program interface. However, a query language has been developed called "SPARQL," which is similar to using SQL for querying relational databases. The SPARQL query language uses the OWL specification to define the syntax and semantics for RDF/OWL queries. SPARQL can be used to express queries across diverse data sources, whether the data are stored natively as RDF or viewed as RDF via middleware. SPARQL contains capabilities for querying required and optional graph patterns along with their conjunctions and disjunctions. SPARQL also supports extensible value testing and constraining queries by source RDF graph. The results of SPARQL queries can be results sets (tables) or RDF graphs. Figure 2.8 shows a SPARQL query to get a list of plans (retrieved against a single case/patient instance).

Like SQL, it is necessary to understand the data schema for SPARQL queries. That schema is the ontology itself.

2.2.5.1 Ontology Editors

There are numerous tools available for developing and utilizing ontologies.[20] The examples in this chapter used Protégé, which is free open source software developed at Stanford University. Protégé supports both "frames" and "OWL." There are currently two versions that are maintained. Version 3.+ supports OWL 1.0 of the standard. There are numerous plug-ins for this version that have been developed at Stanford and contributed by users, which greatly extend the functionality. Version 4.+ supports OWL 2.0 but not frames.

Top Braid Composer[21] is a commercial offering that provides enterprise-level services and is built on an Eclipse platform.

```
SELECT ?PlanName ?PlanDescription ?GeomSiteName ?FileName
WHERE {
?x umro:hasRTPlan ?Plan .
?Plan umro:isRTPlanOf ?GeomSite .
?Plan umro:PlanName ?PlanName .
?Plan umro:PlanDescription?PlanDescription .
?GeomSite umro:GeomSiteName ?GeomSiteName .
?Plan umro:hasPFI_RL ?PFI_RL .
?PFI_RL umro:FileName ?FileName . }
```

FIGURE 2.8 SPARQL query to get a list of plans (for UMRO).

2.2.5.2 Reasoning Tools

As mentioned earlier, ontologies are often developed to provide a logically sound description of a domain. There exist a number of reasoning packages that are commonly used with ontology-based applications.

"Jess" is a rule engine and scripting environment written entirely in Java language by Ernest Friedman-Hill at Sandia National Laboratories in Livermore, California (http://herzberg.ca.sandia.gov/). "Jess" has the capacity to "reason" using knowledge supplied in the form of declarative rules. "Jess" uses an enhanced version of the Rete algorithm to process rules. (The Rete algorithm is a very efficient mechanism for solving the difficult many-to-many matching problem.)[22]

"SWRL" is a Semantic Web Rule Language (SWRL) based on a combination of the OWL-DL and OWL Lite sublanguages of the OWL Web Ontology Language.[23]

For description logic reasoning, "Racer" (Renamed Abox and Concept Expression Reasoner)[24] and "Pellet"[25] are robust servers often used for scalable ontology reasoning providing inference services for sophisticated ontology-based applications.

In Protégé, access to the various tools is either built-in or accessible via plug-ins.

2.2.5.3 Ontology Repository

Over the last few decades, there have been numerous ontologies developed for the biomedical field. A part of the National Institutes of Health's development for the center of biomedical computing has been the creation of a "BioPortal" that now serves as an ontology repository. The BioPortal Web site (http://bioportal.bioontology.org/) provides access to commonly used biomedical ontologies and to tools for working with them. The BioPortal allows one to

- browse the library of ontologies (~300),
- search for a term across multiple ontologies (~6.4 million),
- browse mappings between terms in different ontologies,
- receive recommendations on which ontologies are most relevant for a corpus,
- annotate text with terms from ontologies,
- search biomedical resources for a term, and
- browse a selection of projects that use BioPortal resources.

A short description of a few of the included ontologies that should prove useful to radiation oncology follows.

2.2.6 National Cancer Institute Thesaurus

The National Cancer Institute (NCI) Thesaurus is a vocabulary for clinical care, translational and basic research, and public information and administrative activities. The thesaurus has good coverage for cancer research domains and moderate coverage for more general healthcare research. Especially, good cancer-related drug coverage and chemotherapy regimen coverage. The construction of this ontology is described by Noy.[26] Detailed coding for radiotherapy is included in the ontology (Figure 2.9) but does not include a number of more modern

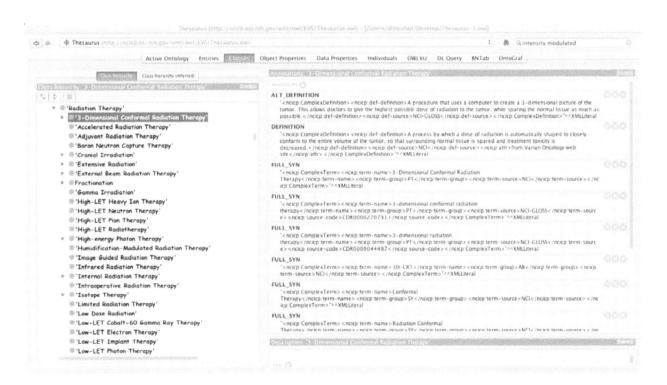

FIGURE 2.9 NCI Thesaurus radiation therapy concepts.

concepts such as intensity-modulated radiation therapy or adaptive therapy. Although the full NCI Thesaurus ontology is available for download, there is also a Web-based server for accessing the ontology through the BioPortal.

2.2.6.1 Ontology for Clinical Research

The BioPortal contains several ontologies targeted toward clinical research and clinical trials. The Ontology of Clinical Research (http://rctbank.ucsf.edu/home/ocre) is an ontology designed to support systematic description of, and interoperable queries on, human studies and study elements. A listing of the Ontology of Clinical Research ontology classes is shown in Figure 2.10.

2.2.6.2 Gene Ontology

The Gene Ontology (http://www.geneontology.org/) project provides an ontology of defined terms representing gene product properties. The ontology covers three domains: "cellular component," the parts of a cell or its extracellular environment; "molecular function," the elemental activities of a gene product at the molecular level, such as binding or catalysis; and "biological process," operations or sets of molecular events with a defined beginning and end, pertinent to the functioning of integrated living units: cells, tissues, organs, and organisms. Clearly, as DNA sequencing becomes readily available for evaluation of radiation therapy patients, identification of genetic profiles will provide evidence for use in individualizing treatments.[27–31]

2.2.7 Foundational Model of Anatomy

The Foundational Model of Anatomy Ontology (FMA)[32] is concerned with the representation of classes or types and relationships necessary for the symbolic representation of the phenotypic structure of the human body in a form that is understandable to humans and is also navigable, parseable, and interpretable by machine-based systems. The FMA is a domain ontology that represents a coherent body of explicit declarative knowledge about human anatomy. It includes a taxonomy (hierarchy) of anatomy, structural abstractions detailing with part-whole and spatial relationships between the entities, descriptions of morphological transformations related to prenatal and postnatal life-cycles, and metaknowledge with rules and definitions used to form these concepts.[33]

One application of interest using the FMA was the "Virtual Soldier Project"—an attempt to determine the extent of bullet penetration damage.[34] Although the FMA provided detailed anatomical descriptions, there was a need in this research to add pathology attributes to those structural descriptions as well as reference tags to the "Visible Human" data set to aid in visualization.[35]

For radiation oncology, there is a growing need to bring together similar types of information to better characterize regions that will be irradiated. Toward this end, the FMA ontology has also been employed by Kalet et al. for radiation therapy planning in an attempt to better define target volumes, identifying primary tumor including lymphatic paths.[36] Their work

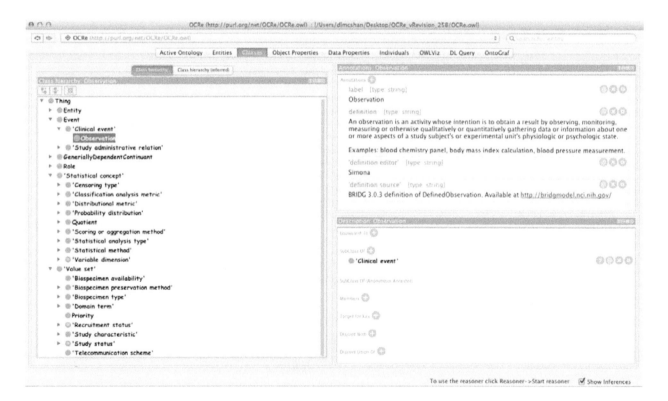

FIGURE 2.10 Ontology for clinical research.

focused on using the FMA representation of the lymphatic system to predict the spread of tumor cells to regional metastatic sites. The application required that (1) the downstream relations associated with lymphatic system components must only be to other lymphatic chains or vessels, (2) the components be at the appropriate level of granularity, and (3) that every path through the lymphatic system must terminate at one of the two well-known trunks of the lymphatic system. In attempt to insure that the FMA would provide this information, software was written to query this information. This research identified a number of weaknesses in the completeness and level of detail formulated in the FMA.

2.2.7.1 DICOM Ontology

The Digital Imaging and Communications in Medicine (DICOM) Ontology (DO) is not yet included in the National Center for Biomedical Ontology BioPortal but is worth mentioning in that this ontology is attempting to reproduce the entirety of the DICOM standard into ontological form. The advantages of doing this would be that the ontology could be queried and reasoned over. Initial work (a Cancer Biomedical Informatics Grid–sponsored project) focused mainly on just parts of the DICOM standard pertaining to images[37] and has not yet tackled DICOM-RT modules. The approach undertaken is toward being able to automate the ontology building process as much as possible by parsing digital (XML) formatted text version of the DICOM standard. Several manual tasks still remain related to harmonization with other terminology standards and resolving some inconsistent definitions for some DICOM elements.

Figures 2.11 and 2.12 mimic the work in DO but extend it to some of the DICOM-RT objects. In particular, showing an example as it would relate to the RT DOSE module.

The ontology contains a real world-class model and DICOM entities. Figure 2.11 follows the DICOM standard summary table A.18.3 in PS 3.3 for the 2011 DICOM standard for the RT DOSE Information Object Definition (IOD) listing the various modules. The RT_Dose_IOD_Module is highlighted and documented in the annotations in the upper right panel. The attribute details of the RT_DOSE module are shown in Figure 2.12. Each element is annotated with the service object pair number. Sequences are further described through macroattributes elsewhere in the ontology.

Using this ontology, DICOM files could be decoded to populate instances of the concepts within this ontology and queried using SPARQL or through an application.

2.2.7.2 Annotate and Image Markup

Another project related to the DO is the Annotate and Image Markup (AIM) ontology, which is a collaborative effort to specify the quantitative and qualitative content that researchers extract from images. The AIM ontology enables semantic image annotation and markup, specifying the entities and relations necessary to describe images. The goal of the AIM project is to enable ontology-based query and mining of images and integration of images with data in other ontology-annotated bioinformatics databases and ultimately enables researchers to link images with related scientific data so they can learn the biological and physiologic significance of the image content.[38]

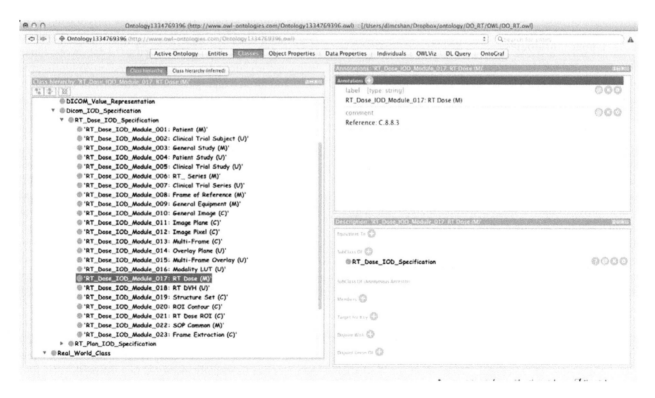

FIGURE 2.11 DO-RT_Dose IOD specification.

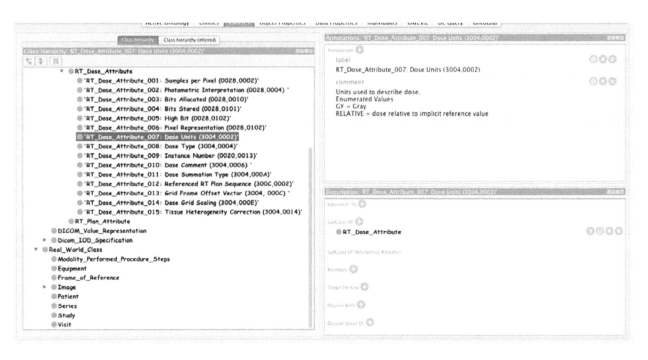

FIGURE 2.12 DO-RT dose attributes.

Radiation oncology heavily depends on the use of images for treatment planning system to derive contours and landmarks, beam placement, and to visualize these objects together with the calculated (or measured) dose distributions. Although much of these data can be stored in the form of RT objects in DICOM, it cannot be directly queried but could be queried using the DO described previously.

2.2.7.3 Upper Ontologies

There are many other ontologies that can prove useful in building new ontologies, providing ways to organize knowledge about specialized domains, such as measurement, units, time, etc., which may be in common for a variety of other domains. These other ontologies are generally referred to as "upper ontologies" (or top-level or foundational ontologies). A number of these are included in the BioPortal: Basic Formal Ontology, General Foundation Ontology, Units of Measurement, and Units Ontology. Other sources of upper ontologies can be found in the OpenCyc project[39] and the Suggested Upper Merged Ontologies[40] developed mainly for computer information systems.

Unfortunately, although the idea of relying on upper ontologies is attractive, there are few of these, if any, that have found very universal acceptance; differences in approach, nomenclature, usages, and sometimes ownership tend to result in developers reinventing or modifying these for use in their own ontologies.[41]

2.2.7.4 Data Access Library

At the University of Michigan Department of Radiation Oncology, we have developed an ontology, we call Data Access Library (DALIB), for data modeling and code generation to create a data access library for data exchange between VMS, Linux, and Windows-based treatment planning and management applications and to support data structure transparency between Fortran and C and C++ code. The ontology was created to identify data elements defined within various data structures that have been identified within our in-house radiation treatment planning system. This system was started close to thirty years ago and is written primarily in Fortran running on an OpenVMS system. The goal was to use the data access library to help migrate code and data to a Linux and/or Windows environment and for developing new code primarily in C++ that could cooperate with legacy Fortran code. Protégé was used to formulate the data model descriptions using classes to describe data structures and class properties to describe data elements. The data model is maintained as an RDF/OWL formatted file.

For program code support, this ontology OWL file is parsed using a Haskel program to build include files for both Fortran and C programs defining the structures defined in the ontology. To facilitate data access between applications, the output of the parsing is also used to generate code for reading and writing RDF files using the same data structures with case-specific instances. This was done with XML stylesheets. To improve access speed, large arrays of numeric data are read/written using XDR formatted files. Figure 2.13 provides a snippet of this ontology. Figure 2.14 shows the generation process.

There are clear advantages of this approach for data interchange, and the data files are based on XML and can be universally accessed. They can be queried using SPARQL, and the data files can be visually inspected by reading the files into the ontology editor. This, of course, implies that these files must be secured and access to these files must be controlled.

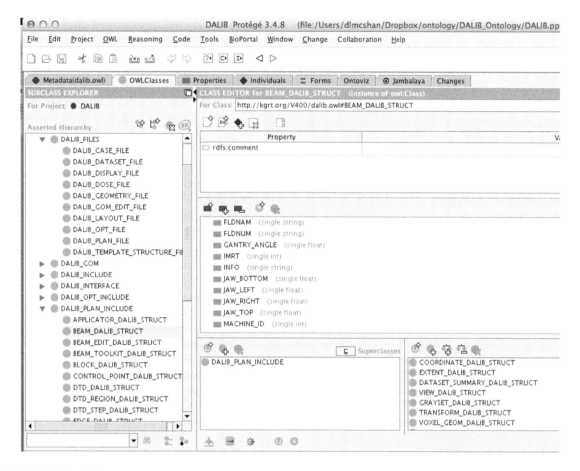

FIGURE 2.13   DALIB ontology.

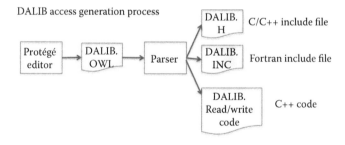

FIGURE 2.14   DALIB data access generation process.

This is mostly a "one-of" ontology in that its contents reflect our own internal data elements, data structure, and terminology but, nevertheless, demonstrates one of the variety of uses of an ontology.

2.2.8 Building Ontologies

There are numerous publications that can help in learning how to build an ontology (knowledge system).[42–45] Typically, these are specific to particular knowledge editor tool or environment and most editor provide user guides and tutorials.

The most important step in building an ontology is to first phrase what questions one would like to have answered by the knowledge system to be built. These questions are sometimes called "competency questions."[45] Such questions help narrow the scope of the supporting ontology and help define the extent and level of granularity of information that needs to be built into the ontology.

When designing a formal ontology, it is helpful to guide that design using objective criteria. Gruber provides the criteria of "clarity" (making sure the ontology is understandable), "coherence" (making sure the ontology is self-consistent), "extensibility" (allowing further specialization), "minimize encoding bias" (avoiding specialized jargon), and "minimal ontological commitment" (defining just enough to be useful). Ontology design, such as most design problems, requires making tradeoffs among these design criteria.[46]

Most successful ontology developments are a team effort including domain specialist, informatics experts (who know the tools and design criteria), software programming expertise (for application-dependent projects), and other stakeholders. Often, large development projects are part of a collaborative effort between institutions. This helps ensure that the design will be clear and consistent and eliminates some of the single developer "encoding bias."

2.3 Discussion

Radiation treatment planning requires abstracting anatomical information including regions to be targeted, treatment planning design and optimization, treatment prescription, treatment delivery management and billing, as well as long-term follow-up on the patient. New techniques, guidelines, and decision strategies are continually being developed and maintained through clinical research to improve the success of treatments.

The sharing of data among systems within a radiation oncology department and the sharing of data among institutions are crucial technologies for the future.

Although data exchange standards such as DICOM are presently the default mechanism for radiotherapy data interchange while providing sufficient detail to describe most geometric and dosimetric quantities, they are not easily extended nor queried.

As mentioned at the beginning of this chapter, there currently does not exist any published ontologies targeted mainly for radiation oncology. Large relational databases and proprietary data stores are being used by commercial companies for treatment planning systems and treatment management systems used in most departments. DICOM is the answer for most questions about how one goes about exchanging data between systems and institutions. Unfortunately, these silos of information exchange omit significant factors that are likely to be helpful in sorting out future radiation oncology research.

Developing an ontology within individual departments would be helpful to support needs, such as protocol guidelines, workflow, organizing and tracking measurement and QA data, and organizing clinical study data and potential many other uses, but that effort is likely beyond the capabilities of most departments to develop. This implies that this technology will need to come from commercial vendors or as a collaborative effort with radiation oncology.

At present, the primary interest in using an ontology for the radiotherapy community will likely be through collaborative research efforts where there is a need to aggregate and share data. Much of that data probably will remain in DICOM files but using the ontology to identify and organize that data (but not necessarily duplicate that data) and using the ontology for management data elements not included elsewhere. Again, the primary reason for using an ontology is to allow querying and reasoning about that data model and instances of that model.

2.4 Conclusion

Ontology development and its applications are rapidly becoming key tools in the armament needed to address the burgeoning biomedical knowledge base and associated patient-specific findings.

Ontologies serve as a mechanism for developing common understandings and communication of knowledge within a given domain. The knowledge model is easily extended, which is often essential for research and discovery.

Obviously, the explicit need for an ontology can be overstated and the argument that standard relational database technology and data modeling tools are sufficient and already predominate in providing efficient support of established needs for data capture, retrieval, and data mining is valid. Nevertheless, modeling changes and defining domain-level understanding within conventional data stores remains problematic, and in that area, ontologies have an advantage. The ability to have that knowledge easily accessible on the Web offers currently unrealized opportunities for sharing and augmenting knowledge within a given domain.

For radiation oncology, ontological-specific considerations (outside of terminology standards) have been limited. (About the first thing I did when learning about this technology in 2005 was to do bibliography search for "ontology" and "radiation oncology." The one hit I got was a misspelling for oncology.) Today, that search will find some legitimate reports, most of which are related to gene expression research for various tumor sites. However, there are other references to work motivating knowledge-based research in the field.[27,36,47–51]

There is a strong argument for developing a radiation oncology-specific ontology that could be coupled with those of other related domains, including medical domain, imaging and annotation, genetic and molecular pathways, clinical research studies, and more pragmatic domains such as workflow and collaborative domains. Access to this level of detail (similar to that illustrated in Figure 2.5) leads to the ability to enlist decision support applications, which combine rules and statistics (Bayesian) into recommendations with traceability to references and annotation found in the ontology. Although seemingly complex, the rate at which information sources about patient's health and status is growing will demand complex solutions to react to and improve patient care.

Acknowledgments

The author acknowledges Mark A. Musen, MD, PhD, and his colleagues in the Stanford BioMedical Informatics Department for allowing him to spend a six-month sabbatical in 2005 to 2006 (winter) learning about ontology-based informatics.

References

1. McKean, E. *The New Oxford American Dictionary*, 2nd edition, Oxford University Press, New York, 2005.
2. Gruber, T. R. (1993). A translation approach to portable ontology specification. *Knowledge Acquisition, 5*(2), 199–220.
3. Gruber, T. R., Where the social Web meets the semantic Web, Keynote Presentation at ISWC, *The 5th International Semantic Web Conference*, 2006.
4. Smith, B. C., *Procedural Reflection in Programming Languages Volume 1*, 1982.
5. *Knowledge Representation-Class Notes for BMI 210/CS 270*, 2005.
6. Woo, C., Construction of Well-Formed Formulas (Version 7). Available for free at http://planetmath.org/construction ofpropositions.html.

7. Lehman, F., *Semantic Networks in Artificial Intelligence*, Oxford, 1992.

8. Minsky, M., A framework for representing knowledge, *MIT-AI Laboratory Memo 306*, 1974.

9. Tu, S. W. & Musen, M. A. (2001). Modeling data and knowledge in the EON guideline architecture. *Studies in Health Technology and Informatics, 84*(Part 1), 280–284.

10. Advani, A. et al. (1999). Integrating a modern knowledge-based system architecture with a legacy VA database: The ATHENA and EON projects at Stanford. *Proceedings of the AMIA Symposium,* 653–657.

11. Musen, M. A. et al. (1996). EON: A component-based approach to automation of protocol-directed therapy. *Journal of the American Medical Informatics Association, 3*(6), 367–388.

12. Tu, S. W. & Musen, M. A. (1996). The EON model of intervention protocols and guidelines. *Proceedings of the AMIA Annual Fall Symposium,* 587–591.

13. Tu, S. W. et al. (2007). The SAGE Guideline Model: Achievements and overview. *Journal of the American Medical Informatics Association, 14*(5), 589–598.

14. Tu, S. W. et al. (2004). The SAGE guideline modeling: Motivation and methodology. *Studies in Health Technology and Informatics, 101,* 167–171.

15. Ram, P. et al. (2004). Executing clinical practice guidelines using the SAGE execution engine. *Studies in Health Technology and Informatics, 107*(Part 1), 251–255.

16. Díez, A. et al., AVICENA, an Ontology for the Design of Executable Clinical Practice Guidelines, Paper presented at 9th International Protégé Conference 2006, Stanford, California, 2006.

17. W3C. OWL Web Ontology Language Reference. W3C Recomendation 10. 2009; OWL Web Ontology Language Reference.

18. W3C. Resource Description Framework (RDF): Concepts and Abstract Syntax; W3C Recommendation 10, 2004. Available at http://www.w3.org/TR/2004/REC-rdf-concepts-20040210/.

19. Oracle Databse Semantic Technologies, 2010. Available at http://www.oracle.com/technetwork/database/options/semantic-tech/semtech11gr2-featover-131765.pdf.

20. Ontology Tools, 2010. Available at http://techwiki.openstructs.org/index.php/Ontology_Tools.

21. Top Braid Composer.

22. Forgy, C. L. (1982). Rete: A fast algorithm for the many pattern/many object pattern match problem. *Artificial Intelligence, 19,* 17–37.

23. SWRL: A Semantic Web Rule Language Combining OWL and RuleML (formally submitted), 2004. Available at http://www.w3.org/Submission/SWRL/.

24. RACER: Renamed Abox and Concept Expression Reasoner. Available at http://www.sts.tu-harburg.de/~r.f.moeller/racer/.

25. Pellet: OWL 2 Reasoner for Java.

26. Noy, N. F. et al. (2008). Representing the NCI Thesaurus in OWL DL: Modeling tools help modeling languages. *Applied Ontology, 3*(3), 173–190.

27. Chanyavanich, V. et al. (2011). Knowledge-based IMRT treatment planning for prostate cancer. *Medical Physics, 38*(5), 2515–2522.

28. Fushimi, K. et al. (2008). Susceptible genes and molecular pathways related to heavy ion irradiation in oral squamous cell carcinoma cells. *Radiotherapy and Oncology, 89*(2), 237–244.

29. Guo, Y. et al. (2012). Identification of genes involved in radio resistance of nasopharyngeal carcinoma by integrating gene ontology and protein-protein interaction networks. *International Journal of Oncology, 40*(1), 85–92.

30. Teng, C. C. et al. (2002). Head and neck lymph node region delineation with 3-D CT image registration. *Proceedings of the AMIA Symposium,* 767–771.

31. Westbury, C. B. et al. (2011). Gene expression profiling of human dermal fibroblasts exposed to bleomycin sulphate does not differentiate between radiation sensitive and control patients. *Radiation Oncology, 6,* 42.

32. Rosse, C. & Mejino, J. L., Jr. (2003). A reference ontology for biomedical informatics: The Foundational Model of Anatomy. *Journal of Biomedical Informatics, 36*(6), 478–500.

33. About Foundational Model of Anatomy, 2012. Available at http://sig.biostr.washington.edu/projects/fm/AboutFM.html.

34. Rubin, D. L. et al. (2006). Using ontologies linked with geometric models to reason about penetrating injuries. *Artificial Intelligence in Medicine, 37*(3), 167–176.

35. Rubin, D. L. et al. (2004). Linking ontologies with three-dimensional models of anatomy to predict the effects of penetrating injuries. *Conference Proceedings–IEEE Engineering in Medicine and Biology Society, 5,* 3128–3131.

36. Kalet, I. J. et al. (2009). Content-specific auditing of a large scale anatomy ontology. *Journal of Biomedical Informatics, 42*(3), 540–549.

37. Kahn, C. E., Jr. et al. (2011). Informatics in radiology: An information model of the DICOM standard. *Radiographics, 31*(1), 295–304.

38. Rubin, D. L. et al. (2009). A semantic image annotation model to enable integrative translational research. *Summit on Translational Bioinformatics, 2009,* 106–110.

39. OpenCyc, 2012. Available at http://opencyc.org/.

40. Suggested Upper Merged Ontology, 2012. Available at http://www.ontologyportal.org/.

41. Schulz, S. et al. (2009). Granularity issues in the alignment of upper ontologies. *Methods of Information in Medicine, 48*(2), 184–189.

42. Horridge, M., *A Practical Guide to Building OWL Ontologies Using Protege 4 and CO-ODE Tools Edition 1.3*, Manchester University, 2011.

43. Noy, N. et al. (2008). Developing biomedical ontologies collaboratively. *AMIA Annual Symposium Proceedings,* 520–524.

44. Noy, N. et al. (2010). The ontology life cycle: Integrated tools for editing, publishing, peer review, and evolution of ontologies. *AMIA Annual Symposium Proceedings, 2010,* 552–556.

45. Noy, N. F. & McGuinness, D. L., *Ontology Development 101: A Gude to Creating Your First Ontology*, Stanford Medical Informatics March, 2001.

46. Gruber, T. R. (1995). Toward principles for the design of ontologies used for knowledge sharing. *International Journal of Human-Computer Studies, 43*, 907–928.

47. Zink, S. (1989). The promise of a new technology: Knowledge-based systems in radiation oncology and diagnostic radiology. *Computerized Medical Imaging and Graphics, 13*(3), 281–293.

48. Buchheit, I. et al. (1997). Optimization of dose distribution of radiation in cancers of the cavum: Association of an expert-system and a mathematical algorithm. *Cancer/Radiothérapie, 1*(1), 74–84.

49. Kalet, I. J. & Paluszynski, W. (1990). Knowledge-based computer systems for radiotherapy planning. *American Journal of Clinical Oncology, 13*(4), 344–351.

50. Olsson, C. E. & Kemp, G. J. L., Standardizing radiation oncology data for future modelling of side effects after radiation therapy, in *Proceedings of the First International Workshop on Managing Interoperability and Complexity in Health Systems*, ACM, Glasgow, Scotland, UK, 2011, pp. 67–70.

51. Wang, D. et al. (2005). A knowledge-based fuzzy clustering method with adaptation penalty for bone segmentation of CT images. *Conference Proceedings–IEEE Engineering in Medicine and Biology Society, 6*, 6488–6491.

3

Web-Based Information Delivery

Charles Mayo
Mayo Clinic

Steven Conners
Mayo Clinic

3.1 Introduction

Transfer of information using the Web is a routine occurrence in our personal and professional lives. If you purchased this book electronically, you likely provided the required delivery and credit card information by manual entry of data using a Web form. If you submitted an article to a journal for publication, then you uploaded files using a Web form. With the growing emphasis on routine aggregation of data to monitor the outcomes of patient treatments, medical physicists will become participants in much more sophisticated transfers of clinical data.

Our radiation oncology information systems and treatment planning system applications operate on internal networks, which are protected by the security protocols of our institutional firewalls. Typically, these are Windows or UNIX/Linux-based applications that exchange data with direct queries to the underlying databases of these applications or by exchanging Digital Imaging and Communications in Medicine (DICOM)-RT files. For aggregation of data from multiple institutions or even multiple departments within an institution, this approach is not viable. A single database that meets all needs of all users is not plausible; therefore, a federated approach that allows exchange of information between independent databases is required. To maintain security and ability of these independent databases to be modified as needed for their primary users (e.g., your department), with minimal impact on the ability of secondary users (e.g., national registries or other departments) to receive that data, direct queries to the database are deemphasized in favor of semi-automated exchange of data in standardized file formats. Direct queries to the database are fast, but in these exchanges flexibility in design of approaches that allow interaction with a wide range of participants with diverse systems is more important.

In this chapter, we will familiarize the reader with the concepts and tools they will encounter in technical discussions of options for Web-based information delivery. We assume that the readers are radiation oncology professionals who have limited or no programming experience but who may be called upon to assist their departments in understanding the implications of Web-based data exchanges. These discussions occur in the context of creating Web-based services to support pooling of data with those of other institutions, in the creation of outcomes databases, and in the development of patient portals to aggregate patient-reported outcomes. Often, these discussions also focus on questions about the implications of resources required to create Web-based solutions for information delivery. A detailed examination of these questions would require much more than a chapter, but we will provide perspective on options and tools.

3.2 Data Exchange Formats

3.2.1 XML

The basis of most communication on the Web is a text file in a format known as Extensible Markup Language (XML). Figure 3.1 illustrates an XML file describing some limited information about a group of linear accelerators. The first line of the file is a standard format declaration that is used to provide information to any program reading the file about the general rules followed by the file. XML data elements, known as tags, are delimited by brackets (Castro 2001; Esposito 2003a,b). Start tags (e.g., <modes>) mark the beginning of information about a data item and end tags (e.g., </modes>) mark the end. Data elements can be nested to define a hierarchy of information. In this example, there are three <accelerator> tags nested in the <accelerators> tag. A data element can be assigned a text value placed between the start and end tags. Here, the first mode of the Machine A accelerator is assigned a value of A-X06. Information can also be defined as an attribute inside of the start tag (e.g.,

```
<?xml version="1.0" encoding="utf-8"?>
<accelerators>
  <accelerator name ="Machine A">
    <mode type ="photon" energy ="6">A-X06</mode>
    <mode type ="photon" energy ="18">A-X18</mode>
  </accelerator>
  <accelerator name ="Machine B">
    <mode type ="photon" energy ="6">B-X06</mode>
    <mode type ="photon" energy ="10">B-X18</mode>
  </accelerator>
  <accelerator name ="Machine C"/>
</accelerators>
```

FIGURE 3.1 Example of an XML file using a free-form approach to describe some of the details of a set of linear accelerators.

name="Machine B"). Attributes can also be used with empty element tags (e.g., Machine C). The location of the "/" character distinguishes an empty element tag from an end tag.

This approach allows a very free form by the user in constructing the file to reflect the user's data. To collaborate with other users or to assure compatibility in naming and format required for use with other programs, it is best to define agreed upon standards for the tags, attributes, and values used to define the shared data set. These are defined by adding an attribute that defines a file on the Web that defines the name space to be used. For example, the Web pages transmitted to your Web browser from the Internet Web server on your universal resource locator (URL) line is a Hypertext Markup Language File (HTML). Figure 3.2 illustrates that the HTML file is an XML file. The user may explicitly declare a namespace attribute used for qualifying the tags (e.g., <html xmlns=http://www.w3.org/1999/xhtml>).

```
<html>
<head>
    <title>List of Accelerators</title>
</head>
<body>
<ul>
    <li>Machine A
        <ol>
            <li>A-X06</li>
            <li>A-X18</li>
        </ol>
    </li>
    <li>Machine B
        <ol>
            <li>B-X06</li>
            <li>B-X10</li>
        </ol>
    </li>
    <li>Machine C</li>
</ul>
</body>
</html>
```

FIGURE 3.2 An HTML file used to display a list of accelerators in a Web browser is a type form of XML file.

However, as the figure illustrates, the declaration is not necessary; the browser is aware of the namespace. The World Wide Web Consortium (W3C) defines the standards that are used on the Web. The value of the xmlns attribute defines a location on the Web where the name space file can be obtained by the program using the XML file.

When examining XML files, you will often note that the information about the name space has been more specifically defined with a refinement of the namespace declaration. It may be split into a generalized definition of data type elements defined in a namespace declaration (e.g., xmlns:xsd="http://www.w3.org/2001/XMLSchema") along with a more specific set of allowed instances of these generalized data types (xmlns:xsi="http://www.w3.org/2001/XMLSchema-instance"). All of these schema files are themselves XML files. Name spaces are not required in XML (or HTML) files. For HTML files, the Web browser application understands implicitly that the rules defining the allowed tags are at the "http://www.w3.org/1999/xhtml" location. There are also alternative XML schema formats, such as RELAX NG (http://relaxng.org/), which are designed to be simpler and more readable than the xsd format maintained by W3C.

The reader should note that standardization of XML schemas to be used by particular groups (e.g., Radiation Oncology) is outside the scope of W3C. At present, there are no widely used XML standards in place for that group. However, as national and multi-institutional efforts to construct and exchange outcomes data mature, such standards may also develop.

XML is a very mature, well-established format for exchanging data on the Web. Tools and standards for use for XML continue to evolve under the direction of W3C. For example, the group is developing standards for encryption of XML. The W3C is evolving security standards for use of XML to support digital signatures, encryption and distribution, and registration of public, cryptographic keys. Display of XML of the information in an XML file can be customized by coupling it to an extensible stylesheet language transformation (XSLT or XSL) file. The XSL file is also an XML file, in which the tags, elements, and attributes are used to map data to display in an HTML file. XSL has been extended with formatting objects to provide standards page layout for publishing large XML documents, including multilingual documents, to HTML, PDF, and other formats. Searching through the hierarchies in an XML file to query the data in it can be carried out with the XML Path Language.

3.2.2 JSON

JavaScript Object Notation (JSON) is a text-based format (http://json.org/) for transmitting data. JSON files can be easily parsed by JavaScript programs. This can be advantageous when designing Web pages that execute code on the client side of a Web page to make the response very fast, for example, in Web pages using asynchronous JavaScript (AJAX). Figure 3.3 shows an example of a JSON file. Standards are in process for developing schema files analogous to those used with XML files for validating content. JSON files are used for data storage of documents

{"accelerators": [{"name":"Machine A",
"modes":[{"type":"photon","energy":6,"value":"A
-X06"},{"type":"photon, "energy":18,"value":"A-
X18"}]},
{ "name":"Machine B" ,
"modes":[{"type":"photon",
"energy":6,"value":"B-X06"},{"type":"photon,
"energy":10,"value":"B-X10"}]},
{ "name":"Machine C"}]}

FIGURE 3.3 Example of a JSON file used to describe a set of linear accelerators.

used by CouchDB (http://couchdb.apache.org/). CouchDB is an open source, forms-based approach to create a database, using JavaScript to query JSON files. As technologies and standards, JSON and CouchDB (Anderson et al. 2010) are much less mature and less widely utilized than XML and relational databases. However, they may be anticipated to become part of the landscape in Radiation Oncology. As with XML, there are no widely utilized standards for data exchange with JSON that are applicable to Radiation Oncology.

3.2.3 DICOM-RT

The National Electrical Manufacturers Association working in conjunction with the American College of Radiology developed a standard for DICOM (http://medical.nema.org/) that was first published in 1993. Subsequently, the DICOM 3.0 standard was extended to include Radiation Oncology (a.k.a. DICOM-RT) with standards for objects defining structures, dose, treatment plans, and brachytherapy. The standards are further extended to include particle beams and pathology. Unlike XML, DICOM files are binary rather than text-based. The tags defining data elements within DICOM files are carefully controlled by the American College of Radiology/National Electrical Manufacturers Association group. DICOM is a mature and widely used format for exchanging data. These standards are utilized to assure interoperability among vended and custom applications.

The binary format of DICOM files accommodates a wide range of data types in a much more compact file than XML. However, reading and writing to DICOM files requires specialized software tools, and much more advanced programming skills, to interface to the files. Programming environments, such as MATLAB (www.mathworks.com) may have built-in DICOM interfaces that facilitate construction of applications to analyze DICOM-RT objects. The Computational Environment for Radiotherapy Research (www.cerr.info) is an example. Other libraries are available for use in production-level languages (e.g., MyDicom in.NET; www.mydicom.net). Open source tool kits for conversion of DICOM data to and from XML are available in the DICOM tool kit from OFFIS (http://support.dcmtk.org/docs/index.html).

3.2.4 NAACCR

The North American Association of Central Cancer Registries (NAACCR) has established text (ASCII)–based data file formats (http://www.naaccr.org/StandardsandRegistryOperations/VolumeI.aspx) for use in exchanging data between cancer registries. The format encapsulates as a wide range of survival, incidence, staging, and other outcomes relevant metrics. The format is utilized by several groups, including National Cancer Institute Surveillance Epidemiology and End Results, Canadian Council of Cancer Registries, and Commission on Cancer of the American College of Surgeons.

3.3 Data Exchange Methods

3.3.1 Web Servers

When a user types into their Web browser a URL, such as http://www.aapm.org/default.asp, they are specifying a file, default.asp, on a server. The server is connected to the Internet with at a unique text address, www.aapm.org, which corresponds to a unique numerical Internet protocol address, 149.28.118.16. The file is accessed and processed according the communication protocol defined in the URL. In this example, the hypertext transfer protocol (HTTP) is used. Other protocols are HTTP secure (HTTPS), file transfer protocol (FTP), FTP secure, and simple mail transfer protocol. The server must be running a Web server application to understand how to process requests specified in the URL. The two most commonly used Web servers applications supporting these protocols are Internet information service (IIS) from Microsoft (http://www.iis.net) and the open source Apache HTTP Project (http://apache.org). HTTP supports methods for data exchange with the Web server; the most frequently encountered are GET and POST. These are used for retrieving and sending data from scripts or using <form> tags in HTML pages (Esposito 2003a,b).

3.3.2 User to Server: Web Forms–Based Data Exchange

Data are often aggregated by having subjects fill out data on a form in an Internet browser. The form will be displayed in HTML; however, the actions associated with processing the data must be carried out by a program running on a server. The data are likely stored in a relational database that runs on another server. Those interactions will also be managed by the program running on the server.

There are several programming languages that may be used to create Web forms. Many run scripts that are compiled and run on the server at the time of the URL request. Common examples are Active Server Pages (ASP; *.asp) and PHP (*.php) (Ullman 2001, 2002). Other scripting languages such as Python or Perl may also be used on ASP or PHP pages Python and Perl. With the ASP.Net language, it is also possible to place the code executing methods invoked by the Web page into a separate file and precompile it to run faster and with better security on the server.

In that case, the underlying code may be written in the same code used for developing non-Web-based applications: C#, C++, Visual Basic, or even in Python (IronPython.Net). It is possible to run pages using these scripting languages on both IIS and Apache Web servers. There are also some open source Web servers written in Python (http://wiki.python.org/moin/).

The Web form, or page, contains controls for specifying data values: radio buttons and combo boxes to specify one among several choices, check boxes to specify several selections, calendars to enter dates, and text boxes or form field to hold text corresponding to a variety of values. When text boxes are used for input of values that follow a specified format (e.g., an ICD-10 code), range validator controls are used to alert the user to errors in data input. The values of the controls, the state of the page, are serialized as a base 64 encoded character string. It is maintained as a hidden form field, so that the programmatic handling if the data are the same as for a visible field but the data are hidden from the user. When the input is complete, the user presses a button to submit the data to the server to be recorded into a database. The state string is sent to the server where it is deserialized, and the server carries out the method invoked by pressing the submit button.

In general, institutions regard transmission of information within the networks and behind the firewalls of that institution as sufficient to ensure the security necessary to protect patient health information (PHI). When the Web forms are used to aggregate PHI, the data encoded in the state string must also be protected as it is passed from client to server on the Internet. There are several steps taken in protecting the string.

To assure that the integrity of the data has not been compromised during transmission, a hash of the state string is created and then appended to the copy sent to the server. A hash is string that has been computed from an input string according to an algorithm that defines a unique mapping from the input string to a new string that is compressed and indecipherable. The server separates the hash from the state string and recomputes the hash for comparison. If they match, this demonstrates that the integrity of the data has not been compromised in the transmission. For example, suppose the value of the variable on the client's Web page was assigned a value of "10.0." If the character for the period were lost in the transmission, this check would alert the server to a problem in the transmission avoiding mistakenly assigning "100" to the value on the server side. This process is called a machine authentication check.

The machine authentication check does not protect the data from being intercepted before it reaches the server and subsequently having the deserialized PHI read by an unscrupulous agent. Web servers and browser applications can be configured to encrypt the view state/hash string before transmission. Typically, the Triple Data Encryption Standard algorithm is utilized. The algorithm uses three 56-bit keys in the encryption. The Triple Data Encryption Standard algorithm is commonly used for encryption in financial exchanges. Encryption of the state string is natively handled by ASP.Net but not by PHP. However, third-party software tools can be purchased and provide this security layer for PHP pages.

In addition, Web sites that receive data often purchase a secure sockets layer (SSL) certificate. The successor to SSL is Transport Layer Security (TLS), but in most conversations SSL is used to generalize the concept of the secure connection. Web sites using SSL certificates have HTTPS rather than http as the prefix to the URL, corresponding to communication through port 443 rather than port 80. The communication between the server and the client is encrypted. The level of encryption depends on the level of SSL certificate purchased. SSL has seen a number of iterations in the last twenty years to build greater security to respond to the ever-increasing capabilities of cyber-criminals: SSL 1.0 (never released), SSL 2.0 (1995), SSL 3.0 (1996), TLS 1.0 (1999), TLS 1.1 (2006), and TLS 1.2 (2008). SSL requires upfront effort and expense to secure, configure, and maintain but should be considered bare minimum protection for Web-based transactions of sensitive information.

3.3.3 File Exchange

Data files, such as DICOM or spreadsheet files, often need to be submitted to a server as part of an exchange. Several years ago, it had been common to use an FTP protocol for this type of submission. Most Web browsers recognize specification of the FTP protocol in the URL (ftp://mysite.com) as flag to invoke a suitable interface for copying the files. These interfaces prompt for user name and password as well as allowing drag and drop of files into the target directory on the server.

However, this approach requires navigating away from the Web page on which the other data are entered. Using ASP.Net, Java, or PHP pages, file upload controls may be invoked that utilize the underlying file system methods of the server's operating system to handle the file transfer. This approach also has the advantage of enabling the Web pages to specify other methods that are used to inspect the contents of the submitted file to validate that it contains the correct information required for the submission. This could not be easily accomplished using the FTP mechanism.

3.3.4 Architecture and Database

A relational database is at the core of most applications used to aggregate data (Fehily 2002). A relational database enables quantifying both the values of fields in the database and the relationships between data elements. Figure 3.4 illustrates view of a subset of tables used in an outcomes database. In the linkages, a one-to-many relationship is indicated with the key symbol at one end (one) and the infinity sign (many) at the other. Rules for the data types (date/time, float, text, Boolean, xml, etc.), default values, allowance for empty (null) values, and many other requirements for fields and records are defined within the database. The ability to quantify and enforce these relationships is a fundamental advantage for a relational database over file-based data records, such as spreadsheets. Relational databases may be accessed by many users at the same time. The database and applications using the database have methods for assuring concurrency of

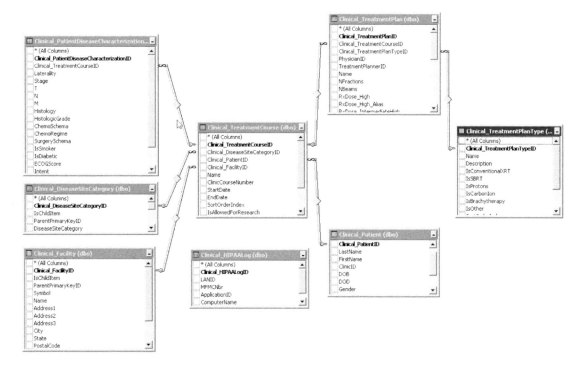

FIGURE 3.4 A database diagram of a set of tables used in an outcomes database to represent basic demographic, course, and treatment plan information. The tables quantify the values of particular data fields each record of the table. The relationships between the information in tables are quantified by the links between fields identified as keys (e.g., Clinical_TreatmentCourseID). Note that in this example the table named Clinical_HIPPALog is used to track access to individual patient records. This is a common requirement for institutional databases containing PHI.

records. That is, if a modification is made to a record by one user, the data seen by the other user will be updated or the user will be alerted to the change. The database application may handle multiple databases and be configured to regulate access and actions for each user to the tables in the database.

There are many relational database applications that are commonly encountered in Web design: MS SQL, MySQL, Postgre, Oracle, and Sybase. Each of the databases uses queries to insert, update, read, and delete records in the database. These queries are written using structured query language (SQL). There may be dialectical differences between the syntax of SQL queries used in the various databases, but the basic approach is the same. As an example, an SQL query that assembles a new table of information from tables in Figure 3.4 to list the patients who's treatment courses were carried out at the facility indicated with the symbol "MCR" could be written as

```
SELECT Clinical_Patient.LastName, Clinical_
Patient.FirstName, Clinical_TreatmentCourse.
ClinicCourseNumber, Clinical_Facility.Symbol
FROM Clinical_Facility
INNER JOIN Clinical_TreatmentCourse
ON Clinical_Facility.Clinical_
FacilityID=Clinical_TreatmentCourse.Clinical_
FacilityID
INNER JOIN Clinical_Patient
ON Clinical_TreatmentCourse.Clinical_
PatientID=Clinical_Patient.Clinical_PatientID
WHERE (Clinical_Facility.Symbol=N'MCR')
```

To execute the query on the database server, the Web server must first connect to it with a connection string that specifies information to authenticate the connection.

When files (e.g., DICOM-RT) are exchanged, how is that handled with the database? Placing large binary files directly into the database is problematic: it significantly increases the size of file and requires separate tracking of the file type. If the data in the file need to be streamed, this slows access. A commonly used alternative it to store the file in a subdirectory on the server and then to store the file path (directory and file name) as a text field in the database. However, in that scenario, if the file is overwritten, moved, or deleted, the database is unaware of the change. Modern, enterprise-level databases introduce field types to cleanly handle files. For example, MS SQL 2008 introduced the FIELDSTREAM data type that provides the best of both approaches. The binary files are stored outside of the database for fast access, but the operating system's file system and the database software are integrated, so that SQL query statements can be used to handle the files and file changes without awareness of the database are prevented.

Protection of patient data within the database is a primary concern. Tools for protecting the data during transmission, such as using an SSL connection, have been discussed. Controlling the access that login accounts may have to the database adds protection. However, additional protection in enterprise-level databases can be added by encrypting the fields containing patient identifying information without substantial reductions in throughput. This means that, even if a copy of the table is

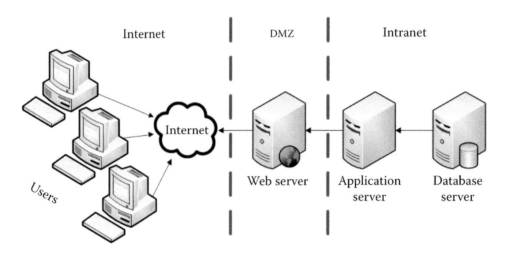

FIGURE 3.5 An example of a network.

stolen (e.g., as an export to a spreadsheet file), the data cannot be read.

Figure 3.5 illustrates a database server as part of a network of servers used to aggregate data. Although the database and Web application (IIS and PHP) may reside on the same server, this is typically avoided in enterprise solutions to preserve high availability and security. The application must be visible to the world, but the database server may reside behind a firewall. Connections to the Web server and logins to the database are restricted to preserve the security of the data. Database fields containing PHI may also be encrypted using built-in standards in the database software.

The power to quantify extensive relationships between data elements and leverage that information to avoid duplication of data elements in different tables of the database can also be viewed as a liability by groups that need to use large volumes of less structured data. The possibility that text-based storage of data can be utilized on the Web without the rigor of a relational database is the premise of NoSQL databases. With this approach, SQL queries are not used to access data. An example of a NoSQL database that has been utilized by some registry groups is Apache CouchDB (http://couchdb.apache.org/).

3.3.5 Server to Server: Web Service

A Web service allows programs running on one server to carry out data exchanges with a program running on different server on the Web. This offers significant advantages. For example, Web forms allow users to manually input data to be exchanged with the server. If entering data into an internal review board–approved national registry from an electronic medical record (EMR), the user would need to look up the needed values in the EMR one by one and then copy them into the form to be exchanged with the server. The staff time associated with manual entry of data quickly becomes a rate-limiting factor in efforts for large-scale aggregation of data into the registry to understand patient outcomes. Use of a Web service enables the user

to run a program on their server that aggregates the submission data from the EMR and then, under the supervision of the user, submit it to the registry. The programs moderating the data exchange between the servers are configured to automate limiting the types of information transmitted, preserving security, conforming to institutional policies for handing of PHI, and validating the entries. By eliminating the onerous manual effort associated with the transfer as well as the errors associated with manual copies of data, the data exchange is made more robust, sustainable, and extensible. Once the Web service architecture and standards are defined, implementing changes (additional data and additional functions) requires only incremental effort.

Web services also provide a means to encapsulate and provide functions to users without requiring them to use a particular Web form. These functions can built into user input forms to enhance their usefulness. For example, the patient demographics page of a registry Web form could potentially request the patient's social security number (SSN) to use as a key in assuring that the entered data are for the correct patient. It would then be desirable to validate the SSN against the records of the Social Security Administration (SSA). The SSA provides a Web service called the Consent-Based Social Security Number Verification Service. The service provides a callable function that accepts an encrypted Name and SSN combination and returns a value indicating if the combination is invalid, valid, or valid but the person is deceased. Access to the service is protected by a username and password embedded in the string sent to the server to request validation. With this approach, the person designing the demographics page for the registry does not need access to or understanding of the databases used by the SSA to provide validation. They need only to write a few lines of code to send the validation request to the SSA and then inform the user of their Web form if the number is valid. By carefully controlling the input and output data, and hiding the details of database queries, security is maintained.

How do Web services work? A simple example will help to illustrate the concepts involved in the function and communication of

Web services. Just as for Web forms, the Web service is invoked by a URL that points to a file on the server. For example in the URL http://www.mockup.edu/ClinicDemo.asmx?op=Eq2GyDVH, points the file ClinicDemo.asmx. In our example, this file contains a single line

```
<%@ WebService Language="C#"
CodeBehind="ClinicDemo.asmx.cs"
Class="WebServiceDemo.ClinicDemo"%>
```

The CodeBehind attributed indicates the file containing the methods that can be invoked on the service. The URL invokes a method in that file named EQ2GyDVH. Figure 3.6 illustrates ClinicDemo.asmx.cs. It exposes just one method, EQ2GyDVH, that accepts as input two single value numbers as well as a collection of paired numbers. The classes, members, and methods exposed in the file use the same programming language as used for other applications, with some minor declaration differences. The example is written in C#.Net, but the principle of extending language tools is the same for a PHP Web service. In this example, the use of the [WebMethod] declaration allows defining which methods are available for use by other servers posting requests to Web server.

The Web server needs to be able to inform other servers on the details of what methods it provides, input parameters and output data, security inputs required, etc. This is accomplished with Web Services Description Language (WSDL). WSDL is,

not surprisingly, an XML file with its schema defined by W3C. Figure 3.7 illustrates a portion of the WSDL file describing the Web service and available methods. It is a complex file. Because the creation, transmission, and reading of the file are handled automatically by the Web server, users do not generally need to become familiar with the syntax to understand the concept.

If the Web page program or even a desktop application invokes a method provided by a Web service, then it must exchange information with the Web service. The communication protocol used for exchanging data between a client and the Web service to invoke methods is an XML-based format known as Simple Access Object Protocol (SOAP). Figure 3.8 illustrates a SOAP message for posting a request to the Web service. The protocol uses tags to identify the Envelope, Body, and an optional Header of the SOAP message. The envelope identifies namespaces used to interpret the message in the body. The Web server coordinates serializing and deserializing the data into XML. Because firewalls protecting computers using the HTTP protocol allow transmission of XML, the SOAP protocol is advantageous. Although information could be more succinctly represented in a custom binary format, it would not be passed by HTTP.

The SOAP message may contain sensitive information such as PHI that should be protected when transmitted from the client to the Web service. If the connection uses an HTTPS connection, the transmissions between the client and the server are encrypted and protected. It is also possible to encrypt the SOAP message using Web services security (WS-Security). This limits

```
using System;
using System.Collections.Generic;
using System.Linq;
using System.Web;
using System.Web.Services;

namespace WebServiceDemo
{
    [WebService(Namespace = "http://cmdemo/webservices/")]
    [WebServiceBinding(ConformsTo = WsiProfiles.BasicProfile1_1)]
    [System.ComponentModel.ToolboxItem(false)]

    public class ClinicDemo : System.Web.Services.WebService
    {
        [WebMethod]
        public List<DVHPoint> Eq2GyDVH(int nfractions,float alphabeta, List<DVHPoint> inputdvh)
        {
            List<DVHPoint> returnvalue = new List<DVHPoint>();
            DVHPoint x = new DVHPoint();
            if(nfractions >0 && alphabeta >0){
                foreach (DVHPoint dvhp in inputdvh)
                {
                    x.dose = dvhp.dose * (1.0f + (dvhp.dose / nfractions) / alphabeta) / (1.0f + 2.0f / alphabeta);
                    x.volume = dvhp.volume;
                    returnvalue.Add(x);
                }
            }
            return returnvalue;
        }
    }

    public struct DVHPoint
    {
        public float dose;
        public float volume;
    }
}
```

FIGURE 3.6 The code used to define the methods publicized in a Web service can easily be coded using the same programming tools used for creation of desktop applications.

```
<?xml version="1.0" encoding="UTF-8"?>
- <wsdl:definitions xmlns:wsdl="http://schemas.xmlsoap.org/wsdl/" targetNamespace="http://cmdemo/webservices/"
  xmlns:http="http://schemas.xmlsoap.org/wsdl/http/" xmlns:soap12="http://schemas.xmlsoap.org/wsdl/soap12/"
  xmlns:s="http://www.w3.org/2001/XMLSchema" xmlns:tns="http://cmdemo/webservices/" xmlns:mime="http://schemas.xmlsoap.org/wsdl/mime/"
  xmlns:soapenc="http://schemas.xmlsoap.org/soap/encoding/" xmlns:tm="http://microsoft.com/wsdl/mime/textMatching/"
  xmlns:soap="http://schemas.xmlsoap.org/wsdl/soap/">
  - <wsdl:types>
    - <s:schema targetNamespace="http://cmdemo/webservices/" elementFormDefault="qualified">
      - <s:element name="Eq2GyDVH">
        - <s:complexType>
          - <s:sequence>
              <s:element name="nfractions" type="s:int" maxOccurs="1" minOccurs="1"/>
              <s:element name="alphabeta" type="s:float" maxOccurs="1" minOccurs="1"/>
              <s:element name="inputdvh" type="tns:ArrayOfDVHPoint" maxOccurs="1" minOccurs="0"/>
            </s:sequence>
          </s:complexType>
        </s:element>
      - <s:complexType name="ArrayOfDVHPoint">
        - <s:sequence>
            <s:element name="DVHPoint" type="tns:DVHPoint" maxOccurs="unbounded" minOccurs="0"/>
          </s:sequence>
        </s:complexType>
      - <s:complexType name="DVHPoint">
        - <s:sequence>
            <s:element name="dose" type="s:float" maxOccurs="1" minOccurs="1"/>
            <s:element name="volume" type="s:float" maxOccurs="1" minOccurs="1"/>
          </s:sequence>
        </s:complexType>
      - <s:element name="Eq2GyDVHResponse">
        - <s:complexType>
          - <s:sequence>
              <s:element name="Eq2GyDVHResult" type="tns:ArrayOfDVHPoint" maxOccurs="1" minOccurs="0"/>
            </s:sequence>
          </s:complexType>
        </s:element>
      </s:schema>
    </wsdl:types>
  - <wsdl:message name="Eq2GyDVHSoapIn">
      <wsdl:part name="parameters" element="tns:Eq2GyDVH"/>
    </wsdl:message>
```

FIGURE 3.7 The WSDL file used to communicate the capabilities of a Web service to clients is an XML file.

```
POST /ClinicDemo.asmx HTTP/1.1
Host: localhost
Content-Type: text/xml; charset=utf-8
Content-Length: length
SOAPAction: "http://cmdemo/webservices/Eq2GyDVH"

<?xml version="1.0" encoding="utf-8"?>
<soap:Envelope xmlns:xsi="http://www.w3.org/2001/XMLSchema-instance"
               xmlns:xsd="http://www.w3.org/2001/XMLSchema"
               xmlns:soap="http://schemas.xmlsoap.org/soap/envelope/">
  <soap:Body>
    <Eq2GyDVH xmlns="http://cmdemo/webservices/">
      <nfractions>int</nfractions>
      <alphabeta>float</alphabeta>
      <inputdvh>
        <DVHPoint>
          <dose>float</dose>
          <volume>float</volume>
        </DVHPoint>
        <DVHPoint>
          <dose>float</dose>
          <volume>float</volume>
        </DVHPoint>
      </inputdvh>
    </Eq2GyDVH>
  </soap:Body>
</soap:Envelope>
```

FIGURE 3.8 HTTP-based Web services exchange data with SOAP messaging. SOAP messages are XML data enclosed in a set of <soap:Envelope> and <soap:Body> tags.

controls access data at the application rather than at the server level. WS-Security uses XML Signature and XML Encryption to protect the exchange. Both are governed by W3C (http://www.w3.org/standards/xml/security). XML Signatures are used to authenticate the content of the XML file. Applications used for creating Web services (e.g., Visual Studio, Java, and PHP) include methods for signing and encrypting SOAP messages. Obviously, these programs may also be used to automate anonymizing patient data.

FTP uses the Transportation Control Protocol to transfer data as a stream of bytes. Because it is possible for a program to trigger a server to initiate an FTP file transfer, Web services may use this approach to transfer binary data files. Although FTP is dated technology, the loose coupling of the upload and processing allows the sender to begin submitting data before the receiver is ready to process the data. It is not tied to (and does not enforce) any particular format. Any validation, processing, error handling, and messaging must be designed outside of FTP. It is only a file exchange.

3.3.6 Near-Term Evolutions

The availability of the Internet and connection speeds has both seen exponential gains over the last decade. Technologies for operating systems, which underpin all of the applications with which users interact, have developed to allow decoupling of the operating systems from an individual server computer. This allows servers to be virtual machines, with the advantage that if the physical hardware on one server fails, the software can seamlessly shift operations to a different server, without interrupting service. Advances in hardware and software technologies that improve the speed and the volume of data that may be managed and assure high availability of servers enable looking to applications to operate as services on the Web. This requires a shift in paradigms for design of applications to operate as interacting of Web services. Service oriented architecture (SOA) is a term that generally refers to Web service design principles that enable applications to evolve from multiple interacting Web services that may exist on a wide range of server types and be written in a wide range of languages. SOA emphasizes defining standards for interfaces that enable the distribution of services across multiple servers.

Thomas Fielding's doctoral dissertation in 2000 introduced Representational State Transfer (REST; http://www.ics.uci.edu/~fielding/pubs/dissertation/rest_arch_style.htm) as an architecture style and set of design principles. These have been embraced by many as important for facilitating the growth of Web services. REST "emphasizes scalability of component interactions, generality of interfaces, independent deployment of components, and intermediary components to reduce interaction latency, enforce security, and encapsulate legacy systems." Applications are sometimes described as RESTful to indicate consistency with these principles. We have seen in the example of Figure 3.6 an XML-based SOAP message used in the communication with a Web service. A RESTful Web service might instead allow exchange of the information in a JSON string that is attached to a URI invoking the Web service. Clarity, consistency, and standardization analogous to W3C are still evolving for REST.

HTTP-based Web services that communicate with SOAP messages are widely used. However, as SOA and REST have emerged, limitations of this approach for developing Web-based applications have become apparent. Windows communications foundation (WCF) was introduced to address these issues. It allows adding support for multiple protocols (e.g., HTTP, Transportation Control Protocol, and namedpipes), transfer of binary as well as text-based data, streaming of data, and additional security protocols. Like Web services, WCF uses WSDL to communicate its services to clients. Although Web services will continue to be an important part of Web-based information delivery, WCF will grow in utilization. Similarly, within the Java and PHP communities, RESTful design of Web services are becoming more frequently encountered. Databases, such as CouchDB, is RESTful.

Another concept gaining in utilization is cloud-based computing. Ability to exchange data on the Web implicitly depends on computer hardware, database, application, and Web servers. There may be substantial variation in demand for server capacity as a result of variability in demand volume from users of the Web site. As users are added, capacity must be added. Maintaining sufficient capacity to accommodate peak demand and to anticipate grow can require substantial technological and staffing resources. The ability to virtualize and cluster servers to maintain highly available systems running in large servers farms creates the ability to provide this capacity as a service. This is the cloud model for infrastructure. A few groups have begun to offer cloud computing services: EC2-Elastic Compute Cloud from Amazon, Azure from Microsoft, OpenStack from RackSpace, and Google Cloud Platform.

When cloud-based solutions are considered as part of database efforts that involve patient information security concerns are heightened. This caution is legitimate. Databases used in cloud architectures used may not be sufficiently mature to provide the same level of protection that would be expected for institutional systems. In the interest of speed, vendors may elect to host the data of many institutions within a single database or they may not encrypt fields containing information that can be used to identify a patient. Choices such as these complicate ability to use these solutions while still assuring institutional information technology (IT) teams that adequate protections are in place.

3.4 Summary

Technologies for input, aggregation, and sharing of data using the Internet while preserving the integrity and security of the data are well established. Thus developing a thorough understanding of the underlying technologies to be used as part of a Web-based information delivery, is important. Institutions are obligated to protect the data of their patients. When the transfer and storage occur outside of the institution, as in cloud-based approaches, then special care must be taken.

When considering options for Web-based information delivery, it is important to consider carefully how general concepts of the problem to be solved are decomposed into well-defined data type and relationships. This decomposition, understanding of available technology options, security requirements for handling of the data, and programming skill sets of team members creating the solution are needed for developing a successful approach to Web-based information delivery.

The annualized cost for establishing a custom Web site with an SSL certificate and SQL database with a commercial Internet service provider can be quite low (e.g., from $200 to $500). However, to assure security necessary to protect PHI, it may be an IT requirement to host the servers within the institutional network. This implies substantially more cost for hardware and professional services but assures protection of PHI. Tools for developing Web sites and services using ASP.Net or PHP enable rapid design and deployment. Significant programming and database are skills need to be provided to set up the services. For small-scale or prototype efforts, these are typically provided either by radiation oncology physics or by IT staff. As efforts expand to require interactions of many users and assure high availability, professional IT staff is the better choice. There are a growing number of vendors providing services for Web-based data exchange, targeted at radiation oncology. The overall cost including hardware and professional services for any of these options may be considerable and will be impacted by the technology decisions and requirements.

Technologies for databases, Web services, and Web hosting continue to evolve for good reason. Developments push to provide better reliability, security, speed, and ease of use. It is impractical to rewrite every system as new technologies are developed, and legacy systems will always be a part of the landscape. However, significant dedication to creating the best applications is required by individuals and groups developing software solutions for Web-based transfer solutions to remain current. When considering options for new solutions, either vended or developed "in-house," inertia can be a cause for concern.

References

Anderson, J. C., et al., *CouchDB the Definitive Guide*, O'Reilly, Sebastapol, California, 2010.

Castro, E., *XML for the World Wide Web*, Peach Pit Press, Berkeley, California, 2001.

Esposito, D., *Applied XML Programming for Microsoft.Net*, Microsoft Press, Redmond, Washington, 2003a.

Esposito, D., *Programming Microsoft ASP.Net*, Microsoft Press, Redmond, Washington, 2003b.

Fehily, C., *SQL*, Peach Pit Press, Berkeley, California, 2002.

Fielding, R. T., *Architectural Styles and the Design of Network-Based Software Architectures*, University of California, Irvine, California, 2000.

Ullman, L., *PHP for the World Wide Web*, Peach Pit Press, Berkeley, California, 2001.

Ullman, L., *PHP Advanced for the World Wide Web*, Peach Pit Press, Berkeley, California, 2002.

4

Electronic Medical Record

Alexis Andrew Miller
Illawarra Cancer Care Centre
and
University of Wollongong

4.1 Introduction

Radiation oncology informatics is an area in its infancy. Although radiation oncologists will agree with the importance and primacy of patient data in the evidence-based medicine era, the reality of their patient's electronic medical record (EMR) data may not mirror this appreciation. Most departments have an EMR, but few would be able to produce an accurate "hands-off" clinical report, that is, a database report that details a patient group based on diagnosis, stage, treatment, and details outcomes without physically looking inside a patient's record. The EMR details the patient's radiation treatment parameters consistently and in fine detail but only because these data are entered "religiously" and are repeatedly quality assured as the basis for safe radiotherapy by a group (dosimetrists and therapists) who work intimately with this data. Why should the diagnosis and stage data be of a lesser quality? Simply because medical staff frequently do not to enter these data in the allocated places with the result that it cannot be used for reporting. Whether this is due to choice, thinking that stating it in a dictated notation is sufficient, or being forced by well-meaning bureaucrats to enter the data in a separate system, the reasons for this failure are numerous (Lapointe & Rivard 2005). Irrespective of reason, the result remains the same: the medical data of millions of patients seen and treated in radiation oncology departments are not available for systematic aggregation and analysis. Yet, radiation oncology is one of, if not the most, suitable medical specialties for an informatics approach.

Readers from an informatics background may be astounded how well structured radiation oncology concepts are. Yet, there is a substantial barrier to the description and manipulation of this knowledge for the informatics domain expert or informatician. The radiation oncology domain knowledge is specific and largely opaque to the informatician. It should be considered that most radiation oncology staff has spent a decade of learning before they are comfortable in dealing with and describing the radiation oncology knowledge structures, so there is a considerable learning curve for the informatician.

Readers from a radiation oncology background may find the informatics concepts difficult, but unless a significant number of radiation oncology professionals come to understand these concepts (Bakken 2001; Greenberg et al. 2010), our information technology (IT) systems will remain dysfunctional. The remaining radiation oncology staff needs to ensure that their data collection behaviors are constant and consistent. It is worth pointing out that informatics is not "computers." Although computers are a useful tool for the rapid, repeated manipulation and correlation of well-structured information, informatics is actually about the structure and use of knowledge, and no one can better appreciate and advise on radiation oncology knowledge than the radiation oncology professionals. Radiation oncology informatics does not need a knowledge structure agreed on by everyone, it needs a knowledge structure that can be demonstrated to work for the major items in the clinical record that are typically used for determining treatment. Although many informatics researchers have tried to extract medical knowledge

37

structures from natural language text, the reverse process (knowledge structure written into natural language) is more likely to demonstrate to radiation oncologists that the knowledge structures defined by informaticians are in fact consistent with the real knowledge structures.

4.2 EMR

The term "electronic medical record" is not precisely defined. The term is interchanged with "electronic health record" in some circumstances, although EMR is more restricted to the portion of the health record that describes the disease and interactions with medical staff. There is general agreement that the term does not indicate control by the patient. Those systems are called patient health records or patient-controlled health electronic records. Being at liberty to define the record, I have chosen to use a wider definition than is usual. For the purposes of this chapter, the EMR is defined as an electronic system used in the radiation oncology department, which contains data that are part of the clinical data set. This is based on the assumption that any data pertaining to disease description, management, and outcome are "medical" in nature and should be recorded and hopefully in a suitable electronic format. The word "medical" therefore is used in a professional sense, that is, the "electronic medical record" is any electronic system that holds medical information pertaining to a patient's clinical management, although the prime focus of the system may not seem to be medical data.

The EMR is also useful for billing, resource management, and scheduling functions within the radiation oncology department, but this discussion begins with the premise that the purpose of the EMR is to store medical knowledge. Much of the following discussion revolves around the way that the knowledge is fragmented for storage and the subsequent uses of this stored knowledge, which may result from the repository.

4.3 What Is Special about Radiation Oncology IT?

Radiation oncology is a disease-centered (cancer) and modality-centered (radiotherapy) tertiary care medical specialty that utilizes radiotherapy in the management of cancer (Miller 2003). This only occupies a small position in a patient's lifelong care. In most health systems, patients will be referred to a radiation oncologist by another secondary care medical practitioner, usually medical specialists (surgeons and internal medicine physicians), who have already obtained a histologic diagnosis and completed preliminary staging investigations. Most of these patients will have a general practitioner/family doctor. Patients with clinical and imaging features of malignancy without a histologic diagnosis are in the minority because occasionally benign disease may be managed with radiotherapy.

Following the establishment of a cancer diagnosis, an accurate description of the extent of the disease (known as "staging") is undertaken. Staging parameters are specific to the cancer

diagnosis. Subsequent decision-making is structured and driven by the disease outcomes reported in the radiation oncology literature. The prevailing model of care is biomedical with a heavy dependence on a detailed understanding of the disease and its response to therapy. Radiation oncology and medical oncology overlap in their disease focus but differ in their modality of therapy.

Radiation oncologists are of the general opinion that patients with cancer should have no delay in the provision of their radiotherapy (Hamilton et al. 2004; Kenny & Lehman 2004). During the delivery of radiotherapy, weekly review of patients is routine, and at completion, patients enter an indefinite period of episodic follow-up, which typically may terminate after several years without recurrence (National Cancer Institute). Thus, the radiation oncologist's care is distinctly episodic, focused on a sequential workflow, and exclusive although undifferentiated by age, gender, and organ system.

The medical knowledge base of radiation oncology is largely explicit and derived from the published literature reports of the application of radiation to biological systems. Clinical judgment is applied within the context of the literature's predicates to ensure optimal outcomes for individuals. Endpoints described as improved survival, local control, and successful palliation are the usual benefits offered to patients. The specific term for this is "evidence-based medicine," which ensures that treatment without benefit is not offered.

Given that patients with cancer have what is commonly a single and well-classified problem, decision-making in radiation oncology follows a deductive path where verifiable and well-documented information is applied to the precise problem of improved outcomes while always reflecting the underlying biomedical model of medicine in operation. The cancer problem is structured. It is defined by the disease (instantiated by a diagnosis and stage), modulated by a series of pertinent prognostic factors known to impact on the disease behavior, and directs the use of therapy options that have already been shown statistically to provide an improved outcome. Other patient factors are considered following this determination, such as patient choice and an assessment of the patient's ability to realize the benefits on offer (Medical Oncology Group of Australia).

The EMR for radiation oncology should reflect these issues if they are to be stored. The need to use accrued data to inform future treatment should be catered for and so must be evident in the knowledge structures on which the design of the EMR stands.

4.4 EMR in Radiation Oncology

There are many electronic systems used in radiation oncology. None of the systems are decorative because they all deal with medical or radiation information for the purposes of safe and effective treatment of patients. Informatics is therefore a crucial but underrecognized component of the radiation oncology system. Although it is tempting to think of two silos, one of medical data and another of radiation data, the reality is that all the medical data can be related to all of the radiation data. This relationship exists because the data are accrued from the treatment of patients. Although it is common that data analysis will group medical or radiation data, increasingly there are reports based on medical categories to look deeper into the radiation data (Peters et al. 2010).

The presence of multiple separate electronic systems in a single radiation oncology department seems to defy this definition; however, this circumstance is largely a historical eventuality where defining a complete system has been harder than a system that addresses core functionality separately. Cooperative and collaborative software development has not been a normal feature of the commercial companies producing medical systems. Although this compartmentalized approach has produced very useful software, it has also resulted in a silo-within-silo eventuality that sees the isodose specification for a plan being stored separately from the clinical data justifying the treatment and the data describing the treatment outcome. It is not impossible to overcome this situation, and because this separation is suboptimal, any discussion about the EMR should point to ways to remedy the deficiency. Indeed, only by seeing a future with these data integrated can the motivation exist to overcome the separation. If the radiation professionals fail to recognize the potential for utilizing the breadth and interconnection of our medical data and to achieve this storage in one site, then we will invariably be left with the underperforming systems we have at present.

Can it be shown that this unintegrated system is not helpful? In 2006, Andrew Jackson from Memorial Sloan-Kettering Cancer Center presented a paper proposing new and different ways of assessing toxicity. The paper used the data of seventy-seven patients treated for lung cancer including their clinical toxicity data. Although this approach was innovative, the patient numbers prevent real conclusions (Jackson et al. 2006). The Digital Imaging and Communications in Medicine (DICOM)-RT container can be stored in a Picture Archiving and Communications System (PACS). Departments can share and collate details from within DICOM-RT files. However, there are no data within these files to allow the selection of the correct group of patients and to correlate the acute and late side effects of treatment. It is almost certain that every radical intent Stage I to IIIa lung cancer treated in the United States at the time of publication had a plan that includes lung dose-volume histogram (DVH) data. With the inclusion of clinical and toxicity data in a DICOM-RT file, the potential pool of patients for lung toxicity studies in stage I to IIIa lung cancer is tens of thousands, not seventy-seven as in Jackson's study.

The magnitude of this task appears daunting after considering the work required to manually assess clinical records to find patients with complete data with the disease, stage, and outcome measures of lung toxicity recorded and then to retrieve, open, and assess the plan DVH. Could this be done more easily? If disease, stage, and treatment toxicities are entered into the EMR at each visit in standardized nomenclature, the searching for the required group does not require that researchers open any patient files. If the treatment plan used for each patient were stored as a DICOM-RT in a PACS, the required DVH data could be reproduced without researchers having to open any DICOM files. In fact, the more tantalizing question is why the clinical, radiation, and outcome data cannot be stored within the DICOM-RT container?

The systems that hold medical data but perform other tasks are addressed throughout this book. The systems already familiar to radiation staff include the electronic radiotherapy system which is more commonly known as the "Record & Verify" (R&V) system (sometimes referred to as the Treatment Management System), the electronic chemotherapy system, and the treatment planning system (TPS). However, other relevant systems also include online data aggregation [e.g., National Radiation Oncology Registrar (Palta et al. 2012)], electronic trial systems, electronic guideline systems, and decision support systems.

4.5 Informatics

Radiation oncology data have multiple uses that cannot be achieved without integration of electronic clinical, radiotherapy (TPS and R&V), and chemotherapy systems. Radiation oncology informatics is about the radiation oncologist's data and knowledge—not so much its storage but particularly its retrieval and reuse to generate new knowledge.

Radiation oncology knowledge is intimately linked to a work process that matches information discovery and flow (Figure 4.1). Because the radiation process is not a free-for-all, the EMR must embed business process mapping and execution tools to constrain work and information flow of radiation professionals in the EMR. Particularly, the EMR needs to support best practice, not infinite and whimsical customization that is the antithesis of quality assurance (QA).

In terms of evolving IT issues, despite its problems with security, all IT initiatives are moving toward Web-based information collection, manipulation, and remote storage. A significant portion of medical and radiation oncologist workload and knowledge now involves images. There are informatics challenges and opportunities integrating radiology with radiation oncology. It is not clear where ROs should store their data—whether this should be in a PACS with images or in a relational database management system (RDBMS) separate from images. Because the ultimate goal is the sharing of stored medical knowledge, the IT decisions involved must include oncologists rather than being made unilaterally by commercial interests.

The efficiency and effectiveness of the radiation oncologists' knowledge will be enhanced by an explicit common terminology and knowledge representation, which is called an "ontology" by informaticians. Only when this is achieved can the integration of genomics be attempted with the clinical data repository. The Iressa story where demographic details predict response to a small-molecule therapy demonstrates that the EMR remains very relevant in the age of genomics (Wu et al. 2007). To date, a problem with the caBIG project (Buetow & von Eschenbach 2006) has been the lack of a corresponding "caSMALL" project working with oncologists to produce a system for routine

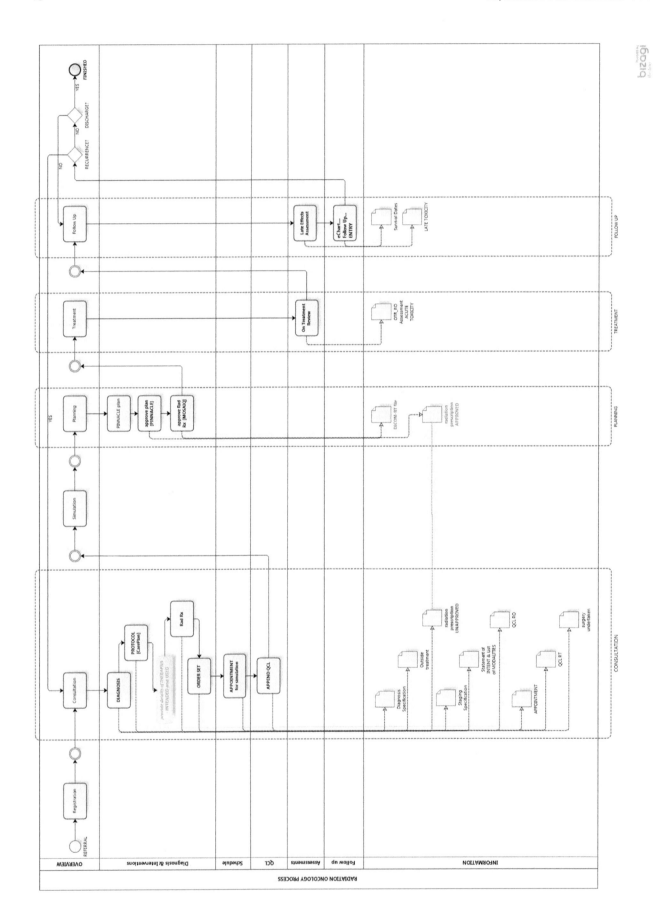

FIGURE 4.1 Radiation oncology workflow map with information production points.

use that provides and coordinates required clinical data. The most effective design and use for an EMR can only result from envisioning the EMR as the informatics tool for research in Oncology. Unfortunately, the design of the EMR is business and IT oriented and does not reliably suit the changing medical data needs of the oncologist. Certainly, collation of data between departments remains a dream or nightmare depending on your task. The current design paradigms and user interface construction methods prominent in proprietary systems have not been helpful in achieving consistent and useful interfaces for radiation oncologists to standardize entry or storage.

Any discussion should also look to the future of the EMR and new data sources and collection methods. The current challenge and opportunity is the technology of mobile devices, whether phone or tablet. With the pervasiveness of this technology, the patient-derived assessment (PDA) is emerging as a useful and plentiful data source, which is useful because Outcomes Modeling requires clinical data, and more and real data will model better. Given the size and processing power of these devices, it is difficult to transfer the EMR entirely to these devices. Different radiation professionals will derive different benefits depending on the peripatetic nature of their workflow.

4.6 Why Store Radiation Oncology Data in an EMR?

The answer to the question of what you are going to do with an EMR has a large influence on how you assess EMRs for purchase, how hard you will push for implementation (Miller & Phillips 2006), and whether you respect the information collected in the system. I start with the concept that my EMR should hold my prospectively obtained clinical data. These data can be superior to any trial system and can be reused (Miller 2006). Therefore, the aim is for my EMR to be able to store my routine clinical knowledge in such a way that I can verify its accuracy and then manipulate it as required to discover knowledge that answers clinically relevant questions without requiring expensive clinical trials. Furthermore, if other oncologists match the pattern of my EMR use, we will be able to pool our data for these analyses.

The notion of data ownership is crucial. The belief that nurses, radiation therapists, or trial coordinators can maintain the veracity of an oncologist's data is fanciful. To verify that the oncologist's knowledge has been disaggregated accurately and entered correctly into the right place requires the understanding of the oncologist. It is a domain expert task. Filled boxes are not the same as boxes filled correctly and consistently. The important question that follows is—What is the expert domain data that only the oncologist must enter, verify, and quality assure? The answer lies in the understanding of the data set without which the oncologist could not adequately manage the patient. Can a patient be managed appropriately by an oncologist or therapist without knowing their name? Of course, the important issue is identifying the patient that can be adequately undertaken with images. Who must have the patient's name then? Only the

clerical staff NEEDS the patient's name for the sending of letters and bills. Similarly, I cannot function as an oncologist if the diagnosis and stage are incomplete. I need them to be accurate and correct. Who then should be responsible for their entry and QA if not me? This, in effect, is what it means to "own" your data (Miller & Phillips 2006). This ownership issue does not address who should be able to look at and use the data.

With these issues in mind, there are two important qualities to the EMR of my choice. First, the way that my knowledge is disaggregated when it is stored inside the database must be predictable and match a formal description of radiation oncology knowledge structures. If this occurs, then I will be able to reaggregate my data into a structure coherent to any radiation oncologist. Unfortunately, this aspect of the EMR is poorly appreciated. Most departments accept the recommendation of the EMR salesman that you can get your medical data back out. There are few publications that demonstrate that this is even possible (Miller 2003), because most studies have just reported on high levels of use (Han et al. 2005; Yu et al. 2010; Colonias et al. 2011). It is commendable that more departments are storing their data electronically. However, if that is all that electronic systems have gained for us, then a lot of expense has been wasted. The second quality is the intuitive nature of the graphical user interface where data are entered, and in this regard, clarity and simplicity are more useful than pretty.

Further to the issues of data structure in storing and reconstituting knowledge, domain experts must develop the skills that allow them to appreciate the biases of software programmers who translate perceived requirements from domain experts into software. If your aim is just to remove paper from your department, then any system that stores unstructured text will suffice. If your aim is to improve billing, then do not be surprised if clinical data structures are sacrificed for a better accounting operation. The software that is written without a true knowledge of required specifications is unlikely to be fit for the purpose of knowledge reuse.

4.7 Radiation Oncology Systems

Many systems in the radiation oncology department hold medical data, and they started as separate systems. These silo systems include the electronic radiotherapy system more commonly known as the R&V system, the TPS, the electronic chemotherapy system, the electronic document repository, and the electronic resource management. Other related systems include electronic systems for online data aggregation, trial management, guideline deployment, and decision support systems. Failure to recognize the breadth and interconnection of medical data will invariably result in an underperforming system.

Early on, computers were employed in radiotherapy for the automated checking of parameters set by humans. When matched and approved ("verify"), the field settings were then recorded ("record"). By and large, this has been successful, although some of the errors in software development were highlighted by the Therac25 radiation accidents in the mid-1980s

(Leveson & Turner 2002). These errors are used in software classes as archetypes of how NOT to program. Notwithstanding the Therac25 debacle where an R&V system malfunctioned, these systems have improved patient safety with a reduction in errors, although errors still occur upstream (Calandrino et al. 1997) and within the R&V (Huang et al. 2002).

The original system consisted of a database with stored treatment and tolerance parameters entered manually by the dosimetrist. Later on, values were transferred from a TPS. At treatment time, hardware intercepts on the gantry detect the settings of the machine and pass the values to the software, which then compares the intended and real parameter values, warning if there is a mismatch. Overriding the warning required human intervention. More recently, the newer R&V systems have the expanded capability of driving the linac without human intervention, hence the change from R&V system to Treatment Management System.

The benefit sought from computerized systems in radiotherapy is the improvement in safety for patients. However, because computerization has increased, the errors have changed rather than having been eliminated. The errors continue to group at the human–computer interface (Patton et al. 2003). Today, it is unlikely that a field will be omitted when using auto-field sequencing, but the probability still remains that the wrong patient may be on the bed. It will not be long before facial recognition software is deployed in the radiation maze to verify the identity and numbers of people entering the bunker. The overdose case in New York in 2005 sounds a strong warning about the problems of electronic systems interfacing with humans (Bogdanich 2010; Roberts & Marsh 2010). Previous examples prove the difficulties occurring at the human-computer interface (Vatnitsky et al. 2001; Borrás 2006; Holmbert 2007).

Similar to radiation treatment delivery, initially radiation planning consisted of manual contours, measured isodose charts, and manual planning at a required SSD of 100 cm. However, the deposition of radiation is predictable and well described by formulas. So early on, it was recognized that computerization would produce a better prediction. The TPS came into being as a way to improve accuracy. Early on, DEC PDP11/77 computers with an array processor to accelerate calculations and a separate graphics processing system were used. They used a variant of the VMS/RT11 operating system called STX. The planning software was written in computer languages such as FORTRAN. These computers were bulky and freestanding, and the outputs were manually transferred to the R&V software. Initially, these systems only permitted "2D planning," that is, the dose calculated in a single plane assuming an equilibrium dose contribution from the planes above and below. A corollary was that the speed of calculation allowed for better algorithms and the use of SSDs that were not 100 cm.

Computerized planning has proven its superiority over manual planning. Technical advances that were enabled include rapid dose calculation at variable SSD (permitting isocentric treatment), reduction in calculation error rates, more accurate predictions of dose deposition (better algorithms, heterogeneity

correction through the use of Hounsfield CT numbers as a measure of electron density), as well as the ability to display dose superimposed over a computed tomography (CT) image. Although not often highlighted, the use of CT images within the TPS has permitted clinicians to describe anatomical (contours) and disease (volume) parameters within a radiation plan, essentially operating as "imaging surgeons" and driving the technical development of inverse planning algorithms to produce highly conformal radiation plans. Seeing the disease volumes drawn over the CT images immediately drives the desire to improve conformality. With the availability of multicore servers and graphical processing units and "the cloud" (which are, in fact, very large data centers that can dynamically recruit computing power as required), dose predictions based on Monte Carlo algorithms are achievable.

Replacement systems were UNIX based, more powerful, and able to transfer output electronically to the R&V software. They permitted the delineation of structures, visualization of target geometry and dose deposition in 3D space, as well as more accurate 3D calculation of dose. As software developed, the ability to "inverse plan" and calculate "dose-volume histograms" was included. Modern systems are based on UNIX, Windows, and Linux using PCs, although there is an increasing trend toward server-based systems as the cost of servers fall and faster copper and optic fiber networks become common. In these systems, the desktop PC becomes a "thin client" exchanging graphics and keystrokes/mouse motion with the server that is able to load balance all its processing demands.

Given the usefulness of the R&V system at the treatment machine, it was an obvious step to add a schedule to manage the workload at the bunker. It was then a small step to transfer this functionality back to the clinic where there was the need to collect and store clinical data such as letters. In the 1990s, the modern EMR in the radiation oncology department emerged in the form of Varian's VaRiS (later reappearing as ARIA), IMPAC's MultiAccess (rebadged as Siemens' LANTIS and later reappearing as MOSAIQ), and MDS-Nordion's OnCentra (later sold to Nucletron) to name the most prominent internationally available examples. These products were expanded to include the R&V for radiotherapy and chemotherapy as well as resource, scheduling, and workflow management and document repositories. All of these systems were based on the business world paradigm of the RDBMS. As a result, one can no longer obtain just an R&V system, it also comes attached to electronic resource management, an electronic document repository, and an electronic clinical database. In many cases, an electronic chemotherapy system is also incorporated. However, until recently, there has been little integration of clinical images as occurs in a PACS.

The RDBMS should be considered in more detail. The RDBMS stores knowledge that has been broken up into smaller pieces of information or data items that repeat frequently. The particular pattern of disaggregation of knowledge for any RDBMS is described in an entity-relationship diagram (ERD). This diagram reflects the presumed relationship of informational entities within the knowledge of the expert domain. For example, if one considers the three entities, *patient*, *diagnosis*, and *therapy*, one could make the proposition that there is a relationship between *patient* and *diagnosis* and between *diagnosis* and *therapy*. I have provided two examples of how an ERD could be structured to reflect this (Figure 4.2). I could relate the *patient* to *diagnosis* and the *diagnosis* to the *therapy* assuming that the *diagnosis* in each relationship is the same entity (Figure 4.2, Example 1). Alternately, I could relate *patient* to *diagnosis* and *patient* to *therapy*. The *patient*-to-*therapy* link comes from arguing that because *patient* is related to *diagnosis*, and *diagnosis* is related to *therapy*, the direct relationship between *patient* and *therapy* is proven (Figure 4.2, Example 2). However, this assumption has consequences. The result of this unlinking is that it will be impossible to use a database report to reliably answer the questions—why did this patient have a mastectomy? Likewise, how many of our patients had their breast cancer treated with a lumpectomy? When the link via a diagnosis does not exist, you can count the number of patients who have breast cancer and the number who also happened to have a lumpectomy

irrespective of whether that operation was undertaken for cancer or not. When you are asking for the number of breast cancer patients who were treated with a lumpectomy, the link is required. For patients with uncomplicated management of a single malignancy, the lack of linking does not cause a problem. For more complicated patients, the linking is necessary as the following real-life example will demonstrate.

This reality is further demonstrated in a patient described in Figure 4.3 who presented with contemporaneous bilateral breast cancer. The two entries reflect the organization in two of the currently available EMRs. The organization in Example 1 (EMR#1) permits directed questions of what happened for each cancer (e.g., *which side had the mastectomy?*) because laterality resides in the diagnosis and the operation is linked to the diagnosis. However, the organization in Example 2 (EMR#2) cannot correctly indicate which side had the mastectomy because of the linkage of the operation to the patient, not to the diagnosis. Even if we included an entry such as "Mastectomy R," it would still

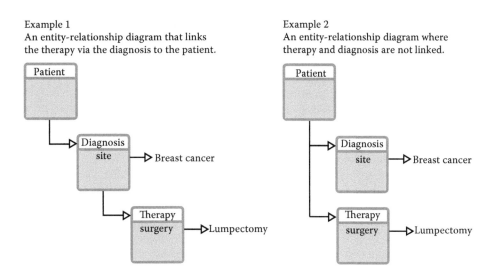

FIGURE 4.2   Possible ERD configurations for the radiation oncology EMR.

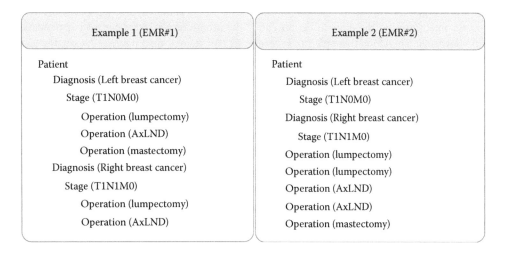

FIGURE 4.3   A real-life example demonstrating the differences in ERD construction.

not be clear that the mastectomy was performed as a result of the diagnosis of cancer. In this example, the lumpectomies were differentiated (R, L) because the system (EMR#2) would not accept two lumpectomies on the same date.

The alteration of "lumpectomy" to "lumpectomy R" introduces a further problem—the problem of free text. The standardization of nomenclature is required if one is to easily analyze stored data such as occurs in the EMR. Variable entries, such as "lumpectomy R," "R lumpectomy," "right lumpectomy," "lumpectomy right," "R lumpictomy" [sic], "quadrantectomy," "partial mastectomy," "R WLE," etc., can result from free text entry, making later analysis impossible unless all possible terms are included. When a system permits free text entry or free text alteration, data integrity suffers.

The RDBMS is well suited to the business world; in fact, it was developed for the business world where the need to manage interrelated standardized lists (e.g., customers, prices, items, and addresses) was difficult using a flat-file structure. The DBMS is therefore based on the knowledge structure of the expert domain that it serves, and, as shown above, this is the vulnerability of the RDBMS. A knowledge structure is required to accurately define the arrangement of tables that make up the particular RDBMS. Put another way, the structure of the radiation oncology EMR's RDBMS is intimately dependent on the knowledge structure of the radiation professionals using the system. Unfortunately, there is currently no specification of radiation oncology knowledge on which to base an EMR for radiation oncology. Although sales representatives describe current proprietary systems such as ARIA and MOSAIQ as complete solutions for radiation oncology departments, these claims are unsubstantiated by studies demonstrating that the data storage mirrors the knowledge structure of radiation oncology. Consideration of one set of functional requirements for the EMR reveals deficiencies (Miller & Phillips 2006).

Of course, how is one to know whether this has already been achieved in the present EMRs? All EMRs can store the medical data even if only as a text document or a scanned image. The way to demonstrate the faithful storage of data according to the expert domain's knowledge structure is to reconstitute the data from storage into something that the domain expert can verify as the original knowledge. Therefore, Example 1 in Figure 4.2 could be reconstituted to "The patient had a lumpectomy as therapy for a diagnosis of breast cancer," whereas Example 2 in Figure 4.2 would read "The patient has a diagnosis of breast cancer. The patient had a lumpectomy." What has been lost from the second example is the reason for the lumpectomy. Although this may not be a problem for a single diagnosis and a single operation, in the patient with multiple diagnoses (benign and malignant) and multiple operations, it becomes difficult to link diagnosis with its therapeutic management. Although it may be tempting to think that data mining techniques can do just as well, the fact remains that Example 1 of Figure 4.2 is the correct representation of medical knowledge in radiation oncology; therefore, the storage system should reflect this, and data mining can focus on finding new knowledge.

4.8 Radiation Oncology Workflow

The knowledge of the radiation oncologist is acquired sequentially during their workflow, which is quite discrete and easily described (Miller & Phillips 2006). In fact, the two are intimately linked, and it is possible to build workflow models (Figure 4.1), which demonstrate this knowledge discovery specifying "no fly" points when data are missing (Potters & Kapur 2012). Although this may sound restrictive, it is normal and good practice and does not prevent one from being a good oncologist. One can confidently state that, at the start of simulation, the diagnosis and stage of the patient must be known because the decision to treat is based on the diagnosis and stage. Similarly, at the start of treatment, the diagnosis, stage, dose, fractionation, field settings, etc., must be known. Paradoxically, the stage of disease is required to determine radiation volumes; however, it is not required for the treatment. So harking back to the issue of data ownership, the patient can be appropriately treated without knowing the stage, although those treating the patient may wish to inform themselves of the stage.

This knowledge flow can be appreciated if one looks at the possible meanings of the data item "10 cm" in radiation oncology. If you say that

$T\_size = 10\ cm$

then you can predict that if

$Field\_size = 10\ cm,$

then treatment will result in a geographical miss because the field has no margin for tumor positioning errors or physical penumbra of the radiation beams. Furthermore, if the patient with this 10 cm tumor is being treated, we can reasonably expect to see that there is a diagnosis and histology also specified. This example also demonstrates some of the QA potentials in the EMR. Where radiation treatment has been delivered, we can reasonably expect that the mandatory data elements relating to site, histology, and stage are complete and consistent. To say that these data exist in the plain text is to deny the EMR its potential. Although the proprietary systems permit customization to make diagnosis and staging data nonmandatory, the result is that the EMR is unable to be efficiently used for research. Although these data can be transferred from natural language into the EMR by data managers and radiation or nursing staff, the clinician who wishes to analyze it later must verify its accuracy. It is less work and of greater accuracy for the data owners to enter it themselves.

The use of the EMR can provide benefits for the clinician as well as for the speed and ease of workflow in the rest of the department. In the paper systems, it was not unusual to record the patient's diagnosis on most pieces of paper generated during the patient's management. The diagnosis was known at the consultation with the physician when they first visit the department. Entry of the diagnosis at this time is likely to be most accurate because the physician has just reviewed all of the clinical information, imaging, and pathologic data.

One of the features of paper systems described above was data redundancy. The diagnosis was entered many times. Unfortunately, the veracity of a diagnosis is not ensured by the frequency of its repetition. If an item is recorded 100 times and one of those is different to the other 99, it is impossible to know which entry is correct without formally inspecting the source data. The perils of copy-and-paste have already been noted with the mindless repetition of previously recorded data (Ash et al. 2004). EMRs should employ QA based on different strategies. There should be a single entry to verify as the truth. There should be several deliberate checks that coincide with use of the data. The entered diagnosis should be specifically reviewed and verified before simulation and before starting treatment. The entered diagnosis should be assessed by a report with inbuilt artificial intelligence for consistency and completeness. Finally, the process is greatly aided by multiple looks from inquisitive staff.

It may be an obvious statement, but it is impossible for an EMR to do its job or to even fail at its job when no data are entered into it. No system can work unless you use it. It is not uncommon for medical staff to expect that a software system will cater to their preferred workflow exactly and, when it does not, to refuse to use it. In all cases, the software system has undergone some degree of workflow engineering and so contains certain inherent assumptions. The one assumption that is prominent is the belief that the data will be entered when it is known. All of the diagnostic and staging items are grouped and are placed at the top or at the start of the EMR. This matches the oncologist's workflow and correctly matches the time when entry of this data is the easiest. The hunt for perfection in an EMR is commendable, but it does not exist. In fact, as the capabilities of computers and software improve, so do the characteristics of perfection.

Much of the data in the EMR will be derived from users who are also the source of the data. There are several points where the entered data represent an interpretation of external data. In these cases, the EMR should hold the primary data. The histopathology of the cancer is an excellent example and reveals a point where technological advance has shifted the IT goalposts. The histopathology entry in the EMR is derived by the oncologist from the histopathology report. Up until now, the histopathology report has been regarded as "primary data." However, advances in slide scanning now make it possible to store the whole image electronically. In comparison with the image, the report is an interpretation and therefore secondary data. Therefore, although the present day EMR is able to hold the histopathology report, the EMR of the future should also be able to hold the scanned histopathology image.

The involvement of commercial interests in the provision of the EMR is a mixed blessing. On one hand, we have EMRs available to store data and undertake the organization of the department. These systems are developed due to a need to be seen to be advancing and using the newer technologies than competitors. On the contrary, commercial companies have their own agenda with software development that does not always match the clinical agenda. Much of this agenda comes from the world of the software developer who believes in customization as a good feature of software. In the radiation oncology EMR, the ability to make diagnosis nonmandatory is not a good feature. The diagnosis is established for each and every patient receiving radiotherapy and chemotherapy. The diagnosis is the basis of patient selection for all clinical reports in the literature and, as such, is fundamental to our workflow. How can it be contemplated that the diagnosis is an optional piece of information?

Balanced against this desire, which has made the field diagnosis optional, is the fixed underlying organizational structures encoded in the software that match the U.S. medical ecosystem. My EMR has a place to enter the name of the "attending" and the "admitting" physicians. Most non-U.S. countries do not use the term "attending" or "admitting." Customization is required in these portions of the EMR, not in diagnosis and staging.

Many departments have also discovered the frustration of promises made by software sales representatives that the company will respond to suggestions and demands for software change, which are invariably made before purchase. Because none of the EMR vendors currently provide application programming interfaces, it is not possible to extend or enhance their systems to overcome these deficiencies. Because the present systems are monolithic and include an R&V that controls machines, the issue of Food and Drug Administration (FDA) approval provides substantial inertia to software alteration. The approval for medical software that controls linear accelerators is mandatory with respect to safety. However, the rest of the software bundle does not require the same level of approval. In fact, apart from consideration that the software was written using a recognized process, like ISO9001, the FDA is not approving the software as "fit for use." The effect of having the entire software suite as "FDA approved" is that there is a veneer of safety produced and that alterations to the software outside the R&V are made more difficult.

4.9 Radiation Oncology IT Future

The future format of the EMR should be considered. Although the future may not necessarily be the placement of all data "on the Web" or "in the cloud," the emerging standard for data is called Web 2.0 or the Semantic Web. Although the terms appear to mandate the involvement of the Internet, it is actually a generic view of how to describe data elements with others that surround it, so that when used it leaves the terms able to be processed electronically. For instance, a Semantic Web definition of diagnosis might contain statements such as "is a specific form of disease," "is confirmed by a histopathologic report," and "is quantified by a stage." The knowledge expressed in these formats proves to be easy to interchange also. You can easily process it into plain text format, an Extensible Markup Language (XML) format, or place it in a database. The conglomeration of the interrelated terms with their relationships specified is called an ontology. Ontology means "study of the thing" and is best illustrated by a family tree where the terms mother, father, stepbrother, cousin, and maternal great-great-grandfather carry information about relationships, gender, and separation. The ontology describes entities and relationships, and in the case of the family tree, only two relationships are important in defining the whole tree—"has_a_gender" and "is_a_parent."

4.10 Radiation Oncology Ontology

The structure of radiation oncology knowledge has not been formally described and verified by domain experts. Although all EMRs are built on some knowledge representation, the normal description of the knowledge structure is derived from a meeting with radiation oncologists that have lasted at best a few days. The failure to include a formal knowledge structure representation has implications for the use of the EMR as a source of knowledge, that is, it is the process of getting the data back out which determines the usefulness of the EMR to the professional. Putting the data into an electronic format provides a benefit only to the department.

The information recorded into the EMR must be linked according to the reasons for its presence. For instance, one current commercial EMR enables the recording of operations, but they are linked to the patient rather than the diagnosis as previously shown in Figure 4.2, Example 2 (EMR#2). Another current commercial EMR uses a linking structure shown in Figure 4.2, Example 1 (EMR#1). The result is that only the data in the EMR#1 can be faithfully reconstructed.

4.11 Novel Data Sources

The new data sources for radiation oncology include Bioinformatics sources such as caBIG and the patients themselves.

4.11.1 caBIG

The integration of individual patient data with Bioinformatics data held on the caBIG is the hope of truly personalized medicine. However, as the recent Iressa data have shown, the patient group responsive to the epidermal growth factor receptor blocker can also be predicted from demographic data (Pereira et al. 2005). It is unknown the extent to which routine demographic and historical data can substitute for laboratory determined values. Certainly, the identification of patients by demographic and symptom features will be far cheaper.

4.11.2 Patient-Derived Assessments

All data collection represents a hierarchy of usefulness and truth. The "gold standard" data source varies depending on the particular data. In diagnosis, the primary data are the images of the *in situ* disease and the histopathologic specimens obtained that determine morphology and grade. The reports describing the radiology and pathology images are therefore secondary data. In the determination of stage, when the oncologist reviews the images, staging represents secondary data.

In the domain of patient history and current health status, the primary data are the patient's response, such as appears in Quality of Life surveys. The quantification of radiation reactions, however, is determined by the oncologist from patient questioning and observation.

In the era of the smart phone, where PDAs can accrue data without oncology staff input, the question should be asked: can a survey by smart phone adequately substitute for patient questioning? (Prusoff et al. 1972; McGee et al. 2002; Zakim et al. 2008). It should be considered that the PDA can be used frequently, can have automated reminders, and can be processed automatically to determine which patients are changing physically, physiologically, and psychologically (Hareendran et al. 2012). Whether or not these PDAs are used to diagnose problems, they can alert radiation staff to a patient who has undergone expected change but requires routine review, or a more rapid unexpected change (e.g., unappreciated radiosensitivity) when specialized review is required.

There are already useful software development kits such as the OpenData Kit. The software can utilize all of the smartphone functionalities, including static and video camera, Global Positioning System, position/motion sensors, microphone, data connection, and a Central Processing Unit capable of signal processing. As an illustration, the software bundle includes three components. An application ("app") for an Android smartphone or tablet that can used by the patient to complete the PDA and that supplies the communication infrastructure to send the completed PDA by wireless, Bluetooth, or data link. Second, an XML file that specifies the PDA structure including specific questions, response types that can include single and multichoice answers, date, time, bar/QR code, image, video, sound, or geopositioning inputs, as well as question order and branching. Finally, there is a Web server to receive the completed PDA. The Web server can use HL7 messaging to automate the storage of these data inside the EMR.

The scope for PDAs covers all areas where medical and nursing staff undertakes assessments. These can be done while at home before assessment. Data derived in this way will be structured and standardized if the assessments built-in software use standardized nomenclature. The continued use of physician and nursing face-to-face assessments will allow the PDAs to be validated.

4.12 Advancing the EMR

The usefulness of the EMR can only be advanced with collaboration between oncology domain experts from all radiation oncology professional groups with informatics domain experts. This will happen more rapidly and more completely when two events occur: (1) when radiation professionals seek to understand their knowledge structures through formal study and undertake consistent and close engagement of informatics domain experts in their place of work and (2) when informatics professionals seek to understand our knowledge structures through formal collaboration seeking to codify the knowledge structures that can be attested as correct because of the consistent and close engagement with radiation professionals. The breadth of the use of the EMR must be addressed by having the portions used by each professional group developed by that group. That is, the portion of the EMR addressing medical data must be built, developed, and altered in collaboration between oncologists and informaticians. Nurses and physicists cannot usurp this role, any more than oncologists can properly advise on treatment machine user interfaces.

All oncologists do not need to be involved in this area, but some must be. The remaining oncologists must follow. The greatest advance in data collection and analysis for radiation oncology will occur with the publication of an ontology of radiation oncology, which is open source and controlled by the radiation oncology profession rather than proprietary concerns. As has occurred with the Integrating the Healthcare Enterprise in Radiation Oncology initiative, some of the questions asked of vendor sales representative would then be "Is your EMR compliant with the Radiation Oncology Ontology? Can you demonstrate that please?"

References

Ash, J. S. et al. (2004). Some unintended consequences of information technology in health care: The nature of patient care information system-related errors. *Journal of the American Medical Informatics Association, 11*(2), 104–112.

Bakken, S. (2001). An informatics infrastructure is essential for evidence-based practice. *Journal of the American Medical Informatics Association, 8*(3), 199–201.

Bogdanich, W. (2010). Radiation offers new cures, and ways to do harm. *The New York Times*, 23, sec. Health. http://www.nytimes.com/2010/01/24/health/24radiation.html.

Borrás, C. (2006). Overexposure of radiation therapy patients in Panama: Problem recognition and follow-up measures. *Revista Panamericana De Salud Pública, 20*(2–3), 173–187.

Buetow, K. H., & von Eschenbach, A. C. (2006). Cancer Informatics Vision: caBIG. *Cancer Informatics, 2*, 22–24.

Calandrino, R. et al. (1997). Detection of systematic errors in external radiotherapy before treatment delivery. *Radiotherapy & Oncology, 45*, 271–274.

Colonias, A. et al. (2011). A radiation oncology based electronic health record in an integrated radiation oncology network. *Journal of Radiation Oncology Informatics, 3*(1), 3–11.

Greenberg, J. et al. (2010). The COPD ontology and toward empowering clinical scientists as ontology engineers. *Journal of Library Metadata, 10*(2–3), 173–187.

Hamilton, C. S. et al. (2004). All delays before radiotherapy risk progression of Merkel cell carcinoma. *Australasian Radiology* (May 2003), 371–375.

Han, Y. et al. (2005). Impact of an electronic chart on the staff workload in a radiation oncology department. *Japanese Journal of Clinical Oncology, 35*(8), 470–474.

Hareendran, A. et al. (2012). Capturing patients' perspectives of treatment in clinical trials/drug development. *Contemporary Clinical Trials, 33*(1) (January), 23–28.

Holmbert, O. (2007). Accident prevention in radiotherapy. *Biomedical Imaging and Intervention Journal, 3*(2) (March), e27.

Huang, E. H. et al. (2002). Late rectal toxicity: Dose-volume effects of conformal radiotherapy for prostate cancer. *International Journal of Radiation Oncology Biology Physics, 54*(5), 1314–1321.

Jackson, A. et al. (2006). The atlas of complication incidence: A proposal for a new standard for reporting the results of radiotherapy protocols. *Seminars in Radiation Oncology, 16*, 260–268.

Kenny, L., & Lehman, M. (2004). Sequential audits of unacceptable delays in radiation therapy in Australia and New Zealand. *Australasian Radiology, 48*(1), 29–34.

Lapointe, L., & Rivard, S. (2005). A multilevel model of resistance to information technology implementation. *MIS Quarterly, 29*(3), 461–491.

Leveson, N. G., & Turner, C. S. (2002). An investigation of the Therac-25 accidents. *Computer, 26*(7), 18–41.

McGee, M. A. et al. (2002). Comparison of patient and doctor responses to a total hip arthroplasty clinical evaluation questionnaire. *The Journal of Bone and Joint Surgery (American), 84*(10), 1745–1752.

Medical Oncology Group of Australia. Understanding the cancer treatment process. http://www.moga.org.au/patients-carers.

Miller, A. A. (2003). Clinical information systems in oncology—Making a difference to patient outcomes. *Health Care & Informatics Review Online* (December).

Miller, A. A. (2006). New informatics-based work flow paradigms in radiation oncology: The potential impact on epidemiological cancer research. *Health Information Management Journal, 34*, 84–87.

Miller, A. A., & Phillips, A. K. (2006). A contemporary case study illustrating the integration of health information technologies into the organisation and clinical practice of radiation oncology. *The HIM Journal, 34*(4), 136.

National Cancer Institute. Follow-up care after cancer treatment. http://www.cancer.gov/cancertopics/factsheet/Therapy/follow-up.

Palta, J. R. et al. (2012). Developing a national radiation oncology registry: From acorns to oaks. *Practical Radiation Oncology, 2*(1), 10–17.

Patton, G. A. et al. (2003). Facilitation of radiotherapeutic error by computerized record and verify systems. *International Journal of Radiation Oncology Biology Physics, 56*(1), 50–57.

Pereira, J. R. et al. (2005). Gefitinib plus best supportive care in previously treated patients with refractory advanced non-small-cell lung cancer: Results from a randomised, placebo-controlled, multicentre study (Iressa Survival Evaluation in Lung Cancer). *The Lancet, 366*(9496), 1527–1537.

Peters, L. J. et al. (2010). Critical impact of radiotherapy protocol compliance and quality in the treatment of advanced head and neck cancer: Results from TROG 02.02. *Journal of Clinical Oncology, 28*(18), 2996–3001.

Potters, L., & Kapur, A. (2012). Implementation of a 'no fly' safety culture in a multicenter radiation medicine department. *Practical Radiation Oncology, 2*(1), 18–26.

Prusoff, B. A. et al. (1972). Concordance between clinical assessments and patients' self-report in depression. *Archives of General Psychiatry, 26*(6), 546–552.

Roberts, G., & Marsh, B. (2010). Fatal radiation—Interactive graphic. *The New York Times.* http://www.nytimes.com/interactive/2010/01/22/us/Radiation.html?ref = health.

Vatnitsky, S. et al. (2001). The radiation overexposure of radiotherapy patients in Panama 15 June 2001. *Radiotherapy & Oncology, 60*(3), 237–238.

Wu, Y. L. et al. (2007). Epidermal growth factor receptor mutations and their correlation with gefitinib therapy in patients with non-small cell lung cancer: A meta-analysis based on updated individual patient data from six medical centers in Mainland China. *Journal of Thoracic Oncology, 2*(5), 430–439.

Yu, P. et al. (2010). Different usage of the same oncology information system in two hospitals in Sydney—Lessons go beyond the initial introduction. *International Journal of Medical Informatics, 79*(6), 422–429.

Zakim, D. et al. (2008). Underutilization of information and knowledge in everyday medical practice: Evaluation of a computer-based solution. *BMC Medical Informatics and Decision Making, 8*(January), 50.

Toward a Terminology for Radiation Oncology

Michael D. Mills
University of Louisville

Robert J. Esterhay
University of Louisville

5.1 Introduction

The modern era has many challenges, not the least of which is the ongoing need to master new terms and concepts. Consider that most English speakers can boast of a 4000-word vocabulary. Shakespeare's vocabulary, perhaps unmatched in the history of English writers, boasted a documented vocabulary of more than 29,000 words, including more than 1700 words he coined that are now used in everyday speech (Khurana 2006). All of this pales in comparison with the geometrical increase in the number of terms and concepts that characterize the advances in life and healthcare technologies representative of the modern era. Medicine and medical technology alone have seen an explosion of terminology and concepts within our generation.

Consider just one example of the history of terminology development. In the early 1960s, a physician specialty group, the College of American Pathologists, realized that they had a problem that threatened their specialty. As health research created new medical terms and concepts that pathologists were obliged to describe in written reports, the danger arose that multiple terms could be used for the same concept, or a single concept could be described using multiple terms, or the same term might be applied to one concept in one area of the nation but mean something quite different in other areas. Using a poorly understood term could improperly communicate essential clinical information about a specific patient or category of patients. In addition, terminology is not static. It changes over time. Terms become obsolete, new terms emerge, and concepts continually need refinement and adjustments with new terms. Pathologists saw clearly the need for a consistent universal terminology such that terms and concepts were used consistently, in the same way everywhere. In addition, pathologists' terminology, including concepts and definitions, would need to exist in a repository that could be accessed, and pathologists would be continually tested on their mastery of these vocabulary and terminology concepts. To meet this need, the Systematized Nomenclature of Pathology was started in 1965, and the concept was later extended into other medical fields. Systematized Nomenclature of Medicine—Clinical Terms (SNOMED CT), the latest iteration of this project, is considered to be the most comprehensive multilingual clinical health and healthcare terminology in the world. SNOMED CT, a joint effort of the National Health Service in England and the U.S. College of American Pathologists to develop an international clinical terminology, was formed in 1999 by the convergence of SNOMED Reference Terminology and the United Kingdom's Clinical Terms Version 3 (formerly known as the "Read Codes"). The U.S. National Library of Medicine (NLM), on behalf of the U.S. Department of Health and Human Services, entered into an agreement with the College of American Pathologists for a perpetual license for the core SNOMED CT and ongoing updates. The contract provides to NLM a perpetual license to distribute SNOMED within the NLM's Unified Medical Language System (UMLS) Metathesaurus for no cost use within the United States by both the U.S. government (federal, state, local, and territorial) and private organizations. The contract also covers updates to SNOMED CT issued by the College of American Pathologists between June 20, 2003 and June 29, 2008. In April 2007, SNOMED CT was acquired by the International Health Terminology Standards Development Organisation (IHTSDO 2009). SNOMED CT now consists of a staggering 370,000 or more unique medical terms and concepts. The UMLS Metathesaurus is even larger, containing all terms from SNOMED CT as well as terms from other organizations, such as the Logical Observation Identifiers Names and Codes (LOINC; http://loinc.org) and Nomenclature Properties and Units (NPU) (http://www.iupac.org). The relationship between SNOMED CT, LOINC, and the UMLS Metathesaurus is diagrammed in Figure 5.1.

FIGURE 5.1 Relationship between the UMLS Metathesaurus, LOINC, and SNOMED CT.

Terminology and concepts contained within the UMLS and SNOMED CT Metathesaurus evolve over time. Additionally, the needs and uses for a Metathesaurus also change, and new projects emerge. The UMLS Clinical Observations Recording and Encoding (CORE) Project was initiated to define a UMLS terminology subset that is useful to report information respecting a patient encounter. Specific encounter tasks include documenting, encoding, and summarizing clinical information. The terminology would be a clinical resource to describe a patient problem list, a discharge diagnosis, or the reason for the encounter. The Project process includes organization and analysis of databases containing terms from participating healthcare institutions. Such clinical databases use controlled terminologies and vocabularies that are evaluated for appropriate precision and frequency of use.

At present, a problem list subset of SNOMED CT under development is based on data sets submitted by seven institutions: Beth Israel Deaconess Medical Center (Boston, Massachusetts), Intermountain Healthcare (Salt Lake City, Utah), Kaiser Permanente (Oakland, California), Mayo Clinic (Rochester, Minnesota), Nebraska University Medical Center (Omaha, Nebraska), Regenstrief Institute (Indianapolis, Indiana), and Hong Kong Hospital Authority (Hong Kong, China). These institutions are large-scale, mixed inpatient-outpatient facilities that cover most major medical specialties (including internal medicine, general surgery, pediatrics, obstetrics, gynecology, psychiatry, and orthopedics). The 14,000 most frequently used terms, which cover ninety-five percent of usage volume in each institution, are mapped to 6800 UMLS concepts. This forms the basis of the UMLS CORE subset. About eight percent of terms cannot be mapped to the April 2009 edition of the UMLS Metathesaurus. The reason is that some terms used by certain specialties are unique to that specialty and no concept has yet been defined for the UMLS. There is a process to add terms and concepts to the UMLS, which should be followed by each specialty to provide a complete terminology for the specialty.

Among the source terminologies in the UMLS, SNOMED CT covers the highest percentage (81%) of the UMLS CORE concepts. Because SNOMED CT is the designated U.S. standard terminology for diagnosis and problem lists, we believe that identifying a CORE subset of SNOMED CT concepts will be useful to users who want to implement SNOMED CT in their clinical systems.

SNOMED is not the only integrated system for standard medical terminology. Other systems include the LOINC and NPU, a project undertaken by the International Union of Pure and Applied Chemistry (http://www.iupac.org). LOINC is a set of universal codes and names to identify laboratory and other clinical observations that facilitates exchange and pooling of clinical results for clinical care, outcomes management, and research. LOINC was initiated by research scientists at the Regenstrief Institute, who continue to develop it with the collaboration of the LOINC Committee. The NPU terminology contains 15,000 strictly defined, named, and coded properties in the field of clinical laboratory sciences. On April 1, 2009, the owners of these three standards that contain laboratory test terminology (LOINC, NPU, and SNOMED CT) began an operational trial of cooperative terminology development and prospective divisions of labor for the generation of laboratory test terminology content. This trial will provide practical experience and important information on opportunities to decrease duplication of effort in the development of laboratory test terminology and to ensure that SNOMED CT works effectively in combination with either LOINC or NPU. A general consensus has been reached that patients and healthcare providers need these standards to work together to minimize risks to patients and to support effective communication, decision support, and health data analysis. The goal, then, is to ensure that SNOMED CT can work effectively in combination with either LOINC or NPU in computer systems that support electronic patient records and health information exchange (LOINC 2009).

Another repository of healthcare terminology worthy of mention was Wiki Healthcare Information Technology, a Wikipedia-based forum and semantic tool for Healthcare Information Technology sponsored by Tolven Health, (http://tolvenhealth.com/). WikiHIT was a California-based partnership of private companies, governmental agencies, nonprofit organizations, private individuals, and public-interest groups that were working together to generate and approve healthcare and life sciences clinical data definitions. WikiHIT enabled and supported widespread exchange of unique and precise meaning or semantic interoperability among healthcare software applications. This semantic infrastructure is needed to develop open-source electronic health record clinical data definitions and advance the adoption of electronic health records around the world. The intent was to use volunteer effort from a large number of contributors to create and vet terminology at a low cost and then offer it to the world in an open-access forum. Therefore, the goal of this effort was to the benefit of the greater community of healthcare organizations, which would be realized most effectively with the broadest participation possible by member agencies. The WikiHIT effort has now been subsumed by the Federal Government by the ARRA HITECH legislation and the terminology standards required by HIT vendors for EHRs to be certified (http://www.healthit.gov/). to share vetted terms and concepts by fostering the development of a standard process for capturing expert knowledge and transforming clinical data sets into open-source clinical data definitions to support automatic coding evaluation.

5.2 Terminology for Radiology

The broad field of radiology surprisingly lagged behind the rest of medicine in terminology development. However, after the year 2000, the explosion of radiology informatics and Picture

Archiving and Communication Systems required radiology to catch up very fast. The Radiological Society of North America and the American College of Radiology recognized this need and enlisted the cooperation and support of other groups. The combined effort is leading to the development of a relatively comprehensive Radiology terminology. The College of American Pathologists (Systematized Nomenclature of Pathology/SNOMED), the NLM (UMLS), and the National Cancer Institute provided cooperation and contributed already existing terms as part of this effort; more than 10,000 unique radiology concepts have been developed and contributed thus far. RadLex [short for Radiology Lexicon] is expected to satisfy the needs of both industry (equipment vendors and software developers) and radiology professionals to serve as the single source for radiology-specific concepts and terms (Flanders 2006; RSNA 2009). For the field of Radiology, RadLex is intended to become the primary lexicon and terminology standard and to unify other standards, such as SNOMED CT and Digital Imaging and Communications in Medicine.

5.3 Terminology for Radiation Oncology

The RadLex project is expected to be a major breakthrough in terms of a comprehensive and robust lexicon for radiology. Radiation oncology, however, did not participate in the development of RadLex and indeed as a specialty has made little progress toward the development of a concept-based terminology. Among the many reasons why this neglected aspect of the radiation oncology professional infrastructure should receive urgent attention are the following:

- Industry terms are dominated by vendor jargon or slang; examples include RapidArc, volumetric modulated arc therapy, step and shoot, dynamic arc, dynamic multileaf collimator, and segmented multileaf collimator.
- New concepts and new terms for those concepts evolve over time; two that have come about in the past 10 years are planning organ-at-risk volume) and internal target volume (ICRU 1999). References for such terms are scattered throughout the literature but do not exist in any structured terminology lexicon.
- Almost every clinical trial proposed by cooperative groups on various aspects of cancer and its treatment (e.g., the Radiation Therapy Oncology Group, the Gynecological Oncology Group, the Cancer Trials Support Unit, and the North Central Cancer Treatment Group) must include explicit definitions of terminology within the scope of the individual clinical trial. Inconsistencies between terms and among the various trials contribute to difficulties comparing data and performing meta-analyses.
- A unified system for naming terms would, in the long run, contribute to a reduction in the cost of delivering radiation therapy and evaluating radiation oncology outcomes. Radiation therapy is considered expensive because its delivery depends on expensive technology (hardware

and software) and personnel. In the absence of hard data in which health benefits, quality of life, and costs are measured and optimized and cost-effectiveness is realized, radiation oncology may find it difficult to compete for payment resources and conduct comparative effective research studies that require a unified naming system.

The challenge for radiation oncology is therefore to develop a "concept-based terminology" from existing sources that supports recent and emerging technologies. Radiation oncology is under the onus to demonstrate the cost-effectiveness of its technologies. To perform cost-effectiveness studies, it will need to have a concept-based terminology. Otherwise, it might be accused of equating different concepts when costs of technologies are compared. Ideally, such terminology would be contributed to the public domain and become adopted through a Standards Development Organization, such as the National Electrical Manufacturers Association, which includes all of the vendors that manufacture and sell radiation oncology products and services. The need is to create a process that provides funding, maintains, and enhances the concept-based terminology database for radiation oncology.

Concept-based terminology has many structural aspects, and these aspects vary depending on the type of the terms and the structure of the repository. One broad example is that any concept should be placed in a relationship with other similar terms and defined uniquely. Is the term describing a structure, a system, or a method? What is its relationship with similar terminology? A simple illustration of how narrower and broader terms relate one to another is presented in Table 5.1. Tomotherapy is a way of irradiating a patient to paint out a precise dose distribution in the body to treat malignant disease. Tomotherapy is a type of Intensity-Modulated Radiation Therapy (IMRT). IMRT includes tomotherapy but also includes other technologies that accomplish similar results for the patient. Tomotherapy may be delivered using one of several different technologies, each of which is conceptually distinct, such as serial tomotherapy (delivered with the Nomos Peacock system) and helical tomotherapy (delivered with the TomoTherapy Hi-Art system).

Another example from LOINC is the structure of the six primary name axes shown in Table 5.2. This table illustrates various

TABLE 5.1 Logical Structure of a Concept-Based Terminology

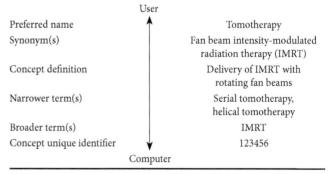

	User
Preferred name	Tomotherapy
Synonym(s)	Fan beam intensity-modulated radiation therapy (IMRT)
Concept definition	Delivery of IMRT with rotating fan beams
Narrower term(s)	Serial tomotherapy, helical tomotherapy
Broader term(s)	IMRT
Concept unique identifier	123456
	Computer

Note: Tomotherapy is a way of irradiating a patient to point out a precise dose-distribution in the body to treat malignant disease (see text).

TABLE 5.2 Six Primary Name Axes in the LOINC Lexicon

Component	Breathing rate, patient diameter, radiation dose	Tomotherapy
Kind of property	Angle, area length, mass, pressure, temperature	Type of
Timing	Point in time, study minimum, maximum, fractionation	Point in Time
System	Head of fetus, lung, linear accelerator	IMRT
Scale	Quantitative, ordinal, nominal (coded), narrative, cgy	Nominal
Method	Stated, measured, estimated, computed tomography, monitor units delivered	3D Treatment Delivery

concepts that must be illustrated to gain an understanding of a term. In this example, tomotherapy names a type of 3D treatment delivery that occurs at a point in time for a patient and represents a type of IMRT. The six axes encourage the lexographer to think about all aspects of the term to assure the definition is complete.

Although a detailed elucidation of the components of registering a term in each of the systems mentioned is beyond the scope of this chapter, the proper assignment of factors such as these to each term requires a trained professional lexographer, someone who knows the domain or field such as radiation oncology and physics as well as the appropriate terminology standards and process for creating terminology. Ideally, a program to establish a radiation oncology terminology should have the following characteristics:

- It should be concept based and include definitions (concepts are illustrated in Tables 5.1 and 5.2).
- It should be part of the UMLS Metathesaurus (most radiation oncology-specific terminology is currently not included in the UMLS).
- It should have a well-established and rigorous process for coding nomenclature (the term should go through a naming process such as outlined in Tables 5.1 and 5.2; then, the term, the definitions, and the results of the naming process should be peer reviewed by a professional lexographer).
- It should be led and supported by a volunteer organization, such as the American Society for Radiation Oncology or the American College of Radiology.
- It should be free of intellectual property concerns.
- Its items should be contributed to the public domain and provided free for worldwide use.
- It should include a means of attribution for the contribution of terms and, ideally, open source and open access.
- It should use open-source mapping and browser tools.
- Actual entry should be restricted to trained professional lexographers.

5.4 A Proposal for Uniform Naming of Tissues

Recently, the Image-Guided Therapy QA Center (ITC) at Washington University in St. Louis developed a process to examine radiation oncology treatment planning data for completeness and consistency. To encourage the use of protocol-specified names of anatomic structures, the ITC established and published a standard list of structure names for clinical trials supported by members of the Advanced Technology QA Consortium. The list of names can be found at http://atc.wustl.edu under the Resources tab, Other Resources. A standard list of organ-at-risk and target volume names was harvested from recent protocols and distributed for comments within the Advanced Technology Integration Steering Committee of the ITC. The principles are listed below:

1. Use of anatomic (Right, Left) is preferred over target-volume-relative (Ipsi, Contra) indicators for laterality in paired organs.
2. Compact labels ("R," rather than "RT" or "RIGHT") should be used where possible.
3. Historical use of indices to distinguish multiple target volumes (PTV1, PTV2, etc.) is inconsistent: PTV1 is used for low-dose TVs in some protocols and for high-dose TVs in others. Thus, target volumes should be labeled with prescription dose level (PTV_5200, PTV_6600, etc.).
4. Names for geometric extensions of contoured structures used for treatment planning [i.e., planning risk volumes (PRVs)] should encode the size of the margin used if it is uniform (e.g., "SPINAL_CORD_PRV5" for a PRV around the spinal cord with a 5 mm margin).
5. Consistency in label "format (delimiters, units, precision)" is important for reliably matching corresponding structures in different patients. Thus, target-volume dose will be expressed as an "integer value" in units of cGy (e.g., "PTV_5240"). This format retains sufficient precision and avoids inconsistencies resulting from the use of (fractional) Gy values (i.e., "PTV52" vs. "PTV_52.4").

A scheme for uniform naming of contoured structures based on these principles is outlined below.

A. Organs at Risk
   1. A list of base names for organs at risk is given in Table 5.3. This list is not exhaustive. It is expected that it will be extended in a consistent manner as new protocols are written.
   2. For paired organs, right or left organs are identified by appending "_R" or "_L" to the base name (e.g., LUNG_L).
   3. For geometric extensions of organs (PRVs) with "uniform margin," a suffix of "_PRVm" is appended to the base name, where m is an integer indicating the size of the margin in mm (e.g., SPINAL_CORD_PRV5). "Nonuniform" PRVs are identified using the suffix "_PRV" (i.e., "without margin size").
B. Target Volumes
   1. Target volumes are constructed using the International Commission on Radiation Units and Measurements designation as the base name.

TABLE 5.3 Base Names for Organs at Risk

Structure Name	Paired?	Structure Name	Paired?
ANAL_CANAL		MAIN_BRONC	
BLADDER		OPTIC_NRV	_L/_R
BRAC_PLX	_L/_R	ORAL_CAVITY	
BRAIN		OVARY	_L/_R
BRAINSTEM		PAROTID	_L/_R
BREAST	_L/_R	PENILE_BULB	
BRONC_TREE		PERINEUM	
CARINA		PHARYNX	
CAUDA_ EQUINA		PITUITARY	
CEREBELLUM	_L/_R	PROSTATE	
CEREBRUM	_L/_R	RECTUM	
CHIASM		RETINA	_L/_R
CN_VII	_L/_R	RIB	
CN_VIII	_L/_R	SACRUM	
COCHLEA	_L/_R	SEM_VES	
CORNEA	_L/_R	SKIN	
DUODENUM		SM_BOWEL	
EAR_MID	_L/_R	SPINAL_CORD	
EAR_EXT	_L/_R	STOMACH	
ESOPHAGUS		SUBMND_SALV	_L/_R
FEMUR	_L/_R	TEMP_LOBE	_L/_R
GLOBE	_L/_R	TESTIS	_L/_R
GLOTTIS		THYROID	
GREAT_VESS		TM_JOINT	_L/_R
HEART		TONGUE	
KIDNEY	_L/_R	TRACHEA	
LG_BOWEL		URETHRA	
LARYNX		VULVA	
LAC_GL	_L/_R		
LENS	_L/_R		
LIPS			
LIVER			
LUNG			
MANDIBLE			

Source: From Bosch, W. R., *Uniform Tissue Names for Use in RTOG Advanced Technology Clinical Trials*, August 19, 2009. Available at http://atc.wustl.edu. Accessed 1 December 2009.

Note: Where paired organs are indicated, laterality is indicated by appending "_L" or "_R" to the base name. Geometric extensions of these structures are indicated by appending "_PRV*m*," where *m* is the nominal margin (mm) used to extend the structure.

2. To distinguish targets receiving different prescription doses, a suffix of "_*d*" is appended to the base name, where *d* is the prescription expressed as an integer in units of cGy [e.g., "PTV_5040" is used for a planning target volume receiving a total prescription dose of 5040 cGy (50.4 Gy)].

3. To distinguish "primary" and "nodal" targets receiving the same prescription dose, append "p" or "n," respectively, to the target volume base name (e.g., "PTVp_5580" or "PTVn_5580").

4. If multiple instances of primary or nodal targets receiving a given prescription dose must be distinguished, an integer may be appended after the "p" or "n" to distinguish the particular target. Thus, PTVn1_5400, PTVn2_5400, etc.

5.5 Conclusions

Patients making treatment decisions need to know the costs and benefits of therapy. As diagnoses and treatments become more rigorous and complex, there is an urgent need to improve communication whether among people, among machines, or between people and machines. Radiation oncology should adopt a unified goal to achieve semantic interoperability by using concept-based terms maintained in a relational database that is free to all. Currently, many in our profession use poorly defined terms or even invent terms for use within a single institution or training tradition. These terms often have no scientific basis to describe how the patient is actually being treated. This problem is larger than any single profession. All professionals, including physicians, physicists, and dosimetrists, bear responsibility for the current state of our terminology. Obviously, the safety of the patients is at issue along with the overall quality and efficiency of the diagnostic and treatment process. In addition to the safety issue, undisciplined terminology affects the overall quality of healthcare. Radiation oncology has neglected to develop a concept-based terminology, which is an important component of infrastructure that would allow cost/benefit studies to be designed and conducted. Today, it is difficult to perform study meta-analysis and almost impossible to design multi-institutional cost/benefit studies because the terminology is not consistent within the industry, and terms are not precise between and among institutions. Future trials may require computers to communicate using terms defined within the UMLS lexicon. Because radiation oncology has no structured terminology and few, if any, terms within existing databases are unique to radiation oncology, the field may find itself at a disadvantage to prove the essential cost/benefit of existing and emerging technologies.

In addition, healthcare reform is based on the adoption of healthcare information technology to improve efficiency, effectiveness, and integration within the healthcare system. Certainly, for radiation oncology to be part of the nation's healthcare reform agenda, it urgently needs to develop a concept-based terminology. Adoption of and stimulus funding for healthcare information technology will be driven by meaningful use of that technology via semantic interoperability or the exchange of meaning. To improve quality and safety and reduce the costs of radiation oncology services, adoption of healthcare information technology with radiation oncology communication and information technology will be required. A concept-based terminology will therefore be required as a fundamental infrastructure if radiation oncology is to move forward in the twenty-first century.

References

Flanders, A. E. (2006). Informatics at RSNA 2006: Major informatics initiatives in a fresh learning environment. *RadioGraphics, 26*, 1259–1261.

International Commission on Radiation Units and Measurements (ICRU), *Prescribing, Recording and Reporting Photon Beam Therapy,* Report 62 (Supplement to ICRU Report 50), Oxford University Press, Oxford, 1999.

International Health Terminology Standards Development Organisation (IHTSDO), *Systematized Nomenclature of Medicine-Clinical Terms (SNOMED CT)*, 2009. Available at http://www.ihtsdo.org/snomed-ct/. Accessed 14 September 2009.

Khurana, S., *Shakespeare Had Some Vocabulary! Simran's Quotations Blog*, 2006. Available at http://quotations.about.com/b/2006/04/22/shakespeare-had-some-vocabulary.htm. Accessed 11 September 2009.

Logical Observation Identifiers Names and Codes (LOINC), *Owners of LOINC, NPU, and SNOMED CT Begin Trial of Cooperative Terminology Development* (press release), 2009. Available at http://loinc.org/news/owners-of-loinc-npu-and-snomed-ct-begin-trial-of-cooperative-terminology-development-joint-press-release.html. Accessed 14 September 2009.

Radiological Society of North America (RSNA), *RadLex: A New Method for Indexing Online Educational Materials*, 2009. Available at http://www.rsna.org/radlex. Accessed 14 September 2009.

U.S. National Library of Medicine, Unified Medical Language System, *The CORE Problem List Subset of SNOMED CT*, 2009. Available at http://www.nlm.nih.gov/research/umls/Snomed/core_subset.html. Accessed 14 September 2009.

WikiHIT, A Collaborative Forum for Advancing Healthcare Information Technology, 2009. Now only available via www.tolvenhealth.com.

Informatics for Accessing Hospital Information Systems

Collin D. Brack
University of Texas Medical Branch

6.1 Clinical Information Systems and the Rise of Electronic Medical Records

We begin by describing the shift away from silos of specialty-specific information systems and toward a unified electronic medical record (EMR).

Modern hospitals are highly complex multispecialty, multidelivery organizations and contain multiple distinct clinical departments and business units. These departments may seem to operate independently with respect to clinical function, yet they are required to standardize certain operations to deliver superior care within a hospital setting. Due to the increasing complexity in healthcare delivery, previous "industrialized" organizational and business models are no longer relevant for hospitals, in which a good or service progresses along a single, preengineered path and results in a predictably consistent outcome. Present hospitals allow patients to coordinate multiple specialist visits, have labs drawn, receive imaging, and have patient charges reviewed all in the same day. The modern hospital, from a business perspective, takes on a federated model and more closely resembles a loose knit group of small business owners who have centralized (or outsourced) certain nonmedical administrative functions. An organizational model of this nature is highly dependent on connectedness and the ability for each department to efficiently exchange information.

Technology plays a prominent role, in this regard, as advances in network infrastructure have allowed affordable and reliable connectedness between departments, and the speed and accuracy at which information is shared between departments continue to improve as adoption increases for clinical data exchange standards.

6.2 Radiation Oncology and the Clinical Information System

How does Radiation Oncology fit into the broader hospital systems framework? Each medical specialty, from Pediatrics to Geriatrics, and each business unit, from Finance to Legal, have specific and unique data collection and information processing needs, which necessitate custom information systems, be it a custom clinical information system (CIS) for pediatric care or a custom billing system designed to track medical procedures and payment schedules. Among the multiple medical specialties, consider the differences in clinical workflow and clinical information between a busy pediatric clinic with fifteen exam rooms and the complex series of clinical and technical processes present in Radiation Oncology. The challenge is that each specialty requires a clinical "expert system" to account for these specialty differences while also satisfying "global" system attributes such as billing, scheduling, and standardized clinical nomenclature. These specialty-based expert systems can be referred to as CIS, which are dedicated systems tailored to a single medical specialty. For example, an Otolaryngology Information System is a custom CIS designed to assist ear, nose, and throat (ENT) physicians with the clinical and data management needs related to ENT procedures and includes specialty-specific information in the form of otolaryngology-based taxonomy and ontology, ENT procedure codes, and data management needs relevant for progressing a patient (referral) through a series of ENT-specific clinical workflows. The CIS is codified in a data model and data structure typically residing in a multidimensional database, which can be retrieved and accessed simultaneously by otolaryngologists and their associated care team through a computer interface.

The "global" systems that unite a hospital's administrative and business functions are the nonclinical systems commonly found in most corporate settings—for example, centralized accounting and financial systems, enterprise asset management systems, and information technology (IT) administration systems for global user access control and system security. These nonclinical departments, or business segments, within the hospital have varying levels of integration yet face similar interface challenges when connecting to a specialty-specific CIS or Hospital Information System (HIS). Examples of nonclinical hospital departments logically segmented by function include Admissions, Health & Safety (ensuring overall hospital safety and security), Information Management, Nutrition, Human Resources, Purchasing, Patient Accounts, Social Work, and Facilities Management. Hospitals commonly first establish connections between these "back-office" systems and, by extension, make efforts to connect specialty systems to the EMR/HIS. In some hospital environments, the radiation oncology information system (ROIS) may require a direct interface to nonclinical systems in situations where a hospital-wide EMR is not present or not fully connected to back-office business systems such as accounts receivable or billing. Regardless of the setting, core challenges in the form of integration, interoperability, and interaction design must be addressed as health systems extend their reach within the hospital and beyond in regional networks (Kuhn et al. 2006).

6.3 EMR as a System of Record

What is the scope of the EMR and is it always the authoritative record in a hospital? The emergence of hospital-wide systems, namely, the EMR, has created a common platform by which nonclinical information system and CIS can exchange information. The EMR, according to the Health Information Management Systems Society (HIMSS 2012), can be defined as "a longitudinal electronic record of patient health information generated by one or more encounters in any care delivery setting. Included in this information are patient demographics, progress notes, problems, medications, vital signs, past medical history, immunizations, laboratory data, and radiology reports." It is this single longitudinal data set that hospitals have decided to contain the authoritative, clinical "system of record." A system of record is

any information management system, be it a paper record filing system or database, which is responsible for housing the original and authoritative "source data." In healthcare, for example, the medical records that a hospital would be required to produce in the event of a court subpoena would be housed within the medical "system of record." As medical information shifts from paper format to electronic, it is important to categorize data and identify source data of this nature for disaster preparedness and legal reasons. It is also important to begin thinking about data with classification in mind because the ease at which electronic information can be replicated and duplicated throughout HIS can lead to data quality issues.

6.4 ROIS System of Record

Within the context of Radiation Oncology, it can be argued that the specialty is so unique in terms of clinical workflow, technology, and information that radiotherapy will always require a dedicated clinical system and is therefore not a good candidate for transitioning its function and source data entirely to a centralized EMR. One of the requirements for implementing a robust ROIS environment is to, in turn, design and implement multiple "interfaces" between the ROIS and the hospital's centralized EMR or HIS. As other specialties shift their information system functionality and source data to EMR, it is important to differentiate and identify which highly technical specialties (like Radiation Oncology and Otolaryngology) to remain outside the purview of a hospital's EMR scope. It is common for a Radiation Oncology department (operating under the management of a hospital system) to have an ROIS connected in a federated model to a HIS or EMR. The Radiation Oncology CIS remains as a dedicated information systems to process and manage patient care and as such will exist as a peripheral system—and contributing a portion of the ROIS clinical data to the EMR. Conversely, the ROIS "sources" certain clinical data from the EMR and other external systems, making the ROIS reliant on sending and receiving clinical data as well as storing the legal, authoritative "system of record" pertaining to the patient's radiation treatment plan, delivery and outcome. Before the advent of centralized EMR systems, integration projects revolved around interconnecting CIS systems. The current push is toward CIS-to-EMR connectivity or specialty representation within the EMR but, for situations where an EMR is not present or only partially deployed, a direct ROIS-to-CIS solution may be required.

6.5 Building Bridges: System Centralization via EMR and HIS

The complexity and interoperability challenges have been outlined above in order that we may appreciate the strides that hospitals have taken over the past twenty-five years to consolidate CIS and address the multiple layers of data redundancy found within CIS. The trends in centralization have arrived in time to meet recent rises in clinical information in terms of volume as well as complexity (in digital form as well as paper-based records). The push to centralize and the need for new data management tools to manage increasing data volume have created additional design challenges for ROIS connectivity projects. Interface projects are commonly referred to as "bridges," and as such, it is helpful to describe these connectivity projects with analogous bridge construction terminology—in which the goal of the project is to construct a bridge between cities. The development of a custom system bridge requires a project manager, blueprints (data schema), a structural engineer (database engineer), permitting (approval from both system custodians), structural testing, go-live, maintenance, surveillance (connectivity monitoring), etc. Interfacing projects, therefore, need to be managed and resourced appropriately by the stakeholders and the interfacing project should be classified as "custom software development." When an ROIS requires connectivity to a hospital-wide EMR, it is best to "approach" the project as if undertaking bids for commissioned software development. With respect to the "traffic" traversing the interface bridge, it has been difficult to achieve industry (i.e., specialty) consensus around the scope (what information to exchange) and format (what form the information will take when traveling between systems). Regardless of industry, data exchange standards require years of governance and consensus to develop, test, adopt, deploy, and sustain. Early demand for clinical system connectedness initially outpaced the supply of well-adopted data exchange standards, but today there are multiple connectivity initiatives and data exchange solutions, which we will review and include the work accomplished by the Integrating the Healthcare Environment (IHE) organization.

6.6 EMR Adoption: Each Hospital Is Unique

Every hospital is in varying stages of EMR adoption, functionality, and implementation. When preparing for an ROIS-to-EMR connectivity project, the first step is to embark on a discovery stage to determine the current deployment state of the EMR system. Some EMR deployments can last up to three years depending on the level of disruption the new software causes. HIMSS Analytics has created a model describing common EMR "adoption levels" and tracks more than 5000 U.S. hospitals progress from Stage 0 (in which no ancillary information systems exist) to Stage 7 with "Complete EMR; CCD transactions to share data; Data warehousing; Data continuity with ED, ambulatory, OP" and is accessible online (http://www.HIMSSanalytics.org/stagesGraph.asp). Stage 1 contains foundational components of an EMR and includes the following three ancillaries systems: Laboratory Information System (LIS), Radiology Information System, and Pharmacy. This wide variability of EMR adoption, technology infrastructure, and software systems within hospitals has encouraged user-centered, nontechnical approaches for solving deployment and integration issues by encouraging closer involvement between physicians, end users, and system developers (Urda et al. 2012).

6.7 In-Bound and Out-Bound ROIS Data Feeds

The next four sections provide high-level examples of ROIS bridges. This section lists in-bound and out-bound system data points commonly found between an ROIS-to-EMR interfacing project. Other CIS, the EMR, and nonclinical external systems have varying levels of data dependencies, which rely on an ROIS capable of sending and receiving data. The following list of data dependencies are not exhaustive but provide a high-level data inventory to be considered when designing a fully integrated ROIS.

6.7.1 In-Bound Data That Flow Into the ROIS

- Patient registration and demographic data (radiation oncology patients are referrals and have already been registered in HIS/EMR)
- Patient admission and status (admissions, discharges, and transfers)
- Laboratory results
- ICD-9/ICD-10 codes, hospital-specific procedure codes, and coding bundles
- Imaging: positron emission tomography, magnetic resonance imaging, and computed tomography imaging data that originate outside the radiation oncology department
- External clinical assessments (treatment protocols and assessments required for use in clinical trials)
- Oncology-specific physician reports/notes (treatment summaries and physician notes related to chemotherapy regimens and surgical outcomes)

6.7.2 Out-Bound Data That Flow Out of the ROIS

- Patient registration status (patient discharges and transfers)
- Outbound billing and code capture
- Physician documentation, notes, and treatment summary reports
- Laboratory requests (computerized physician order entry)
- Export of radiation oncology medical imaging for procedures and care outside radiation oncology (neuro-oncology procedures and surgical planning)
- Export of radiation oncology medical imaging to institutional PACS for archival purposes
- Export relevant clinical trial data
- Export relevant cancer registry data (diagnosis and staging information)

These data flows between the ROIS and the hospital-wide EMR are commonly sourced from the hospital's LIS, Enterprise PACS, Cancer Registry Systems, Charge Automation and Billing Systems, and Medical Oncology Information Systems. Describing patient information in terms of "patient information objects" is also helpful when designing a strategy to determine

ROIS data exports for historical archival purposes. Mandal et al. identify the following Radiation Oncology patient information objects: Patient profile, Patient history, Disease status, Diagnostic (Dx) data, Treatment plan, Simulation data, Treatment planning data, Physics data, Treatment (Tx) data, Treatment verification data, Radiation biology data, Follow-up information, and Record/billing/discharge (Mandal et al. 2008).

6.7.3 Ancillary Information System Interface Example: The LIS

Ancillary Information Systems include Laboratory, Radiology and Pharmacy Information Systems. A modern LIS can be equipped with an industry-standard [Health Level Seven (HL7)] order automation interface that allows it to receive and processes lab orders from an ROIS equipped with a complementary HL7-based interface. These HL7 interfaces are simply "request queues" in which the systems place individual requests, or messages, in a format that the LIS can understand. The LIS HL7 interface can be considered to be a "fulfillment center" that simply processes requests and delivers the results back to the requesting system. The entire exchange is automated, whereby the LIS programmatically delivers the laboratory results back to the ROIS via these preconfigured HL7 interfaces.

In order for the ROIS to participate in this exchange, it must assume the following laboratory-specific roles or features: "order entry, order placing, placer order management and follow-up, order result tracking, management of patient biologic history, specimen calculation, specimen identification" (IHE 2012). A workflow diagram describing the data flows is available online (http://wiki.ihe.net/index.php?title=Laboratory_Testing_ Workflow). The LIS, as specified in the IHE "Laboratory Testing Workflow," is responsible for the following functionality: order reception, specimen calculation, specimen identification or specimen acceptance, order check, scheduling, filler order management, production of work lists, clinical validation and interpretation of results, and result reporting. The ROIS can be equipped with a modular LIS, which performs a subset of traditional LIS-specific roles; however, if both systems have appropriate HL7 interfaces, laboratory automation should be designed into the clinical automation workflow.

6.7.4 Nonclinical System Example: Billing and Charge Capture

Another key part of the HIS revolves around tracking billable medical procedures, known as charge capture. The hospital "billing system" is designed to capture relevant charges from each department and provides a centralized method of tracking, analyzing, correcting, submitting, and resubmitting medical insurance claims for payment. Submission claims are prepared by medical billing and coding professionals to ensure that the actual medical procedure performed by the hospital is expressed (or codified) in an industry standard classification (ICD-10), so that insurance companies have a standard method

of interpreting the level, sophistication, and volume of medical procedures. A "code capture system" or "charge capture system" must be in place in order for a billing department to accurately submit and receive medical insurance payment for radiotherapy procedures. First, the ROIS contains tables of records listing all relevant procedure codes. Each procedure code is also tied to the system's ICD-9 or ICD-10 database. Every medical procedure that is scheduled within the ROIS is tied to these records, and when a scheduled procedure is completed, systems commonly refer to the event as being "captured." Captured charges then trigger an ROIS billing interface that sends information about the procedure to the hospital's billing system.

Billing reports can then be generated listing all captured procedures for given date ranges and other relevant search criteria. These reports can be generated from preexisting ROIS reports or a set of custom reports can be designed to query the relevant data against the ROIS database. Industry standard reporting software (e.g., Crystal Reports from SAP or Microsoft's SQL Server Reporting Services) can connect to the ROIS system to design and run custom billing reports, and ROIS vendors provide consulting services to create custom reports as well as provide training for report writing analysts. Charges can be captured and tracked programmatically, known as automated code capture, through an HL7 billing interface. In larger settings, the ROIS can be fully equipped with a direct claims submission and processing engine, whereby the ROIS submits captured codes directly to insurance carriers or clearinghouses. However, automated code capture environments must still rely on reports to validate submissions, discover lost or undercoded procedures, and assist billing staff with rejected claims.

6.7.5 External Interfacing Example: Data Repositories

The Clinical Data Management System or Clinical Trials Management System is a repository for clinical trials data and has the ability to source and report clinical data to the clinical trial sponsor. The Clinical Data Management System or Clinical Trials Management System can monitor or report against relevant clinical trial data within the ROIS. The two systems can exchange data via a custom standards-based interface [for HL7 or Digital Imaging and Communications in Medicine (DICOM)– based data] or a custom report can be designed to query against the ROIS. Ad hoc queries are primarily designed within the context of the ROIS by modifying pre-existing reports or templates. The industry also refers to the default set of reports as "canned reports."

Cancer Registries collect, manage, analyze, and provide information to clinicians and other healthcare providers, public health officials, administrators, and scientists for a variety of purposes to assess and manage the burden of cancer. A large portion of the Cancer Registry data is sourced from Oncology-based Information Systems and is used by public health officials for epidemiologic reporting and cancer incidence trending. The ROIS contains legally mandated reportable diagnoses of cancer cases and can report these data to a cancer registry (or registry intermediate) through standard reporting, whereby a series of reports are executed and the results compiled into the appropriate format or programmatically via an interface engine.

6.8 System Integration Standards and Organizations

Successful integration within healthcare requires both data standards and data connectivity "best practices." The following sections describe the current state of health information standardization relevant to Radiation Oncology as well as efforts made by organizations that demonstrate real-world connectivity and document integration best practices.

6.8.1 HL7 in Radiation Oncology

HL7 was founded in 1987 to create an international industry messaging standard for CIS to exchange data. It has grown to become the most widely used interoperability standard in healthcare and provides a common framework for CIS to exchange, request, and retrieve electronic health information. The most recent version of the HL7 standard is version 3 and includes a Reference Information Model and a new Extensible Markup Language–based messaging syntax designed to support clinical workflows for every medical specialty. The Reference Information Model is an information model, or framework, that is the common source for the information content of the HL7 standard (Kagadis & Langer 2011). Each specialty within this HL7 framework includes a functional profile that describes the health data requirements, specialist definitions, and clinical workflow use-cases germane to the specialty. The Radiation Oncology functional profile describes the clinical Radiation Oncology role as "The Radiation Oncologist uses a Radiation Oncology–specific Electronic Health Record (RO-EHR) that is designed to support the specific requirements for calculating and documenting radiotherapy. The RO-EHR controls, verifies, and records all aspects of each individual radiation treatment. As part of the overall oncology treatment team, the Radiation Oncologist uses their RO-EHR to communicate with the oncology EHR systems. The RO-EHR also communicates with Treatment Planning Systems, Treatment Management Systems, Treatment Delivery Systems and image viewing systems" (ANSI/HL7 2010). RO-specific use-cases and oncology definitions are further described in the "HL7 EHR-S Ambulatory Oncology Functional Profile" maintained by the HL7 "Ambulatory Oncology Profile Task Group" and "EHR Work Group" and are available online (http://wiki.HL7.org/images/0/05/Ambulatory_Oncology_EHR_Functional_Profile.doc).

6.8.2 IHE

IHE is described online (http://www.ihe.net/about/ihe_faq.cfm) as "an initiative by healthcare professionals and industry to improve the way computer systems in healthcare share information. IHE promotes the coordinated use of established standards such as DICOM and HL7 to address specific clinical need in support of optimal patient care. Systems developed in accordance with IHE communicate with one another better, are easier to implement, and enable care providers to use information more effectively." The primary goal of the IHE is not to develop data exchange standards (Lenz et al. 2007) nor is the initiative a certifying body (Siegel & Channin 2001); instead, the goal of the IHE is to demonstrate, validate, and support real-world connectivity between CIS, HIS/EMR, ancillary systems, and other healthcare systems that have adopted data exchange standards such as HL7/DICOM. Every medical specialty is represented and software vendors are encouraged to assist in "Interoperability Showcases" and "Connectathons," which occur at industry trade group conferences and health IT conferences. These events allow the IHE to showcase their common language process-based approach that overcomes individual vendor architectural differences (Vegoda 2002). More information about IHE is available online (http://www.ihe.net).

6.8.3 IHE-RO

IHE divides specialties into clinical "domains" where subject matter experts (clinical and technical experts) contribute to the identification of "integration and information sharing priorities" and "vendors of relevant information systems develop consensus, standards-based solutions to address them." IHE domain information is available online (http://www.ihe.net/Domains/index.cfm). The Radiation Oncology Domain is sponsored by American Society for Radiation Oncology and is a rich source of real-world connectivity outcomes and challenges faced by the ROIS/HIS integration.

6.8.4 IHE-RO Technical Framework

The IHE RO Domain has published two "technical framework" volumes, in final text version, in 2011. The work contained in these volumes represents the most comprehensive analysis of system interoperability between ROIS and systems, which support the HL7 and DICOM standards. Both volumes were a collaborative effort by American Society for Radiation Oncology, American Academy of Pain Medicine, Radiological Society of North America, HIMSS, and IHE International, Inc., and are each available in full-text below.

Volume 1 (RO TF-1): Integration Profiles document specifies "implementations of [health IT] standards that are designed to meet identified clinical needs. They enable users and vendors to state which IHE capabilities they require or provide, by reference to the detailed specifications of the IHE Radiation Oncology Technical Framework" and is accessible online (http://www.ihe.net/Technical_Framework/upload/IHE_RO_TF_Rev1-7_Vol1_2011-0509.pdf).

Volume 2 (RO TF-2): Transactions document builds upon Volume I, which provided an overview of IHE functionality and described how clinical roles and needs are organized into "Integration Profiles." Volume II (IHE International 2011) contains "detailed technical descriptions of each IHE transaction including the clinical problem it is intended to address and the IHE actors and transactions it comprises" and is accessible online (http://www.ihe.net/Technical_Framework/upload/IHE_RO_TF_Rev1-7_Vol2_2011-0509.pdf).

6.9 Primacy of Oncology Connectivity

The focus of this chapter was on health information connectivity and compatibility with the larger HIS; however, smaller institutions should redirect their initial efforts around integrating oncology-specific information systems. Oncology integration will have the greatest benefit to the cancer patient by means of efficient clinical oncology workflow and information quality and integrity. As such, connectivity and accessibility challenges between oncology-based CIS should be given precedence and focus on interconnections between Medical Oncology, Radiation Oncology, Surgical Oncology, and Clinical Trials.

Radiation Oncology is a subset of three principle oncology-based specialties: Medical, Surgical, and Radiation Oncology. As such, the ROIS has principle oncology-based data dependencies and data overlap between their Medical and Surgical Oncology Systems. Understanding these oncology-based interdependencies is arguably more important than selecting a handful of ROIS data to export to a central HIS or EMR. A tight interconnection among Medical, Surgical, and Radiation Oncology disciplines is the hallmark of a modern interdisciplinary cancer care system in which all oncology specialties work in tandem to provide a seamless continuum of care. Additional cancer-specific information systems emerge to join all the oncology specialties, namely, Cancer Registry System and Clinical Trials Information System. Upon successful integration of these oncology-specific systems, the "next" stage of connectivity can begin in which these systems gain greater efficiency by connecting to the broader HIS.

References

IHE International, *IHE Radiation Oncology Technical Framework Vol II*, IHE International, 2011.

Kagadis, G. C. & Langer, S. G., *Informatics in Medical Imaging*, Taylor & Francis, Boca Raton, Florida, 2011.

Kuhn, K. et al. (2006). Expanding the scope of health information systems. IMIA Yearbook of Medical Informatics. *Methods of Information in Medicine, 45*(Suppl 1), S43–S52.

Lenz, R. et al. (2007). Semantic integration in healthcare networks. *International Journal of Medical Informatics, 76*, 201–207.

Mandal, A. et al. (2008). Development of an electronic radiation oncology patient information management system. *Journal of Cancer Research and Therapeutics, 4*, 178–185.

Siegel, E. L. & Channin, D. S. (2001). Integrating the healthcare enterprise: A primer. *RadioGraphics, 21*, 1339–1341.

Urda, D. et al. (2012). Addressing critical issues in the development of an oncology information system. *International Journal of Medical Informatics, pii*, S1386–5056.

Vegoda, P. (2002). Introducing the IHE (Integrating the Healthcare Enterprise) concept. *Journal of Healthcare Information Management, 16*, 21–24.

II

Information and the Radiation Oncology Process

7

Information Flow through the Radiation Oncology Process

Luis Fong de los Santos
Mayo Clinic

Michael G. Herman
Mayo Clinic

7.1 Introduction

Current demands of the healthcare industry require the use of information systems (IS) and information technologies (IT) to deliver care that is safe, patient centered, timely, efficient, and equitable (Committee-on-Quality-Health-Care-in-America 2001; Chaudhry et al. 2006). Nevertheless, as IT and IS evolve and become essential components of clinical practice, the complexity of developing and implementing systems and applications that support clinical information flow and workflow increases. Additionally, replacing electronic systems that merely reproduce the previous paper-based workflow can potentially lead to unsafe processes and inefficient information flow. Therefore, comprehensive understanding and definition of the components of the modern radiation oncology practice, the underlying processes, and the social infrastructure of a specific organization or clinical environment is essential for the successful development, evaluation, and implementation of IT/IS applications that will support safe, effective, and efficient information flow and workflow (Berg 1999; Ammenwerth et al. 2003, 2004; Garde et al. 2006).

The main goal of this chapter is to address the complexity and interdependencies of the various components of modern radiation oncology practices and their impact on information flow and workflow. Understanding these components and the role that each of them plays in the practice will facilitate the development and use of IT/IS applications to improve information flow and workflow. Section 7.2 describes a typical paper-based system (i.e., paper charts), notes its advantages and disadvantages, and compares it with modern electronic-based systems. Section 7.3, the core of the chapter, provides in-depth descriptions of the components of the modern practice and the role that each component plays in the information flow and workflow of a department. Section 7.4, the final section, is intended to provide an overview of current and well-established methodologies to

improve workflow and information flow and to help practices cope with the continuous evolution of their own processes and technological advances.

7.2 Old Paradigm—Paper Charts in Clinical Care

Historically, the paper chart has been the essential tool for paper-based IS supporting the clinical care process and practice workflow. Despite the evolution of IT and information management systems, the paper chart is still a fundamental component in the management and transfer of clinically relevant information for many practices worldwide (Saleem et al. 2009). Some of the arguments in favor of the paper chart are that it is tangible, fast, easy to use, and modify (needing only a pen), very adaptable for the specific needs and processes of the practice, and, at least in the short run, a low-cost tool. Another reason for the persistence of paper charts is the high comfort level that users have with the status quo. Users firmly believe that it is safer and more efficient to maintain their current system than it would be to change their processes and practice by introducing electronic information management systems. Practices using paper-based systems have developed a false sense of efficiency due to the fact that many of the routine and time-consuming information management activities are done by several layers of administrative personnel. As demands on the field increase and the amount of information grow, this alternative becomes highly ineffective and costly. Overall, paper-based systems are not conducive to building efficient and safe communication environments in modern practices or to supporting real-time and team-based decision-making, medicolegal protection, quality improvement, and outcome analysis. Moreover, paper-based systems do not facilitate multidisciplinary patient-centered, technologically advanced care, where the care team may be numerous and dispersed in location. Morgan (2008), in a chapter on identifying and understanding clinical processes, uses the analogy of the paper day-planner and the electronic organizer: "Those using a paper day-planner will initially be faster than those using an electronic organizer at simple activities such finding addresses, penciling in appointments, and entering new contacts and phone numbers. It is not until information management becomes more complex, of greater volume, and required by more than one user that the electronic organizer's advantages become clear." The current demands and complexity of the radiation oncology practice and the healthcare industry make it impossible to deliver high-quality care that is efficient, effective, safe, and at competitive cost using paper-based information management systems.

IS and IT applications in radiation oncology have come a long way and have followed the evolution and development of electronic medical records. These applications have been developed by vendors, users, or, in some cases, a combination of both, with the goal of supporting the processes and needs of the modern radiation oncology practice. Among the challenges confronted by both developers and users is the fact that the practice and infrastructure can vary tremendously from department to department. Developers, without any clinical experience or guidelines, tend to create applications that may isolate features and functionality, with rigid processes and workflow that collectively make the application cumbersome, counterproductive, and potentially useless. The users, accustomed to their paper-based system, wrongly believe that the IS will completely mimic their previous processes and workflow, therefore developing unrealistic expectations that lead to frustration and rejection of the system. In many cases, without a systematic approach, users blindly select applications and IT applications that provide excellent functionality for a subset of the process but lack integration, connectivity, and a view of the overall operation. This leads to inefficient storage of and access to information and the formation of information or functional "silos," all of which defeat the purpose of having a system capable of integrating essential information throughout the entire episode of care. Finding an electronic-based information management system that will meet all clinical and administrative needs and processes, allow connectivity among multiple systems and technologies, and support multiple and different workflows as well as efficient, safe, and effective transfer of information remains a considerable challenge (Jha et al. 2009). A key element in developing such a system (or combination of systems) is to define and understand the processes and workflow, infrastructure, and information flow of the modern radiation oncology practice, as described in the following section of this chapter.

7.3 Modern Radiation Oncology Practice

The introduction of computer-based systems into the practice of radiation oncology has simplified and streamlined some tasks that in the past had been time-consuming, difficult, or unsafe

and provided new capabilities that a few decades ago were practically impossible. IS in radiation oncology was first introduced in the late 1970s as the interface between the treatment unit and the verification and recording process of patient treatments (Rosenbloom et al. 1977; Fredrickson et al. 1979; Morrey et al. 1982). The role of such systems continues to evolve and develop into the clinical element that provides the integration and overall management of the information and technology needed throughout the patient treatment and subsequent follow-up in a highly complex and demanding patient care setting (Herman et al. 1997). As the technology and systems evolve, the complexity is shifting to the domain of processes, information management, and technology infrastructure and connectivity. The modern radiation oncology practice is a complex and continuously evolving system in itself. Simplifying that complexity requires developing a model that broadly represents the modern radiation oncology practice. One proposed hierarchical model classifies the practice into five major and interlinked components (Fong & Herman 2007) (Figure 7.1):

1. Peopleware (Lorenzi & Riley 1995)
2. Processes
3. Information
4. Software
5. Hardware

The first of these components is the "people" or "individuals" that create and are part of the clinical practice, develop the clinical tools, or receive the clinical care; this component is known as "peopleware." A subset of these individuals, based on clinical needs and goals of the practice, develop clinical "processes" that will maintain an efficient and safe workflow. Clinical processes generate, use, and exchange "information" throughout the whole episode of care. With the introduction of IT/IS tools and applications into the practice, the underlying component that manages the exchange of information flow and workflow is "software." Finally, software requires a reliable "hardware" infrastructure

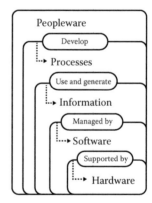

FIGURE 7.1 Modern radiation oncology practice comprises five major components: peopleware, processes, information, software, and hardware. Successful development, implementation, and use of IS and IT applications requires proper consideration of the hierarchy as well as synergy among each of the components.

and stable connectivity that allows continuous information flow and workflow. Each of these components is relevant in itself, but synergy among all the components is the only way to achieve high standards of care where workflow and information flow are safe, efficient, and effective. These elements are described in depth below.

7.3.1 Peopleware

The concept of peopleware was first used in the context of software engineering by Neumann (1977) and later popularized in a book by DeMarco and Lister (1999) called "Peopleware: Productive Projects and Teams." Recently, Lorenzi (Lorenzi & Riley 1995) and Gremy (Gremy et al. 1999; Gremy 2005) introduced this concept in the healthcare setting. Peopleware refers to all the aspects related to the role of people as individuals and as a group in the development, implementation, management, or use of software and hardware. The success of IS and IT with subsequent information flow and process dynamics is heavily influenced by the people who will design, use, and manage those systems. Practices that do not considerer human factors in the design and implementation of information management systems cannot reasonably expect to avoid unanticipated and potentially catastrophic consequences in their products and processes.

Peopleware should be integral to the process throughout the conception, realization, and evaluation of health IS. Samaras and Horst (2005) identified three groups representing the peopleware at different stages of a system cycle: users, producers, and developers. The users group represents the needs and functions of patients, clinicians, operators, and clinical managers. The producers group consists of manufacturers, assemblers, and maintainers. The developers group contains the designers, testers, product managers, and executives. Each of these three groups has its own role in the system, but with overlapping needs and interdependencies. Clinical users are and should be the driving force in the development of health IS; nevertheless, constant negotiation between developers and producers will guarantee effective and efficient product development and functionality. This is easier said than done, because the needs and goals of each group may be divergent or in conflict in some cases. This conflict can potentially be alleviated by recalling that the health industry, unlike other industries, has one common goal: the needs of the patient. Developing and using tools and applications with the mindset that "patients' needs come first" provides a lens through which the groups can focus their efforts and achieve their individual goals by fulfilling one common mission.

The radiation oncology setup is characterized by a highly diverse group of users, each having a defined role throughout the episode of care. In the most traditional definition, the clinical users involved in the management and delivery of radiation therapy (World Health Organization 2008) are radiation oncologists, medical physicists, treatment planners (also known as dosimetrists), radiation therapists (also known as radiation therapy technologists or therapeutic or therapy radiographers), nurses (nurse practitioners and radiation oncology nurses),

administrative staff, and IT/IS staff. Clear and detailed description of the role and responsibilities of each of the members of the care team is an essential component on the development of efficient workflow and information flow (some of those responsibilities will be shown in the following section when describing the clinical and administrative processes).

The practice of radiation oncology is a moving target in itself as technologies and systems as well as the interactions among them become more complex. Nevertheless, successful implementation of these systems in a clinical setup requires more than a good understanding of the technology alone but also the interaction between IT and the corresponding human players. Evaluation of IT applications thus has to consider the environment in which IT is used; in other words, the peopleware. Human-centered methods and technologies specifically developed for healthcare environments are essential for the successful development of health IS that will increase efficiency and productivity; are easy to learn and use, thereby increasing user adoption, retention, and satisfaction; and at the same time reduce medical errors, development time, and overall cost.

7.3.2 Processes

"An organization is only as effective as its processes" (Rummler & Brache 1995). Processes are fundamental components of any organization, especially the healthcare industry. Work in any medical practice, including the practice of radiation oncology, consists of many common clinical and administrative activities (writing a prescription, scheduling a treatment, developing a plan, etc.). Each activity occurs daily and is performed by one or more members of the staff based on series of steps. Each activity in combination with the steps to complete it and the individual(s) who will be responsible for it is referred as a process. The work that is accomplished as processes move from one step to another is referred to as the workflow (Workflow Management Coalition 1999). Processes are the fundamental elements of all the work in a practice, especially in the era of informatics. In the analogy of the paper day-timer and the electronic organizer, the paper day-timer is very accommodating when it comes to stepping out of the common workflow (e.g., adding an appointment; users can always add Post-it notes anywhere at any time). In contrast, the electronic organizer most likely has a few very well-defined ways to perform and complete a task or add information to it. Health IS are the same way; they have the potential to improve workflow, but only if the proper process and workflow analysis precedes their development and implementation. Moreover, automating an already unsafe and risky process will definitely put a patient at a greater risk.

One of the most significant process improvements that can be made as a result of implementing a health information system is in the area of information flow, but the processes supporting the information flow have to be carefully planned and designed. During the development of a process or workflow, it is important to identify what information is needed for each step of the process and the earliest point at which that information can be captured. Paper-based systems are, as mentioned before, a convenient tool because they provide visual cues of where and what work is pending. For example, a paper chart waiting on the physicist's desk means that the plan needs to be checked. A well-defined workflow with corresponding processes has to be carefully planned when the visual cues are no longer there and the workflow is replaced by a list of tasks on a screen. In an electronic-based system, the information is no longer tangible in the same sense as it is in the paper-based system, opening opportunities for mistakes that would not be obvious with a paper chart (e.g., if the system allows opening multiple patient records in the same workstation, there is a high risk of entering information for the incorrect patient). Information flow changes need to be carefully planned, maintaining all of the required review and auditing steps so efficiency does not compromise safety.

Cancer treatment is a complex, multistage process involving the treatment of a great variety of conditions using multiple modalities, technologies, and related professional expertise. The overall workflow involves multiple transfers of data between different professional groups and across work areas. The number of pieces of information belonging to an individual patient is massive when considering the administrative and clinical information that is generated, used, and collected during the entire episode of care in support of clinical, educational, and research activities. Figure 7.2 shows an integrated cancer treatment workflow, emphasizing the radiation oncology process.

Any radiation oncology practice follows basic business and clinical processes. Effectively managing those processes requires that the relevant information for each process be available at the time it is needed. Understanding those processes and defining the information needed and generated as well as identifying the users involved is a fundamental step toward building effective and efficient information flow channels. Brooks et al. (Brooks 1999; Brooks et al. 1998) defined seven basic processes that operate within the clinical setting to support the primary radiation oncology workflow (patient registration, consultation, departmental scheduling, departmental charting, setup for treatment simulation, treatment delivery, and administrative services). In addition to the processes proposed by Brooks, it is important to consider two additional clinical processes essential in the workflow of any radiation oncology practice: treatment planning and quality assurance (QA). Each of these processes is described below, with the focus on the users involved and the information they require and generate.

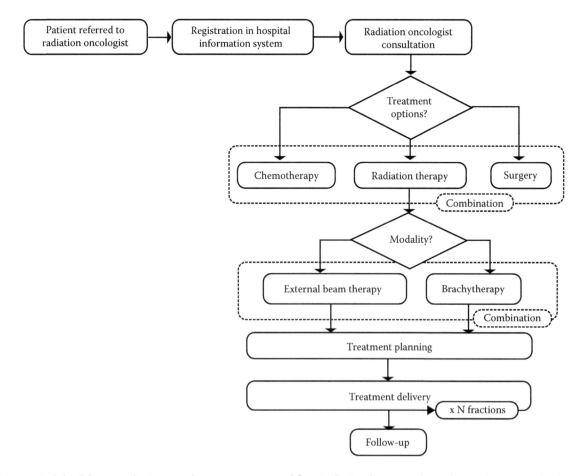

FIGURE 7.2 High-level diagram of an integrated cancer treatment workflow. Radiation therapy can be used in combination with other treatment options (e.g., surgery and chemotherapy), requiring that information be shared among multiple disciplines and modalities.

7.3.2.1 Patient Registration Process

In a clinical environment, this process typically consists of acquiring all of the relevant demographic information to reliably identify the patient to the clinic. Information gathered during this process includes that pertaining to patient identity, demographics, primary and secondary insurance coverage, employer, family members and contact information, referring physician, and primary oncologist. All of this information is important for further communication and proper organization of the patient data. Given the current demands on the healthcare system, practices are focusing more on providing care that is both clinically sound and cost-effective. When pertinent information is gathered during the registration process, the clinic manager can determine whether the institution or department is providing cost-effective care based on the reimbursement model implemented.

7.3.2.2 Consultation Process

The clinical encounter is at the "core" of the healthcare delivery process. During the initial clinical encounter, information is exchanged between the patient and the radiation oncologist, and decisions regarding the therapeutic plan are made. The episode of care in radiation therapy occurs over several consecutive days for several (e.g., 6–8 weeks); further, assessments of outcome and complications could take from several months to years. Therefore, clinical staff (e.g., physicians, nurses, physician assistants, and social workers) meet with the patient on a regular basis (e.g., weekly) and long-term follow-up encounters are required as part of a comprehensive episode of care. It is during the patient consultation process, at the different stages of the episode of care, where the clinical staff generates and gathers the relevant information to support the decision as to the specific treatment modality, to monitor and assess clinical outcome and complications, and to provide feedback to the patient. The quality of the information collected during this process will facilitate clinical, research, and quality management (QM) tasks (e.g., outcome analysis, QM reporting, claims for reimbursement, and legal inquiries). Collecting high-quality information during this process is a challenging task, especially for clinical outcome assessment. Many practices are still following the old paper-based paradigm, where the format of the information is nonstandard and unstructured. In moving forward to the informatics era, practices must adopt systems and processes that use international codes and standards [e.g., the *International Classification of Diseases* (World Health Organization Web site: http://www.who.int/classifications/icd/en/) and the Response Evaluation Criteria In Solid Tumors (Therasse et al. 2000)] as well as structured and template-based data entry to facilitate efficient data query and reporting as well as analysis of large amounts of information.

7.3.2.3 Departmental Scheduling Process

The scheduling process of a radiation oncology practice involves coordinating and managing the time of different staff members as well as different resources (e.g., simulator units, treatment units, and examination or treatment rooms). For the patient, the scheduling process should be transparent and punctual; the care provider or a treatment unit needs to be available at the scheduled time. Developing scheduling processes that will control the patient's schedule as well as manage the time of the clinical team and the resources used is essential to guarantee efficient workflow throughout the episode of care. As the patient moves through the treatment process, the scheduling system coordinates all of the appointments and resources to be used at each step, avoiding conflicting designation of resources and care providers (i.e., one resource cannot be assigned to more that one patient at time, and a single person cannot be at two places at once). As the complexity of a practice increases, the scheduling process should be extended beyond the schedule of appointments and should play a role in coordinating all the underlying processes and tasks occurring at the different stages of the development, evaluation, and delivery of treatment, providing a means of communication between the staff members and allowing capture of procedures codes for billing purposes at task completion. In most settings, one system serves as the "master" scheduler and other clinical or departmental systems are "slaved" to that master scheduler. One-time data entry and connectivity among systems should be pursued to avoid conflicts and data duplication across the different scheduling systems.

7.3.2.4 Departmental Charting Process

The chart is the centralized location of the patient data. Historically, the standard patient chart consisted of a binder containing documentation of the patient's medical history, results from physical examinations and pathology studies, nursing notes, and records of treatment—in other words, all the relevant data for a single patient. Familiarity with the paper chart, and the processes that surround its creation and maintenance, is one of the main factors in the reluctance of many practices to move from paper-based documentation to an electronic version (e-chart). The charting process starts during the first clinical encounter, when the radiation oncologist makes the decision to treat and defines the diagnosis, prescription, and treatment modality. The charting process, like the scheduling process, continues throughout the entire care episode. One of the advantages of e-charts in workflow during the charting process is their portability. Instead of having a single copy of the information, which can be at only one place at any given time, the e-chart allows multiple applications at different stages of the patient care process to be accessible, viewable, and modifiable in multiple locations by multiple users (assuming those users have been granted proper user rights in the system). In addition to data on patient care management and treatment, the e-chart also provides accessibility to the images (setup and ports) taken during the treatment course, facilitating the review process and communication between physician and therapists. Internal audits (i.e., chart checks) of the plan and the treatment are part of the charting process. Different practices implement this part of the charting process differently, but access to all of the elements and parameters of the treatment before, during, and after treatment is necessary. Additional information such as electronic signatures, time stamps, monitoring of overrides, and historical tracking of changes facilitates the process and helps address random and systematic issues during the overall care process.

7.3.2.5 Setup for Treatment Simulation Process

During the simulation process, the patient is "set up" in the anatomical position that will be used during the course of the treatment, and a customized immobilization system is created to ensure reproducibility of the setup during treatment delivery. A conventional X-ray-based (2D) or computed tomography–based (3D) simulator is used to generate a set of images, which the physician then uses to locate the target and define the isocenter (in the case of external-beam radiation therapy) to guide placement of marks on the patient's skin by a set of lasers. The therapist describes the patient setup using notes, sometimes in combination with pictures; this information is part of the patient chart. The information collected during this process has a substantial influence on patient treatment, because incorrect or unclear information could potentially cause a systematic error that would be carried out during the whole treatment delivery.

7.3.2.6 Treatment Planning Process

Treatment plans are developed from information generated during the simulation process in combination with other diagnostic imaging modalities (e.g., computed tomography scans, magnetic resonance imaging, or positron emission tomography). During this plan-development process, the number, orientation, type or modality, and characteristics of the radiation beams (or in the case of brachytherapy, the radiation sources) are chosen to achieve the therapeutic goal (Fraass et al. 1998). Modern practices commonly use computerized treatment planning systems as an integral part of the treatment planning process to define the target and volumes of nearby critical structures, determine the beam directions and shapes, and calculate and evaluate the corresponding dose distribution. Recent progress in information transfer standards (e.g., those provided by the Digital Imaging and Communications in Medicine [DICOM] and DICOM-Radiation Therapy [DICOM-RT]) have improved connectivity, information flow, and workflow of the planning process (Law & Liu 2009) and allowed communication with imaging and delivery devices as well as departmental IS so that the planning information becomes part of the patient record.

7.3.2.7 Treatment Delivery Process

During the treatment process, the prescribed radiation dose is delivered to the target location while the dose to healthy tissue is minimized. Efficient and effective interaction between users and a radiation delivery system is an essential element for guaranteeing that the treatment process will run smoothly and that the intended dose will be delivered. The process begins with the transfer of the treatment plan parameters to the delivery system. This information is typically transferred automatically from an IS (commonly known as a record and verify [R&V] system) to

the delivery system. Therapists are then required to check that the parameters and accessories needed for the treatment are correct. Information about beam localization and patient setup is facilitated by the R&V system and used by the therapists to position the patient. Modern delivery systems include image guidance systems to localize the target (or a surrogate of the target) and double check the placement of the beam aperture (in the case of external-beam treatment). These systems have several layers of software and hardware verification systems to check that the prescribed parameters are being used and to interrupt the treatment if one or more of those parameters is outside a defined tolerance. At the end of the treatment session, all of the treatment delivered parameters, including the radiation dose delivered and the cumulative dose to date, are recorded as part of the patient chart. Sophisticated treatment modalities such as intensity-modulated radiation therapy have made R&V systems a necessary tool for the treatment delivery process, and they have also introduced new processes in the overall radiation therapy workflow (e.g., patient-specific QA processes). Despite the technological advances on the delivery process, it is important to point out that no R&V system is completely independent of well-educated and well-trained treatment teams.

7.3.2.8 Administrative Services

Among the numerous administrative services in any department, four processes should be emphasized because of their impact on maintaining a practice and their relevance to providing high-quality care. Those processes are billing, patient education, report generation, and retrospective (tumor registry) or prospective (clinical trials and protocols) data collection and analysis. The billing process is essential to maintaining a financially healthy clinical practice. As mentioned above, linking the billing process to the scheduling process and code capture upon completion of tasks and activities can improve efficiency and reduce charging and coding errors. Integration of the clinical and billing systems will also eliminate inefficient manual data entry. It is the responsibility of the care team to provide information to patients throughout the entire care episode. IS can automatically generate necessary information specific to the patient's health issue, letters, and reminders or reports to patients and a variety of health plans. It is becoming increasingly common for practices to evaluate their own progress; this is achieved by generating internal overall performance reports from their own systems. If meaningful, consistent, and well-structured information were collected during all stages of the care episode, then it would be simple to run reports to evaluate and improve the quality, safety, and efficiency of the practice.

7.3.2.9 QA Process

This is a fundamental and integral process of the practice of radiation oncology. The current paradigm of QM and QA in radiation therapy is based on measuring and assessing the performance of radiation therapy devices and equipment in terms of predefined tolerances to guarantee satisfactory overall treatment delivery (e.g., within 5% of the prescribed dose) (ICRU 1976). Guidelines for QA and QM have been historically provided by national and international agencies and organizations. A new paradigm for QA and QM, led by the American Association of Physicists in Medicine Task Group 100, is shifting the focus from predefined tolerances to process mapping and failure mode and effect analysis (Huq et al. 2008). The goal of this new paradigm is to overcome the complexity of implementing and using new technologic advances while simultaneously improving clinical outcomes and the overall value of clinical care (where value = quality/cost). Based on this premise, overall information flow and workflow, as well as understanding the variability of processes and infrastructure on the practice of radiation oncology, should be considered part of the QA process. New developments in radiation oncology information management systems should consider implementing tools, applications, and corresponding infrastructure to support this new QA/QM paradigm as part of the overall workflow. Additionally, the clinical care team should become familiar with the tools for process analysis and failure mode and effect analysis so that they can develop, implement, and monitor safe and efficient processes (Pawlicki et al. 2005; Ford et al. 2009).

These nine clinical processes are fundamental elements present in any radiation oncology practice. Nevertheless, the clinical goals, needs, and infrastructure of each practice should drive the definition of their own underlying subprocesses, tasks, and activities. The final section of this chapter (Section 7.4: Improving Information Flow and Workflow) includes a series of references to guide managers and the overall care team in developing efficient, safe, and effective processes for their own practices.

7.3.3 Information

The third component of the modern radiation oncology practice is information. As technology progresses and the accuracy of treatment and patient care improve, the field of radiation oncology is generating and using more and more information. Since the introduction of image guidance, the practice has evolved from weekly review of megavoltage-based treatment ports to daily localization and verification using 2D-, 3D-, and 4D-based diagnostic quality imaging. Other novel diagnostic imaging modalities, dose calculation algorithms, and treatment modalities are also being used, thereby increasing the amount of information needed for and generated by those systems. One of the major challenges arising with technologic advances is that the information generated throughout the process is commonly isolated and scattered among multiple systems, therefore impeding the efficient and effective management and use of that information. The introduction of information standards has facilitated improvements in workflow and information management. At present, three initiatives have been undertaken to develop standards for communication among IS and diagnostic and treatment devices: DICOM, Health Level Seven (HL7), and Integrating the Healthcare Enterprise (IHE). It is important for our field to understand the role and characteristics of each of these standards in the context of improving the information flow through the radiation oncology process.

7.3.3.1 DICOM

The goal of the DICOM standard is to achieve compatibility between imaging, therapeutic, and information management systems in healthcare environments (e.g., cardiology, dentistry, endoscopy, mammography, ophthalmology, orthopedics, pathology, pediatrics, radiation therapy, radiology, and surgery) (DICOM Strategic Document 2009). At the application layer, the services and information objects of the DICOM standard address five primary areas of functionality: (i) transmission and persistence of complete objects such as images, waveforms, and documents; (ii) query and retrieval of such objects; (iii) performance of specific actions; (iv) workflow management (i.e., support of worklists and status information); and (v) quality and consistency of image appearance. DICOM-RT, which follows the same standard DICOM query-retrieve model used in radiology (DICOM Supplement 11 1997; DICOM Supplement 29 1999), defines seven information objects: RT Image, RT Dose, RT Structure Set, RT Plan, and the three components of the RT Treatment Record (RT Beams Treatment Record, RT Brachy Treatment Record, and RT Treatment Summary Record; Figure 7.3) (DICOM Part 3 2008). All of the information and images needed and used during the episode of care in radiation therapy can be converted to the DICOM-RT standard and integrated into the information management system, thus improving connectivity across systems and the flow of information and work within a department. Additionally, effective use and management of the information collected at different stages of the episode of care can also support the treatment decision-making process and improve overall clinical care (Liu et al. 2007; Law & Liu 2009; Law et al. 2009). Current commercial treatment and information management systems in radiation oncology are taking advantage of the DICOM-RT standard to improve communications both among their systems and with other systems.

7.3.3.2 HL7

HL7 is a not-for-profit organization accredited by the American National Standards Institute dedicated to providing a comprehensive framework and related standards for the exchange, integration, sharing, and retrieval of electronic health information that supports clinical practice and the management, delivery, and evaluation of health services (HL7 Web site: http://www.hl7.org/). HL7 specifies several flexible standards, guidelines, and methodologies by which various healthcare systems can communicate with each other. Such guidelines or data standards are a set of rules that allow information to be shared and processed in a uniform and consistent manner. Currently, HL7 is developing Conceptual Standards (HL7 RIM), Document Standards (HL7 CDA), Application Standards (HL7 CCOW), and Messaging Standards (e.g., HL7 v2.x and v3.0). Messaging standards are particularly important because they define how information is packaged and communicated from one party to another. Such standards set the language, structure, and data types required for seamless integration from one system to another.

| RT objects | | | | | | |
RT dose	RT image	RT plan	RT structure set	RT beams treatment record	RT brachy treatment record	RT treatment summary record
General image [C]	General image [M]	RT general plan [M]	Structure set [M]	RT general treatment record [M]	RT general treatment record [M]	RT general treatment record [M]
Image plane [C]	Image pixel [M]	RT prescription [U]	ROI contour [M]	RT patient setup [U]	RT patient setup [U]	RT treatment summary record [M]
Image pixel [C]	Contrast/bolus [C]	RT tolerance tables [U]	RT ROI observation [M]	RT treatment machine record [M]	RT treatment machine record [M]	
Multi-frame [C]	Cine [C]	RT patient setup [U]	Approval [U]	Measured dose reference record [U]	Measured dose reference record [U]	
RT dose [M]	Multi-frame [C]	RT fraction scheme [U]		Calculated dose reference record [U]	Calculated dose reference record [U]	
RT DVH [U]	Device [U]	RT Beams [C]		RT beams session record [M]	RT brachy session record [M]	
Structure set [C]	RT image [M]	RT brachy application [C]		RT treatment summary record [U]	RT treatment summary record [U]	
ROI contour [C]	Modality LUT [U]	Approval [U]				
RT dose ROI [C]	VOI LUT [U]					
Overlay plane [U]	Approval [U]					
Multi-frame overlay [U]						
Modality LUT [U]						

(Left axis label: RT object modules)

FIGURE 7.3 DICOM-RT objects and corresponding modules. Depending on its usage, RT Object Modules can be defined as Mandatory [M], Conditional [C], and User Optional [U]. The DICOM-RT standard has all the necessary elements to represent the information and data needed and generated during the radiation oncology process. Thus, facilitating the connectivity across systems and improving the information flow and workflow during the episode of care. DVH = dose-volume histogram, ROI = region of interest, LUT = look-up table (DICOM Part 3, 2008).

7.3.3.3 IHE

IHE is an initiative undertaken by healthcare professionals and industry to improve the way computer systems in healthcare share information [IHE Web site: http://www.ihe.net/; Wiki IHE Web site: http://wiki.ihe.net/]. The goal of IHE is not to develop new standards but rather to use current well-established standards (e.g., DICOM and HL7) to address specific clinical needs. IHE for Radiation Oncology focuses on addressing information sharing, workflow, and patient care in practice of radiation oncology.

7.3.4 Software

At one point or another, clinical processes and information are controlled and managed by pieces of software and IS. For most of the applications used during the regular workflow of a department, software is a predefined entity that users learn to use and (in most cases) accept over time. The tasks around existing software are mainly training, operation, and maintenance. New needs are constantly arising, and vendors and users are constantly developing new applications. On the one hand, vendors tend to tailor their applications to address the needs

of users through the development of competitive and marketable innovations; users, on the other hand, develop applications to address internal needs and issues with or limitations of the applications provided by vendors. Despite the application and use of several such systems and pieces of software, they share a common element that must be considered during the development process—they all have human safety implications and they all interface with human users. Therefore, it is vital to consider a human-centered approach to the conceptualization, design, and development of such systems (Chapanis 1996).

The software-developing process marks the tipping point that will define the impact of the application in the flow of information and work during its operation. Using a systems-engineering, human-centered approach, Samaras and Horst (2005) proposed a model representing the software development and life-cycle process (Figure 7.4). The development process starts with the initial conceptualization of the system. The assessment of users' needs is probably the most complex activity during software development and depends on the assumption that the correct target audience has been selected. Some techniques to collect the user needs include interviews, questionnaires, ethnomethodologic studies, brainstorming, problem-domain storyboarding, prototyping, literature review, ergonomic laboratory research,

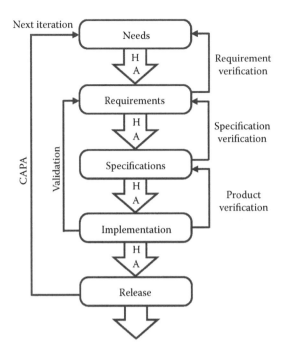

FIGURE 7.4 Software development and life-cycle process proposed by Samaras & Horst (2005). This model follows a systems-engineering, human-centered approach and consists of validation and verification modules and feedback loops with incremental hazard analysis (HA) and postdeployment corrective and preventative actions (CAPA). (From Samaras, G. M., and Horst, R. L., *Journal of Biomedical Informatics, 38*, 61–74, 2005. With permission.)

and evolutionary software development techniques. Once the user needs have been determined, the next task is to translate them into requirements such as response time, storage capacity, load balancing, data backup and disaster recovery, system availability, and ease of use. Properly formulated requirements should be described using common language statements (i.e., "in English") so that they can be clearly understood by users, design teams, and product managers. Properly formulated requirements must be traceable to specific user needs and verifiable by a predefined operational quantity (e.g., effectiveness, efficiency, and user satisfaction). Once the requirements are properly established and verified, the next task is to translate those requirements into engineering design specifications. Such specifications represent the actual product design and quantitative product attributes with associated units and tolerances. The next step is product implementation, which might include several cycles of preproduction development until a defined goal is reached and the system or application is finally released. The software development process includes several feedback loops: validation testing (implementation vs. requirements), verification testing (of requirements, specifications, and implementation), incremental hazard analysis, and postdeployment corrective and preventative actions. Users and developers should not underestimate the importance of the software development process. Proper implementation of this process will considerably improve the safety,

efficiency, and effectiveness of applications supporting the clinical workflow.

7.3.5 Hardware

The modern practice of radiation oncology involves the use of several sophisticated pieces of technology. Figure 7.5 illustrates the IT/IS aspects of the modern radiation oncology practice with corresponding hardware/technology components and connectivity infrastructure. The information flow, workflow, and processes of modern practices are highly dependent on those technologies and the infrastructure to support them. The successful and effective use of those technologies and hardware has three main aspects: network, storage, and support, each of which is described briefly below. Neglecting or undermining the importance of any of these aspects will negatively affect the performance and efficiency of the information flow and processes supported by the corresponding technologies.

7.3.5.1 Network

It is now very common that an institution provides the necessary network infrastructure to support its internal systems. Nevertheless, there might be cases where information needs to be shared outside the institutional network. In those cases, closer attention needs to be given to key performance network parameters (e.g., bandwidth, packet loss rate, latency, and jitter). Security is an essential aspect of the network, data communications, and infrastructure management (Beaver & Herold 2004; van der Linden et al. 2009). Data managers and system administrators need to understand five basic principles in securing any information system: (i) proper identification of the user and computer used by the system; (ii) authentication of the user and the system to be used; (iii) access control, ensuring that the users, systems, and process are allowed to use specific resources; (iv) confidentiality, ensuring that sensitive or private information is protected from unauthorized disclosure; and (v) integrity of the data and systems configuration and attribution, functions that allow tracking of information and actions with users or computers involved.

7.3.5.2 Storage

The introduction of new image-guided modalities for assessment and daily target localization has increased the need for storing large amount of data (on the order of terabytes per year). This issue is not new to radiology practices, the data management needs of which are typically managed through the use of Picture Archiving and Communication System (PACS). Similar PACS are now being designed by vendors and researchers to support the workflow and data management needs of radiation oncology (see also Chapter 9 of this text). IS, as mentioned before, possess several advantages over paper-based records; nevertheless, it is not uncommon that all electronic information is stored in a centralized location (e.g., server and data storage devices); therefore, corruption or malfunction of the system could potentially cause catastrophic data loss. As any practice becomes more

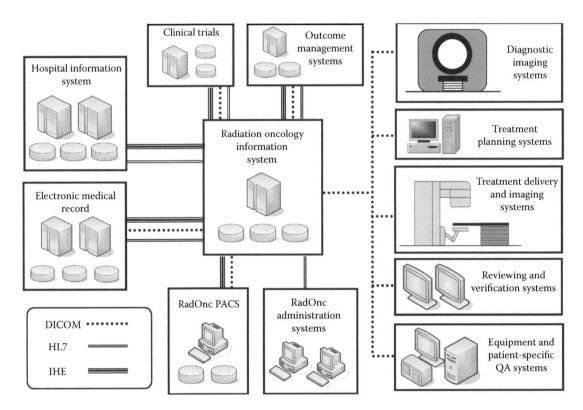

FIGURE 7.5 Hardware components and connectivity infrastructure of the modern radiation oncology practice. Highly available data storage and efficient information flow are highly dependent on having and maintaining reliable, safe, and stable infrastructure and connectivity across all the systems inside and outside the radiation oncology department.

dependent on IT/IS applications, it is extremely important (as part of storage and network management) to have in place daily data backup processes (preferably automated) and fault tolerance high-availability clusters (Schmidt 2006; Bowman 2008) that will guarantee access to the information at any time and under conditions where the main system fails.

7.3.5.3 Support

Given the complexity of the systems, applications, and devices used in the modern practice of radiation oncology, no single person is capable of solving and addressing all of the possible problems that may arise within a system and between systems. This underscores the critical need for IT/IS support. Defining the right amount of IT/IS support is not a simple task. Siochi et al. (2009) proposes using a combination of number of patients and amount of services (archive storage size, number of network nodes, and number of CPUs involved for each specific computing task, etc.) as a measurement for assigning cost-effective and efficient IT/IS supporting resources. These resources can be drawn internally from the institution and/or department (IT staff/physicists) as well as from the vendor (using comprehensive service agreements).

7.3.6 Summary

The previous sections provide detailed descriptions of each component of a modern radiation oncology practice: peopleware,

processes, information, software, and hardware. The development and use of IT/IS applications to the practice of radiation oncology offer enormous opportunities for advancing the field, providing the care team with powerful tools to assist and manage the delivery of safe, high-quality, and cost-effective care. Nevertheless, this great potential comes with new and difficult challenges. Overcoming those challenges and maintaining effective and efficient information flow throughout the radiation oncology process requires a thorough understanding of each component, both in itself and in its relation to the specific needs, goals, environmental restrictions, and infrastructure of a particular practice.

7.4 Improving Information Flow and Workflow

One obvious question that comes to mind is how can an individual (physician, physicist, administrator, and vendor) use all of the concepts proposed here to develop and use IT/IS to maintain effective and efficient information flow and workflow in a particular practice? Traditionally, administrators and project managers focus on the last two components of the radiation oncology practice, software and hardware, seeking the best technology that would bring the highest marketability and profit while potentially neglecting the fact that people will be using this technology and that processes will need to be changed or adapted to maintain efficient information flow and workflow. In short, this mindset emphasizes tools above the final product itself, which in this case the final product is the best possible benefit to the patient. Lorenzi and Riley (1995) proposed that, for a system to work effectively, an organization needs to develop three sets of skills: technical, project management, and people and organizational skills. Technical skills refer to the knowledge, experience, and abilities necessary to create and maintain a successful system. Project management skills include the knowledge, techniques, and skills to successfully manage internal changes in complex IS and practices while maintaining safe and efficient processes and information flow. Last but definitely not least, people and organizational skills are becoming absolutely necessary, particularly as practices become more complex. These are the skills that are needed to meet the often conflicting needs of various stakeholders and to build a team approach and guarantee the successful development, implementation, and use of a particular system. Developing these skill sets as part of the healthcare team environment is the starting point for developing successful, safe, and efficient practices in which IS and technologies can effectively support information flow and workflow.

The next level is to build process-centered practices (Smith & Flarey 1999) in which the work-environment paradigm shifts from task-oriented piecemeal endeavours to management of core processes. In this paradigm, leaders, managers, and supervisors manage processes, not departments. They encourage everyone on the team to work in an organized and coherent fashion, making clear to each member the details of each process and their corresponding roles. Patient care is provided in a fully integrated environment in which each member participates and sees himself or herself as partnering with many different people and specialties. Superior process performance is achieved by having a superior practice design, the right people, and the right environment in which to work. It is becoming increasingly imperative to identify the organization's mission and vision and to ensure that that mission and vision are appropriately supported by the key processes. One of the main objectives of this chapter was to identify the core processes that characterize the modern radiation oncology practice and to describe the role of each team member in the overall episode of care. These core processes provide the foundation upon which practices can organize and build their own processes, based on their specific goals and infrastructure.

It is clear that maintaining efficient and effective information flow and workflow depends greatly on the processes supporting them. Moreover, processes are not and should not be stagnant entities. As practices evolve, processes should evolve in parallel to maintain or improve efficiency and efficacy. But improving or developing a process first requires detailed knowledge of that process. Therefore, practices need to develop the necessary skill set to embark on a continuous process improvement journey. Process mapping, a concept highlighted below, provides the necessary tools to start that journey.

The complexity of the process mapping concept precludes covering the details in couple of paragraphs and is beyond the scope of this chapter. Detailed descriptions of process mapping

and its applications can be found in the following published literature:

- *Process Mapping, Process Improvement and Process Management* by Dan Madison, Paton Press (2005)
- *The Basics of Process Mapping* by Robert Damelio, Productivity Press (1996)
- *Workflow Modeling: Tools for Process Improvement and Application Development* by Alec Sharp, Artech House Publishers (2008)
- *Improving Performance: How to Manage the White Space in the Organization Chart* by Geary A. Rummler and Alan P. Brache, Jossey-Bass Publisher (1995)

Getting one's practice processes mapped is the initial step before one can move toward improving or changing a process or a practice's information flow or workflow. The next step is to choose the quality improvement methodology that fits best the specific practice goals. These methodologies are well-defined, systematic approaches to improve processes, systems, and infrastructure to reduce waste, eliminate inefficiencies, and ultimately provide the highest-quality patient care. Among the methodologies proven useful in the healthcare industry are the following (Mayo Clinic Quality Academy 2009):

- "Change Management"—intended to support organizational and personal transitions from a determined current state to a desired future state to achieve and ultimately sustain the desired business vision and strategy. The objective in change management is to minimize the impact that the change will have on the organization and business performance while maximizing the business benefits of change;
- "Lean"—a continuous improvement methodology that focuses on improving customer service, performance, and quality by reducing waste, process variation, and imbalance. The process helps to identify the value stream or streams and highlight waste that can then be eliminated. Lean uses a set of tools that assist in the systematic identification and steady elimination of waste. As waste is removed from the value stream, quality improves while process times and costs are reduced;
- "Plan-Do-Study-Act (PDSA)"—a shorthand for testing a change by developing a plan to test the change (Plan), carrying out the test (Do), observing and learning from the consequences (Study), and determining what modifications should be made to the test (Act). This is the scientific method used for action-oriented learning;
- "Six Sigma"—a focused, systematic, statistically based process improvement methodology that aims to eliminate or reduce defects in a product or process by reducing process variability. Six Sigma involves the use of facts, data, statistical analysis, and the understanding of customer needs to design processes that have high reliability and low defect rates; and
- "Theory of Constraints"—which is based on the premise that the rate of throughput is limited by at least one constraining process (i.e., a bottleneck). Only by increasing

throughput (flow) at the bottleneck process can overall throughput be increased. This method is useful when a constraint is preventing the system as a whole from performing at a higher level. It can be applied to both internal processes and supply chain.

The field of radiation oncology is complex and constantly evolving. Current needs and goals of the health industry demand improvements in workflow and information flow to provide patient care of the highest standards as those standards continue to evolve and become more complex. IS have the potential to achieve that goal, but only if the synergistic dependency of the components of the modern radiation oncology practice (i.e., peopleware, processes, information, software, and hardware) is considered during the development, implementation, evaluation, and everyday use of the system.

References

Ammenwerth, E. et al. (2003). Evaluation of health information systems—Problems and challenges. *International Journal of Medical Informatics, 71*, 125–135.

Ammenwerth, E. et al. (2004). Visions and strategies to improve evaluation of health information systems—Reflections and lessons based on the HIS-EVAL workshop in Innsbruck. *International Journal of Medical Informatics, 73*, 479–491.

Beaver, K., & Herold, R., *The Practical Guide to HIPAA Privacy and Security Compliance*, Auerbach Publications, Boca Raton, Florida, 2004.

Berg, M. (1999). Patient care information systems and health care work: A sociotechnical approach. *International Journal of Medical Informatics, 55*, 87–101.

Bowman, R. H., *Business Continuity Planning for Data Centers and Systems: A Strategic Implementation Guide*, Wiley, 2008.

Brooks, K., Radiation oncology information management system, in *The Modern Technology of Radiation Oncology*, Van Dyk, J., Ed., Medical Physics Publishing, Madison, Wisconsin, 1999, pp. 509–520.

Brooks, K. W. et al., Advanced therapy information management systems: An oncology information systems RFP toolkit, in *Imaging in Radiation Therapy*, Hazle, J. D., & Boyer, A. L., Eds., Medical Physics Publishing, Madison, Wisconsin, 1998.

Chapanis, A., *Human Factors in Systems Engineering Systems Engineering and Management*, Wiley-Interscience, New York, 1996.

Chaudhry, B. et al. (2006). Systematic review: Impact of health information technology on quality, efficiency, and costs of medical care. *Annals of Internal Medicine, 144*, 742–752.

Committee-on-Quality-Health-Care-in-America, *Crossing the Quality Chasm: A New Health System for the 21st Century*, National Academy Press, Washington, District of Columbia, 2001.

DeMarco, T., & Lister, T., *Peopleware: Productive Projects and Teams*, Dorset House Publishing Company, New York, 1999.

DICOM Supplement 11 (1997). Digital Imaging and Communications in Medicine (DICOM) Supplement 11. Radiotherapy Objects. Available at ftp://medical.nema.org/medical/dicom/final/sup11_ft.pdf. Accessed 9 May 2010.

DICOM Supplement 29 (1999). Digital Imaging and Communications in Medicine (DICOM) Supplement 29. Radiotherapy Treatment Records and Radiotherapy Media Extensions. Available at ftp://medical.nema.org/medical/dicom/final/sup29_ft.pdf. Accessed 9 May 2010.

DICOM Part 3 (2008). Digital Imaging and Communications in Medicine (DICOM) Part 3. Information Object Definitions. Available at ftp://medical.nema.org/medical/dicom/2008/08_03pu.pdf. Accessed 9 May 2010.

DICOM Strategic Document (2009). Digital Imaging and Communications in Medicine (DICOM) Strategic Document. Available at http://medical.nema.org/dicom/geninfo/Strategy.pdf. Accessed 9 May 2010.

Fong, L., & Herman, M. (2007). TU-C-AUD-09: Comprehensive assessment methodology for radiation oncology information systems (abstract). *Medical Physics, 34*, 2551.

Ford, E. C. et al. (2009). Evaluation of safety in a radiation oncology setting using failure mode and effects analysis. *International Journal of Radiation Oncology Biology Physics, 74*, 852–858.

Fraass, B. et al. (1998). American Association of Physicists in Medicine Radiation Therapy Committee Task Group 53: Quality assurance for clinical radiotherapy treatment planning. *Medical Physics, 25*, 1773–1829.

Fredrickson, D. H. et al. (1979). Experience with computer monitoring, verification and record keeping in radiotherapy procedures using a Clinac-4. *International Journal of Radiation Oncology Biology Physics, 5*, 415–418.

Garde, S. et al. (2006). CSI-ISC—Concepts for Smooth Integration of Health Care Information System Components into established processes of patient care. *Methods of Information in Medicine, 45*, 10–18.

Gremy, F. (2005). Hardware, software, peopleware, subjectivity—A philosophical promenade. *Methods of Information in Medicine, 44*, 352–358.

Gremy, F. et al. (1999). Information systems evaluation and subjectivity. *International Journal of Medical Informatics, 56*, 13–23.

Herman, M. G. et al. (1997). Management of information in radiation oncology: An integrated system for scheduling, treatment, billing, and verification. *Seminars in Radiation Oncology, 7*, 58–66.

Huq, M. S. et al. (2008). A method for evaluating quality assurance needs in radiation therapy. *International Journal of Radiation Oncology Biology Physics, 71*, S170–S173.

ICRU, *Determination of Absorbed Dose in a Patient Irradiated by Beams of X- or Gamma-Rays in Radiotherapy Procedures, ICRU Report 24*, International Commission on Radiation Units and Measurement, Bethesda, Maryland, 1976.

Jha, A. K. et al. (2009). Use of electronic health records in U.S. hospitals. *New England Journal of Medicine, 360*, 1628–1638.

Law, M. Y. Y., & Liu, B. (2009). Informatics in radiology DICOM-RT and its utilization in radiation therapy. *Radiographics, 29,* 655–667.

Law, M. Y. Y. et al. (2009). DICOM-RT-based electronic patient record information system for radiation therapy. *Radiographics, 29,* 961–U3.

Liu, B. J. et al. (2007). Image-assisted knowledge discovery and decision support in radiation therapy planning. *Computerized Medical Imaging and Graphics, 31,* 311–321.

Lorenzi, N. M., & Riley, R. T., *Organizational Aspects of Health Informatics: Managing Technological Change: Computers in Health Care,* Springer-Verlag, New York, 1995.

Mayo Clinic Quality Academy (2009). Mayo Clinic Quality Academy. Available for internal institutional use at http://mayoweb.mayo.edu/quality-learning/index.html. Accessed 9 May 2010.

Morgan, M., Identifying and understanding clinical processes, in *Electronic Health Records,* Carter, J. H., Ed., American College of Physicians, 2008, pp. 169–192.

Morrey, D. et al. (1982). A microcomputer system for prescription, calculation, verification and recording of radiotherapy treatments. *British Journal of Radiology, 55,* 283–288.

Neumann, P. G., "Peopleware in Systems." in Peopleware in Systems. Cleveland, OH: Assoc. for Systems management, 1977, pp. 15–18.

Pawlicki, T. et al. (2005). Statistical process control for radiotherapy quality assurance. *Medical Physics,* 2777–2786.

Rosenbloom, M. E. et al. (1977). Verification and recording of radiotherapy treatments using a small computer. *British Journal of Radiology, 50,* 637–644.

Rummler, G. A., & Brache, A. P., *Improving Performance: How to Manage the White Space in the Organization Chart,* Jossey-Bass, San Francisco, 1995.

Saleem, J. J. et al. (2009). Exploring the persistence of paper with the electronic health record. *International Journal of Medical Informatics, 78,* 618–628.

Samaras, G. M., & Horst, R. L. (2005). A systems engineering perspective on the human-centered design of health information systems. *Journal of Biomedical Informatics, 38,* 61–74.

Schmidt, K., *High Availability and Disaster Recovery: Concepts, Design, Implementation,* Springer, 2006.

Siochi, R. A. et al. (2009). Information technology resource management in radiation oncology. *Journal of Applied Clinical Medical Physics, 10,* 16–35.

Smith, S. P., & Flarey, D. L., *Process-Centered Health Care Organizations,* Jones & Bartlett Publishers, Gaithersburg, Maryland, 1999.

Therasse, P. et al. (2000). New guidelines to evaluate the response to treatment in solid tumors. *Journal of the National Cancer Institute, 92,* 205–216.

van der Linden, H. et al. (2009). Inter-organizational future proof EHR systems: A review of the security and privacy related issues. *International Journal of Medical Informatics, 78,* 141–160.

Workflow Management Coalition (1999). Workflow Management Coalition: Terminology & Glossary. Available at http://www.wfmc.org/Glossaries-FAQs/. Accessed 9 May 2010.

World Health Organization, *Radiotherapy Risk Profile—Technical Manual,* in Barton, M., & Shafiq, J., Eds., World Health Organization, 2008.

8

Integrating Radiology with Radiation Oncology: Challenges and Opportunities

George C. Kagadis
University of Patras

Steve G. Langer
Mayo Clinic

Peter Kijewski
Memorial Sloan-Kettering Cancer Center

Paul G. Nagy
The Johns Hopkins University

8.1 Introduction

Acquiring and analyzing data are processes of great importance and complexity in healthcare. These processes are the basis on which specific patient problems are categorized. The data acquired in this procedure may vary from simple narratives, numerical values, or images to more complex forms such as magnetic resonance spectroscopy images and genetic polymorphism analyses. All such data are characterized by uncertainty due to their nonuniform acquisition while the archiving of these data was, until recently, based on various hard-copy types. Nowadays, it seems unreasonable for a modern healthcare enterprise to operate without integrated information technology (IT) systems. IT systems can be invaluable for organizing the acquisition, processing, management, and distribution of a vast abundance of information from a variety of sources (Kagadis et al. 2008, 2010, 2012).

Medical informatics deals with aspects of data generation, distribution, visualization, manipulation, storage, transmission, integration, security, and overall management (Huang 2005; Shortliffe & Cimino 2006). Although medical informatics evolved initially in radiology, the application of informatics in managing data in radiation oncology departments is increasing greatly, as is its acceptance.

This chapter discusses the various aspects of integrating radiology with radiation oncology in terms of IT. Section 8.2 deals with the history of the Picture Archiving and Communication System (PACS) industry. Section 8.3 deals with the current status of and future trends in the field of radiation oncology. Section 8.4 outlines some specifications for developing a "common language" for the integration of radiology and radiation oncology.

8.2 History of the PACS Industry

The current maturity of the PACS industry in radiology is an excellent example of the diffusion of innovation model (Reiner et al. 2002). This model describes the social and economic stages of a technology's penetration in a market as well as the social archetypes of consumers of that technology. The model was created by psychologists who were studying the adoption of planting hybrid corn seed in Iowa in the early 1940s (Ryan & Gross 1943). This group identified five groups of technology adopters, each of which had different social and economic priorities: "innovators, early adopters, early majority, late majority, and traditionalists" (Figure 8.1). This model has been used to describe the adoption of virtually all technologies, from videocassette recorders to smartphones. Each stage can be identified by market penetration as well as business and technical innovations. The remainder of this section consists of a brief history of the PACS industry, presented in terms of the diffusion of innovation model, with the goal of identifying current opportunities and future challenges

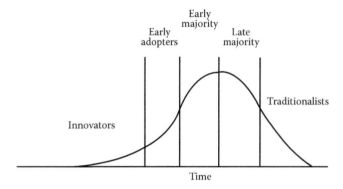

FIGURE 8.1 Diagram of the diffusion of innovation model (Ryan & Gross 1943).

for this industry. Consumers decide at what point of maturity to invest in technology. The rational economic model describes the decision point as the point at which a consumer perceives the benefits of a technology outweighing the risk of obsolescence.

8.2.1 Innovator Phase (1980s to 1992)

The innovator phase of the PACS industry dates back to the late 1980s with the creation of dedicated proprietary vendor networks to display and transmit images. This phase involved the least risk-averse group trying to determine whether it was even possible to read medical images electronically. Early pioneers had to address three challenges. The first was the display of medical images in ways that could approach the resolution and brightness of film mounted on view boxes. The second challenge was the storage of medical images in an environment in which the cost of one terabyte of online storage was $10 million. The third challenge was developing the standards required to support the storage and transfer of images. The American College of Radiology, along with the National Electrical and Manufacturing Association, created the eponymous "ACR/NEMA" file formats for the common storage of medical images but did not address a means to transfer them. The pioneers of the innovator phase implemented these systems in laboratory and academic settings.

8.2.2 Early Adopter Phase (1993 to 1997)

The clear trigger for the early adopter phase was the release of the Digital Imaging and Communications in Medicine (DICOM) 3.0 standard in 1993. The DICOM 3.0 standard provided a flexible binary run length encoded information model for the common storage of medical images along with a means to query and retrieve DICOM images on a DICOM local area network. The Radiological Society of North America funded reference implementations of the DICOM standard, which greatly aided the adoption of this standard by commercial vendors. The U.S. Department of Defense and U.S. Department of Veterans Affairs specified DICOM as a requirement for their purchase approval

of all PACS equipment (DeMoor et al. 1993). The DICOM standard allowed a "best-of-breed" model, whereby PACS vendors could integrate with competing modality vendors. The cost of the technology behind PACS also continued to decline, with one terabyte of storage costing approximately $1 million in 1993. Another key factor in driving the early adopter phase was the presence of investigators demonstrating the utility of PACS for improving the productivity of radiologists and technologists and minimizing the referring physician's time. Studies demonstrated that, under certain conditions, radiologists could improve their productivity by up to fifty percent (Reiner et al. 2001), technologists could improve their productivity by up to forty percent (Reiner et al. 2002), and clinicians could save up to fifty minutes a day by not having to visit the radiology department (Dreyer et al. 2005). Although the early adopters were not overly cost conscious, these reports on the effectiveness of the technology were critical for the next phase, that of the early majority.

8.2.3 Early Majority Phase (1998 to 2002)

The trigger for the early majority phase was the introduction by several new vendors of a Web-based, enterprise-level PACS that used the hospital's own network and delivered an intuitive user interface to clinicians throughout that hospital network. The development of Web-based systems such as these introduced PACS to a whole new class of physicians. Return on investment became achievable, if not revenue neutral, with the use of commercial workstation and server components. The cost of a terabyte of online storage continued to drop to $100,000. PACS began to be adopted outside of academic medical schools by private practices that wanted to improve their productivity and extend their reach. Such practices used teleradiology to extend their geographic coverage and become a competitive differentiator. At that time, the state of computed tomography (CT) had advanced to the extent that it became disadvantageous to read hundreds of axial slices on film-based view boxes. Stack mode computer-based viewing showed significant improvements in the speed and efficiency of image review. PACS vendors began to offer fee-for-service models of payment that removed the capital purchase price as a barrier for entry into the PACS market.

8.2.4 Late Majority Phase (2003 to 2008)

The tipping point in the PACS industry, the point at which fifty percent of the market had adopted this technology, came in 2003 boosted by innovations in the liquid crystal display (LCD) market. LCDs proved advantageous compared with cathode ray tube display technology in cost, weight, brightness, uniformity, and heat generation. Commercial LCDs could be calibrated to the DICOM grayscale display function, which ensures that the images are displayed with the same characteristics across monitors of different capabilities. LCDs allow better ergonomics and space design in image reading rooms. LCDs were also widely adopted throughout hospitals during that time in areas of high ambient light and tight space constraints. By 2003, the cost of a terabyte of storage had dropped to $10,000, and return on investment for PACS became clear. Another clear sign of the maturity of the industry was the standardization of the competencies required to be a successful PACS administrator. In 2006, the American Board of Imaging Informatics released a certification program for imaging informaticists. That board was created by the Society of Imaging Informatics in Medicine and the American Registry of Radiologic Technologists after years of working on defining a common body of knowledge for PACS administrators.

8.2.5 Traditionalist Phase (2009 Onward)

In 2008, a market research publication reported that the penetration of the PACS market had reached ninety percent (PACS Industry Study 2008). Although the accuracy of this report is a matter of debate, most agree that the PACS market is now mature. A radiology practice without a PACS is considered uncompetitive in terms of recruiting new radiologists and competing for contracts that require rapid report turnaround times. Contracts between hospitals and radiology practices now require report turnaround times of a few hours and all cases to be turned around within twenty-four hours. These performance requirements are not feasible without IT support. Small facilities in rural areas are using PACS to help radiologists cover large geographic areas. Practice approaches such as NightHawk, preliminary reads only, and teleradiology have emerged that leverage PACS as part of their core business model.

A further sign of the underlying technology maturity of PACS is storage. The cost of a terabyte of online storage, in keeping with Moore's Law of having in price every two to three years for the same capacity, had dropped to $1000 in 2009. Many of us who have witnessed the transformation of radiology over the past twenty years still find it difficult to grasp the 10,000-fold decrease in the cost of storage and its impact on our industry. Similar beneficial—and disruptive—changes have occurred in processing power, networking, and display technologies. The cost of hardware has dropped to such a point that it is advantageous to duplicate a PACS architecture as an option for fault tolerance and business continuity. Complete hardware redundancy should be a requirement for hospitals with users throughout the enterprise and operations that run evenings and weekends. Many hospitals have found that even perfect disaster recovery plans to preserve data are wholly insufficient in terms of preserving operations: although the data may be intact, operations can be disrupted for hours and even days while hardware is being replaced and the system is being rebuilt. Such disruptions have been shown to be traumatic to all diagnostic imaging operations. Today, hospitals are putting much more emphasis on business continuity plans in addition to mere disaster recovery.

The largest differentiator in today's PACS market lies in the vendor's ability to interoperate with other clinical systems. In the 1990s, the term interoperability was typically used to describe the ability of PACS to send and receive DICOM images. Today, the connotation of interoperability has grown to encompass much more, as we discover the value of integrating PACS with more information systems throughout and beyond a hospital, so that radiologists are provided with additional relevant information to further improve their diagnostic abilities. Interoperability based on open standards such as DICOM, Health Level Seven (HL7), and Integrating the Healthcare Environment (IHE) becomes extremely important to facilitate a successful migration between vendors. Like most software applications, the lifetime for PACS software is approximately five years. Many early-generation architectures predated DICOM standards for the storage of key image selections as well as annotations and presentation states, effectively locking them in a proprietary architecture.

8.3 Current Status and Future Trends in Radiation Oncology

8.3.1 Enterprise IT

A trend in hospital enterprise IT is a transition from systems that are exclusively dedicated to serving a particular clinical department toward systems that absorb an increasing proportion of departmental IT functions into systems that serve the entire enterprise. There are several reasons for this direction of development. Among them is that each clinical department performs a specialized function that is only part of a patient's care; therefore, departmental IT systems must be part of the workflow of the larger enterprise. As one example, communication of information between departmental systems and enterprise systems is essential for building and maintaining electronic medical records (EMRs).

An example of a function within this departmental enterprise communication is Admission, Discharge, and Transfer (ADT). The ADT information generated and modified by an enterprise system is distributed to all departmental systems. For example, a patient's medical record number is generated by an enterprise system and distributed to all departmental systems. Similarly, any change in a patient's demographic information is automatically made known to all departmental systems.

More recent enhancements in enterprise workflow include means by which physicians can enter orders directly into the enterprise IT system—otherwise known as computerized physician order entry (CPOE). Orders to perform a particular clinical procedure are entered into the system and then transmitted to the appropriate departmental system. At the Memorial Sloan-Kettering Cancer Center (MSKCC), CPOE is a component of the Sunrise Clinical Manager system (Eclipsys, Atlanta, Georgia). However, any institution considering such a tool should be aware that implementing an enterprise-level CPOE that works well at the level of the individual physician is an extremely difficult undertaking and requires a very high level of coordination among all stakeholders. At this time, a common problem associated with use of CPOE is the significant discrepancy between the system as delivered by the vendor and what the physician actually needs for rapidly and conveniently entering an order for an a appropriate procedure, particularly in view of ever-increasing patient loads.

Means of scheduling patient procedures is another example of a function that may be part of a departmental IT system but may be moved into an enterprise-wide scheduling system so that an institution can coordinate all patient activities no matter where a particular procedure is performed. At the MSKCC, the Cadence EWS (Epic, Verona, Wisconsin) has been implemented for all patient scheduling. At present, departmental systems receive information generated by the system but cannot modify that information.

As is apparent from the previous sections, multiple systems from different vendors continuously exchange data to maintain workflow throughout the enterprise. Data formats and transmission protocols may be incompatible. Moreover, data may have to be queued for later transmission if a particular subsystem fails. For these reasons, the MSKCC has installed an interface engine through which all data must flow (SeeBeyond eGate integrator, acquired by Oracle). A simplified diagram of the associated enterprise workflow is shown in Figure 8.2.

8.3.2 Enterprise PACS

Aside from the practical reason for an enterprise IT workflow (i.e., the fact that multiple clinical departments contribute to a patient's care), there are technological reasons for transitioning specific functions from departmental to enterprise systems as well. A key example is the imaging infrastructure being built out as imaging changes from film-based to digital systems such as PACS. The practice of radiation oncology increasingly depends on digital imaging for its clinical activities, and it would make little sense to duplicate much of the basic imaging capabilities in radiology for radiation oncology. The MSKCC has operated an institution-wide PACS for more than ten years (GE Healthcare Centricity PACS, version 3.0.5, General Electric Corporation, Milwaukee, Wisconsin). An indicator of the growth of the facility during this time span is the rate of image accumulation, which now is at 1.5 million

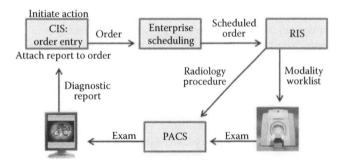

FIGURE 8.2 A workflow action is initiated via a CPOE in a CIS. The order is scheduled and transmitted to the RIS. The specific procedure to be performed is transmitted to the PACS and a worklist item is presented to the technologist. After images are acquired, the examination results are "pushed" into the PACS and then read by the radiologist. The workflow ends when the diagnostic report is attached to the order placed by the clinician.

images per week. The technology infrastructure was recently upgraded in anticipation of the opening of a large breast imaging center in September 2009 and the Center for Image-Guided Intervention in April 2010.

Within six months of startup, the PACS at the MSKCC had grown to serve the entire enterprise with access to radiological images and diagnostic reports. As film use has declined, the ability to view images on display systems had to be made available throughout the enterprise, including clinic work areas, operating rooms, conference rooms, offices, and physicians' homes. To meet this need, a high performance wide-area network was developed to link seven distinct sites with the radiology information infrastructure.

8.3.3 Principal PACS Components

The central subsystem within any PACS is an image management system that controls image acquisition, storage, routing, archiving, and display. In the MSKCC system, high performance access to all images is provided by seventy terabytes of network-attached storage (Hewlett Packard and Microsoft). A PACS Web application permits images to be viewed on any hospital PC, with the viewer having direct access to images for the current patient selected in the enterprise clinical information system (CIS). DICOM query and send functions serve workstations with specialized applications such as treatment planning and 3D visualization and analysis. However, significant effort has been devoted to moving specialized image processing, including 3D viewing, onto dedicated PACS workstations with more extensive functionality compared to the PACS Web viewer. A combination of magneto-optical disks, ultra-density optical drives, and hard drives (Centera by EMC^2, Hopkinton, Massachusetts) is used for archival storage. About 140 image acquisition devices use modality worklists (MWL) to avoid duplicate data entry and significantly reduce the risk of misidentifying patients or image studies. Because the MSKCC is a tertiary referral center, a large volume of images that are imported into its PACS were acquired at other institutions. To facilitate the importation process, CD readers and film digitizers have been installed with additional functions to edit imported information and "push" the exams into PACS (GEHC RA600; General Electric Corporation).

Even considering the extensive imaging capabilities built into the PACS installation, the overriding motivation for elevating PACS to the status of an enterprise facility is business continuity, which encompasses both high availability and disaster recovery. The MSKCC maintains duplicate PACS and ancillary facilities in two data centers separated by about eleven miles that are connected by means of multiple ten gigabit Ethernet links and include duplicate seventy terabytes of online image storage. In the event of a malfunction at the primary site, operations are switched to the failover site. This type of system redundancy is considered essential for a large digital imaging facility. In fact, future enhancements of this failover system are planned with the goals of both more rapid recovery of system functions and minimizing potential loss of data.

8.3.4 Radiation Oncology and PACS

As is clear from the description of PACS in the previous section, radiation oncology can derive substantial benefit from taking advantage of an enterprise-level digital imaging installation that has been built out and refined over several years. The most obvious example is the management of images acquired on the same type of devices in radiology and radiation oncology, such as CT, magnetic resonance, or positron emission tomography (PET)/CT scans. At the MSKCC, a PET/CT scanner in radiation oncology has been linked into the same enterprise workflow as multiple PET/CT scanners in radiology. The existing imaging installation and enterprise workflow can be used without change. The effort during the implementation would center primarily on making sure that the radiation oncology staff has a thorough understanding of the workflow that has already been implemented for radiology.

Workstations dedicated to specific tasks provide convenient access to images managed by PACS via the use of DICOM query or DICOM send functions. Convenient export of complete examinations or selected images into DICOM-formatted files is a common request. Figure 8.3 shows one possible configuration for integrating radiation oncology imaging systems with PACS to take advantage of facilities already in place. (The implementation of this integrated facility at the MSKCC is not yet complete.)

8.3.5 Future Developments

Notwithstanding the extensive imaging infrastructure already in place for radiology, much work is left to be done to integrate radiation oncology into the enterprise imaging workflow. The system at the MSKCC is no exception. Significant additional effort is needed to complete this integration. Some important issues requiring attention by vendors are outlined in the following sections.

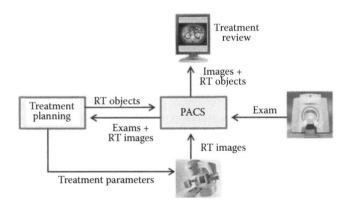

FIGURE 8.3 Images acquired on scanners are acquired using the workflow shown in Figure 8.2. DICOM is used to exchange images between the PACS and treatment planning. Images acquired during treatment can be stored in PACS. Access to information for treatment review is also via DICOM, with the required application installed on the PACS workstation. RT, radiation therapy.

8.3.5.1 Domain Partitioning

With radiation oncology objects such as RT Image, RT Dose, RT Structure Set, RT Plan, and RT Beams Treatment Record (DICOM, Part 3, p. 494) stored in PACS, radiology and radiation oncology departments can share the entire imaging backend infrastructure. However, users of an imaging system would prefer to see, as their default, only those images that are relevant for their clinical practice. For example, a radiologist assigned to reading CT images would want to see only studies that are acquired specifically for diagnosis. The same principle applies for radiation oncologists who are reviewing CT images for treatment planning; however, in addition to the treatment planning images, a radiation oncologist may also want to see diagnostic studies to aid in target delineation. Convenient and rapid switching between these alternate views of imaging studies is an important feature that is not difficult to implement but is not yet available in the MSKCC PACS.

8.3.5.2 Layered Annotations

Radiologists and radiation oncologists generate annotations for different reasons. Such annotations may use the same storage facility, but they must be viewable separately depending on the physician group that generated the annotations. Use of the grayscale softcopy presentation state is one way to incorporate such a feature.

8.3.5.3 Nonviewable Objects

Radiation oncology typically generates data sets that are not viewable by the PACS workstation application. A potential solution is to automatically activate an application with the specific functions required for Radiation Oncology objects. Such a feature has been implemented for radiology when accessing data stored specifically for volume visualization.

8.3.5.4 Context Sharing among Applications

Applications targeted for various radiologic specialties may need to be purchased from either different vendors or largely autonomous divisions of the same vendor. This situation has led to the proliferation of workstations dedicated to specific applications. At the MSKCC, an attempt has been made to reverse this trend by loading applications onto the PACS workstation and sharing patient and examination context among applications. This has been done for diagnostic reporting, volume visualization and analysis, and reading of PET/CT scans.

A similar approach can be considered for radiation oncology. Applications specific for radiation oncology can be loaded onto the PACS workstation, and communication between the PACS application and the radiation oncology applications is then established. Ideally, images should be included as shared objects, but such an implementation is more difficult than sharing such items as the current patient and active study. So-called "virtual machines" can be of value for applications that require a completely different software environment, for example, a different operating system. However, additional effort is usually needed for integration among different applications. At present, all context-sharing implementations at the MSKCC are proprietary. However, use of the clinical context object workgroup, an HL7 standard protocol designed to enable disparate applications to synchronize in real-time and at the user interface level, may prove valuable for allowing applications to present information at the desktop or portal level in a unified way.

Large hospital enterprises are increasingly taking IT functions out of clinical departments and moving them into enterprise-wide systems. Examples include CPOE, enterprise-wide scheduling, and CIS. Image management is no exception. The required infrastructure for imaging is too complex, too expensive to implement, and too quickly changing to duplicate for each clinical specialty. The MSKCC has made some progress in this direction, but much effort will be needed in the future. For example, specific features to accommodate multiple clinical specialties must be incorporated into PACS, as described in the following section. Considerable progress has been made in incorporating various features for cardiologic practice, but radiation oncology is more demanding with respect to the required features; therefore, progress in making expensive enterprise resources for imaging available to radiation oncology has thus far been less than satisfactory.

8.4 Specifications for a "Common Language" for Integrated Radiology and Radiation Oncology

8.4.1 Imaging Components and Institutional Workflow

When a patient enters a medical center to be seen in a radiology or radiation oncology department, several tasks must be performed that are similar in both departments:

(a) The patient must be registered into the hospital. The actor that does this is often called Registration/ADT (Reg/ADT) and may actually be a part of the Hospital Information System (HIS).
(b) Once the patient is registered, physicians create imaging or radiation oncology orders in the HIS.
(c) The resulting orders must be acted upon by scheduling systems within each department.
(d) Imaging studies are then obtained (diagnostic examinations in radiology or planning/verification studies in radiation oncology).
(e) Images are reviewed for quality assurance and results are compiled (reports or treatment verification).
(f) Images and results are archived.
(g) Images and results are published to the HIS.

The two subsections that follow describe how these tasks play out in their respective arenas with the equipment typical of that environment.

8.4.2 Workflow between the Institution and the Radiology Department

The principal components in a radiology department are the following:

(a) Radiology Information System (RIS)
(b) MWL provider, which provides worklists to imaging modalities
(c) Imaging modalities (e.g., CT and magnetic resonance)
(d) PACS
(e) Separate image archive and manager (optional)

As mentioned in Section 8.4.1, the patient's experience begins upon being "admitted" into the medical center. If the patient is unknown to the hospital, the patient is interviewed and their personal information forms the basis for their "registration" in the hospital (Reg/ADT). At this point, the patient is assigned a patient identifier (PID), which forms the key used to search for results (labs, pathology, and radiology reports) in the EMR. From this point, patients are "transferred" to the radiology department. This transfer is captured in the Reg/ADT so as to track the patient's location within the medical center. The information system components used in the radiology department are then used as follows:

(a) The order from the referring physician arrives from the HIS and is serviced by the RIS. The RIS determines what resource is needed (e.g., a CT suite) and locates a free time slot for it. The RIS then sends this information back to the HIS so the patient is provided with a single list of the scheduled examinations in the medical center.
(b) The patient arrives at the imaging department at the scheduled time. A receptionist enters the patient into the RIS. This entry triggers several events; the CT suite is now reserved for this patient; a message is sent to the Reg/ADT system so that the patient's location is tracked; and the order information is sent to the MWL server for subsequent use at the scanner.
(c) The patient is escorted to the CT suite and the examination is begun. Using the PID, the scanner queries the MWL for examination information, thereby preventing the introduction of human error during data entry at the scanner.
(d) The MWL responds with the order information. This is usually identified with a study ID of some kind, often called the "accession" number.
(e) The scanner creates and marries images to the order information specified by the accession number. When the imaging study is complete, it is sent to the PACS.
(f) Upon reaching the PACS, the study is held as "pending" until quality assurance can be performed by a technologist knowledgeable in the procedure that has just been performed. If the study is complete and of sufficiently high quality, then it is released for interpretation.
(g) The radiologist interprets the images on a PACS workstation and then dictates a report that is transcribed back into the RIS. When both the images and report are marked as final, they are available for publication to the EMR.

The preceding section presents a brief introduction to the institutional and radiology components that implement the workflows required for an imaging examination. The protocols used and further details of how the workflow is accomplished are described in the following section.

8.4.3 Workflow between the Institution and the Radiation Oncology Department

Components of the workflow between the institution and the radiation oncology department at the MSKCC are very similar to that presented in the previous sections. A treatment to be performed starts with an entry in the CPOE system and is then scheduled via the enterprise-wide scheduling system. The scheduled order is then passed to the Oncology Information System (Aria, Varian Medical Systems, Palo Alto, California). Treatment dictations are forwarded to an enterprise repository as the permanent record. However, the actual implementation of this workflow is not as complete as it is in radiology. Rather, legacy systems that had been developed in-house are gradually being phased out to conform to the standardized hospital workflow. Aspects of that standardized workflow are described in the section that follows.

8.4.4 Standard Protocols and Integration Use Cases

To this point, we have introduced a lengthy list of acronyms and abbreviations, including Reg/ADT, HIS, EMR, RIS, MWL, PID, and PACS. However, each of these agents can also be considered to be a compound made up or more fundamental "atomic" software actors, with all of the attendant potential for integration problems among and between them. In 1997, the Radiological Society of North America, the Healthcare Information and Management Systems Society, several academic centers, and several medical imaging vendors embarked on a program to solve integration issues across the breadth of healthcare informatics—the IHE initiative (Channin 2000; Channin et al. 2001). The IHE initiative has prompted in-depth analysis of most of the workflows used in medical centers, and these analyses have been used to extract an idealized set of actors that accomplish specific tasks. These actors are combined and communicate with standard protocols to enable the specific workflow required. Today, IHE spans a broad variety of implementation domains, including anatomic pathology, cardiology, eye care, IT infrastructure, laboratory analysis, patient care coordination, patient care devices, quality assurance, radiation oncology, and radiology.

Notably, IHE does not attempt to specify new standards but rather gives guidance on how to implement existing standards in a coherent way to successfully complete the required workflows. The "use cases" developed to accomplish this guidance are referred to as "integration profiles" and have been designed to work with existing standards. For instance, the vast majority of integration profiles rely on DICOM (http://medical.nema.org) and HL7 (http://www.hl7.org). DICOM is the native tongue among imaging devices (PACS and imaging modalities), whereas HL7 is the native language of ordering and result systems (e.g., EMR and HIS).

When this chapter was written, the following "integration profiles" had been defined for the practice of radiology:

(a) *Scheduled Workflow*: The steps in a normal, planned order

(b) *Charge Posting*: The steps that permit an examination to be properly billed

(c) *Patient Information Reconciliation*: A means of updating patient demographics if they were originally incorrect

(d) *Import Reconciliation Workflow*: A means of "coercing" images from an outside medical center to agree with the PIDs used at the current site

(e) *Presentation of Grouped Procedures*: A means of building associations among, for example, several studies of different body parts so that Display/Report Actors will open all of those studies together

(f) *Post Processing Workflow*: A means of managing worklists and examination status for advanced processing and computer-aided design steps beyond initial image acquisition

(g) *Reporting Workflow*: A means of managing worklists and report status for reporting workflow

(h) *Mammography*: A means of creating, storing, managing, and using mammographic objects

(i) *Nuclear Medicine*: A means of creating, storing, managing, and using nuclear medicine objects

(j) *Consistent Presentation of Images*: A means of ensuring the consistent appearance of images across hard-copy and soft-copy systems

(k) *Evidence Documents*: Workflow steps and worklists for managing computer-aided design and report objects (e.g., DICOM Structured Reporting)

(l) *Key Image Notes*: Methods for flagging and managing significant images within an examination

(m) *Simple Image and Numeric Reports*: A means of creating and managing simple reports with an optional image

(n) *Access to Radiology Information*: A means of consistent access to images and reports

Similarly, the following "actors" have been defined for the practice of radiology:

(1) Modality
(2) Charge Processor
(3) Department System Scheduler/Order Filler
(4) Display
(5) Document Consumer
(6) Document Repository
(7) Evidence Creator
(8) Image Archive
(9) Image Document Source
(10) Importer
(11) Patient Demographics Supplier
(12) Performed Procedure Step Manager
(13) Portable Media Creator/Importer
(14) Post Processing Manager
(15) Print Composer/Server
(16) Report Creator/Mgr/Reader/Repository
(17) Image Display
(18) Image Manager
(19) Image Document Consumer

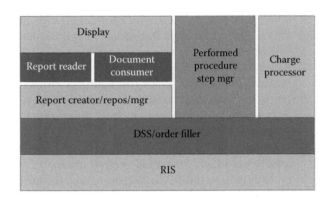

FIGURE 8.4 Actor components from the IHE initiative that could be included in an RIS. DSS, department system scheduler.

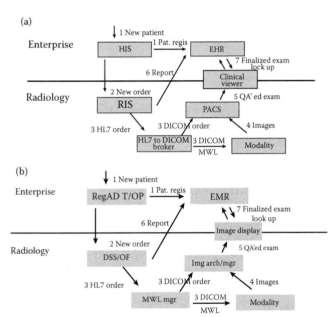

FIGURE 8.5 (a) In this view, the process of scheduled workflow to image a patient is illustrated within the context of conventional information systems as described in Section 8.4.2. (b) In this view, the conventional systems have been mapped onto the corresponding IHE actors. DSS/OF, department system scheduler and order filler (parts of an RIS); MWL Mgr, MWL manager.

A visual representation of an RIS, expressed in terms of IHE actor components, is presented in Figure 8.4. Figure 8.5 illustrates the text-based description of workflow processes (Section 8.4.2) from a conventional information systems viewpoint (Figure 8.5a) and in terms of IHE actors (Figure 8.5b).

8.5 Conclusions

Medical informatics is a fertile field that is continuously growing in importance and relevance for all medical physicists, and the pace of that growth continues to increase as well. Until now, medical imaging has covered the entire spectrum of biological organization from genotypes to phenotypes, combining the

knowledge gained from various imaging modalities into a unified framework (PACS Industry Study 2008). Patterns emerging from each imaging modality reveal different information of a certain clinical problem (anatomy, physiology, metabolism, etc.).

IT systems are indispensible in current radiology and radiation oncology departments. Both domains are image intensive and both use a variety of parameters that need to be recorded for future reference. Medical decision-making systems will continue to evolve, integrating medical imaging metadata into their structure along with laboratory data and patient history. The future of PACS, as an example, will undoubtedly be characterized by provision of greater opportunities and reliability. Future generation PACS will not only be able to store and deliver higher quality images but also be tightly integrated with other hospital systems to provide an integrated electronic health record for all patients. The role of medical physicists in this process is strategic and is analytically discussed in the first chapter of this book.

References

Channin, D. S. (2000). M:I-2 and IHE: Integrating the healthcare enterprise, year 2. *Radiographics, 20,* 1261–1262.

Channin, D. S. et al. (2001). Integrating the healthcare enterprise: A primer. Part 3. What does IHE do for ME? *Radiographics, 21,* 1351–1358.

DeMoor, G. J. E. et al., Eds., *Progress in Standardization in Health Care Informatics,* IOS Press, Amsterdam, 1993.

Dreyer, K. J. et al., *PACS: A Guide to the Digital Revolution,* 2nd edition, Springer, New York, 2005.

Huang, H. K. (2005). Medical imaging informatics research and development trends—An editorial. *Computerized Medical Imaging and Graphics, 29,* 91–93.

Kagadis, G. C. et al. (2008). Anniversary paper: Roles of medical physicists and health care applications of informatics. *Medical Physics, 35,* 119–127.

Kagadis, G. C. et al., Overview of medical imaging informatics, in *Advances in Medical Physics,* 3rd edition, Hendee, W. R. et al., Eds., Madison, Wisconsin, Medical Physics Publishing, 2010.

Kagadis, G.C., & Langer, S.G., Eds., *Imaging in Medical Diagnosis and Therapy: Informatics in Medical Imaging,* 1st edition, CRC Press, Boca Raton, Fl, 2012.

Picture Archiving and Communications System (PACS) [medical devices industry study], TriMark Publications, 2008.

Reiner, B. I. et al. (2001). Radiologists' productivity in the interpretation of CT scans: A comparison of PACS with conventional film. *AJR American Journal of Roentgenology, 176,* 861–864.

Reiner, B. et al. (2002). Changes in technologist productivity with implementation of an enterprise-wide PACS. *Journal of Digital Imaging, 15,* 22–26.

Ryan, B., & Gross, N. C. (1943). The diffusion of hybrid seed corn in two Iowa communities. *Rural Sociology, 8,* 15–24.

Shortliffe, E. H., & Cimino, J. J., Eds., *Biomedical Informatics: Computer Applications in Health Care and Biomedicine,* 3rd edition, Springer, New York, 2006.

9

Volker Steil
University of Heidelberg

Stuart Swerdloff
Elekta AB

Fred Röhner
University of Freiburg

Frank Schneider
University of Heidelberg

Frederik Wenz
University of Heidelberg

Gerald Weisser
University of Heidelberg

Frank Lohr
University of Heidelberg

Image Management in Radiotherapy

9.1 Introduction

Radiotherapy (RT) workflow revolves around image data. What makes workflow in RT different from that in other specialties is that RT images are typically associated with, and thus visualize and document, a specific step in the treatment delivery chain. It is therefore necessary to establish an image documentation and archiving system that properly represents these associations between images and treatment steps.

Several recent developments in RT have rendered this endeavor more complex:

- More sophisticated treatment techniques such as dynamic paradigms, volume-modulated arc therapy, TomoTherapy, and particle therapy, which can only be successfully managed in a digital environment and which have extensive documentation requirements, are being introduced into clinical routine at a very fast pace, soon all but completely replacing "conventional" RT. In particular, these techniques generate plan data that may be very different from plan data that have been generated to date as 4D or adaptive strategies may require more than one structure set on a computed tomography (CT) data set or more than one CT data set, each one with a new structure set.

- Such techniques with typically isotropically steep gradients can exploit their full potential only under the auspices of precision image-guided RT (Boda-Heggemann et al. 2011). In this context, a large amount of image data is generated in association with specific treatment steps. With increasing frequency, volume data are acquired, which

require substantial amounts of image storage capacity. The current acceleration of complex therapy techniques based on improved treatment machine control systems, faster collimators, and the increasing number of filter-free beam lines will lead to these techniques being used in more than fifty percent of all external beam treatments. Constantly improving local and systemic treatments result in longer survival. This longer survival applies not only to patients treated with curative intent. Patients treated for metastases also live longer and will have an increasing number of repeat treatments of new metastatic lesions or retreatments of lesions treated with a subradical dose in palliative intent as a consequence. Complex treatments will not be limited to curative intent situations but will be used increasingly for treatment of metastases. RT departments and practices are increasingly forming network structures, either as a peer network or as a central unit with satellites. Image storage and administration cannot be limited to a local setup in these situations. Storage periods in RT are extensive (well in excess of ten years in most countries) and therefore require an image management and archiving strategy that extends substantially into the future.

Expanding upon previous articles and documents discussing electronic data management in RT (Starkschall 1997; Herman 1999; Steil et al. 2007, 2009; Swerdloff 2007; Heinemann et al. 2009), this chapter will specifically discuss requirements and strategies for image management and archiving along the RT workflow.

9.2 Medical Images, DICOM Standard

The majority of image information in medicine is already generated, dealt with, and communicated in a filmless fashion. Analog image information is likely to completely disappear in due course. All image management will therefore be based on the Digital Imaging and Communications in Medicine (DICOM) standard because all relevant manufacturers of imaging or image processing systems have adopted or implemented this standard. Adoption of this standard ensures connectivity and, for most pure image data and some modalities, also interoperability of various systems in a clinical environment. Both the data format for short- or long-term storage and the communication protocol for data exchange are defined in the standard. This forms the basis for digital image archiving in what is known as Picture Archiving and Communication System (PACS).

To provide information about the DICOM functionality of a particular imaging or image processing system, manufacturers have to publish an exact description of all their DICOM-compatible systems (DICOM Conformance Statement). If such a statement exists, the system's interoperability with other systems is, however, not a given. Interoperability of two systems can only be assessed by comparing both conformance statements. The DICOM standard itself is constantly expanding. The DICOM Working Groups add supplements to the DICOM standard every year. More than 100 supplements have already been published. Image processing units such as radiation treatment planning systems (TPS) may add additional information such as regions of interest (ROI), surface definitions, landmark points, image registration parameters, etc. Some manufacturers also add vendor-specific information representing certain proprietary functions in what is known as private tags. These possible variations of the standard already at the image data level may pose a problem when exchanging data or visualizing data sets with software tools other than the system that generated these data (Ackerly et al. 2008).

9.3 DICOM Data in RT

Within the DICOM standard, the specific requirements of RT are dealt with in DICOM objects (Law & Liu 2009). Objects may comprise information about image data, geometric and mathematical information (spatial registration objects), or treatment-specific information. DICOM-RT objects (RT for "radiation therapy") contain only treatment data and reference the respective underlying images. The following RT objects are defined and widely used (see also Figure 9.1):

- RT-Dose
- RT-Structure Set
- RT-Plan
- RT-Ion Plan
- RT-Image
- RT-Treatment Record
  - RT-Beams Treatment Record
  - RT-Brachy Treatment Record
  - RT-Ion Beams Treatment Record
- RT-Treatment Summary Record

These DICOM-RT objects are constantly supplemented with new objects as a consequence of the continual development of the DICOM standards by the DICOM Standards Committee (Working Group 7, Radiotherapy). New objects currently being defined and established in the DICOM supplement 147 ("Second Generation RT", not yet published but available in draft) are, for example, RT Patient Positioning, RT Replanning (Adaptive

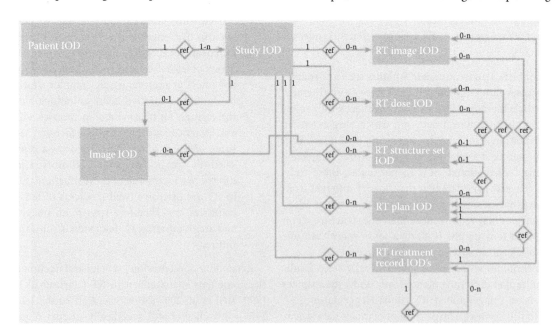

FIGURE 9.1 Data structure of DICOM Data with RT objects, modified after: Figure7.2c DICOM information model (Radiotherapy), in: (Dicom Standards Committee Working Group 7, 1999).

Planning), RT Plan Review etc. Changes defined in this supplement will also enable workflow management based on the "unified procedure step" defined in Supplement 96. Within the framework of this process, existing objects will also be modified or replaced. This may affect downward compatibility. From a user's perspective, this means that visualization tools will have to be able to visualize both first- and second-generation objects for long archiving periods.

9.4 Generating DICOM-RT Objects in Association with the Treatment Chain

Within the treatment chain "treatment planning CT (TP-CT)→treatment planning→RT treatment," an image data set is created and continuously expanded with the respective above-mentioned DICOM objects (Law et al. 2009).

9.4.1 Diagnostic Imaging Data

Various image data are integrated into the treatment planning process. Among these are conventional X-ray data, CT, magnetic resonance imaging, positron emission tomography (PET), PET-CT, and even 4D PET-CT. This supplementary diagnostic image material is either generated and visualized separately from the TP-CT or it is already generated respecting treatment planning requirements and is then matched or fused to the TP-CT. The legal situation regarding the need to archive, directly or indirectly, diagnostic data used for target volume definition is not clear and in any way country specific. In the author's opinion, however, archiving of the written report that is indicative of the image source should suffice in the treatment context, because the diagnostic images are stored at the point of origin for at least ten years anyway.

9.4.2 TP-CT

The TP-CT, acquired in the treatment position, forms the calculation basis of treatment planning.

Images acquired in this context are initially conventional DICOM image data, with their data structure being no different from images acquired in a diagnostic context. When virtual (CT) simulation is performed based on these data, the first DICOM-RT objects such as RT-Structure (outer contour) or the isocenter as part of RT-Plan are added to the data set. These objects are, however, not yet adequate for treatment (e.g., there are outer contours only and no other ROI); there are geometric plan data only as generated by virtual simulation, but the plan does not yet contain dose. Therefore, a new RT-Structure set will be created later in the process, and finally, a plan containing all required elements including dose will be generated. Then, this modified data set is transferred to the TPS. Already at this first interface, differences in the DICOM structure of both systems can result in difficulties processing the above-mentioned partial objects by the TPS.

9.4.3 Treatment Planning

This already modified CT DICOM data set is then supplemented with additional objects during treatment planning. Among these are the Structure Set (based on segmentation of target volumes and organs at risk) then RT-Plan (objects that are created when defining treatment fields) and finally RT-Dose (which is created during dose calculation). In addition, at this stage for the first time, treatment-specific image material [digitally reconstructed radiographs (DRR)] is created as part of the RT-Image object. After the treatment planning process is finished, the object's RT-Structure Set, RT-Plan, and RT-Dose are complete and the DRRs represent the first part of the RT-Image object. This high concentration of DICOM information at this stage suggests that this might be a good point in time along the workflow to centrally archive the DICOM data generated so far. Technically, however, this approach primarily links these DICOM data to a specific application (the TPS) and its respective dedicated database with a relatively low level of standardization in the current implementation of RT-Image. This means that access to these data is only certain as long as this particular TPS application is supported. This may be problematic, given the long archiving periods mandatory for RT. The Integrating the Healthcare Environment–Radiation Oncology (IHE-RO) initiative, enabling standard compatibility test procedures, may reduce these issues. A second problem of archiving plan data in a dedicated TPS-archive is that links to other classes of information and particularly other image data cannot easily be established.

Standards of practice as well as regulations in some jurisdictions require the final treatment plan to be approved by a licensed physician and physicist. Typically, in addition to this final approval, the individual steps in the treatment planning process are recorded with reference to the respective personnel, normally based on written signatures.

Conventional simulation complicates the workflow significantly (Shakeshaft 2010). Due to its diminishing role in clinical RT, it is not discussed in detail.

9.4.4 Radiation Treatment

During the first treatment with a new treatment plan, the spatially correct application of dose is recorded by acquiring images. Until recently, mainly 2D portal images recorded on film were the mainstay of this documentation. Film has all but been replaced by electronic portal image detectors (EPID) and supplemented by both ultrasound- and CT-based volume imaging in the vicinity of the treatment device. Images are also regularly acquired during the treatment course to provide constant quality assurance of the overall treatment process. These images have to be assessed by radiotherapists and physicians and their approval requires an individual's signature. All these image data finally have to be archived together with these approval signatures. DICOM provides a hierarchy of Unique Identifiers (UIDs) for each individual Service Object Pair (SOP) Instance. This hierarchy consists of Patient ID, Study Instance UID,

Series Instance UID, and finally SOP Instance UID. With this hierarchy, specific SOP Instances are identified. The treatment record also has the PLAN SOP instance UID and should therefore reference the image data so that they can be tied together and the treatment step context of the image acquisition can be recorded. In analogy to the situation already discussed for TPS, the respective image data are typically created in a format that is proprietary to the individual vendor's software of the imaging device. Historically, these data were also often stored in a proprietary format but more recently are provided in DICOM format and also processed as such by a Treatment Management System/Oncology Information System (TMS/OIS).

This short overview of the treatment delivery chain has indicated the most relevant issues to be addressed within the framework of RT image management:

- DICOM data sets have to be handled in conjunction with their respective RT objects and an unequivocal relationship between data elements and signatures has to be established and maintained, particularly because direct signing of objects is problematic.
- Among the various modules that process and modify images, data are transferred to the respective next step by a DICOM export. It is of relevance that a DICOM export actually sends a "copy" of the data to the next module and the original data remain at the respective source. It is therefore important to make sure that no uncontrolled data redundancy is generated, that it is always clear which data set is currently "valid" (a consequence of DICOM inadequately supporting versioning), and that problems such as creating different ID-Numbers for identical patients are avoided, not to mention possible alteration or corruption of an object without the UID being changed. As long as this cannot yet be formally dealt with using the abovementioned second-generation DICOM objects, the only remedy is a stringent handling of the nomenclature of different data versions by the user.
- As indicated above, these combined image and object data are normally associated and can be accessed with the software application that was involved in their generation and are stored in the respective proprietary database. Although these data may actually be stored as DICOM-RT data, as a consequence of private tags and the architecture of the application, etc., they are usually not freely available "without" the application of origin. Even nonmandatory but standard elements are sometimes not supported by some applications, creating unnecessary and annoying interoperability failure. This aspect is of utmost relevance for long-term archiving of these data.
- Complex techniques, particularly 4D or adaptive strategies may cause problems. For example, they create more than one structure set on a CT data set or more than one CT data set, each one with a new structure set. During treatment planning, multiple plan versions create different RT-Plan and RT-Dose objects, which are then transferred downstream the treatment delivery chain (Boosts, Replanning) or that remain on the TPS (different plan versions). If these partial treatment data are only available at the respective module of origin, an integral visualization/recording/documentation of all treatment related data is difficult.

9.5 Strategies for Image Management

Figure 9.2a–c outline the most relevant strategies and structures to establish a filmless environment in an RT department.

Two fundamentally different approaches to RT image management shall be discussed, with different intermediate strategies existing in clinical practice:

1. Application-based image management and archiving. When this approach is implemented, images are created, modified, approved/signed, visualized, and archived within one dedicated software application and its associated hardware (Figure 9.2a). A modification is providing centralized archiving with a diagnostic PACS (Figure 9.2b). With this approach, associating image data with numerous DICOM objects and with the respective steps in the treatment delivery chain is virtually impossible.
2. Central image management and archiving with a DICOM-RT PACS/TMS/OIS. With this strategy, images are also created and modified with a dedicated application but are then transferred, following the DICOM Standard, to a central storage location, where a TMS/OIS creates and manages the relationship between images, treatment steps, and clinical data (Figure 9.2c).

Implementing a local, RT-only PACS not administered by a TMS/OIS (Yamada et al. 1999) is an intermediate solution, needs a supplemental workflow administration, and is therefore not discussed in detail.

"Application-based image management" currently enables access to image data and additional elements with all functionalities. An association of these data to patient data or individual steps in the treatment delivery chains is not established. Archiving is not centralized but is performed on the respective proprietary database. This requires numerous independent redundancy systems to provide sufficient data safety. In addition, this approach requires extensive administration because the data can only be transported into the future with a multitude of system migrations.

Only "central image management" in a DICOM environment (ideally fully DICOM-RT compliant) that is governed by a TMS/OIS and that is decoupled from the application of data origin supports the current treatment processes in their full complexity. In addition, the central storage location can be protected against downtime and data loss with acceptable efforts. Limiting data as much as possible to the DICOM-RT standard facilitates migration, provides data safety into the future, and is therefore mandatory for real filmless archiving of all image and related data.

(a)

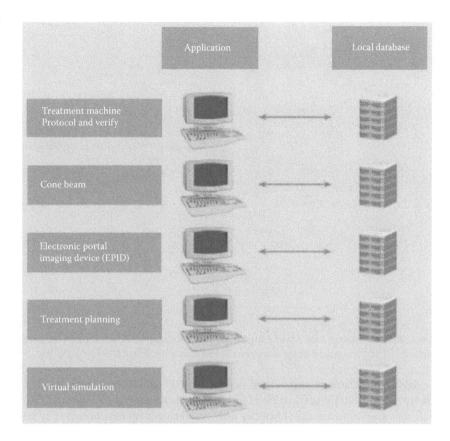

(b)

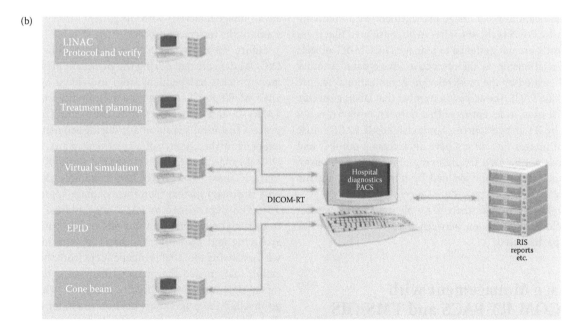

FIGURE 9.2 Different strategies for establishing a filmless department. (a) Application-based image management. (b) Image management based on a central PACS. (c) Centralized image management based on a DICOM-RT PACS/TMS/OIS.

(c)

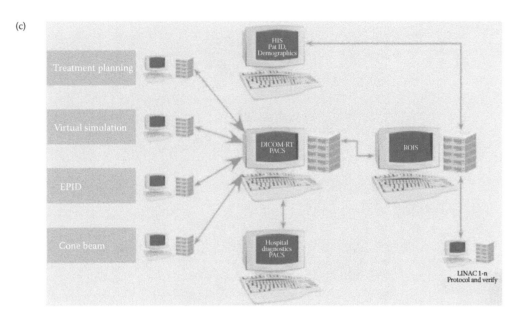

FIGURE 9.2 (**Continued**) Different strategies for establishing a filmless department. (a) Application-based image management. (b) Image management based on a central PACS. (c) Centralized image management based on a DICOM-RT PACS/TMS/OIS.

A central, patient-related, and treatment step–related image archiving approach independent from the application of image origin is therefore preferable despite occasional difficulties with visualization of certain data elements. Following the DICOM format for data modifications and data storage is mandatory.

When application-based image management with archiving in a diagnostic PACS is chosen, it has to be considered that these systems mostly are not designed to manage DICOM-RT objects with their relationship to the respective image data. Another problem is created by the most relevant organizational feature in a diagnostic PACS, the accession number that labels a specific diagnostic session in its entirety. This number, however, is not supported by RT image sources. Some diagnostic PACS refuse handling of images that do not have an accession number and can therefore not be used for archiving of RT data right away. This generally can only be avoided by implementing a strict (DICOM) worklist-based procedure step environment.

In the following section, therefore, only image management based on close collaboration between a TMS/OIS and an RT PACS will be discussed.

9.6 Image Management with DICOM-RT PACS and TMS/OIS

Although ultimately of utmost importance, for the daily workflow, archiving of DICOM data with correct representation of their relationships is rather an afterthought when compared with functional handling of these data along the treatment delivery chain. Image management based on various individual applications generating and storing images (e.g., using the proprietary software of the EPID system for comparison of EPID

with DRR) is therefore frequently observed. It is, however, preferable to integrate image handling/processing with the clinical and treatment data provided by a Record and Verify Radiation OIS (TMS/OIS) (Starkschall 1997; Law & Liu 2009; Law et al. 2009; Shakeshaft 2010). This is the only way to directly associate all image materials with the patient history and the current status of the treatment process.

Figure 9.3 illustrates a potential workflow that unites all DICOM data created during the treatment planning and treatment process independent from individual applications in a DICOM-RT PACS. Image data are initially imported into the TMS/OIS and associated with clinical and treatment data (a process that ideally is automatic during acquisition that already happens in the context of a treatment plan). Image data and DICOM objects are then transferred to the RT PACS (copied to the RT PACS and deleted in, for example, the TMS), leaving only the image history (comments, image approval, etc.) in the TMS/OIS (represented by PACS node address and UID). This is a crucial step ensuring a clean architecture for long-term archiving and providing the basis for viewing these images with a viewing tool that is independent from the original application that created the images. This means that all plan and image data are always directly accessible without having to go through the original application when, for example, an old dose distribution has to be assessed when a patient is to be reirradiated. In addition, in an RT PACS, as in many PACS, original non-DICOM objects, such as Word, JPG, MP3, AVI, and MPG files, can be saved. This should be realized as a DICOM Structured Record modality or as a PDF/A document inside a DICOM image object. In any case, proprietary formats such as MS Word (which is, to further complicate things, version dependent) should be strictly avoided.

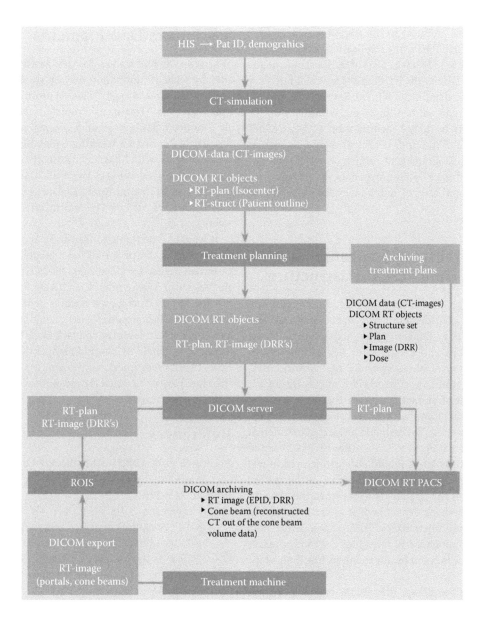

FIGURE 9.3 Image workflow optimized for an RT PACS.

The more information and images that are prevalent digitally in a department, the more imminent becomes the question of how to properly electronically approve and sign these data. This has already been discussed in the literature. Some individual DICOM objects (RT-Plan, RT-Ion Plan, and RT-Structure Set) explicitly record an approval status. The RT-Plan object, for example, contains data on dose prescription, patient setup, fractionation, and treatment technique but also an approval status (approved, unapproved, rejected). There is even the possibility to create a logical relationship between treatment plans (RT-Plan relationship: prior, alternate, predecessor). These aspects of RT objects are, however, regrettably not uniformly supported by all TPS vendors; therefore, the archiving safety in PACS is doubtful.

A possible approach to establish a signature trail along the treatment planning workflow that enables safe archiving is suggested as follows:

- Treatment plans to be actually treated are preselected and, if the system supports plan approval, approved in the TPS. This protects the treatment plans with the version number selected for treatment against further modifications. Only approved treatment plans are then transferred to and imported into the TMS/OIS (RT-Plan). A PDF file [specifically PDF/A, which has become an official standard (ISO-19005-1:2005) to prevent future readability problems] of the respective treatment plan is created by the TPS that contains, in a user-defined fashion, all data that are necessary to unequivocally identify the original plan (version number, etc.) and is imported into the TMS/OIS where it is approved and locked by a signature with login and password (simple electronic signature) or (technically already possible) with a higher-level electronic signature protocol. This plan-PDF in the TMS/

OIS is then the "pointer" to the respective DICOM-RT data set in the RT PACS. The implication is that the PDF is linked to the SOP Instance UID of the RT Plan, and the SOP Instance UID of the RT Plan is the actual reference and link to the data in the RT PACS. Only DICOM-RT data of approved treatment plans are imported into the RT PACS. This facilitates identification of DICOM-RT data in the RT PACS and, if critical plan data such as dose-volume histograms and sample dose distributions are also part of the PDF, improves data safety by providing an additional level of redundancy for these critical plan data.

9.7 Digital Archiving Environment–Required Hardware Infrastructure

Archiving with a high level of data safety requires a hardware infrastructure with considerable performance and a certain minimum redundancy. Otherwise, regular operations are prone to interruptions and a significant risk of data loss is incurred. At the center of such a system is the network infrastructure with a minimum port bandwidth of one gigabit per second and the central storage hardware that consists of at least two independent storage servers and storage units (such as Storage Area Networks). These should be located in different fire sections of the department to reduce the risk of server and thus data loss. These two units should also be mirrored in real time with mirroring software provided as part of the storage solution. This reduces the risk for data loss when the primary server is damaged and the department is operational after a minimum time of nonfunctionality. To further enhance data safety, it is also recommended to introduce a second, cheaper line of redundancy, which can be, for example, based on a tape backup system.

When calculating the storage dimensions of the entire system, the image production of the department has to be estimated as accurately and as far into the future as possible. The storage should be scalable to accommodate a potentially growing archive; an estimate of the required scalability should be attempted. Particularly critical are volume image data that are generated with increasing frequency. A typical TP-CT requires a storage space of approximately seventy-five megabytes, whereas a cone beam-CT requires between seventy-five megabytes (reconstructed CT data set) and one to two gigabytes (raw data for research purposes, using the established redundant storage system though typically not in standard format). Planar images and RT-specific DICOM objects are less relevant in comparison (Swerdloff 2007).

Finally, the constant requirement for updating the hardware and software basis of all operations should be kept in mind as well as the need for maintenance agreements that offer short response times. Maintaining an up-to-date hardware basis typically mandates replacement cycles of five years or less. This has to be taken into account for budget calculations.

9.8 Future Developments

When the DICOM standard is stringently followed, all relevant therapy-related information may be stored in an RT PACS. If objects that already exist ("Plan Approval") or are being developed, such as "Treatment Protocol," are reliably established in a uniform fashion among systems and vendors (DICOM Standards Committee Working Group 7), there will be considerable progress with imaging and archiving the actual workflow in a standardized storage device such as PACS. This renders archiving of all treatment-related data less dependent from individual software such as TPS and therefore increases archiving safety.

It would be desirable to transfer long-term image archiving to the primary hospital PACS at a predefined time to limit the resources needed for a dedicated RT PACS. Given that diagnostic PACS also have to provide archiving for extended time periods for certain data subsets such as pediatric images, the long archiving periods needed for RT should not be a problem. It is as of yet unclear, however, if a lack of certain functionalities of a general hospital PACS would hamper the overall image workflow. The desired functionalities and the overall data integrity therefore have to be tested for every individual image workflow/PACS architecture.

References

Ackerly, T. et al. (2008). Radiotherapy DICOM packet sniffing. *Australasian Physical & Engineering Sciences in Medicine, 31*, 243–251.

Boda-Heggemann, J. et al. (2011). kV cone-beam CT-based IGRT: A clinical review. *Strahlentherapie und Onkologie, 187*, 284–291.

DICOM Standards Committee Working Group 7, Radiotherapy Extensions, *Radiotherapy Treatment Records and Radiotherapy Media Extensions*, 1999.

DICOM Standards Committee Working Group 7, Radiotherapy Extensions, *Second Generation Radiotherapy*, in draft.

Heinemann, F. et al. (2009). Department and patient management in radiotherapy. The Freiburg model. *Strahlentherapie und Onkologie, 185*, 143–154.

Herman, M. G., *Computer Networking and Information Systems in Radiation Oncology*, 1999. Available at www.aapm.org/meetings/99AM/pdf/2755-16806.pdf.

Law, M. Y. & Liu, B. (2009). Informatics in radiology: DICOM-RT and its utilization in radiation therapy. *Radiographics, 29*, 655–667.

Law, M. Y. et al. (2009). Informatics in radiology: DICOM-RT-based electronic patient record information system for radiation therapy. *Radiographics, 29*, 961–972.

Shakeshaft, J. (2010). Picture archiving and communications system in radiotherapy. *Clinical Oncology, 22*, 681–687.

Starkschall, G. (1997). Design specifications for a radiation oncology picture archival and communication system. *Seminars in Radiation Oncology, 7*, 21–30.

Steil, V. et al. (2007). Picture Archiving and Communication Systems as an Element of Digital Information and Image Management in Radiation Oncology. Touch Briefings 2007. *European Oncological Disease, 1*(1), 102–105.

Steil, V. et al. (2009). Patient-centered image and data management in radiation oncology. *Strahlentherapie und Onkologie, 185*, s1–s7.

Swerdloff, S. J. (2007). Data handling in radiation therapy in the age of image-guided radiation therapy. *Seminars in Radiation Oncology, 17*, 287–292.

Yamada, T. et al. (1999). Image storing system for radiation therapy (radiation oncology information system: ROIS) as a branch of diagnostic PACS; Implementation and evaluation. *Computerized Medical Imaging and Graphics, 23*, 111–117.

10

Connectivity

R. Alfredo Siochi
*University of Iowa
Hospitals and Clinics*

Stuart Swerdloff
Elekta AB

10.1 Introduction

The need for connectivity in radiation oncology may seem obvious to the practitioner, but one could imagine an environment where almost no connectivity is needed. If all systems involved in the radiation oncology process were instead subsystems inside a conglomerated application service, and all data were stored in a single data store of some kind, then the "apparent" connectivity would take the form of computer terminals accessing a server. One could also hearken back to the days of the paper chart. However, those subsystems would need to agree on how data were to be kept in the data storage and somehow identify the state of various subprocesses. Even with an all-paper environment, one has to agree on storage locations and formats so that the consumer of a written document knows where to find the needed information and how to interpret it. Regardless of the nature of the data and equipment, there is a need for the subprocesses and subsystems to communicate.

This communication requires interfaces based on agreed-upon conventions. In practice, these define the boundaries between subprocesses within radiation oncology. They provide a means of communicating the data and state from one system to another. A connection between two systems is the minimum level of communication in which the two systems have reached the point of acknowledging each other's presence and agreeing in principle that some (particular) kind of interface will be shared between them. One can list today's set of systems, but the list could readily change as new products are brought to market.

Although the current practice of radiation oncology includes numerous well-known subsystems, it is worthwhile to consider that the boundaries surrounding certain subprocesses already vary over geographical regions and can shift over time. For example, the computed tomography (CT) simulation subprocess may be performed in the radiology department. The images are then sent to the radiation oncology department so

that dosimetrists can import them into their treatment planning system. In a larger regional hospital, a patient could decide to get their treatments closer to home at some point in the middle of their treatment. All of the current treatment data would then need to be transferred to the local clinic.

The subprocesses are, in general, defined by the roles of the various staff members in radiation oncology. For example, radiation therapists acquire images and send them to dosimetrists for treatment planning. In almost all cases in current practice, the format and interpretation of these images is governed by the Digital Imaging and Communications in Medicine (DICOM) standard (National Electrical Manufacturers Association 2011a). Figure 11.1 in the next chapter not only provides an overview of the process from a resource perspective but also indicates the flow of data in the radiotherapy process and the locations where interfaces are required (arrows with dashed lines).

If these subsystems incorrectly apply the standard, data may be transferred correctly but may be incorrectly interpreted and misrepresented by a subsystem, leading to potential treatment errors (Siochi et al. 2011). The issue of connectivity is one of safety and efficiency, because each subprocess relies on the correct use of the interface. Both consumers and producers of data must abide by the rules of the interface to ensure that the meaning of the data is preserved through the entire chain of subprocesses.

In recent years, the rapid increase in the number of image-guided radiation therapy systems has made connectivity an even more important issue. Whereas two devices (call them A and B) require only two communication pathways (A-to-B and B-to-A), three devices require six pathways and four devices require twelve pathways. The addition of a single device adds potentially twice as many pathways as there are existing devices in the system. Although not all these devices have to communicate with each other, it is not uncommon to have one device communicating with several others. Some of these communications may occur sequentially, which provides opportunities for correcting errors, but some could occur in parallel, propagating errors very quickly throughout the system. On the contrary, parallel or centralized operations could prove to be more efficient. Parallel operations could also create redundant data sets that could be used for data recovery. These considerations create a large number of possible permutations for the connections of subprocesses, devices, and subsystems.

It is not only the type of device that matters, but also the actual number of logical units. Although a department may have only one CT or magnetic resonance (MR) scanner, it may have several treatment planning system computers, several treatment management system (TMS) workstations, several linear accelerators, and even more desktops for electronic medical record (EMR) applications. Many of the desktop and workstation computers may be running several applications that connect to different locations. They all have to be configured correctly to ensure connectivity. This is not just a matter of two or three or even ten devices; we are potentially dealing with tens of devices for smaller clinics and perhaps hundreds of devices for large centers.

An additional challenge is that the subprocesses may be spread across wide distances, as in the case of satellite clinics being supported remotely. Here, the issue of connectivity also deals with the physical connections and underlying technologies that support the transfer of data over several network nodes. This is also a network configuration issue, because two devices that are physically adjacent to each other may actually connect through very long physical distances, making several hops over multiple network nodes.

Such communications can be unattended in some cases but must be monitored by humans in others. The variety of the timing of activities and communication and the availability of the personnel needed to achieve or monitor communication also drive the need for central storage or, if not central, then at least "easily and remotely accessible data."

Fortunately, modern systems have achieved a certain level of uniformity in terms of the means of communicating between them using computer networking. The network model includes a certain amount of abstraction, making it much simpler for the applications involved while at the same time providing a richer set of choices for those deploying the physical systems and attempting to connect them.

10.2 Communication Layers

The transfer of data in modern networks involves the seven-layer Open Systems Interconnection (OSI) model. The various layers move bits from physical systems through the network and finally appear as data in the application. The physical infrastructure in a network supports the first three layers, which can be considered as "Media" layers (also referred to as lower layers), where electrical signals transport bits that are arranged into frames containing sender and receiver Media Access Control addresses. These frames are then grouped into packets with additional

information about the sender and receiver Internet Protocol (IP) addresses. They are moved into the "Transport" layer, where they are grouped into segments to improve the flow of data. The next set of layers is the "Host" layers (the upper layers), where sessions are established between receiving and sending computers, and data are transformed (representation, compression and encryption) and finally presented to the application. More details about the OSI model can be found in any standard networking textbook (Wetteroth 2002).

The upper layers are of primary interest in radiation therapy. Much of the data are transferred via the DICOM standard. This standard relies on Transmission Control Protocol/IP and its underlying layers to physically and logically transfer the data. However, the layers above the "Transport" layer must conform to the standard, with attention to Parts 3 to 8 and 10 (National Electrical Manufacturers Association 2011c–i). The standard specifies not only the data formats of the radiotherapy objects but also the communication that takes place between application entities (AEs). For DICOM applications to send messages to each other, they must be aware of the IP addresses, ports and AE titles of all the other applications with which they wish to communicate. Before DICOM data are transmitted, AEs will perform a series of handshakes (associations) to make sure that they support the types of services being requested. When a DICOM transmission fails, it could fail either because the AE ports and IP addresses have been incorrectly configured (in which case the receiving AE may not actually be aware of the existence of the sending AE) or the services offered by the receiving AE are not compatible with those requested by the sending AE. A more subtle connectivity issue involves the successful transmission of data with the inaccurate interpretation by an AE resulting in wrong information. This error deals with the lack of adherence to formats and conventions for the various DICOM objects. For a more detailed explanation of these data formats, see Chapter 11 of this book and the references cited therein.

The lower OSI layers are used in Lower Layer Protocols and, along with file transfers, are the most commonly used transport mechanisms for Health Level Seven (HL7) messages. However, there are multiple protocols that are higher level, and indeed HL7 refers to the seventh OSI layer, the application layer. HL7 messages are typically used to exchange demographic information among medical applications. They can also exchange schedules, allergies, orders, insurance information, and Admit, Discharge, Transfer (ADT) messages (Benson 2012). Currently, there is insufficient support for radiotherapy-specific HL7 messages, and radiotherapy-specific EMRs exchange a limited number of HL7 messages for demographics and scheduling with Hospital Information System (HIS) applications (e.g., MOSAIQ ADT messages exchanged with EPIC or GE).

10.3 Services and Objects

Entities that consume services and objects in radiotherapy need to adhere to a number of standards to connect. The primary standards of interest have already been mentioned: HL7 and

DICOM. There are a number of other "Legacy" objects that are formatted according to the Radiation Therapy Oncology Group (RTOG) and Radiotherapy Treatment Planning (RTP)-Link or RTP-Connect file formats. These are purely file format conventions and data transmission proceeds according to any convenient means for file transfers.

The most common entities within a radiotherapy department are imaging devices, planning systems, treatment management and delivery systems, on-treatment patient monitoring systems, archives [e.g., Picture Archiving and Communication System (PACS)], the EMR, and/or the HIS. Whereas communications with the HIS are typically limited to HL7 messages that are used for workflows between departments or domains (HIS to TMS or EMR), communications among the other systems are usually done via DICOM and deal with the information that designs and delivers the radiation treatment.

This dichotomy represents the two broad classes provided by these entities: services and objects. Services generally deal with demographics (e.g., ADT) and scheduling or workflow. These are generally provided through HL7, but there are DICOM-based services. These services are more common in radiology departments, although they are starting to find their way into radiotherapy. The Modality Worklist for image acquisition is becoming available on the TMS rather than relying on the Radiology Information System, especially in "Oncology only" facilities. The Modality Performed Procedure Step essentially checks off completed items on the Modality Worklist, although it is not commonly supported by the TMS and is generally only used in radiology departments. Combining these two DICOM objects leads to a new approach known as the Unified Procedure Step. However, this is not only a mere combination of services but also a redesign that addresses more complex workflows and accommodates multiple inputs and outputs. It can be used for treatment delivery and it is anticipated to be used for general workflow communication and control for treatment planning and other activities in the department.

The other class deals with objects, their storage, and interpretation. DICOM not only specifies a communication model but also object formats, as described in Parts 3, 6, and 10 of the standard. The object is represented in binary format, with both Big-Endian and Little-Endian byte orders supported. For data that are represented by several bytes, the Big-Endian format puts the most significant bytes first, whereas the Little-Endian format does the reverse. For example, the number 2049 is represented as "08 01" in Big-Endian hexadecimal notation, whereas the Little-Endian format gives "01 08." When the byte order is not specified, Little-Endian is the implicit default. If the communication stream is written to a file, additional metadata are included to help decode the file, essentially replacing the association step. Digital signatures and encryption are also available for the objects. Chapter 11 goes into more detail about the structure of the DICOM file.

The HL7 format could be easier to read than a DICOM file, because HL7 files are in ASCII text. They are line oriented, with individual elements separated by the pipe character "|". The text consists of messages and segments. The lines consist of three letter codes (e.g., PID for the patient ID) followed by the data for that code (medical record number, name, date of birth, gender, address, and telephone number). In version 3.0 of the standard, Extensible Markup Language (XML) will be available, making the information even more readable.

The RTOG format (Radiation Therapy Oncology Group 1999) was primarily intended to support the RTOG 3D Clinical Trial Quality Assurance (QA). It preceded DICOM and has been superseded by DICOM. The RTOG 3D QA group now utilizes DICOM and has been, and continues to be, deeply involved in the development of the DICOM Standard and the efforts for Integrating the Healthcare Enterprise in Radiation Oncology (IHE-RO).

The RTOG format continues to be supported. Older, unconverted RTOG data may exist in the archives of many radiotherapy departments, and some familiarity with the format will be helpful. The format consists of binary (raw data) and ASCII (metadata). It uses implicit file naming schemes and directory structures to indicate the relationships among the data. The objects that are supported are those that are consumed or produced by treatment planning systems. Some examples are CT, MR, and ultrasound images, structures (target volumes, organs), beam geometry, digital film images (digitally reconstructed radiographs and portal images), dose distributions, dose-volume histograms, and brachytherapy seed geometry. The RTOG Web site (http://www.rtog.org) provides more information about their plans for DICOM and RTOG file format support.

Before DICOM, in the early days of Record and Verify systems, a proprietary format known as RTP-Link (the current version is known as RTP-Connect) was used to send treatment parameters from the planning systems to the V&R systems (and currently to the TMS). These are ASCII files, with each line representing an object such as a field or an MLC definition. Each line is terminated with a Cyclic Redundancy Check checksum, which is a function of all the preceding characters in the line. The objects were related to each other in a hierarchy that included course, prescription, setup, treatment fields, and dose points. Newer versions of the standard (IMPAC Medical Systems, Inc. 2012) have support for linking DICOM images, via their Unique Identifier (UID), to a given field. The specifications can be found online (http://www.elekta.com/dms/elekta/elekta-assets/Elekta-Software/pdfs/technical-references/LED17001.pdf).

10.4 Achieving Results

Most users have an expectation that the vendors will create products that communicate with each other, and for the most part, the first six layers of the OSI work well. The problems begin at the application level, where data are converted to information. For example, the data might contain the number zero to represent a gantry angle. The information we seek is whether this means our beam is pointing toward the floor (IEC 1217 convention) or toward the ceiling (native coordinate system on older Varian accelerators).

When applications assume the existence of certain data elements that are not required by a standard, they will either fail the transmission or fill in default data. In some cases, when vendors have a suite of applications, they will use private tags to fill in what they perceive to be gaps in the DICOM standard. In other situations, they rely on the user to manually populate the missing information.

One method that people have used to assess whether applications will work well with each other is to examine the DICOM conformance statements of both AEs. Part 2 of the standard explains how to write a conformance statement (National Electrical Manufacturers Association 2011b). Such documents specify the information objects, service classes, communication protocols, and media storage application profiles that are supported by the AE. Because a number of elements in an information object may be optional, the conformance statement should indicate how they support such elements.

The AE should also describe what their role is in the communication, that is, are they consumers of a service [Service Class User (SCU)], providers [Service Class Provider (SCP)], or both? Do they support an object and its corresponding service? [Service Object Pair (SOP)]. Or they may be SCPs for some objects and SCUs for other objects. For example, a CT scanner sends images to a PACS archive for storage. In this example, the service is "Storage," the SCU is the CT scanner, the SCP is the PACS archive, and the SOP is "CT Image Storage." Several storage SOPs combine the storage service with a specific object (e.g., MR Image Storage, MR Spectroscopy Storage, NM Image Storage, and Ultrasound Multi-frame Image Storage), where the archive is the SCP and the imaging device is the SCU. A treatment planning system could also be an SCU sending RT Dose objects to an archive. The archive still plays the role of the SCP, and the SOP is "Radiation Therapy Dose Storage." An archive could also be an SCU, when it is transferring data to another archive, which plays the role of the SCP. Although object storage is the predominant type of SOP, there are other SOPs such as patient query and retrieve services that could either "FIND," "MOVE," or "GET" all the objects (multiple types of images, reports, and dose distributions) for a given patient.

The specific combinations of roles, services, and objects must be described along with caveats. This type of information is also provided by the AE during an association step with another AE to determine if they can communicate with each other. In effect, the users create a mental picture of the association step for every object and service they are interested in and for every pair of AEs that they envision participating in data exchanges.

Such a task is technically demanding and time consuming. Most users do not have the luxury of verifying that the conformance statements will bear out in practice and would prefer that vendors guarantee the connection. This desire for plug-and-play radiotherapy systems gave birth to the IHE-RO efforts that abstract these DICOM association and conformance concepts by developing use cases. These use cases specify how products should interact with each other to complete a given process.

The process is broken down into well-defined tasks that are performed by "actors." The dialogue between the actors represents a transaction. These are tested between various pairs of actors at events known as "connectathons." Judges determine whether the transfer of information between a pair of actors has preserved the information (e.g., correct orientations and contours appearing in the correct coordinate system). The results of these events are incorporated into the vendor's IHE Integration Statement. Radiotherapy clinics can request these statements from the vendor, and the vendor can indicate the integration profiles they fulfill.

Consider the Simple Treatment Planning Integration Profile as an example. It specifies the data transfer of DICOM images and RT objects. In this profile, the list of actors is given in the order of the data transfers: (1) The CT scanner acquires images and exports them; (2) the Contourer imports the images, allows the user to create volumes of interest, and exports images and contours; (3) the Geometric Planner imports the images and contours, places treatment beams on the patient, and exports a plan containing the beam geometry; (4) the Dosimetric Planner imports all the data exported by the Geometric Planner, calculates dose, and exports the plan and dose distributions; (5) the Dose Displayer imports everything from the Dosimetric Planner and presents the dose distribution to the user; and (6) the Archive exports and imports all of the objects generated by all of the preceding actors.

The previous steps might use a mixture of different vendors. For example, a GE CT scanner could send the CT images to a Coherence Dosimetrist Virtual Simulation System, which then transfers the treatment beams and contours to a Philips Pinnacle planning system. The Pinnacle workstation sends the treatment data to the Candelis RT Viewer to review the dose distributions and treatment plan, and finally the data could be archived on a VelocityGrid RT-PACS. However, the entire chain of events can only be successfully completed if each of the vendors fulfills the role of the actor. Examining each of the vendors' integration profiles will indicate whether the right set of actors have been assembled to complete the task.

The current difficulty with connectathons is that a very large number of permutations have to be tested. Each vendor that wants to test a particular actor will have to work with another vendor that has the corresponding actor that will either receive or send the data that the actors work with. As there are several vendors and several actors, the number of communication pathways can be very large. A single vendor may not be able to test a particular actor against all the other existing vendors, but it can still pass the connectathons. Although the requirements for passing as a particular actor are strict, there is no guarantee that all possible combinations of vendors will work, but it greatly reduces the likelihood of interoperability problems.

There are several other profiles that have been tested at connectathons: Multimodality Registration for Radiation Oncology, Advanced RT Objects Interoperability, and Integrated Positioning and Delivery Workflow. A number of use cases are under development at this writing. Those on the short list include the Integrated Patient QA Checker, Treatment Delivery Device Data Integration, On-line Image Review, and Flattening Filter Free. Many other use cases have been suggested, and the IHE-RO Wiki (http://wiki.ihe.net/index.php?title=Radiation_Oncology) contains a list of those under consideration for future development.

10.5 Troubleshooting

When connectivity problems arise, one must determine the OSI level where the issue is occurring. In general, the physical infrastructure and networking in a modern radiotherapy department is robust enough that the issues rarely occur at the lower levels. Configuration issues tend to be a more common problem, where administrators of the AEs forget to enter the correct IP addresses or ports or fail to specify which services the DICOM AE should support. DICOM conformance statements and successful connectathons only indicate that the AE has the functionality, but it may not be enabled by default. Another possible problem may arise at the transport layers due to different AEs existing on different networks with incompatible security requirements. There may also be unique requirements for remote sites (e.g., VPN, Webex, log-me-in, and citrix server farms) that may present basic connectivity infrastructure problems regardless of the nature of the data being exchanged. For these cases, some of the middle and lower OSI levels may have to be probed. Such problems are almost always an issue of configuration and deployment, with additional constraints due to political IT boundaries between departments and vendors. Clear lines of communication among all those who participate in the data exchange are necessary to create simpler solutions that satisfy all requirements (Siochi et al. 2009a). Medical physicists should take the lead in resolving this issue (Siochi et al. 2009b) because they are familiar with all the radiotherapy data exchange requirements.

Once all devices are communicating with each other, subtler connectivity issues involving the interpretation of the data must be tested (Siochi et al. 2011). If coordinate conventions are not followed, systematic errors, such as contours appearing in a different location from the image or wrong-sided treatments, are possible. In some cases, incorrect UIDs may be listed in a data set, and some objects might be incorrectly associated with other objects. Some systems will fail if they are unable to retrieve the objects specified by these related UIDs. For example, DICOM-RT structure sets indicate the UIDs of the CT images that were used to create the contours. When these data do not load properly, there may have been a sequence of data transfer operations that was not correctly followed. Performing end-to-end tests will not only verify that the wide variety of devices in the department can connect successfully to support a workflow but also confirm that the data transfer operations within the workflow produces the correct representation of information. Before the clinical release of new devices, their integration within the workflow should be tested so that appropriate modifications to the workflow can be implemented to support the correct sequences of data transfer that result in the successful transmission of DICOM objects.

10.6 Conclusion

The increase in the number of devices and new products will always present new challenges for connectivity. The amount of data being exchanged has grown tremendously with the introduction of image-guided radiation therapy, where multiple images are being acquired for a single course of treatment. Manual data entry is no longer an option, and our medical records should take full advantage that electronic systems have to offer. The requirements for archiving are in the tens of terabytes range and will probably continue to grow as more data about our radiotherapy treatments are desired. The current best option for driving interoperability is IHE-RO. Connectivity issues are being resolved for the basic data transfers in radiotherapy, allowing IHE-RO and the participating vendors to anticipate use cases for their products and proactively test them, with the goal of being plug-and-play as they hit the market.

References

Benson, T., *Principles of Health Interoperability HL7 and SNOMED*, Springer, New York, 2012.

IMPAC Medical Systems, Inc., *RTPConnect Radiotherapy Treatment Planning Import/Export Interface Specification*, 2012.

National Electrical Manufacturers Association, *Digital Imaging and Communications in Medicine (DICOM) Set*, National Electrical Manufacturers Association, Rosslyn, Virginia, 2011a.

National Electrical Manufacturers Association, *Digital Imaging and Communications in Medicine (DICOM) Part 2: Conformance*, National Electrical Manufacturers Association, Rosslyn, Virginia, 2011b.

National Electrical Manufacturers Association, *Digital Imaging and Communications in Medicine (DICOM) Part 3: Information Object Definitions*, National Electrical Manufacturers Association, Rosslyn, Virginia, 2011c.

National Electrical Manufacturers Association, *Digital Imaging and Communications in Medicine (DICOM) Part 4: Service Class Specifications*, National Electrical Manufacturers Association, Rosslyn, Virginia, 2011d.

National Electrical Manufacturers Association, *Digital Imaging and Communications in Medicine (DICOM) Part 5: Data Structures and Encoding*, National Electrical Manufacturers Association, Rosslyn, Virginia, 2011e.

National Electrical Manufacturers Association, *Digital Imaging and Communications in Medicine (DICOM) Part 6: Data Dictionary*, National Electrical Manufacturers Association, Rosslyn, Virginia, 2011f.

National Electrical Manufacturers Association, *Digital Imaging and Communications in Medicine (DICOM) Part 7: Message Exchange*, National Electrical Manufacturers Association, Rosslyn, Virginia, 2011g.

National Electrical Manufacturers Association, *Digital Imaging and Communications in Medicine (DICOM) Part 8: Network Communication Support for Message Exchange*, National Electrical Manufacturers Association, Rosslyn, Virginia, 2011h.

National Electrical Manufacturers Association, *Digital Imaging and Communications in Medicine (DICOM) Part 10: Media Storage and File Format for Media Interchange*, National Electrical Manufacturers Association, Rosslyn, Virginia, 2011i.

Radiation Therapy Oncology Group, *Specifications for Tape/Network Format for Exchange of Treatment Planning Information*, RTOG 3D QA Center, St. Louis, Missouri, 1999.

Siochi, R. A. et al. (2009a). Information technology resource management in radiation oncology. *Journal of Applied Clinical Medical Physics, 10*(4), 3116.

Siochi, R. A. et al. (2009b). Point/counterpoint. The chief information technology officer in a radiation oncology department should be a medical physicist. *Medical Physics, 36*(9), 3863–3865.

Siochi, R. A. et al. (2011). A rapid communication from the AAPM Task Group 201: Recommendations for the QA of external beam radiotherapy data transfer. AAPM TG 201: Quality assurance of external beam radiotherapy data transfer. *Journal of Applied Clinical Medical Physics, 12*(1), 3479.

Wetteroth, D., *OSI Reference Model for Telecommunications*, McGraw-Hill Telecom Professional, McGraw-Hill, New York, 2002.

Information Resources for Radiation Oncology

R. Alfredo Siochi
University of Iowa
Hospitals and Clinics

11.1 Introduction

Radiation oncology processes require information to manage a patient's treatment appropriately. The information can be part of a pool of experience derived from the literature and from clinical trial databases. This information can drive the determination of the treatment strategy, and it dictates the information remaining to be gathered through imaging, treatment planning, and treatment delivery.

Different imaging modalities provide different types of information: physiologic, functional, and anatomical. Some of the information that accompanies these images is related to the images themselves (e.g., the type and activity of an isotope used for positron emission tomography, a patient's body weight) and can be found in the headers of the image files. Treatment planning systems may not be able to extract these data in any meaningful way, and third-party tools or in-house solutions are often needed to derive additional information for the planning process.

A similar data analysis or transformation step occurs between treatment planning and beam delivery when the physicist performs patient-related quality assurance (QA) and, in some cases, tests the validity of data sets used for image guidance. To ensure that the planned dose distributions match the delivered dose distributions, a QA step checks that the meaning of the data is preserved as the data are transformed between separate systems with different proprietary conventions and implementations of standards.

Here, we see the general definition of information at work: information is meaningful, well-formed data (Floridi 2009).

Information resources can be described within the framework of this definition. Resources may be the information itself, gleaned from community knowledge in the form of well-explained images and text, or resources may exist as bits suitable for machine communication and require decoding for human use. The tools for viewing, reading, editing, and transforming the data are also information resources, for without these tools, the data do not provide information. The act of decoding and encoding the information may produce well-formed data (e.g., text strings that meet some notation convention), but those data may be meaningless to a human who does not know or remember the applicable standards and conventions. Data dictionaries, metathesauri, and medical terminology systems can therefore be considered information resources, and they are often used by applications in their attempts to deliver meaning, not just data, to the user.

Tools that facilitate the search for information are also resources: search engines, literature databases, radiation therapy (RT) Web portals, and professional society Web sites are good starting points for obtaining treatment protocols, file formats, standards, data dictionaries, file readers, viewers, and editors.

Figure 11.1 shows a business process modeling notation (White 2008) diagram for the radiation oncology treatment process and indicates in the data messages and annotation fields the information resources that are used in each component of the process. The diagram is not intended for process analysis; rather, it provides an overview of the resources we will explore in this chapter and where they fit in the context of the radiation oncology process. The resources will be described for each component of the process, along with the tools required to use them. The

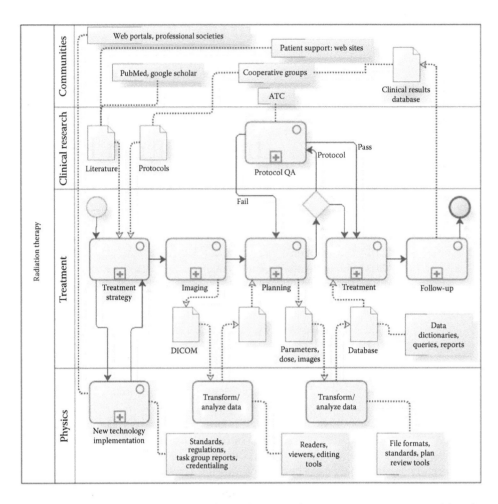

FIGURE 11.1 A business process modeling notation diagram of the radiation oncology treatment process. Some of the information resources are indicated as data messages. Additional resources are indicated by using the annotation fields connected to the processes they support.

remainder of the chapter devotes sections to the components of the process illustrated in Figure 11.1. If the reader is interested in a resource rather than a process, Figure 11.1 provides a visual index to help find the chapter sections that discuss the resource.

11.2 Designing the Treatment Course

The determination of a patient's course of RT involves histories, symptoms, and examinations coupled with the physician's experience and knowledge. When the patient data fit a well-established pattern, typical treatment strategies can be applied. For example, men with prostate cancer can be sorted into low-, intermediate-, and high-risk groups depending on Gleason score, prostate-specific antigen level, and T category (Mohler et al. 2007). Each risk group has a specific schedule of doses and radiation beam arrangements.

Although this may seem like a straightforward decision-making process, this experience had to be gleaned from clinical research. Physicians improve their practice by searching clinical trial results, clinical practice guidelines, the scientific literature, and open clinical trials that may be suitable for their

patients. Expanding on the previous example of men with prostate cancer, physicians might further subdivide the risk groups to determine if a particular patient will benefit from adjuvant androgen deprivation therapy (Beasley et al. 2008). Or they may decide that the results of a published hypofractionation study (Kupelian et al. 2005) are encouraging enough to try the technique with their own patients. Finding a clinical trial is a safe way of implementing new techniques because the participants must conform to a protocol and submit their data for QA. For example, local control rates were better for patients with Ewing's sarcoma who did not deviate from the protocol specified in the Pediatric Oncology Group study POG-8346 (Donaldson et al. 1998). Research-oriented practices may set up their own protocols to try out new technology (Soares et al. 2005). The results of some of these clinical trials become part of searchable databases from which other researchers may be able to perform data mining.

Much of the information described here can be found by Internet searches. Table 11.1 shows the uniform resource locators (URLs) of some relevant Web sites. Results of studies funded by the National Institutes of Health (NIH) can be downloaded

TABLE 11.1 Organizations Providing RT Information Resources

Organization/Resource	URL
NIH	http://www.nih.gov/
Clinical Trials	http://www.clinicaltrials.gov/
NLM	http://www.nlm.nih.gov/
NCBI	http://www.ncbi.nlm.nih.gov
Literature databases	http://www.ncbi.nlm.nih.gov/Literature
PubMed	http://www.ncbi.nlm.nih.gov/pubmed
NCI	http://www.cancer.gov
Clinical Trials Search	http://www.cancer.gov/clinicaltrials/search
Clinical Trials Results	http://www.cancer.gov/clinicaltrials/results
PDQ	http://www.cancer.gov/cancertopics/pdq/cancerdatabase
NCCN	http://www.nccn.org
Cooperative Groups	See Table 11.2
Standards, Guidelines, Regulations	See Table 11.3
ATC	See Table 11.4
DICOM	See Table 11.5
Treatment Database Management Systems	See Table 11.6

for free in portable document format (PDF) through PubMed Central, a database that contains whole research articles. PubMed, the citation database from the U.S. National Library of Medicine (NLM), will return citations and abstracts for articles matching the search criteria. The citations are primarily from Medline, the NLM bibliographic database whose records are indexed with the NLM Medical Subject Headings (MeSH), a controlled vocabulary of medical terminology. Medline's records cover the life sciences with an emphasis on biomedicine.

The service can be accessed from the Web site of the National Center for Biotechnology Information (NCBI). NCBI offers other services such as a bookshelf with online biomedical textbooks, a compendium of human (OMIM) and animal (OMIA) genes, molecular databases, data mining tools, and even programming utilities to interact with Entrez, the search engine that works with all the literature, molecular, and gene databases of the NCBI. The NCBI also has links for consumer health (NLM's Medline Plus), providing physician directories, dictionaries, encyclopedias, and medical information for patients. Both Medline Plus and the NCBI have links to the NIH clinical trials Web site, where patients and physicians can search for open clinical trials.

A database of cancer-related open clinical trials can also be searched on the Web site of the National Cancer Institute (NCI). The NCI also provides the results of clinical trials. Physicians can study the results of the trial to determine if the proposed treatment would be clinically beneficial to their patients and to assess their ability to comply with trial protocols. Additional information on specific clinical trials can be found on the Web site of the appropriate groups that are part of the NCI-sponsored cooperative group program (Table 11.2). (Table 11.2 is not meant to be an exhaustive list; the result of a Google search for the phrase

"oncology group" contains many other cooperative groups that have not been included in that table.)

In addition to trial protocols, clinical practice guidelines from the National Comprehensive Cancer Network (NCCN) can be adopted to improve patient care. Guidelines for clinical oncology have been developed for all the major disease sites and are published online. Also, the physician data query (PDQ) is a convenient comprehensive cancer database from the NCI that contains peer-reviewed summaries of cancer treatment options, directories of professionals, and registries of open and closed clinical trials.

Institutional or multi-institutional trials that were not registered with the NCI's clinical trials reporting program, or have funding from sources other than grants from the NIH, might not have results that are readily available. Although the results of such trials may be published and included in Medline, the articles might not be free PDF downloads. The information may also be in textbooks that are not part of the NCBI bookshelf. In these cases, academic centers may have electronic access to the journals or books in question. A Google search sometimes returns excerpts of textbooks through Google Scholar. Some articles can be read for free through the local library by using online services like AccessMyLibrary. Most of these resources can be found through other Internet searches. If all else fails, one can find hard copies through libraries or from bookstores. However, electronic versions are more powerful, because the entire text can also be searched for relevant information.

Physicians can use all these resources to devise multiple treatment strategies to provide patients with options. Increasing numbers of patients are well informed before the consultation and have visited Web sites such as Medline Plus, WebMD, various hospital Web sites, or Web sites found through their favorite

TABLE 11.2 Organizations Participating in the NIH Cooperative Group Program

Organization	URL
American College of Radiology Imaging Network	http://www.acrin.org
American College of Surgeons Oncology Group	http://www.acosog.org
Cancer and Leukemia Group B	http://www.calgb.org
Children's Oncology Group	http://www.childrensoncologygroup.org
Eastern Cooperative Oncology Group	http://www.ecog.org
European Organisation for Research and Treatment of Cancer	http://www.eortc.be
Gynecologic Oncology Group	http://www.gog.org
National Cancer Institute of Canada, Clinical Trials Group	http://www.ctg.queensu.ca
National Surgical Adjuvant Breast and Bowel Project	http://www.nsabp.pitt.edu
North Central Cancer Treatment Group	http://ncctg.mayo.edu
RTOG	http://www.rtog.org
Southwest Oncology Group	http://www.swog.org

Internet search engine. They may already have formed opinions about various treatment options and may have a strong preference. Other patients rely on the consultation for their primary source of information, and physicians can provide them with additional information, including the appropriate Web sites for patients to visit. The actual treatment course is determined after discussions with the patient.

11.3 Implementing Treatment Techniques

At times, radiation oncologists may favor a particular treatment strategy but settle for another option because of a lack of appropriate equipment, processes, and training. In some cases, this eventually leads to a request to implement the treatment technique, and medical physicists investigate its feasibility (Purdy 2007). Some techniques may already be well established at other centers, and task group reports may even be available to provide guidelines [e.g., American Association of Physicists in Medicine (AAPM); Table 11.3]. Other techniques may truly be "cutting edge" and collaborations with equipment manufacturers are needed. In either case, a thorough understanding of the use of the equipment, the clinical workflow, and its connectivity with other devices is essential.

Such information is typically found in the user's manual, but it can be helpful to contact the manufacturer and find technical staff who developed the equipment. Many manufacturers are involved in RT, and an Internet search can often reveal contact information. When the equipment has already been purchased, the sales representatives can often find the right technical people to work with the institution's medical physicists.

However, manufacturers generally specify the standards met by their equipment, and medical physicists should be familiar with those that affect their clinical workflow. Standards indicate conventions for coordinate systems of treatment units, required safety interlocks, file formats, performance requirements, and communication protocols, among other things (International Electrotechnical Commission 1998, 2000, 2007, 2008a; 2008b; Lillicrap et al. 1998; Jones 1999; Health Level Seven, Inc. 2005, 2007; National Electrical Manufacturers Association 2007e; Van Dyk 2008). Such standards also improve patient and operator safety (International Electrotechnical Commission 1998, 2000, 2007, 2008a, 2008b) and facilitate interoperability and communication among devices (Oosterwijk 1998, 2001, 2008; Health Level Seven, Inc. 2005, 2007; National Electrical Manufacturers Association 2007e; Law & Liu 2009; Law et al. 2009). The relevant standards bodies and their Web sites are listed in Table 11.3.

When the technique to be implemented is part of a protocol, some type of credentialing may be required. The Advanced Technology Consortium (ATC) is a group of organizations that participate in this process (see Table 11.4) and support specific cooperative group clinical trials. The Image-Guided Technology Center handles the protocol QA for the Radiation Therapy Oncology Group (RTOG), National Surgical Adjuvant Breast and Bowel Project, Gynecologic Oncology Group, European Organisation for Research and Treatment of Cancer, National Cancer Institute of Canada, Japan Clinical Oncology Group, and New Approaches to Brain Tumor Therapy, whereas the Quality Assurance and Review Center manages the QA for the Children's Oncology Group, Cancer and Leukemia Group B, and American College of Surgeons Oncology Group protocols. The Radiological Physics Center provides dosimetry services in support of these activities, such as sending a head phantom for testing the dosimetric accuracy of intensity-modulated RT plans (Molineu et al. 2005) to institutions seeking credentials for protocols requiring such credentialing. RTOG provides much of the technical support and research development for the exchange (Radiation Therapy Oncology Group 1999) and analysis of the clinical data.

Generally, credentialing bodies will not determine whether an institution satisfies government regulations. Rather, the medical physicist should determine if any policies and procedures need to be adopted for regulatory compliance. U.S. Federal law regulating the medical use of byproduct material can be found on the Nuclear Regulatory Commission Web site (United States Nuclear Regulatory Commission 2007) (Table 11.3). Each state in the United States also has its own requirements regarding the

TABLE 11.3 Organizations Providing Relevant Radiotherapy Standards, Guidelines, and Regulations

Group	URL and References
AAPM	http://www.aapm.org
Task Group Reports	http://www.aapm.org/pubs/reports
International Electrotechnical Commission	http://www.iec.ch[a]
National Electrical Manufacturers Association	http://medical.nema.org[b]
Health Level Seven, Inc.	http://www.hl7.org[c]
RTOG	http://itc.wustl.edu/exchange_files/tapeexch400.htm[d]
US Nuclear Regulatory Commission	http://www.nrc.gov[e]
International Atomic Energy Agency	http://www.iaea.org[f]

[a] International Electrotechnical Commission 1998, 2000, 2007, 2008a, 2008b.
[b] National Electrical Manufacturers Association 2007e.
[c] Health Level Seven, Inc. 2005, 2007.
[d] Radiation Therapy Oncology Group 1999.
[e] United States Nuclear Regulatory Commission 2007.
[f] International Atomic Energy Agency 1996a, 1996b.

TABLE 11.4 Member Organizations of the ATC

Organization	URL
ATC	http://atc.wustl.edu
Image-Guided Therapy Center	http://itc.wustl.edu
RTOG Dosimetry Group	http://www.rtog.org
Radiological Physics Center	http://rpc.mdanderson.org/rpc
Quality Assurance Review Center	http://www.qarc.org
Washington University in St. Louis School of Medicine	http://medschool.wustl.edu

medical use of radiation (e.g., the Iowa Administrative Code 2008). International guidelines may also apply, such as those provided by the International Atomic Energy Agency (1996a, 1996b).

11.4 Imaging

The resources described thus far are existing documents and services from organizations that serve the population of cancer patients. As we start the patient-specific treatment process, we collect data that turns into information (defined previously as meaningful, well-formed data). Our resources must now include tools for processing the data, beginning with patient images that flow through the clinical process using the Digital Imaging and Communications in Medicine (DICOM) standard (National Electrical Manufacturers Association 2007e).

The DICOM standard specifies a transfer protocol as well as the format of the data set being transmitted (National Electrical Manufacturers Association 2007d). The transfer protocol is based on a layer that sits on top of the Transmission Control Protocol/Internet Protocol layer. The format of the data set deserves closer inspection. Figure 11.2 depicts a DICOM data set (National Electrical Manufacturers Association 2007b) that consists of several data elements. The data elements at the beginning of the data set include patient demographics and information about the image (e.g., the number of rows and columns). The data element that contains the image, the pixel data, appears at the end of the file. Each data element is composed of three or four data fields. The first data field in each element provides a tag

with two sixteen-bit values for the group and element numbers. This defines what the data element represents. For example, the pairing of group and element values 7FE0 and 0010 (hexadecimal notation) represents the pixel data. The descriptive names for the standard tags are listed in the DICOM data dictionary (National Electrical Manufacturers Association 2007c), indexed in ascending tag order (see Figure 11.3). The next data field indicates how the value is represented [VR (Value Representation)], that is, as one of the data types defined in DICOM PS 3.5 (National Electrical Manufacturers Association 2007b). In some transfer syntaxes, VR is optional, because it can be implicitly determined from the VR that is defined in the data dictionary. In the case of pixel data, the VR is typically "OW" (Other Word), a stream of words (two-byte values). The third field, VL (Value Length), indicates the number of (even) bytes that comprise the actual piece of data (possibly padded) that appears in the last part, the Value. For pixel data, VL is the product of the number of rows, columns, and bytes per pixel in the image, whereas the Value field is the stream of words that represent the grayscale values (one pixel per word) in the image. For some types of data elements, the value field could contain nested data elements in a sequence. The first three data fields in a data element serve as a header that allows one to interpret the bytes in the value field.

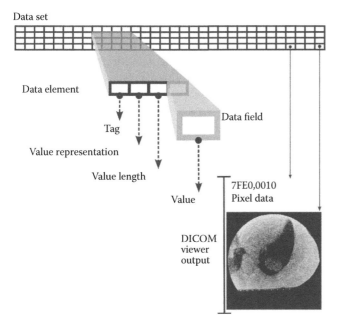

FIGURE 11.2 A DICOM data set for an image. The data set consists of data elements that in turn consist of data fields. The last data element is the pixel data. The example viewer output was created with ImageJ (Rasband 1997–2009).

```
Group,Element:Item Name:VR:VM:RET(IRED)
                    ⋛
0010,1005:Patient's Birth Name:PN:1:
                    ⋛
0010,1030:Patient's Weight:DS:1:
                    ⋛
0010,2110:Contrast Allergies:LO:1-n:
                    ⋛
0028,0002:Samples per Pixel:US:1:
0028,0003:Samples per Pixel Used:US:1:
0028,0004:Photometric Interpretation:CS:1:
0028,0005:Image Dimensions:US:1:RET
                    ⋛
0028,0010:Rows:US:1:
0028,0011:Columns:US:1:
                    ⋛
7FE0,0010:Pixel Data:OW or OB:1:
FFFA,FFFA:Digital Signatures Sequence:SQ:1:
FFFC,FFFC:Data Set Trailing Padding:OB:1:
FFFE,E000:Item:NA:1:
FFFE,E00D:Item Delimitation Item:NA:1:
FFFE,E0DD:Sequence Delimitation Item :NA:1:
```

FIGURE 11.3 Excerpts from a DICOM data dictionary. The first line indicates the format of each of the remaining lines. VR, values representation; VM, value multiplicity; RET, retired dictionary record. This particular dictionary was created by using the Microsoft Word version of DICOM PS 3.6 and applying some simple rules for table to text conversion and character replacement. (From National Electrical Manufacturers Association, *Digital Imaging and Communications in Medicine (DICOM) Part 6: Data Dictionary*, National Electrical Manufacturers Association, Rosslyn, Virginia, 98, 2007.)

The fields of the data elements are transferred via the DICOM protocol in this order, and the data elements are transferred in ascending tag order, with a few exceptions. DICOM files are essentially a transcript of the DICOM transmission and follow the same rules. However, files have a header called the file meta information to deal with the absence of a DICOM association stage that occurs at the beginning of the data transfer between two DICOM application entities.

DICOM image data sets consist of several data elements. This complexity is hidden from the user, as long as all DICOM application entities play by the rules. When connectivity problems occur, the physicist generally turns to a set of DICOM tools that allows images to be viewed and text-based data elements to be read and corrected. These tools have also facilitated the extraction of other information that the dosimetrist may need but that the planning system ignores during the DICOM transfer. An example software tool is ImageJ (Rasband 1997–2009), an image processing application that uses a number of plug-ins for DICOM (Barboriak et al. 2005). Table 11.5 provides a sampling of Web sites that provide DICOM tools, many of which are open-source or freeware. Performing an Internet search with the keyword "DICOM" yields many more Web sites.

All of these DICOM software applications make use of the PS 3.6 data dictionary (National Electrical Manufacturers Association 2007c) to identify the data elements they expect to receive. However, they do not transmit all the data elements in the data dictionary; the standard specifies which ones are required based on the type of information object being transmitted, but some data elements are optional. DICOM communications and file formats cannot be parsed by anticipating a standard set of data elements in a specific order; the only requirement is that the data elements appear in ascending tag order (with the exception of item and sequence delimiters). These tags serve as keys

for looking up values in the data dictionary to determine the data type (VR) and to help programmers define the appropriate internal data structure for representing information to the user [i.e., the translation between the real world and the DICOM world (Pianykh 2008)]. They can also be useful for those who troubleshoot connectivity problems. Some data dictionaries are hard coded, and others are external files that can be read and edited (e.g., the one used by Matlab's image processing toolbox). The standard data dictionary has also been supplemented by some private data dictionaries. Although these private dictionaries may be proprietary, DICOM readers can ignore private data because their tags must follow a certain pattern. If the VR is explicit (i.e., the VR data field is not optional), then it is possible that the value field could be decoded and information could still be displayed to the user. DICOM Reader applications that have editable data dictionaries may allow the addition of private groups, assuming that the manufacturer provides them.

Figure 11.3 shows an excerpt from a data dictionary. The keys in this case are the tags (the first item in each record). Each record has a description of what the data means, followed by the data type (VR) and the value multiplicity (VM), which indicates how many items of the same type are in the value field. For example, in Figure 11.3, the contrast allergies record supports the entry of multiple long string (LO) items (1 up to n) separated by a delimiter, thus allowing the operator to indicate several allergies. Some items have also been retired (RET) and this is indicated in the last field.

The data types are defined in DICOM PS 3.5 (National Electrical Manufacturers Association 2007b). A DICOM application should be able to determine the data type for a data element by using the dictionary entries; hence, the VR data field can be optional in a data element. The VR definitions also have a syntax; for example, a PN (Person's Name) data type is not just a text field but rather a string containing a specific delimiter, "^", that separates up to five different name fields that follow a specific order and a set of rules. This data type can only be used for a person's name.

Knowing these basic rules can help in analyzing the output of DICOM readers. Some readers will include the data element item name, whereas others may not. Using a simple word editor like Notepad to open a DICOM file will reveal a lot of text elements as well as numerical data that must be encoded as a string. Very simple troubleshooting tricks are possible, as described by Pianykh (2008). It is even possible to insert data elements that other applications require to work properly. Some development environments can also open files and display the bytes in hexadecimal form, making it easier to find the group and element bytes that comprise a data element tag. An Internet search using the phrase "hex dump" will yield a number of freeware tools that serve the same purpose.

TABLE 11.5   A Sampling of DICOM-Related Resources

Resource	URL
DICOM homepage	http://medical.nema.org/dicom/[a]
Free DICOM Viewers	http://www.idoimaging.com/index.shtml
Segmentation and Registration toolkit	http://www.itk.org
Visualization toolkit	http://www.vtk.org
Medical Imaging Interaction toolkit	http://mitk.org
DICOM toolkit	http://dicom.offis.de/dcmtk
ImageJ	http://rsbWeb.nih.gov/ij[b]
DICOM Editor	http://mircwiki.rsna.org
DICOM-RT viewer	http://sites.google.com/site/hscheurigdicomrt/
DICOM viewer	http://www.mccauslandcenter.sc.edu/mricro/ezdicom/
DICOM Server	http://www.xs4all.nl/~ingenium/dicom.html

[a] National Electrical Manufacturers Association 2007b, 2007c, 2007d, 2007e.
[b] Rasband 1997–2009; Barboriak et al. 2005.

11.5   Treatment Planning

The creation of radiation oncology treatment plans involves the use of images from many modalities. Volumes of interest, such as tumors and critical organs, are derived from these images and are used to determine beam orientations, treatment field shapes, dose-volume histograms, and the RT equipment parameters that deliver the prescribed dose distribution. The tools described in the previous section can help provide additional information that dosimetrists might need to create the treatment plan. In some cases, the volumes of interest may be created elsewhere along with some treatment beams (e.g., virtual simulation) and exported to the planning system. Such objects have been incorporated into the DICOM standard as DICOM-RT information object definitions (IODs) (National Electrical Manufacturers Association 2007a). Increasing numbers of treatment planning systems support these IODs.

The adoption of DICOM-RT provides a framework for distributing RT department processes. For example, physicists do not have to review a treatment plan in the dosimetry area; they can import DICOM-RT files into plan review workstations that may run several DICOM-RT-compatible applications (Spezi et al. 2002; Deasy et al. 2003). In another example, the Computational Environment for Radiotherapy Research (Deasy et al. 2003) has been used for performing part of the treatment-plan review in a paperless clinic (Siochi et al. 2009). Such clinics rely on an electronic patient record, and some researchers (Law et al. 2009) have demonstrated the use of DICOM-RT as the underlying structure.

DICOM-RT can also be used for data submission to the ATC. The conformance statement for the DICOM part 10 file set reader can be found via links on the ATC Web site. This statement describes what IODs the file set reader application supports and what IODs are expected. The applications that created the file set for submission must support these IODs as well. The conformance statements for these applications can be found on their manufacturer's Web sites.

These conformance statements help to determine what connectivity issues might exist between applications. In many cases, the treatment planning system database or data files are separate from the treatment delivery database. Transformations of data are sometimes necessary to ensure compliance and proper interpretation of delivery parameters (Siochi et al. 2009). Many corrections can be done directly on the database, but this is time-consuming, especially when there are a lot of plan parameters to correct. Some centers have written in-house applications to deal with these issues (Fraass et al. 1995; Siochi et al. 2009). Their programmers make use of the file formats specified by the manufacturers [e.g., RTP-Connect (IMPAC Medical Systems Inc. 2007); do a web search for LED17001.pdf]. Some manufacturers require an account (linked to the purchase of equipment) to access these formats.

In addition to resource tools for analyzing the information flow between imaging, planning, and treatment, there are also image-based tools that help people research the concepts behind preparing a treatment plan, similar to those described in Section 11.2. There are situations where a physician, dosimetrist, or physicist might recall reading about a particular dose distribution or image analysis method, but their memory of it

is based on the images. They may want to compare those images and doses against what is on their treatment plan. They could use Web-based search engines for biomedical images such as GoldMiner (Kahn & Thao 2007) and Yale Image Finder (Xu et al. 2008). GoldMiner retrieves images from figures in open-access radiology content over the Internet [e.g., *Radiographics*, *American Journal of Roentgenology*, *British Journal of Radiology*, *Radiology*, and the European Association of Radiology's online database (http://www.EURORAD.org)]. The search terms are found in figure captions. GoldMiner uses the Unified Medical Language System (UMLS) metathesaurus to perform concept-based searching as well, increasing the number of relevant hits compared with a keyword search. Yale Image Finder does something similar, except that it also searches text in the images. Yale Image Finder uses the articles in PubMed Central as the domain for the search. In both applications, the user can browse through the results with thumbnails of the images and select links to the journal articles. Besides these applications, several other content-based image retrieval tools have been developed to handle a variety of medical literature and image databases (Antani et al. 2004, 2007; Lehmann et al. 2004; Muller et al. 2004; Hsu et al. 2007; Deserno et al. 2009).

One area in which such retrieved images might be used is in contouring the volumes of interest in a treatment plan. They can provide visual examples and guidelines of contours, similar to the ATC atlas of contours and tissue naming conventions that were adopted to improve the uniformity of dose volume histogram reporting.

11.6 Treatment Delivery

Treatment management systems use a database of the treatment parameters to control the delivery of radiation. When each fraction of radiation is delivered, the database records the positions of beam-defining elements, the delivered dose, variations from desired parameters, tolerances, and overrides for parameters that were out of tolerance. This allows physicists to perform chart reviews and administrative personnel to run reports and keep track of billing.

Automating these chart reviews to detect deviations from anticipated patterns would involve reading the database or at least exporting the records in question to an application that can analyze them. Billing audits could also be automated to some extent. Unfortunately, the built-in tools provided by manufacturers have not been flexible enough to be customized to the needs of different departments. In-house database programmers, when available, are able to automate the creation of reports that provide the needed information. To do so effectively, they need the data dictionary of the database. Like the DICOM standard, the data dictionary describes the type and length of each data element. However, it does not simply provide a key (tag) and some attributes associated with it. Rather, it describes all the tables, records, and fields, along with their relationships.

A typical hierarchy of data elements in an RT database is shown in Figure 11.4. This is not meant to be an exhaustive

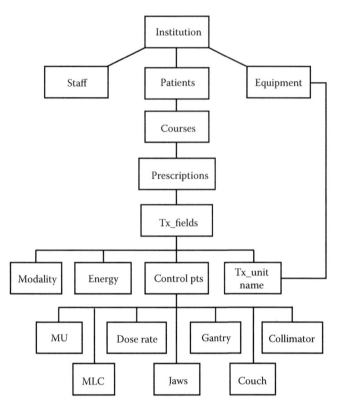

FIGURE 11.4 Relationships among some of the data elements in an RT treatment database.

diagram; rather, it provides a glimpse into the relationships among the data elements and tables in an RT database. Data dictionaries defining these relationships are available from the manufacturer, usually through extended service support agreements. Table 11.6 lists resources for some of the major treatment management systems.

Even without access to the manuals for the data dictionary, several simple queries can be made by using tools that provide a basic view of the database tables and relationships. The fields have usually been given names that are intuitive for users with radiation oncology domain knowledge. For example, using the Pervasive SQL Database Management System with the Lantis database allows tables for the patients, their prescriptions, and

TABLE 11.6 Treatment Database Resources

Resource	URL
Data Dictionary Creator	http://www.codeplex.com/datadictionary
Pervasive	http://www.pervasive.com
MySQL	http://www.mysql.com
Mosaiq/Multi-Access	http://www.elekta.com/mosaiq
Aria/Varis	http://www.varian.com/us/oncology/radiation_oncology/aria
American Medical Association CPT codes	http://www.ama-assn.org/ama/pub/physician-resources/solutions-managing-your-practice/coding-billing-insurance/cpt.page?

their treatment fields to be viewed, and queries can be formed by selecting items from the various tables as variables for the query. This type of graphical query builder provides the user with lists of data elements rather than relying on the user's knowledge of the database structure (more experienced users might prefer to directly type their queries for the more common data elements).

Although the structure of the database can be reverse engineered in a limited manner with tools such as Data Dictionary Creator, the metadata may not be complete enough to understand how the database maps onto the real-world elements. Fortunately, the databases encountered in RT have had rather intuitive names, so basic operations can be performed. Usually, tools such as the SQL Server Management Studio are used by database developers to create the manuals that describe their data dictionaries. Similar tools can be found through an Internet search for "create data dictionary." For more complex queries, one may need support from the database developers. Not all data dictionaries will contain extended properties, so the database may not exactly be self-documenting.

Exploring a database with queries leaves it open to the possibility of unintended changes. This can be extremely dangerous with a treatment parameter database. For example, accidentally changing the number of monitor units for a treatment field could lead to mistreatment of a patient. Typically, lower-level interactions with the database are left to medical physicists with the appropriate training, the manufacturer's technical support staff, or dedicated IT staff in the radiation oncology department under supervision of a medical physicist. Higher-level interactions with the database are performed through dedicated applications (such as Mosaiq and Aria) and are access-limited according to the security clearance given to the user.

These dedicated applications also automate code capture using the Current Procedural Terminology (CPT) codes from the American Medical Association. An Internet search for "CPT billing codes" yields several online tools that help in searching the CPT database. There are also patient-oriented sites that explain the meaning of the codes and discuss billing issues. These sites also provide information about other coding systems such as the International Statistical Classifications of Diseases and Healthcare Common Procedure Coding System for Medicare billing, which is based on CPT.

11.7 Follow-Up

The treatment database management system would be an ideal database to search when creating the data required for cancer registries. Unfortunately, this is not the only database used by cancer clinics; rather, clinics tend to rely on more global hospital databases that may not accurately reflect the content of the individual departmental treatment databases (not just RT but medical oncology as well). Although such databases are more global, they are also less detailed. The data reported to state and national cancer registries are primarily for epidemiologic studies, so the level of detail may be sufficient.

Guidelines and tools for data submission are available through each participating state's cancer registry as well as the national registry. Data dictionaries (Thornton & O'Connor 2009), data exchange standards (Havener 2009), record descriptions (Havener 2009), and quality standards (Hofferkamp 2009) are provided by the North American Association of Central Cancer Registries. The standards can be retrieved from the "Cancer Data Standards" link on their home page (http://www.naaccr.org). They also provide links to the homepages of their member cancer registries. The National Program of Cancer Registries, through the Centers for Disease Control and Prevention, provides the Registry Plus suite of applications that help collect and process data. The Surveillance, Epidemiology and End Results (SEER) coding rules (Johnson & Adamo 2007) explain how to convert a variety of types of information (neoplasm description, tumor staging, demographics, treatment, and follow-up) into code strings. Coding instructions for collaborative staging (Collaborative Staging Task Force of the American Joint Committee on Cancer 2006) are also used to maintain uniformity throughout various cancer registries.

The data submitted to the cancer registries end up in various state publications as well as the SEER database, which provides cancer incidence and survival statistics for the United States. A knowledge of these statistics can help health professionals spot outliers in their clinics and possibly identify aggressive cancers. Such information could change the treatment strategy decision.

The more detailed treatment data generally are submitted to cooperative groups only if the patients are part of the cooperative group protocol. However, investigators of prospective clinical trials, regardless of whether the trial is NIH funded or part of a cooperative group, may still be required to register through the protocol registration system on clinicaltrials.gov (http://clinicaltrials.gov/ct2/manage-recs/fdaaa).

Much of the data preparation for cancer registries typically involved tedious manual entry, so vendors of radiation oncology treatment management systems have developed tools to facilitate the communication between the cancer registry and the radiation oncology database. For example, the cancer registry application, "METRIQ," an oncology management database developed by IMPAC (http://www.elekta.com/metriq), uses "Mosaiq Connect" to import data from the radiation oncology EMR database, "Mosaiq." Varian's efforts include their radiation oncology database, Aria, their HL7 Information Exchange Manager and their cancer registry partner OncoLog (http://www.varian.com/us/oncology/radiation_oncology/aria).

11.8 Research

Data submitted to cooperative groups or cancer registries end up in databases that researchers can access. The SEER data are available through the Internet or can be shipped on disc once the appropriate agreements have been signed (http://seer.cancer.gov/data). The RTOG has a data-sharing policy document on their Web site that outlines the procedures for data requests. Similar data-sharing policies and procedures are available on the Web sites of the other cooperative groups. Using the phrase "oncology cooperative group data sharing policy" in an Internet search produces a convenient summary of links to the various policy documents.

Searching through these databases may also require a metathesaurus, a compendium of concepts, their relationships, and their linked terminologies from many other thesauri and vocabularies. The UMLS (Nelson et al. 2002) metathesaurus, available through the NLM (http://www.nlm.nih.gov/pubs/factsheets/umlsmeta.html), is a database of biomedical terms from several source vocabularies (e.g., biomedical literature, code sets, and lists of controlled terms). It indicates the relationships among the terms and their relevant concepts, making it possible for computer applications to "interpret" the user's medical language. PubMed, for example, makes use of UMLS. Several authors have used UMLS in oncology informatics research for image searching (Cerveri et al. 2000; Frankewitsch & Prokosch 2001; Kahn & Thao 2007; Kahn & Rubin 2009; Kammerer et al. 2009), natural language processing (Cognetti & Cecere 2003), knowledge servers (Sherertz et al. 1995; Tuttle et al. 1996; Mitchell et al. 2003), and tumor documentation standards (Altmann et al. 1999; Cognetti & Cecere 2003). A subset of terms contained in UMLS that is a standard for electronic health information exchange is the Systematized Nomenclature of Medicine–Clinical Terms; it is used within the DICOM standard.

A thesaurus such as MeSH from the NLM can also be helpful when selecting search terms. Researchers have used it to find relevant articles and databases of clinical data to perform reviews as well as retrospective studies (Vicini et al. 1999; Ganswindt et al. 2008). Others have analyzed MeSH to characterize cancer information systems (Rada 2006). A browser for MeSH terms is available for performing online searches, and an electronic copy can be downloaded from http://www.nlm.nih.gov/mesh.

A super-set thesaurus of MeSH terms, EMTREE, is used by EMBASE/Excerpta Medica, a database of biomedical literature from Elsevier. A significant number of journals in EMBASE are not in Medline, and researchers have used both databases to improve their search results (Felber 2000; El-Ghobashy & Saidi 2009). Many of the databases mentioned in Section 11.2 have also been searched to increase the number of clinical trials used in a meta-analysis of data (Au et al. 2003). Additional databases used in recent radiation oncology reviews (Hart et al. 2005; Tsao et al. 2005; Antonarakis et al. 2007; Mintz et al. 2007; Viani et al. 2007, 2009a, 2009b; James et al. 2008; Sanghera et al. 2008) are CANCERLIT, Cochrane Library, Cumulative Index to Nursing and Allied Health (CINAHL), Latin American and Caribbean Health Sciences (LILACS), Allied and Complementary Medicine (AMED), Current Contents, and the Proceedings of the American Society of Clinical Oncology, American Society for Radiation Oncology, and European Society for Therapeutic Radiation Oncology. Search strategies, keywords, and filters have been described by Felber (2000). URLs for the various databases are given in Table 11.7.

TABLE 11.7   Biomedical Literature Databases

Database	URL
AMED	(through OVID)
CANCERLIT	http://www.cancer.gov/search/cancer_literature
CINAHL	http://www.ebscohost.com/cinahl
Cochrane	http://www.cochrane.org
EMBASE	http://www.embase.com
LILACS	http://bases.bireme.br/cgi-bin/wxislind.exe/iah/ online/?IsisScript=iah/iah.xis&base=LILACS&lang=i
Medline	http://www.ncbi.nlm.nih.gov/pubmed
OVID	http://www.ovid.com/site/catalog/Catalog_DataBase.jsp

11.9 Discussion

Many of the resources described in this chapter can be accessed conveniently through Web portals. For example, the American Society for Radiation Oncology (ASTRO) has a Web page (http://www.astro.org) with links to PubMed, NCI, NIH, RTOG, the PDQ database, clinical trials database, NCCN practice guidelines, Health Services/Technology Assessment Text (a clinical guidelines database), the Cochrane Collaboration, and the Agency for Healthcare Research and Quality. Several other links are provided for coding and billing, online journals, and clinical practice tools. Other features, like online access to the *International Journal of Radiation Oncology Biology Physics* (commonly known as the *Red Journal*), are available only to members. The AAPM also has a homepage (www.aapm.org) with links to the homepages of related organizations (ASTRO and its European counterpart ESTRO, American College of Radiation Oncology, American Brachytherapy Society, NCI, RTOG, and Society for Radiation Oncology Administrators). A link to a list of radiation oncology resources is also provided on the AAPM Medical Physics Resource page (which has links to an online cancer guide, the Association of Cancer Online Resources, the Radiological Physics Center, The University of Texas M. D. Anderson Cancer Center, the ATC, the National Electrical Manufacturers Association, a list of DICOM resources, education materials, search engines, medical physics departments, several mailing lists, and other medical physics–related standards and reference data). The Association of Cancer Online Resources (http://www.acor.org) is also a portal, but it is geared toward cancer patients and their families. They have links to clinical trials, mailing lists, reference materials, patient support, and publications.

Most of the information resources described in this chapter are available online. However, URLs can change, and links can break. The lists of resources created here were found by using standard search engines such as Google. Although the searches performed were rather general, more detailed information can be found by using Internet searches more effectively, as described by several authors (Gatlin 1998; Wu & Li 1999; Bin & Lun 2001; Day 2001; Younger 2005; Yu & Kaufman 2007). However, the quality of the information should be evaluated by those with the appropriate domain knowledge, especially when the information is geared

toward consumers. Attempts at automating the identification of inaccurate information with criteria have not been successful, at least in the case of breast cancer information (Bernstam et al. 2008). Health information consumers may also acquire accurate information, but they may not be able to determine its relevance or completeness (Skinner et al. 2003; Morahan-Martin 2004). Few, if any, Web-based assessment tools are available to help consumers identify reliable information (Lorence & Greenberg 2006). Healthcare practitioners should be aware that their patients may have found inaccurate or incomplete information on the Web and should be prepared to assess their misconceptions and provide them with the proper Web sites and guidance on information retrieval.

Many of this chapter's information resources were documents retrieved through the Internet. Several of those resources are regulations provided by organizations like the ones in Table 11.3. Information like this may seem static, without utility as data. However, applying informatics methods to these regulations (e.g., encoding them and creating data dictionaries, controlled vocabularies, and thesauri) can bring them to life. For example, many regulations exist because of accidents. There is an implicit analysis that occurs, similar to that of Failure Modes and Effects Analysis (FMEA). Hence, these regulations could be used as data to evaluate the regulatory compliance of a FMEA for a process. Some researchers have combined FMEA and informatics concepts to reduce medical errors (Lenz et al. 2005; Crane & Crane 2006).

Such errors happen more frequently when data are entered in several places or transformed through human interaction. RT processes that involve many manual steps could benefit by automating information capture. This is especially true in brachytherapy. Treatment management systems for brachytherapy either do not exist or are at a very immature state. Most brachytherapy treatment records are on paper. As a result, the informatics processes and resources available for external beam treatments are not particularly applicable to brachytherapy at this time. Automated deliveries, such as those with high-dose-rate systems, have the potential for providing databases that contain the actual source positions. Some researchers have even looked into image-guided robotic systems to automate the delivery of low-dose rate radiation for prostate cancer by using needles (Davies et al. 2004; Wei et al. 2005; Hu et al. 2007; Yu et al. 2007). Perhaps the use of such systems will increase the amount of electronically captured brachytherapy treatment information. The capture of plan information is a little more mature; the DICOM-RT IODs have modules for brachytherapy that can be used for submission to the ATC, and several brachytherapy planning systems support DICOM-RT. DICOM-RT also has modules for brachytherapy treatment records, but this capability is currently underutilized.

Increasing the amount of electronically available information sources in radiation oncology will require a shift to the paperless and filmless environment, creating an informatics-based workflow paradigm (Miller 2006). The DICOM-RT modules, along with DICOM Structured Reports, have the potential to serve as

IODs for an electronic chart (Law et al. 2009). This has the potential to provide more well-defined information in a standard format, which in turn can be leveraged for more robust outcomes analyses and to improve the state of evidence-based medicine.

References

Altmann, U. et al. (1999). A model for integration and continuous development of standards for tumour documentation using relational database techniques and extensible markup language. *Studies in Health Technology and Informatics, 68,* 895–898.

Antani, S. et al. (2004). Content-based image retrieval for large biomedical image archives. *Studies in Health Technology and Informatics, 107,* 829–833.

Antani, S. K. et al. (2007). Geographically distributed complementary content-based image retrieval systems for biomedical image informatics. *Studies in Health Technology and Informatics, 129,* 493–497.

Antonarakis, E. S. et al. (2007). Survival in men with nonmetastatic prostate cancer treated with hormone therapy: A quantitative systematic review. *Journal of Clinical Oncology, 25,* 4998–5008.

Au, H. J. et al. (2003). Systematic review of management of colorectal cancer in elderly patients. *Clinical Colorectal Cancer, 3,* 165–171.

Barboriak, D. P. et al. (2005). Creation of DICOM—aware applications using ImageJ. *Journal of Digital Imaging, 18,* 91–99.

Beasley, M. et al. (2008). Expanded risk groups help determine which prostate radiotherapy sub-group may benefit from adjuvant androgen deprivation therapy. *Radiation Oncology, 3,* 8.

Bernstam, E. V. et al. (2008). Commonly cited Website quality criteria are not effective at identifying inaccurate online information about breast cancer. *Cancer, 112,* 1206–1213.

Bin, L. & Lun, K. C. (2001). The retrieval effectiveness of medical information on the Web. *International Journal of Medical Informatics, 62,* 155–163.

Cerveri, P. et al. (2000). Remote access to anatomical information: An integration between semantic knowledge and visual data. *Proceedings of the AMIA Symposium,* 126–130.

Cognetti, G. & Cecere, L. (2003). E-oncology and health portals: Instructions and standards for the evaluation, production organisation and use. *Journal of Experimental & Clinical Cancer Research, 22,* 677–686.

Collaborative Staging Task Force of the American Joint Committee on Cancer, *Collaborative Staging Manual and Coding Instructions, Version 01.04.00, NIH Pub. No. 04-5496,* U.S. Department of Health and Human Services, Bethesda, Maryland, 2006.

Crane, J. & Crane, F. G. (2006). Preventing medication errors in hospitals through a systems approach and technological innovation: A prescription for 2010. *Hospital Topics, 84,* 3–8.

Davies, B. L. et al. (2004). Brachytherapy—An example of a urological minimally invasive robotic procedure. *International Journal of Medical Robotics, 1,* 88–96.

Day, J. (2001). The quest for information: A guide to searching the Internet. *Journal of Contemporary Dental Practice, 2,* 33–43.

Deasy, J. O. et al. (2003). CERR: A computational environment for radiotherapy research. *Medical Physics, 30,* 979–985.

Deserno, T. M. et al. (2009). Ontology of gaps in content-based image retrieval. *Journal of Digital Imaging, 22,* 202–215.

Donaldson, S. S. et al. (1998). A multidisciplinary study investigating radiotherapy in Ewing's sarcoma: End results of POG #8346. Pediatric Oncology Group. *International Journal of Radiation Oncology Biology Physics, 42,* 125–135.

El-Ghobashy, A. E. & Saidi, S. A. (2009). Sentinel lymph node sampling in gynaecological cancers: Techniques and clinical applications. *European Journal of Surgical Oncology, 35,* 675–685.

Felber, S. H. (2000). Searching for evidence-based oncology: Tips and tools for finding evidence in the medical literature. *Cancer Control, 7,* 469–475.

Floridi, L., Philosophical conceptions of information, in *Formal Theories of Information: From Shannon to Semantic Information Theory and General Concepts of Information, Lecture Notes in Computer Science,* Sommaruga, G., Ed., Springer, New York, 2009, 13–53.

Fraass, B. A. et al. (1995). A computer-controlled conformal radiotherapy system. IV. Electronic chart. *International Journal of Radiation Oncology Biology Physics, 33,* 1181–1194.

Frankewitsch, T. & Prokosch, U. (2001). Navigation in medical Internet image databases. *Medical Informatics and The Internet in Medicine, 26,* 1–15.

Ganswindt, U. et al. (2008). Adjuvant radiotherapy for patients with locally advanced prostate cancer—A new standard? *European Urology, 54,* 528–542.

Gatlin, L. (1998). How to make Internet searches easier: Tips for effective use of Web search engines. *American Journal of Orthodontics and Dentofacial Orthopedics, 114,* 355–357.

Hart, M. G. et al. (2005). Surgical resection and whole brain radiation therapy versus whole brain radiation therapy alone for single brain metastases. *Cochrane Database of Systematic Reviews,* CD003292.

Havener, L. A., *Standards for Cancer Registries Volume I: Data Exchange Standards and Record Descriptions. Version 11.3,* North American Association of Central Cancer Registries, Springfield, Illinois, 2009.

Health Level Seven, Inc., *The Clinical Document Architecture,* Health Level Seven, Inc., Ann Arbor, MI, 2005.

Health Level Seven, Inc., *Application Protocol for Electronic Data Exchange in Healthcare Environments,* Health Level Seven, Inc., Ann Arbor, MI, 2007.

Hofferkamp, J., *Standards for Cancer Registries Volume III: Standards for Completeness, Quality, Analysis, Management, Security and Confidentiality of Data,* North American Association of Central Cancer Registries, Springfield, Illinois, 2009.

Hsu, W. et al. (2007). SPIRS: A framework for content-based image retrieval from large biomedical databases. *Studies in Health Technology and Informatics, 129,* 188–192.

Hu, Y. et al. (2007). Hazard analysis of EUCLIDIAN: An image-guided robotic brachytherapy system. *Conference Proceedings—IEEE Engineering in Medicine and Biology Society, 2007*, 1249–1252.

IMPAC Medical Systems, Inc. (2007). RTPConnect radiotherapy treatment planning import/export interface specification.

International Atomic Energy Agency, *International Basic Safety Standards for Protection Against Ionizing Radiation and for the Safety of Radiation Sources*, International Atomic Energy Agency, Vienna, 1996a.

International Atomic Energy Agency, *Radiation Protection and the Safety of Radiation Sources*, International Atomic Energy Agency, Vienna, 1996b.

International Electrotechnical Commission, *Medical Electrical Equipment—Part 2-1: Particular Requirements for the Safety of Electron Accelerators in the Range 1 MeV to 50 MeV*, International Electrotechnical Commission, Geneva, Switzerland, 1998, 131.

International Electrotechnical Commission, *Medical Electrical Equipment—Requirements for the Safety of Radiotherapy Treatment Planning Systems*, International Electrotechnical Commission, Geneva, Switzerland, 2000, 59.

International Electrotechnical Commission, *Medical Electrical Equipment—Medical Electron Accelerators—Functional Performance Characteristics*, International Electrotechnical Commission, Geneva, Switzerland, 2007, 96.

International Electrotechnical Commission, *Medical Electrical Equipment—Medical Electron Accelerators—Guidelines for Functional Performance Characteristics*, International Electrotechnical Commission, Geneva, Switzerland, 2008a, 68.

International Electrotechnical Commission, *Radiotherapy Equipment—Coordinates, Movements and Scales*, International Electrotechnical Commission, Geneva, Switzerland, 2008b, 73.

Iowa Administrative Code, *Therapeutic Use of Radiation Machines*, 2008.

James, M. L. et al. (2008). Fraction size in radiation treatment for breast conservation in early breast cancer. *Cochrane Database of Systematic Reviews*, CD003860.

Johnson, C. H. & Adamo, M., *The SEER Program Coding and Staging Manual 2007*, National Institutes of Health, National Cancer Institute, Bethesda, Maryland, 2007.

Jones, D. (1999). Radiotherapy equipment standards from the International Electrotechnical Commission. *British Journal of Radiology, 72*, 623.

Kahn, C. E., Jr. & Rubin, D. L. (2009). Automated semantic indexing of figure captions to improve radiology image retrieval. *Journal of the American Medical Informatics Association, 16*, 380–386.

Kahn, C. E., Jr. & Thao, C. (2007). GoldMiner: A radiology image search engine. *AJR American Journal of Roentgenology, 188*, 1475–1478.

Kammerer, F. J. et al. (2009). Design of a Web portal for interdisciplinary image retrieval from multiple online image resources. *Methods of Information in Medicine*, 48.

Kupelian, A. et al. (2005) Hypofractionated intensity-modulated radiotherapy (70 Gy at 2.5 Gy per fraction) for localized prostate cancer: Long-term outcomes. *International Journal of Radiation Oncology Biology Physics, 63*, 1463–1468.

Law, M. Y. & Liu, B. (2009). Informatics in radiology: DICOM-RT and its utilization in radiation therapy. *Radiographics, 29*, 655–667.

Law, M. Y. et al. (2009). Informatics in radiology: DICOM-RT-based electronic patient record information system for radiation therapy. *Radiographics*.

Lehmann, T. M. et al. (2004). Content-based image retrieval in medical applications. *Methods of Information in Medicine, 43*, 354–361.

Lenz, R. et al. (2005). Demand-driven evolution of IT systems in healthcare—A case study for improving interdisciplinary processes. *Methods of Information in Medicine, 44*, 4–10.

Lillicrap, S. C. et al. (1998). Radiotherapy equipment standards from the International Electrotechnical Commission. *British Journal of Radiology, 71*, 1225–1228.

Lorence, D. P. & Greenberg, L. (2006). The zeitgeist of online health search. Implications for a consumer-centric health system. *Journal of General Internal Medicine, 21*, 134–139.

Miller, A. (2006). New informatics-based work flow paradigms in radiation oncology: The potential impact on epidemiological cancer research. *HIM Journal, 34*, 84–87.

Mintz, A. et al. (2007). Management of single brain metastasis: A practice guideline. *Current Oncology, 14*, 131–143.

Mitchell, K. J. et al. (2003). A knowledge-based approach to information extraction from surgical pathology reports. *AMIA Annual Symposium Proceedings*, 937.

Mohler, J. et al. (2007). Prostate cancer. Clinical practice guidelines in oncology. *Journal of the National Comprehensive Cancer Network, 5*, 650–683.

Molineu, A. et al. (2005). Design and implementation of an anthropomorphic quality assurance phantom for intensity-modulated radiation therapy for the Radiation Therapy Oncology Group. *International Journal of Radiation Oncology Biology Physics, 63*, 577–583.

Morahan-Martin, J. M. (2004). How Internet users find, evaluate, and use online health information: A cross-cultural review. *CyberPsychology & Behavior, 7*, 497–510.

Muller, H. et al. (2004). A review of content-based image retrieval systems in medical applications-clinical benefits and future directions. *International Journal of Medical Informatics, 73*, 1–23.

National Electrical Manufacturers Association, *Digital Imaging and Communications in Medicine (DICOM) Part 3: Information Object Definitions*, National Electrical Manufacturers Association, Rosslyn, Virginia, 2007a, 1040.

National Electrical Manufacturers Association, *Digital Imaging and Communications in Medicine (DICOM) Part 5: Data Structures and Encoding*, National Electrical Manufacturers Association, Rosslyn, Virginia, 2007b, 108.

National Electrical Manufacturers Association, *Digital Imaging and Communications in Medicine (DICOM) Part 6: Data Dictionary*, National Electrical Manufacturers Association, Rosslyn, Virginia, 2007c, 98.

National Electrical Manufacturers Association, *Digital Imaging and Communications in Medicine (DICOM) Part 10: Media Storage and File Format for Media Interchange*, National Electrical Manufacturers Association, Rosslyn, Virginia, 2007d, 34.

National Electrical Manufacturers Association, *Digital Imaging and Communications in Medicine (DICOM) Set*, National Electrical Manufacturers Association, Rosslyn, Virginia, 2007e.

Nelson, S. J. et al., The Unified Medical Language System (UMLS) project, in *Encyclopedia of Library and Information Science*, Kent, A. & Hall, C. M., Eds., Marcel Dekker, Inc., New York, 2002, 369–378.

Oosterwijk, H. (1998). DICOM versus HL7 for modality interfacing. *Journal of Digital Imaging, 11*, 39–41.

Oosterwijk, H. (2001). Defining DICOM (Digital Imaging and Communications in Medicine) requirements. *Radiology Management, 23*, 41–44.

Oosterwijk, H. (2008). DICOM questions, answered. *Radiology Management, 30*, 33–39.

Pianykh, O. S., *Digital Imaging and Communications in Medicine (DICOM): A Practical Introduction and Survival Guide*, Springer, Berlin, 2008, 383 pp.

Purdy, J. A. (2007). From new frontiers to new standards of practice: Advances in radiotherapy planning and delivery. *Frontiers of Radiation Therapy and Oncology, 40*, 18–39.

Rada, R. (2006). Characterizing cancer information systems. *Journal of Medical Systems, 30*, 153–157.

Radiation Therapy Oncology Group, *Specifications for Tape/Network Format for Exchange of Treatment Planning Information*, RTOG 3D QA Center, St. Louis, MO, 1999.

Rasband, W. S., *ImageJ*, U.S. National Institutes of Health, Bethesda, Maryland, 1997–2009.

Sanghera, P. et al. (2008). Chemoradiotherapy for rectal cancer: An updated analysis of factors affecting pathological response. *Clinical Oncology (Royal College of Radiologists), 20*, 176–183.

Sherertz, D. D. et al. (1995). Accessing oncology information at the point of care: Experience using speech, pen, and 3-D interfaces with a knowledge server. *Medinfo, 8 Part 1*, 792–795.

Siochi, R. A. et al. (2009). Radiation therapy plan checks in a paperless clinic. *Journal of Applied Clinical Medical Physics, 10*, 2905.

Skinner, H. et al. (2003). How adolescents use technology for health information: Implications for health professionals from focus group studies. *Journal of Medical Internet Research, 5*, e32.

Soares, H. P. et al. (2005). Evaluation of new treatments in radiation oncology: Are they better than standard treatments? *Journal of the American Medical Association, 293*, 970–978.

Spezi, E. et al. (2002). A DICOM-RT-based toolbox for the evaluation and verification of radiotherapy plans. *Physics in Medicine and Biology, 47*, 4223–4232.

Thornton, M. & O'Connor, L., *Standards for Cancer Registries Volume II: Data Standards and Data Dictionary, Record Layout Version 12*, North American Association of Central Cancer Registries, Springfield, Illinois, 2009.

Tsao, M. N. et al. (2005). The American Society for Therapeutic Radiology and Oncology (ASTRO) evidence-based review of the role of radiosurgery for malignant glioma. *International Journal of Radiation Oncology Biology Physics, 63*, 47–55.

Tuttle, M. S. et al. (1996). Toward reusable software components at the point of care. *Proceedings of the AMIA Annual Fall Symposium*, 150–154.

United States Nuclear Regulatory Commission, *Medical Use of Byproduct Material*, 2007.

Van Dyk, J. (2008). Quality assurance of radiation therapy planning systems: Current status and remaining challenges. *International Journal of Radiation Oncology Biology Physics, 71*, S23–S27.

Viani, G. A. et al. (2007). Breast-conserving surgery with or without radiotherapy in women with ductal carcinoma in situ: A meta-analysis of randomized trials. *Radiation Oncology, 2*, 28.

Viani, G. A. et al. (2009a). Whole brain radiotherapy with radiosensitizer for brain metastases. *Journal of Experimental & Clinical Cancer Research, 28*, 1.

Viani, G. A. et al. (2009b). Brachytherapy for cervix cancer: Low-dose rate or high-dose rate brachytherapy—A meta-analysis of clinical trials. *Journal of Experimental & Clinical Cancer Research, 28*, 47.

Vicini, F. A. et al. (1999). The role of androgen deprivation in the definitive management of clinically localized prostate cancer treated with radiation therapy. *International Journal of Radiation Oncology Biology Physics, 43*, 707–713.

Wei, Z. et al. (2005). 3D TRUS guided robot assisted prostate brachytherapy. *Medical Image Computing and Computer-Assisted Intervention, 8*, 17–24.

White, S. A., *BPMN Modeling and Reference Guide: Understanding and Using BPMN*, Future Strategies, Inc., Lighthouse Point, Florida, 2008.

Wu, G. & Li, J. (1999). Comparing Web search engine performance in searching consumer health information: Evaluation and recommendations. *Bulletin of the Medical Library Association, 87*, 456–461.

Xu, S. et al. (2008). Yale Image Finder (YIF): A new search engine for retrieving biomedical images. *Bioinformatics, 24*, 1968–1670.

Younger, P. (2005). The effective use of search engines on the Internet. *Nursing Standard, 19*, 56–64; quiz 66.

Yu, H. & Kaufman, D. (2007). A cognitive evaluation of four online search engines for answering definitional questions posed by physicians. *Pacific Symposium on Biocomputing*, 328–339.

Yu, Y. et al. (2007). Robotic system for prostate brachytherapy. *Computer Aided Surgery, 12*, 366–370.

Radiogenomics: The Future of Personalized Radiation Therapy?

Bryan Allen
*University of Iowa
Hospitals and Clinics*

12.1 Introduction

The prescribed radiation dose depends on both the cancer type and the radiation sensitivity of the surrounding normal tissue. Normal tissue radiation dose constraints are often set to ensure that less than five percent of patients will experience a serious radiation-induced toxicity within five years of completing treatment (Emami et al. 1991; West & Barnett 2011). Unfortunately, one size does not necessarily fit all when describing an individual's sensitivity to radiation. To protect the minority who may develop radiation-induced toxicity, the majority may not receive the radiation dose necessary for tumor control. Because cancer survivorship continues to dramatically increase (Travis et al. 2012), with approximately half of all cancer patients receiving radiation therapy (Delaney et al. 2005), it is prudent to begin to develop markers to predict for radiation response and radiation toxicity. Radiogenomics investigates how genetic variation influences radiation response in cells, tissues, and individuals.

12.2 Single Nucleotide Polymorphism and Genome Wide Association Studies

Previous studies have established a correlation between radiation sensitivity and genes involved in DNA repair pathways (Kleinerman 2009), apoptosis (Hendry & West 1997), cellular metabolism and free radical scavenging (Aykin-Burns et al. 2011), and cytokine formation (Anscher et al. 2010). Furthermore, numerous studies have suggested an association between single nucleotide polymorphisms (SNP) and radiation toxicity (Suga et al. 2007; Burri et al. 2008; Werbrouck et al. 2009; Alsbeih et al. 2010; Mangoni et al. 2011). A SNP is a single base-pair change that is observed in at least one percent of the population and is an easily assayable measure of genetic variation. However,

a major challenge for radiogenomic studies is to obtain a sufficient number of patients with adequate documentation of radiation toxicity history in addition to having quality samples to perform predictive assays. Thus, unfortunately, most studies that assessed SNPs and radiation toxicity had limited patient numbers and were difficult to replicate (Andreassen et al. 2006). However, rapidly developing technology has reduced the cost of genotyping and sequencing, enabling researchers to study thousands of patients (Barnett et al. 2009) and perform genome-wide association studies (GWAS) to assess for markers associated with radiation toxicity (Kerns et al. 2010; Michikawa et al. 2010; Niu et al. 2010). Because GWAS must include such large numbers of patients and samples to have significant statistical power, an international radiogenomics consortium was established in 2009 (West et al. 2010). Additionally, the consortium provides a link between the existing collaborative groups, including RAPPER (Burnet et al. 2006), Gene-PARE (Ho et al. 2006), and RadGenomics, as well as other small groups (West & Barnett 2011) that are attempting to identify genetic variants associated with developing radiation-induced normal tissue toxicity (West et al. 2010). Participation in this research endeavor often requires access to a significant number of tissue samples that may be difficult to accumulate and obtain at a single institution. A list of tissue collection centers and a brief description of how to obtain access to these tissues are provided in this chapter.

Once genes of interest and/or SNPs have been identified, sequence and expression analysis may be performed and compared to assess genetic variation influences on radiation response by using available databases provided by the National Institutes of Health (http://www.ncbi.nlm.nih.gov/guide/all). For example, the Database of Short Genetic Variations (dbSNP) provides a searchable public domain database of SNPs, microsatellites, small insertions, and small deletions for both benign and clinical mutations. The SNP database may be searched from the dbSNP homepage (http://www.ncbi.nlm.nih.gov/SNP/) or by using Entrez SNP (http://www.ncbi.nlm.nih.gov/sites/entrez?db=snp). Investigator identified SNPs may also be submitted to the dbSNP database as either genomic DNA or cDNA sequences. Once a SNP and surrounding sequence has been submitted, it is assigned a unique submitted SNP ID number. If several SNPs are mapped together, then it is called a "reference SNP cluster" or refSNP.

12.3 Tissue Collection and Distribution Centers

12.3.1 United States

The Cooperative Human Tissue Network (CHTN) (http://chtn.nci.nih.gov) is a National Cancer Institute–supported prospective collection and distribution service of malignant and benign tissues and fluids obtained from six academic institutions. Anatomic sites include, but are not limited to, breast, uterus, ovary, colon, stomach, esophagus, bladder, kidney, prostate, liver, pancreas, spleen, lung, pharynx, oral cavity, brain, spinal cord, peripheral nerves, muscle, skin, and soft tissue. To receive a sample, investigators must complete and submit a CHTN application and a Data Use Agreement and provide a copy of an Institutional Review Board letter of approval. Tissues are distributed to the widest group of investigators possible, with priority established in the following order: (1) peer-reviewed funded investigators, (2) investigators developing new projects in academic and nonprofit centers, and (3) other investigators including those in for-profit research centers. Because it is a prospective service, the collection and distribution of tissue samples may be unpredictable and is dependent on surgery and autopsy schedules. CHTN is also associated with providing slides from tissue microarrays of both malignant and nonmalignant tissues (http://chtn.nci.nih.gov/tissue-microarrays/). Tissue arrays currently available include the human nervous system and ovarian carcinoma, whereas previous available tissue arrays included normal human tissues, colorectal carcinoma, and breast cancer.

The Biopathology Center (BPC) maintains both the Children's Oncology Group and the Gynecological Oncology Group tissue banks (http://www.nationwidechildrens.org/biopathology-center-resources). Investigators may request slides from tissue microarrays from a variety of pediatric tumors, including alveolar rhabdomyosarcoma, embryonal rhabdomyosarcoma, rhabdoid tumor, low-grade glioma, medulloblastoma, Ewing's sarcoma, osteosarcomas, clear cell sarcoma of the kidney, neuroblastomas, Wilms' tumors, and Hodgkin's lymphoma. Available gynecologic slides from tissue microarrays are available for ovarian carcinomas and endometrial carcinomas. Available tissue microarrays, number of unique cases, number of controls, and number of blocks may be found on their Web site (http://www.nationwidechildrens.org/biopathology-center-resources). If needed, additional specimens from the BPC may be obtained after completing a Tissue Update Form that is available on their Web site. To obtain access to these arrays, the investigator needs to complete an application and Data Use Agreement as well as provide a project description, Institutional Review Board approval, and possible material transfer agreements. In addition, several cores and other resources are available at the BPC. The cell line core will extract DNA from saliva, whole blood, and skin biopsy samples as well as prepare DNA and cell lines for archiving. The biostatics core assists researchers with study design, data analysis, and interpretation of statistical results. The Flow Cytometry Core assists researchers for flow analysis and/or cell sorting in addition to providing access to FlowJo analysis software for data analysis. The epidemiology core will assist researchers with selection of the appropriate study design, selecting appropriate populations and control groups, as well as assisting with the selection of appropriate data sets for secondary analysis and performing basic statistical analysis and interpretation of your data. Instructions on how to order services from these cores is provided on their Web site.

12.3.2 Europe

The Organisation of European Cancer Institutes TuBaFrost is a searchable catalogue of European tissue banks including frozen tissues, blood samples, cell lines, xenografts, and tissue microarrays associated with cancer. Currently, there are more than three million samples in the catalogue. The resource is available to investigators whose institutions are part of the Organisation of European Cancer Institutes or participates in the ErocaPlatform project or cooperates with such members.

12.4 Future Direction

Radiogenomics faces many obstacles to accomplish its aims of developing an assay(s) that can predict an individual's toxicity to radiation therapy. Current problems include normalizing differences in radiation techniques between different centers, normalizing different radiation toxicity scoring approaches, determining the best time point to assess toxicity (acute versus late), and determining the best way to combine and standardize data from multiple studies. Theoretically, this assay should increase the therapeutic ratio of radiation in individual patients by allowing dose escalation in radioresistant patients as well as identifying radiation-sensitive patients and treating them with alternative means.

12.5 Conclusion

Radiogenomics is the process of studying how genetic variability influences sensitivity to radiation therapy. GWAS are increasing the understanding of how genetic variation may influence radiation efficacy and toxicity. The genetic markers of radiation sensitivity discovered today may potentially be used in the future to individualize patient radiation treatment.

References

Alsbeih, G. et al. (2010). Association between normal tissue complications after radiotherapy and polymorphic variations in TGFB1 and XRCC1 genes. *Radiation Research, 173*(4), 505–511.

Andreassen, C. N. et al. (2006). Risk of radiation-induced subcutaneous fibrosis in relation to single nucleotide polymorphisms in TGFB1, SOD2, XRCC1, XRCC3, APEX and ATM—a study based on DNA from formalin fixed paraffin embedded tissue samples. *International Journal of Radiation Oncology Biology Physics, 82*(8), 577–586.

Anscher, M. S. et al. (2010). The negative impact of stark law exemptions on graduate medical education and health care costs: The example of radiation oncology. *International Journal of Radiation Oncology Biology Physics, 76*(5), 1289–1294.

Aykin-Burns, N. et al. (2011). Sensitivity to low-dose/low-LET ionizing radiation in mammalian cells harboring mutations in succinate dehydrogenase subunit C is governed by mitochondria-derived reactive oxygen species. *Radiation Research, 175*(2), 150–158.

Barnett, G. C. et al. (2009). Normal tissue reactions to radiotherapy: Towards tailoring treatment dose by genotype. *Nature Reviews Cancer, 9*(2), 134–142.

Burnet, N. G. et al. (2006). Radiosensitivity, radiogenomics and RAPPER. *Clinical Oncology (The Royal College of Radiologists), 18*(7), 525–528.

Burri, R. J. et al. (2008). Association of single nucleotide polymorphisms in SOD2, XRCC1 and XRCC3 with susceptibility for the development of adverse effects resulting from radiotherapy for prostate cancer. *Radiation Research, 17*(1), 49–59.

Delaney, G. et al. (2005). The role of radiotherapy in cancer treatment: Estimating optimal utilization from a review of evidence-based clinical guidelines. *Cancer, 104*(6), 1129–1137.

Emami, B. et al. (1991). Tolerance of normal tissue to therapeutic irradiation. *International Journal of Radiation Oncology Biology Physics, 21*(1), 109–22.

Hendry, J. H. & West, C. M. (1997). Apoptosis and mitotic cell death: Their relative contributions to normal-tissue and tumour radiation response. *International Journal of Radiation Oncology Biology Physics, 71*(6), 709–719.

Ho, A. Y. et al. (2006). Genetic predictors of adverse radiotherapy effects: The Gene-PARE project. *International Journal of Radiation Oncology Biology Physics, 65*(3), 646–655.

Kerns, S. L. et al. (2010). Genome-wide association study to identify single nucleotide polymorphisms (SNPs) associated with the development of erectile dysfunction in African-American men after radiotherapy for prostate cancer. *International Journal of Radiation Oncology Biology Physics, 78*(5), 1292–1300.

Kleinerman, R. A. (2009). Radiation-sensitive genetically susceptible pediatric sub-populations. *Pediatric Radiology, 39 Suppl 1,* S27–S31.

Mangoni, M. et al. (2011). Association between genetic polymorphisms in the XRCC1, XRCC3, XPD, GSTM1, GSTT1, MSH2, MLH1, MSH3, and MGMT genes and radiosensitivity in breast cancer patients. *International Journal of Radiation Oncology Biology Physics, 81*(1), 52–58.

Michikawa, Y. et al. (2010). Genome wide screen identifies microsatellite markers associated with acute adverse effects following radiotherapy in cancer patients. *BMC Medical Genetics, 11,* 123.

Niu, N. et al. (2010). Radiation pharmacogenomics: A genome-wide association approach to identify radiation response biomarkers using human lymphoblastoid cell lines. *Genome Research, 20*(11), 1482–1492.

Suga, T. et al. (2007). Haplotype-based analysis of genes associated with risk of adverse skin reactions after radiotherapy in breast cancer patients. *International Journal of Radiation Oncology Biology Physics, 69*(3), 685–693.

Travis, L. B. et al. (2012). Second malignant neoplasms and cardiovascular disease following radiotherapy. *Journal of the National Cancer Institute, 104*(5), 357–370.

Werbrouck, J. et al. (2009). Acute normal tissue reactions in head-and-neck cancer patients treated with IMRT: Influence of dose and association with genetic polymorphisms in DNA DSB repair genes. *International Journal of Radiation Oncology Biology Physics, 73*(4), 1187–1195.

West, C. M. & Barnett, G. C. (2011). Genetics and genomics of radiotherapy toxicity: Towards prediction. *Genome Medicine, 3*(8), 52.

West, C. et al. (2010). Establishment of a radiogenomics consortium. *International Journal of Radiation Oncology Biology Physics, 76*(5), 1295–1296.

III

Informatics for Teaching and Research

13

Teaching Support

Joann I. Prisciandaro
University of Michigan

13.1 Introduction

Studies have shown that there is more to the learning process than simply presenting material to an audience. In 1956, a committee of college examiners led by Benjamin Bloom identified three domains of educational activities: the cognitive, the affective, and the psychomotor [1]. The cognitive domain is traditionally emphasized in education. Within this domain, Bloom et al. [1] identified six hierarchical levels of educational goals (the taxonomy of educational objectives): knowledge, comprehension, application, analysis, synthesis, and evaluation (Figure 13.1).

At the foundation of the cognitive domain is knowledge. Knowledge simply involves the recall of information previously presented to an audience. At this level, due to their limited exposure to the new material, the learner does not have a full appreciation and understanding of the material. With time, the learner begins to comprehend the material and can summarize and translate this material. The next step in the learning process is application, translating this newly acquired knowledge into practice. Once the learner can begin applying this idea, he/she can analyze the elements of the knowledge received and recognize unstated assumptions, such as identifying patterns or organization of the material. At the next level, synthesis, the learner can begin to piece together elements of the information that have been conveyed to formulate new ideas and hypotheses. Lastly, at the pinnacle of Bloom et al. [1], educational hierarchy is evaluation, the ability of the learner to make individual value judgments regarding the knowledge they have received.

The purpose of Bloom et al. [1] taxonomy is to assist educators in developing curricula based on their intended educational objectives and to appropriately evaluate students based on these objectives. In the information age, there are many different venues available to convey knowledge to our students; however, it is important to consider whether these forums are the most appropriate based on our educational objectives and expectations.

13.1.1 Classroom-Based Learning

Traditionally, teaching has been performed face-to-face in a classroom setting. This technique is still largely used today because face-to-face discussions allow individuals, namely, peers, an opportunity to easily interact, socialize, and exchange their ideas [3]. In addition, and probably more importantly, we continue to use this form of learning because we are accustomed to it. However, there are inherent issues with classroom-based learning, including limitations for self-directed and self-paced learning, and the customization of the learning experience for individual learners [3]. Some educators have tried to address these issues by altering their approach to learning from the classic instructor-centric to a learner-centric model.

13.1.1.1 Instructor-Centric Learning

The most traditional forum for teaching is an instructor-centric model. The instructor maintains the reins of the learning environment, that is, the content of the course, the pace in which the material is presented, and its mode of delivery [4]. In this educational approach, several assumptions are typically made. First, all students have a similar level of knowledge in a given topic. Second, knowledge is imparted from instructor to student [5]. Third, students will progress at a similar pace [6].

13.1.1.2 Learner-Centric Learning

As many educators can attest, the assumptions made in an instructor-centric or purely lecture-based educational approach

121

Bloom's taxonomy for thinking

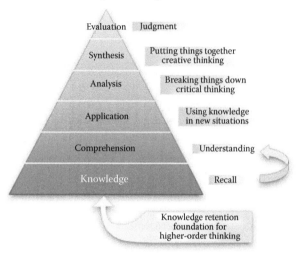

FIGURE 13.1 Bloom's taxonomy of the cognitive domain. (From http://www.ehow.com/how_5108411_apply-blooms-taxonomy.html, retrieved January 5, 2011.)

are flawed. Many institutions complement lectures with smaller breakout sessions, such as recitations, to give students the opportunity to ask questions and work at their own pace. Increasingly more educators are advocating a learner-centered approach in which education is perceived as a shared responsibility [5–7]. In this model, students are expected to play an integral and active role in their education, whereas the instructor acts as a facilitator of the learning process. The rationale for a learner-centered educational model is based on five premises [6]:

1. "Learners are distinct and unique."
2. One must take into account the differences among learners to provide all learners "with the necessary challenges and opportunities for learning and self-development."
3. Learning occurs best when the material covered is relevant and meaningful to the learner, and the learner is actively engaged.
4. "Learning occurs best in a positive environment."
5. "Learners are naturally curious" and interested in learning about their world.

The learning environment in a learner-centric model may be independent, collaborative, cooperative, or competitive [4]. Eric Mazur presents an example of a collaborative approach to education in "Peer Instruction" [8]. In Mazur's approach, the lecture format is not entirely abandoned, but it represents a considerably smaller fraction of the overall class time. Students are expected to prepare for their class by completing pre-class reading assignments. Once in class, students are presented with a series of short presentations on key concepts. After each presentation, several short conceptual questions are posed to the students. The students are given time to discuss the questions and their answers. Students are then given time to try to convince their neighbors that their answers are correct by presenting their rationale

followed by additional time to revise their answers. The answers are then tallied, and an explanation of the correct answer is presented by either the instructor or a student. Although this approach may require some additional preparatory work compared with the traditional lecture format, Mazur contends that the course material becomes more accessible to students if you engage them in the learning process.

13.1.2 Electronic and Blended Learning

With the widespread availability of the Internet, many educators are turning toward electronic means to enhance the educational experience of their courses. The advantage of electronic learning (e-learning) is that it allows students to learn anytime, anywhere, as long as they have access to the Internet [3]. There are nominally two e-learning styles, asynchronous and synchronous. Asynchronous e-learning may involve facilitated or self-directed instruction [3]. Facilitated asynchronous training involves a facilitator (typically the instructor) and a group of learners who interact and communicate online but not in real time. Self-paced e-learning does not involve interactions between the instructor, the learner, or his/her peers. Typically, information is disseminated through the review of prepared learning modules, which can be reviewed at one's own pace. On the contrary, synchronous e-learning involves real-time interactions between the instructor and the learner via the Internet [3].

Electronic and the traditional classroom-based learning (whether it be instructor or learner-centric) need not be mutually exclusive. The extent in which the electronic element is incorporated may be considered as a continuum ranging from no computer or Internet aids involved in the learning process, to hybrid learning where classroom time is reduced with the addition of online learning (Web enhanced/supplemented), to fully online learning (Web dependent), such as distance learning [9].

To maximize the learning experience, instructors may consider a blended learning approach. Although instructors may have different views and approaches to blended learning, it is commonly defined as a combination of different instructional strategies such as classroom and e-learning [3, 10]. Regardless of the extent in which one hopes to incorporate electronic media into their instruction, it is imperative that educators become familiar with these tools and their potential impact.

13.2 Information Technology Resources

There are a multitude of technological resources available that may be used for educational purposes. The extent by which these technologies are incorporated into educational endeavors may vary from instructor, based on their IT access and comfort using these tools. However, a brief description of these tools is provided below.

13.2.1 Learning Management Systems

A Learning Management System (LMS) is a software application that administers instructor-based or online courses and keeps track of the progress of the target audience [11], which may range from students, employees, customers, and business partners, depending on the content of the material presented. As more institutions look for online solutions to enhance the learning process (i.e., by offering online courses, providing supplemental study material or support), there is an increasing need for LMS applications. Several signatures of a robust LMS are its ability to centralize and automate administration, assemble learning content, and consolidate training initiatives [12]. Presently, two types of LMS applications are available: proprietary and open source.

Proprietary LMS applications are systems that involve encrypted source code. Therefore, users are legally restricted from modifying the code on their own (e.g., BlackBoard and WebCT) [13]. As such, some argue that proprietary software provides an added level of security; however, even proprietary applications are not free from software bugs or glitches. In contrast, open-source LMS applications allow end users to modify or customize the source code (e.g., Sakai and Moodle). There are advantages and disadvantages to both proprietary and open-source LMS, and each institution will need to weigh its options carefully before deciding which LMS option to select. Ultimately, the decision may be dictated by cost. After considering the LMS applications that are available, one must ask oneself whether an out-of-the-box product (possibly with additional customization performed by the vendor) will sufficiently meet one's needs or whether the institution has the resources to manage and customize a LMS system, in addition to providing end-user support (i.e., is there a dedicated IT team that can manage the program) [14]. Although licensing fees are associated with proprietary LMS software, depending on the size of the institution and the available IT resources, this may be more cost-effective than maintaining an open-source LMS application.

When considering LMS applications for medical physics educational programs, namely, residency programs, cost, functionality, and software availability play a significant role in influencing which LMS application may be used. Although many universities provide software through their Graduate Medical Office, oftentimes they are designed for physician residency programs and are either unavailable or may be ill suited for medical physics residency programs [15]. A review of four commercially available LMS applications has been performed by Zacarias et al. [15] specifically for their utility in affiliated medical physics residency programs. Based on their study, Typhon Group software (Typhon Group LLC, Metairie, Louisiana) was considered a viable option.

13.2.2 Web Conferencing

Web conferencing or Web seminars (webinars) are an online, interactive meeting forum that allows multiple participants to communicate over the Internet either directly using a Web application or by downloading an application to the participants' personal computers before the meeting [16]. Web conferencing allows participants an opportunity to share audio, video, and files in real time. This tool has become popular in both the academic and business sectors, because it provides a useful and convenient means of communicating while minimizing the cost in time and funds to travel to a central location. At present, there are a number of vendors that provide Web conferencing solutions including Adobe Acrobat Connect, Citrix GoToMeeting, Elluminate Live!, IBM Lotus SametimeUnyte Meeting, Microsoft Office Live Meeting, WebEx MeetMeNow, Saba Centra, and ILinc [17]. Although most of these products are similar in nature, when determining the right product for one's use, one should compare their cost and functionality.

13.2.3 Social Networking

Social networking is "a system of social interactions and relationships" [18] among a group of individuals connected by a commonality such as a friendship, kinship, belief, or interest [19]. Although the concept of social networking has been around far longer than the Internet, the Internet has created new and seemly limitless potential for communication and collaboration between individuals who typically would not have the opportunity to meet or interact [20]. Users typically post their profiles and personal networks on social networking sites such as Facebook, Twitter, Windows Live Profile, LinkedIn, and MySpace. The rationale behind social networking sites is to provide individuals a new means of communicating and to display their acquaintances and how they are connected. By sharing one's profile and connections, their acquaintances can better understand the individual and possibly make new connections to other individuals that share a common interest. In essence, social networking sites have created virtual communities, changing our primary means of communication and moving our personal lives into a public forum [21].

According to a recent report published by comScore, social networking is the most popular global, online activity, accounting for nineteen percent of all time spent online [22]. In October 2011, it was found that "nearly 1 in every 5 minutes spent online is now spent on social networking sites," thus indicating the growth of social networking as truly a "global cultural phenomenon" [22]. With the pervasiveness of such sites, one should not underestimate their role in fostering personal communication and in changing the approach toward fields such as marketing and education.

Students today are "digital natives"; they have grown up with the Internet and largely incorporate social networking into their daily routines [21, 23, 24]. Because of this, to successfully engage students, educators must adapt. One approach is to tap into social networking sites and revise our approach to education. Integrating social networking sites into course curricula may enhance, and promote collaborative learning, and improve both faculty-student and student-student interactions [21].

13.2.4 Podcasts

A podcast is an audio recording that has been posted to a Web site for download to digital audio players such as an iPod or any other MP3 player [25] or one's PC. The term "podcast" was coined as a combination of "iPod" and "broadcast." Although podcasts were originally audio, they have evolved to include images (photofeed) and video (vodcasts/vidcast) [26]. Because their content is pre-recorded, podcasts can be paused, fast-forwarded, and rewound [25].

Podcasts have become a popular media for businesses to promote products and for the average person to share amateur content on social networking Web sites [27]. However, podcasts are also gaining popularity in the academic setting, as instructors and students seek alternative teaching aids. A host of Web sites share educational videos and lectures, typically free of charge, such as Academic earth, iTunesU, TeacherTube, and yovisto.

13.2.5 Wikis

A wiki (Hawaiian for "quick") is an open and collaborative Web site in which any visitor may participate [28, 29]. The site content can be edited or created by any participant; consequently, the material can be continuously revised. The beauty of a wiki is its ability to track the history of the document as it is revised, and if necessary, allow an author to restore an older version of the document. The first wiki site (WikiWikiWeb) was created by Ward Cunningham in 1995, and since then, numerous wiki Web sites and programs have been developed. In the realm of education, wikis have a great deal of potential for improving collaborations between instructors, students, and between instructors and their students. Wikis may be shared between educators to develop and improve lesson plans and notes, between students to develop study guides and develop and share information related to group projects, and between an instructor and his students to convey and/or correct any misconceptions related to a lesson plan, class notes, or a newly covered concept [30].

13.2.6 Blogs

A blog (a portmanteau of "Web" and "log") is a Web site containing an online journal with regular posts chronicling an individual's activities and/or commentary of events. Unlike a typical Web site, a blog allows members of the public to upload their comments on a specific post. Since blogs first appeared in the 1990s, they have evolved beyond just serving as personal journals. Although personal blogs are still very popular, group blogs, authored by multiple contributors, are becoming increasingly popular. Blogs vary in style and content and cover the gamut of social and political issues [31]. As an instrument for education, blogs can serve as a means of communication between instructors and students, parents, and peers in addition to providing a forum for students to communicate with one another and learn [32]. There are a number of blogging platforms available online, such as Blogger.com and WordPress.com, as well as EduBlogs. org, which is advertised as an education blogging platform focused on providing educational resources for both teachers and students.

More recently, microblogging has been gaining popularity. Compared with traditional blogs, microblogs are brief posts (typically 140 to 200 characters) and can easily be written or received on a variety of portable electronic devices, including cell phones [33]. Currently, one of the most popular microblogging services available is Twitter. Posts on this service are referred to as tweets, and, microblogs are commonly also known as tweets.

13.2.7 Rubrics

A rubric (scoring rubric) is an explicit set of criteria or scoring guidelines used to clarify expectations and assess an individual's performance on a given project [34, 35]. A rubric contains a description of expectations at each level and the corresponding gradation level [36]. Rubrics can be used for grading purposes in addition to help students gain an appreciation for the quality of their own work in comparison with their peers and techniques for improvement [37]. Rubrics may be used by instructors to evaluate students, students to evaluate peers, and as a self-assessment tool. The two common rubric types are analytic and holistic. An analytic rubric is process oriented; it separates a project into individual activities and assigns a grading scheme to each component [36]. On the contrary, holistic rubrics are more product oriented; the grading scheme rates a project in its entirety [36].

Rubrics have become popular with educators as a means of communicating expectations, providing feedback on works-in-progress and grading final product [37]. There are numerous rubrics Web sites that provide examples and/or templates including rubistar, irubric, rcampus, teAchnology, and squidoo.

13.3 Educational Resources

The development of the World Wide Web in the 1990s has revolutionized many aspects of human society and our system of socialization. Not only has the Internet provided us with a means of accessing instant, electronic information and entertainment resources, it has also provided us with a new means of communication. By pursuing various search engines, Web sites can easily be identified on a desired topic(s) of interest. For this chapter, a search of available educational resources was performed. A summary of interesting and useful educational resources is provided next; however, the author would like to remind readers that this list is by far not comprehensive due to the ever-changing and evolving nature of the Internet.

13.3.1 MIT Open Course Ware [38]

The Massachusetts Institute of Technology (MIT) Open Course Ware (OCW) is a free and openly licensed digital publication of educational material. Starting in 2000, the MIT faculty proposed the OCW concept as a means to advance knowledge and educate students. The OCW Web site allows students and faculty members at MIT and other institutions access to MIT syllabi, lecture notes, assignments and solutions, exams and solutions, and multimedia content (i.e., video lectures). Starting in 2002, MIT published course material for fifty courses online, and as of 2010, this number has expanded to 2000 courses.

13.3.2 OCW Consortium [39, 40]

In 2002, a worldwide community of 250 higher education institutions and associated organizations joined together to form the OCW consortium. This nonprofit organization is committed to advancing OCW sharing and expands global educational opportunities. Collectively, the consortium has published material from more than 13,000 courses. Similar to the MIT's OCW, the intention of publishing this material is to assist instructors, especially those with limited resources, in teaching a given topic and as a study guide or refresher for students.

13.3.3 Creative Commons [41]

Creative Commons (CC) is a nonprofit organization that was founded in 2001 to provide a legal and technical infrastructure to maximize "digital creativity, sharing, and innovation [41]." Customizable intellectual property licenses and tools are available free of charge on the CC Web site to assist artists, writers, programmers, and others define acceptable use of their work [42]. Lawrence Lessig, one of the founders of CC, has likened this to a "some rights reserved" approach to copyright [42].

The commons are a "body of work freely available to the public for legal use, sharing, repurposing, and remixing" [41]. The commons are classified based on a series of projects or areas of interest. At present, these projects include Culture, Education, Science, and Government. By selecting one of these projects, you are provided with a list of resources available in the public domain in that area of interest.

13.3.4 Medical Physics Educational Resources [43]

13.3.4.1 List Servers

A list server is an E-mail management software application that handles subscription requests for a mailing list and distributes messages, newsletters, and other posts from members of the list to the entire list [44, 45]. List servers provide a useful forum for discussion and exchange of ideas. To participate (send or receive a post) a user must subscribe to a specific list server. Mailings are received by participants of the list either in real-time as they are received or batched into what is typically called a digest and distributed to the group on a regular basis (i.e., daily, weekly, or monthly). The first and one of the most popular list servers available is LISTSERV [46]. However, other list server programs such as Majordomo and GNU mailman are also available.

There are a number of official mailing lists that are geared specifically for medical physicists. These include the following:

- DXIMGMEDPHYS—Diagnostic imaging medical physicist list

- MEDPHYSUSA—The American medical physics mailing list
- MEDPHYS—The global medical physics mailing list
  - Unlike MEDPHYSUSA, this mailing list is intended to be a forum of communication for the international medical physics community [47].
- MEDPHYSBOARDPREPARATION—Medical physics board preparation study group
  - This list server is intended to attract and assist medical physics trainees that are in the process of taking their medical physics board exams [48].

13.3.4.2 Professional Organization Web Sites

1. AAPM [49]

The American Association of Physicists in Medicine (AAPM) is a nonprofit organization dedicated to the scientific, educational, and professional advancement of physics in medicine. To achieve this end, the AAPM Web site has a host of resources available to its members. Links are available to journals such as *Medical Physics*, the *Journal of Applied Clinical Medical Physics*, and *Physics Today*. In addition, the AAPM has more than 100 reports summarizing the finding of specially convened task groups and working groups on clinical, scientific, professional, and educational topics available to download free to its members.

The AAPM's Virtual Library maintains copies of recorded presentations that were given at annual AAPM meetings or specialty medical physics meetings. The recordings include streaming video and/or audio of the presenter, slides, and/or audio transcriptions. The vast material contained on this site is well worth a visit, especially when one considers the breadth of the field of medical physics and the rapid introduction of new technology and procedures into the clinical setting.

For an additional fee, the Online Learning Center (OLC) is available for AAPM members to aid in the Maintenance of Certification (MOC) process. The OLC offers online self-assessment modules (SAMs) presentations and quizzes, allowing users to obtain online SAMs credits.

The AAPM Educators Resource Guide provides resources such as curriculum guidance, online modules, and references to medical physicists involved in education. Although the guide is currently organized into five categories, at present, the most developed is the Physics Education for Diagnostic Radiologists and Residents. This guide provides links to the AAPM/Radiological Society of North America (RSNA) physics tutorials [50], PowerPoint presentations, and online modules, such as Dr. Perry Sprawls' online resources for learning and teaching the physical principles of medical imaging [51].

There are efforts under way to improve the online didactic content available on the AAPM Web site. Several subcommittees and task groups have been charged with this endeavor, including the Online Learning Services Subcommittee and Task Group 206.

2. ABRF [52, 53]

The American Board of Radiology Foundation (ABRF) is an independent, nonprofit organization whose mission is to demonstrate, enhance, and continually improve accountability in the use of medical imaging and radiotherapy. To achieve this end, the ABRF has begun hosting annual summits and has worked to develop a professionalism series.

The intention of the annual summit is to address national healthcare challenges such as addressing overutilization of medical imaging (2009) and improving patient care through electronic communication in imaging (2010). A summary of the summits, in addition to select presentations, is available for free downloads on the ABRF Web site.

The professionalism series resulted in the development of online ethics and professionalism modules. The modules were developed by a team of experts and the content has been peer reviewed. The development of these modules was financially supported by the ABRF, the Academy of Radiology Research, the AAPM, the American Board of Radiology, the American College of Radiology, the American Radium Society, the American Society for Radiation Oncology (ASTRO), and the RSNA and are available free to members of any of these organizations. The modules are self-guided and include tests and practicums to allow users to assess their comprehension of the material. At present, ten modules are available covering the following topics:

• Attributes of Professions and Professionalism
• Physician/Patient/Colleague Relationships
• Personal Behavior and Employee Relationships
• Conflicts of Interest
• Ethics of Research
• Human Subjects Research
• Research with Animals
• Relationships with Vendors
• Publication Ethics
• Ethics of Education

3. ASTRO [54]

The American Society for Radiation Oncology (ASTRO) is an independently managed organization that is dedicated to improving the quality of patient care through education, clinical practice, research, and advocacy. Although the vast majority of its members are radiation oncology physicians, medical physicists make up approximately seventeen percent of ASTRO's total membership [55]. Consequently, the majority of the educational resources available on the ASTRO Web

site are designed for physicians. With this said, there are still some online educational tools that are relevant to medical physicists. These include SAMs, webinars, and virtual meetings, which are available for a nominal fee.

4. ESTRO [56]

The European Society for Radiotherapy and Oncology (ESTRO) is an international organization dedicated to the advancement and support of education, research, and networking across all areas of radiation oncology. ESTRO's membership has an international and multidisciplinary flair, spanning five continents and including specialists involved in all aspects of the multimodality treatment of cancer. To support the educational needs of their members and the radiotherapy and oncology community, ESTRO has published a series of useful resources on their educational portal. This includes curricula, guidelines, publications, and e-learning tools. There are two online publications that may be most interesting for medical physicists. The first is the GEC-ESTRO Handbook of Brachytherapy [57]. The Handbook of Brachytherapy provides readers with a comprehensive summary of clinical presentations and brachytherapy techniques used to treat various anatomical sites by the most innovative European teams, basic principles of physics, radiobiology, and imaging. The second is the ESTRO physics series. The series consists of ten booklets on an array of clinically used medical physics topics such as in vivo dosimetry, quality assurance, monitor unit calculations, and intensity-modulated radiation therapy.

To minimize travel and expense, ESTRO also offers a series of e-learning resources. ESTRO's Application for Global Learning (EAGLE) provides video and audio material, presentations, text, and virtual classroom sessions. In addition, the Fellowship in Anatomic Delineation and Contouring (FALCON) and the Tutorial for Image Guided External Radiotherapy (TIGER) provide presentations, lessons, and online course to help guide and validate the contouring techniques of professionals such as physicians, physicists, and dosimetrists.

5. IAEA

The International Atomic Energy Agency (IAEA) is an international organization that seeks to promote safe, secure, and peaceful use of nuclear technology [58, 59]. The IAEA's mission is guided by the interests and needs of its member states [58]. Of particular interest to the medical community is the Division of Human Health within the Department of Nuclear Sciences and Applications. The objective of the Division of Human Health is to address the needs of member states related to the use of nuclear technology to prevent, diagnosis, and treat health-related issues [60]. To address these needs, the Division is further subdivided into four sections: (1) Nuclear Medicine, (2) Applied Radiation Biology and Radiotherapy, (3) Dosimetry and Medical Radiation Physics, and (4) Nutritional and Health-Related Environmental Studies. Each section Web site provides useful links, a frequently asked questions page, and free resources related to the medical application of nuclear technology. This includes a number of educational resources such as the following:

- Radiation Oncology Physics: A handbook for teachers and students [61]
- Radiation Oncology Physics Slides [62]
- Radiation Biology: A handbook for teachers and students [63]
- Distance learning course in radiation oncology for cancer treatment [64]

Another section of interest on the IAEA Web site is the Radiation Protection of Patients Web site [65]. The intention of this Web site is to disseminate information to help healthcare professionals safely use radiation in medicine. A discussion on standards and prevention of errors is provided for professionals specializing in areas such as radiology, radiotherapy, nuclear medicine, interventional fluoroscopy, and interventional cardiology. Other publications such as safety guides, safety reports, technical documents, and radiologic accidents are available for free downloads under their additional resource page [66]. In addition, this site provides free training modules on the following topics [67]:

- Diagnostic and interventional radiology
- Radiotherapy
- Nuclear medicine
- Prevention of accidental exposure in radiotherapy
- Cardiology
- Positron emission tomography/computed tomography
- Pediatric radiology
- Digital radiology

6. ICRP [68]

The International Commission on Radiological Protection (ICRP) is an independent international organization that provides recommendations on the safe use of ionizing radiation. These recommendations are intended to provide assistance to the appropriate regulatory agencies of individual countries to develop radiologic protection standards, legislation, guidelines, programs, and codes of practice.

The ICRP has published more than 100 reports regarding radiation protection that are available electronically for a fee or to subscribers of the Annals of the ICRP. In addition, this material may be downloaded either free or at a discounted rate to developing countries. In addition to their reports, the ICRP has a number of summary recommendations, guides and explanatory notes, and educational material available free to download from their Web site. These downloads include presentations summarizing the following ICRP reports:

- 84: Pregnancy and medical radiation
- 85: Avoidance of radiation injuries from medical interventional procedures
- 86: Prevention of accidents to patients undergoing radiation therapy
- 87: Managing patient dose in computed tomography
- 93: Managing patient dose in digital radiology
- 112: Preventing accidental exposures from new external beam radiation therapy technologies

7. ICRU [69]

The International Commission on Radiation Units and Measurements (ICRU) is an organization that establishes "internationally accepted recommendations on radiation related quantities and units, terminology, measurement procedures, and reference data for the safe" use of ionization radiation for medical and scientific applications. The ICRU continually reviews these areas and has focused on the development of recommendations for four programs: (1) diagnostic radiology and nuclear medicine, (2) radiation therapy, (3) radiation protection, and (4) radiation in science. To date, approximately ninety reports have been published and many are available electronically for a fee or to subscribers of the Journal of the ICRU.

8. NCRP [70]

The National Council on Radiation Protection and Measurements (NCRP) was chartered by the U.S. Congress in 1964 to develop and disseminate information, guidance, and recommendations on radiation protection and measurements. The Council was originally known as the Advisory Committee on X-ray and Radium Protection, which was established in 1929. Although chartered by Congress, the Council is a nongovernmental, public service organization that cooperates with other national and international organizations concerned with radiation protection. The NCRP has published more than 150 reports that are available for a fee either in hardcopy or electronically. In addition, the NCRP has also published commentaries, annual meeting proceedings and presentations, symposia proceedings, and lectures, which are also available for a fee.

9. RSNA [71]

The RSNA is a professional organization committed to excellence in patient care through education and research. To promote its educational mission, the RSNA Education portal was developed. Currently, RSNA Education is divided into eight subcategories of interest.

1. Online education
2. Maintenance of certification
3. Continuing professional development
4. Education center store

5. Resources for residents
6. Resources for medical students
7. Resources from RSNA annual meeting

Most relevant to medical physics education are the sections titled Online Education, Maintenance of Certification, Continuing Professional Development, Resources for Residents, and Resources from RSNA Annual Meeting. Online education offers in excess of 300 programs to members and nonmembers. For instance, refresher courses are available for viewing free of charge to members and registered nonmembers. However, one must be a member of the RSNA to earn CME credits. The Maintenance of Certification section is designed to assist members in their MOC process. This section provides links to the ABR, RSNA material that would allow one to earn CME credits, online SAMs, and tools to help one get started with practice quality improvement programs in radiology. Continuing professional development provides information pertaining to professional development such as a list of relevant imaging-based workshops and conferences, links to education material, and webinars designed to instruct SAMs faculty how to appropriately design a SAMs session. Resources for residents provides links to the RSNA/AAPM physics modules, the Web-Rad-Train radiation biology study guide [72], an online course on the business of radiology, and physics teaching files for radiology residents [73]. Lastly, resources from the RSNA annual meeting supplies links to sessions, award-winning exhibits and contributors, and electronic handouts from RSNA refresher courses.

10. UNSCEAR [74]

The United Nations Scientific Committee for the Effects of the Atomic Radiation (UNSCEAR) was established in 1955 by the General Assembly of the United Nations. Its mandate is to collect data and assess and report on the levels and effects of ionizing radiation. The Committee is composed of scientists representing twenty-seven countries. Since 1955, the Committee has published twenty reports, which are available for free downloads, in addition to short briefings of their reports for the media and general members of the public.

13.3.4.3 Miscellaneous Web Sites

1. EMERALD, EMIT, EMITEL [75–78]

In 1995, following the European Conference on Post-Graduate Education in Medical Radiation in Budapest, delegates from several European Union (EU) universities and hospitals engaged in a pilot project called European Medical Radiation Learning Development (EMERALD). The goal of this project was to develop training material, such as curricula and e-learning modules, to improve the training of medical physicists in the area of diagnostic radiology, nuclear medicine, and radiotherapy [75]. Shortly after the development of the initial modules, a second EU sponsored project was initiated and titled European Medical Imaging Technology (EMIT). Similar to EMERALD, the EMIT project was initiated to develop medical physics training material but, in this case, focusing on diagnostic ultrasound and magnetic resonance imaging [75]. Both projects continue to be developed and enhanced. At present, EMERALD II, a new and larger consortium and project has been organized. The training materials and workbooks are available online free of charge from the EMERALD II website (http://emerald2.eu/cd/Emerald2/).

The EMERALD II Web site also contains several other useful educational links, including EMITEL and MEP. The European Medical Imaging Technology e-Encyclopaedia for Lifelong Learning (EMITEL) is a free medical physics electronic encyclopedia and multilingual dictionary [79]. The dictionary can be used to define and translate terms into anyone of twenty-nine languages. The Medical Engineering and Physics (MEP) portal currently provides a link to two e-books. The first e-book, "Medical Radiation Physics from a European Perspective," is the conference proceedings from the 1994 European Conference on Post-Graduate Education in Medical Radiation, which describes the status of medical physics education in a number of European countries. The second e-book, "Medical Physics and Engineering Education and Training Part I," was published in 2011. This e-book is a collection of papers from educational conferences summarizing the experience of medical physics education and training both within and outside of the EU, including the experience of educators in African, Asia, and South American countries. The two e-books are available for free downloads online, and a third e-book, a continuation of the 2011 publication, is planned to be released in 2012 to 2013 [80].

2. RRTL [81–83]

The Remote Real-Time Learning (RRTL) project was developed with the expressed interest of developing a means of providing medical physics didactic training to physicists living in small communities and/or remote regions of the world, where assess to formal, classroom-based learning might be limited or nonexistent. Through a collaborative effort between the Department of Medical Physics at the Toronto-Sunnybrook Regional Cancer Centre and the Department of Radiology at the University of Malaysia, a pilot project was initiated and is ongoing. The objective of the RRTL project is to develop a simple, widely accessible, and cost-effective method to allow real-time interactive education and consultation via the Internet. This project has been piloted with a class of medical physics graduate students at the University of Malaysia, and the results are promising, thus given credence to the viability and economic feasibility of real-time interactive remote education.

3. Stanford Dosimetry Training Tool [84–86]

To address the national shortage of medical dosimetrists, the Stanford Dosimetry Training Tool (DTT) was made available to registered users. The site was supported by a National Cancer Institute training grant received by Arthur Boyer and then at Stanford University. This Web-based training program consisted of a series of twenty-four didactic modules covering topics such as clinical oncology, anatomy, fundamental physics, imaging, and treatment planning. To assess the competency of the user, a pre- and post-quiz was administered. Unfortunately, due to a lack funding, the DTT program was retired in June 2011.

4. Computer-Based Learning Modules for Clinical Dosimetry [87–90]

A new computer-based learning initiative is under way at the Department of Radiation Oncology at the University of North Carolina School of Medicine. A team headed by Dr. Robert Adams has developed a series of educational modules that would allow students to develop and hone practical, hands-on skills using a treatment planning system (PLanUNC). The modules consist of didactic and interactive components and are designed to allow students to apply their newly acquired knowledge in topics such as target delineation, dose calculations, electron dosimetry, treatment planning (beginning with two field plans), beam modifiers, and anatomic planning considerations [88, 89]. Short quizzes are given to the students pre- and post-module completion to assess their level of comprehension and retention [89]. The modules are currently geared to radiation therapy and dosimetry students; however, their utility for radiation oncology and medical physics residents will be evaluated at a later date. The Computer-Based Learning Modules for Clinical Dosimetry is expected to be available online in mid or late-2013.

5. VERT [91–98]

In response to a shortage of clinical resources and equipment time available for the training of staff and students in radiotherapy, a team of two computer scientists and one medical physicist from the University of Hull and the Princess Royal Hospital embarked on a research project in 2001 that culminated in a product known as Virtual Environment for Radiotherapy Training (VERT) (© Vertual, East Yorkshire, UK) [91, 98]. VERT is an immersive, life-size, virtual environment that allows staff and students to enter a virtual linac suite and practice setting up a virtual patient and using radiotherapy equipment [91, 92]. The virtual

linac modeled in VERT has all of the movement and the majority of the functionality of a true linac. Users may choose from anatomic datasets provided by the vendor or import anatomized images via a Digital Imaging and Communications in Medicine-Radiotherapy import tool to model a patient. In addition to providing a visualization of the radiotherapy treatment vault, VERT includes [91]:

- Anatomic views of the patient on the patient table, including an internal view of patient to investigate the relationship of isocenter with the planning target volume and organs at risk
- Visualization of the treatment beams
- Visualization of the radiation dose distribution (i.e., isodose surfaces)
- Collision detection
- Automated skin marking tools
- Tools to quantify setup errors

Recently, the scope of the VERT system has been expanded to include a range of medical physics equipment such as a [97]

- Scanning water phantom
- Solid water QA block/ion chamber
- Light/radiation coincidence phantom
- Laser alignment phantom
- Water-based calibration phantom

VERT can be operated in demonstrator or virtual simulator mode depending on the desired utility [91]. In simulator mode, the 3D glasses worn by the user interfaces with the computer/projection system and tracks the position of the observer. As a result, the system creates an observer's eye view of the virtual linac suite and "allows the user to 'walk around' an object, viewing it from different aspects" [93].

The VERT system consists of a 3D projector, screen, 3D glasses, high-performance computer(s), audio system, and linac hand pendant(s) that controls the linac and treatment couch [87, 93]. Although VERT can be tailored to accommodate the space available at a given facility, to truly appreciate the virtual experience, VERT should be projected at a life-size scale [93, 96].

At the time of writing, VERT has been installed in eighty-one facilities within fourteen countries worldwide. These installations include twenty-eight fully immersive VERT systems and fifty-three seminar style systems [96].

13.4 Conclusions [43]

The advent of the World Wide Web and multimedia technology has transformed how instructors and learners approach education [99]. Not only has the Web provided us with a means of accessing instant, online information resources, it has also provided us with a new means of communicating and interacting. Online educational tools offer novel and compelling instructional resources and provide instructors and learner with new educational venues [99]. Furthermore, today's students are "digital natives" [24]; they are technologically versed and skilled. Thus, as these technologies continue to evolve and become increasingly integrated into our culture and society, it is important that we, as educators, adapt and tailor our approach to education to better suit the needs of today's learners [99].

Acknowledgments

The author thanks Marisa Conte for her assistance in identifying useful educational links and resources for this chapter. Additionally, the author thanks Dr. George Starkschall for the opportunity to write this chapter, and for his many helpful comments and suggestions.

References

1. Bloom, B. S., *Taxonomy of Educational Objectives: The Classification of Educational Goals*, Longsmans, Green and Co., Inc., New York, 1956.
2. eHow. Available at http://www.ehow.com/how_5108411_apply-blooms-taxonomy.html. Retrieved January 5, 2011.
3. Hofmann, A. (2008). Developments in blended learning. *Economics and Organization of Future Enterprise, 1*(1), 55–62.
4. Griffiths, J. et al. Available at http://ehlt.flinders.edu.au/education/DLiT/2002/environs/suyin/overview.html. Retrieved January 5, 2011.
5. Huba, M. E. & Freed, J. E., *Learner-Centered Assessment on College Campuses, Shifting the Focus From Teaching to Learning*, Allyn & Bacon, Needham Heights, Massachusetts, 2000.
6. McCombs, B. L. & Whisler, J. S., *The Learner-Centered Classroom and School*, Jossey-Bass Publishers, San Francisco, 1997.
7. Blumberg, P., *Developing Learner-Centered Teaching: A Practical Guide for Faculty*, John Wiley & Sons, Inc., San Francisco, 2009.
8. Mazur, E., *Peer Instruction*, Prentice Hill, Inc., Upper Saddle River, 1997.
9. E-learning, in Wikipedia. Available at http://en.wikipedia.org/wiki/E-learning. Retrieved January 5, 2011.
10. Driscoll, M. (2002). Blended learning: Let's go beyond the hype. *E-learning, 54* (March).
11. PCMag, LMS. Available at http://www.pcmag.com/encyclopedia_term/0,2542,t=learning+management+system&i=46205,00.asp. Retrieved December 4, 2010.
12. Ellis, R. Field Guide to Learning Management Systems Learning Circuits. Available at http://www.astd.org/NR/rdonlyres/12ECDB99-3B91-403E-9B15-7E597444645D/23395/LMS_fieldguide_20091.pdf. Retrieved 2009.
13. Mondani, P. Open source vs. proprietary: What does it mean to a small business?. Available at http://pennymondani.com/2010/08/open-source-vs-proprietary-what-does-it-mean-to-a-small-business/. Retrieved December 12, 2010.
14. Gupta, M., in G-cube solutions, Vol. 2010.
15. Zacarias, A. S. & Mills, M. D. (2010). Management of an affiliated physics residency program using a commercial software tool. *Journal of Applied Clinical Medical Physics, 11*(3), 265–275.
16. PCMag, Web conferencing. Available at http://www.pcmag.com/encyclopedia_term/0,2542,t=Web+conferencing&i=54287,00.asp. Retrieved December 6, 2011.
17. Mosson, A. (2009). *Focus, 2011*.
18. Oxford English Dictionary Online, Oxford University Press, 2011.
19. Social network, in Wikipedia. Available at http://en.wikipedia.org/wiki/Social_network. Retrieved January 1, 2012.
20. Weaver, A. C. & Morrison, B. B. (2008). Social networking. *Computer, 41*(2), 97–100.
21. Lester, J. & Perini, M. (2010). Potential of social networking sites for distance education student engagement. *New Directions for Community Colleges, 2010*(150), 67–77.
22. comScore. It's a social world: Top 10 need-to-knows about social networking and where it's headed. Available at http://www.comscore.com/it_is_a_social_world, 2011.
23. Fisher, L., in Simply Zesty (2011), Vol. 2012.
24. Richardson, W., *Blogs, Wikis, Podcasts and Other Powerful Web Tools for Classrooms*, Corwin Press, Thousand Oaks, CA, 2006.
25. Harper, L., Podcasting power for the people, Online Newshour. Available at http://www.pbs.org/teachers/connect/resources/4531/preview/. Retrieved December 30, 2011, 2005.
26. PCMag, podcast. Available at http://www.pcmag.com/encyclopedia_term/0,2542,t=podcast&i=49433,00.asp. Retrieved December 30, 2011.
27. iPodder, Podcasting. Available at http://www.ipodder.org/video_podcasting. Retrieved December 30, 2011.
28. Matias, N., What is a Wiki?. Available at http://www.sitepoint.com/what-is-a-wiki/. Retrieved December 6, 2011.
29. Cunningham, W., Wiki Wiki Web. Available at http://c2.com/cgi/wiki?WikiWikiWeb. Retrieved January 1, 2012.
30. Cunningham, W., Wiki in education. Available at http://c2.com/cgi/wiki?WikiInEducation. Retrieved January 1, 2012.
31. Bortree, D. S., in Half an Hour (2009), Vol. 2011.
32. Carvin, A., in learning.now (PBS Teachers, 2006), Vol. 2011.
33. Search Mobile Computing, Microblogging. Available at http://searchmobilecomputing.techtarget.com/definition/microblogging. Retrieved December 6, 2011.
34. TLT Group, Rubrics: Definition, tools, examples, references. Available at http://www.tltgroup.org/resources/flashlight/rubrics.htm. Retrieved January 3, 2012.
35. Teachers First, What are rubrics. Available at http://www.teachersfirst.com/lessons/rubrics/what-are-rubrics.cfm. Retrieved January 3, 2012.
36. Jackson, C. W. & Larkin, M. J. (2002). Rubric: Teaching students to use grading rubrics. *Teaching Exceptional Children, 35*(1), 40–45.

37. Andrade, H., Available at http://rubistar.4teachers.org/index.php?screen=WhatIs. Retrieved January 3, 2012.

38. MIT, MIT Open Courseware: Our history. Available at http://ocw.mit.edu/about/our-history/. Retrieved December 12, 2010.

39. MIT, About the Open Course Ware consortium. Available at http://ocw.mit.edu/about/ocw-consortium/. Retrieved December 12, 2010.

40. OSW Consortium, Open Course Ware Consortium. Available at http://www.ocwconsortium.org/. Retrieved December 12, 2010.

41. Creative Commons, Creative Commons. Available at http://creativecommons.org. Retrieved April 3, 2012.

42. Plotkin, H., in SFGate (San Francisco Chronicle, San Francisco, 2002).

43 Prisciandaro, J. I. (2013). Review of online educational resources for medical physicists. *Journal of Applied Clinical Medical Physics*, submitted for publication.

44. PCMag, List server. Available at http://www.pcmag.com/encyclopedia_term/0%2C2542%2Ct%3DLISTSERV&i%3D46180%2C00.asp. Retrieved January 10, 2012.

45. Search SOA, List server. Available at http://searchsoa.techtarget.com/definition/list-server. Retrieved January 10, 2012.

46. LISTSERV, in Wikipedia. Available at http://en.wikipedia.org/wiki/LISTSERV. Retrieved January 7, 2012.

47. AAPM, AAPM Medical Physics Resource Page. Available at http://www.aapm.org/links/medphys/default.asp#lists. Retrieved January 10, 2012.

48. Kulasekere, R., MedPhysBoardPreparation. Available at MedPhysBoardPreparation. Retrieved January 10, 2012.

49. AAPM, AAPM, from http://www.aapm.org/. Retrieved January 28, 2012.

50. RSNA Education, RSNA/AAPM Physics Modules. Available at http://www.rsna.org/education/physics.cfm. Retrieved January 22, 2012.

51. Sprawls, P., Online Resources for Study, Review, Reference and Teaching Physics and Technology of Medical Imaging. Available at http://www.sprawls.org/resources/. Retrieved January 22, 2012.

52. ABR, The American Board of Radiology. Available at www.theabr.org. Retrieved January 30, 2012.

53. ABRF, The American Board of Radiology Foundation. Available at www.ABRFoundation.org. Retrieved January 30, 2012.

54. ASTRO, ASTRO. Available at https://www.ASTRO.org/. Retrieved January 31, 2012.

55. ASTRO Membership Department, personal communication by J. I. Prisciandaro, February 14, 2012.

56. ESTRO, ESTRO. Available at http://www.estro.org/Pages/default.aspx. Retrieved January 31, 2012.

57. GEC ESTRO, The GEC ESTRO Handbook of Brachytherapy, edited by Gerbaulet, P. et al. (ESTRO, Brussels, 2002). Available at http://www.estro-education.org/publications/Documents/GEC%20ESTRO%20Handbook%20of%20Brachytherapy.html. Retrieved February 1, 2012.

58. IAEA, About the IAEA. Available at http://www.iaea.org/About/index.html. Retrieved January 10, 2012.

59. Wikipedia, International Atomic Energy Agency. Available at http://en.wikipedia.org/wiki/International_Atomic_Energy_Agency. Retrieved January 10, 2012.

60. IAEA, Division of Human Health. Available at http://www-naWeb.iaea.org/nahu/. Retrieved January 10, 2012.

61. IAEA, Radiation Oncology Physics: A handbook for teachers and students, edited by Podgorsak, E. B. (IAEA, Vienna, 2005). Retrieved from http://www-naWeb.iaea.org/nahu/dmrp/syllabus.shtm.

62. Podgorsak, E. B. & Hartmann, G. H. Slides to Radiation Oncology Physics Handbook. Available at http://www-naWeb.iaea.org/nahu/dmrp/slides.shtm.

63. IAEA, Radiation Biology: A handbook for teachers and students (IAEA, Vienna, 2010), Retrieved January 10, 2012, from http://www-pub.iaea.org/MTCD/publications/PDF/TCS-42_Web.pdf.

64. IAEA, Distance learning course in radiation oncology for cancer treatment. Available at http://www.iaea.org/Publications/Training/Aso/register.html. Retrieved January 10, 2012.

65. IAEA, Radiation Protection of Patients. Available at https://rpop.iaea.org/RPOP/RPoP/Content/index.htm. Retrieved January 10, 2012.

66. IAEA, Radiation Protection of Patients—Publications. Available at https://rpop.iaea.org/RPOP/RPoP/Content/AdditionalResources/Publications/index.htm. Retrieved January 10, 2012.

67. IAEA, Radiation Protection of Patients—Free material. Available at https://rpop.iaea.org/RPOP/RPoP/Content/AdditionalResources/Training/1_TrainingMaterial/index.htm. Retrieved January 10, 2012.

68. ICRP, ICRP. Available at http://www.icrp.org/. Retrieved January 17, 2012.

69. ICRU, ICRU. Available at www.icru.org. Retrieved January 20, 2012.

70. NCRP, NCRP. Available at www.ncrponline.org. Retrieved January 22, 2012.

71. RSNA. Available at http://www.rsna.org/. Retrieved January 30, 2012.

72. Hall, E., Web-Rad-Train. Available at http://www.columbia.edu/~ejh1/Web-rad-train/. Retrieved January 30, 2012.

73. RSNA Education, Physics Teaching File for Radiology Residents. Available at http://www.upstate.edu/radiology/education/rsna/. Retrieved January 30, 2012.

74. UNSCEAR, UNSCEAR. Available at www.unscear.org. Retrieved January 20, 2012.

75. Tabakov, S., et al. (2005). Development of educational image databases and e-books for medical physics training. *Medical Engineering & Physics 27*, 591–598.

76. EMERALD, EMERALD, EMIT, EMITEL. Available at http://www.emerald2.eu/. Retrieved February 3, 2012.

77. Stoeva, M. & Cvetkov, A. (2005). e-Learning system ERM for medical radiation physics education. *Medical Engineering & Physics 27*, 605–609.

78. Tabakov, S. (2008). e-Learning development in medical physics and engineering. *Biomedical Imaging and Intervention Journal, 4*(1), e27.

79. EMITEL, EMITEL e-Encyclopaedia of Medical Physics and Multilingual Dictionary of Terms. Available at http://www.emitel2.eu/emitwwwsql/index-login.aspx. Retrieved February 14, 2012.

80. MEP, Medical Physics and Engineering Education and Training Part I, edited by Tabakov, S., et al. (Abdus Salam International Centre for Theoretical Physics, ICTP, Trieste, 2011). Available at http://emerald2.eu/mep/e-book11/ETC_BOOK_2011_ebook_s.pdf. Retrieved February 14, 2012.

81. Woo, M. & Ng, K. (2003). A model for online interactive remote education for medical physics using the Internet. *Journal of Medical Internet Research, 5*, e3.

82. Woo, M. & Ng, K. (2005). Using the Internet for real-time medical physics education. *Medical Physics, 32*, 2416.

83. RRTL, Remote Real-Time Learning. Available at http://www.neteinfach.com/rrtl/index.htm. Retrieved February 3, 2012.

84. Boyer, A., Web-based Dosimetry Training Tool. Available at www.dosimetrytrainingtool.com. Retrieved February 2, 2012.

85. Boyer, A. & Kaylor, S. (2005). The dosimetry training tool (DTT) project. *Medical Physics, 32*(6), 2103.

86. Keall, P. et al. (2007). Five year-report on a Web-based interactive dosimetry training tool. *Medical Physics, 34*(6), 2400.

87. Adams, R. D., personal communication by J. I. Prisciandaro, April 26, 2012.

88. Zeman, E. M. personal communication by J. I. Prisciandaro, April 24, 2012.

89. Schreiber, E. et al. Development of a Web-based dosimetry training tool for therapy and dosimetry education, AAPM 54th Annual Meeting Innovations in Medical Physics Education Session, 2012.

90. Adams, R. D., Curriculum and Modules for Computer-Based Dosimetry Instruction, ISRRT World Congress and CAMRT Annual General Conference, 2012.

91. Phillips, R. et al. (2008). Virtual reality training for radiotherapy becomes a reality. *Studies in Health Technology and Informatics, 132*, 366–371.

92. Hall, J. (2008). Virtual vision of the future. *Synergy News, 3* (November).

93. Beavis, A. W. et al. (2007). Radiotherapy training tools for yesterday's future. *Imaging & Oncology*, 30–35.

94. Beavis, A. et al. (2007). Is virtual reality training really a reality? *SYNERGY: Imaging & Therapy Practice*, 22–24 (September).

95. Shah, U. & Williams, A. (2010). How to use VERT for interactive CT anatomy for post-registration training. *SYNERGY: Imaging & Therapy Practice*, 12–14 (July).

96. Antons, J. et al. personal communication by J. I. Prisciandaro, April 26–May 10, 2012.

97. Beavis, A. et al., The Development of a Virtual Reality Dosimetry Training Platform for Physics Training, AAPM 54th Annual Meeting Innovations in Medical Physics Education Session, 2012.

98. Vertual, VERT. Available at http://www.vertual.co.uk/. Retrieved May 10, 2012.

99. AAMC Institute for Improving Medical Education, Effective use of educational technology in medical education: Colloquium on Educational Technology: Recommendations and Guidelines for Medical Educators, March 2007.

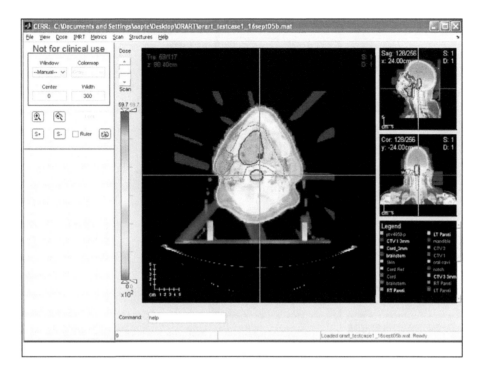

FIGURE 16.3 Basic CERR view window shows transverse, sagittal, and coronal slice views along with the selected dose distribution. This plan was submitted to the authors by the University of Michigan and is distributed with CERR as part of the Operations Research Applications in Radiation Therapy Toolbox (Deasy et al. 2006). The view is highly customizable (see the Wiki link for more details). Different color bar color maps are available. The dose color-wash color bar itself can be interactively rescaled to cut-off cold dose regions. Isodose lines are also available. Window snapshots can be conveniently captured and automatically placed into a Web page, a feature the developers often use.

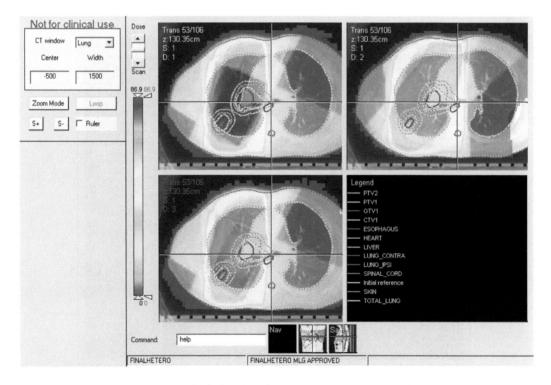

FIGURE 16.5 CERR can conveniently compare multiple dose distributions registered to the same slices. An interactive, drag-able, dose profile tool is available.

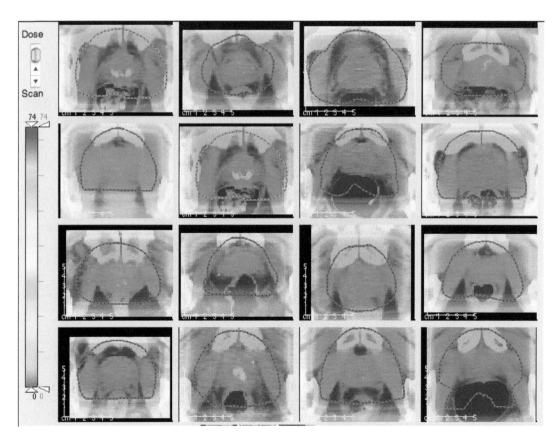

FIGURE 16.9 Cohort review tools. This tool allows CERR to directly compare multiple plans for the same patient or different patients. Stepping through a slice moves all images in unison. Without such a tool, it is difficult to adequately appreciate changes in treatment plans within a population.

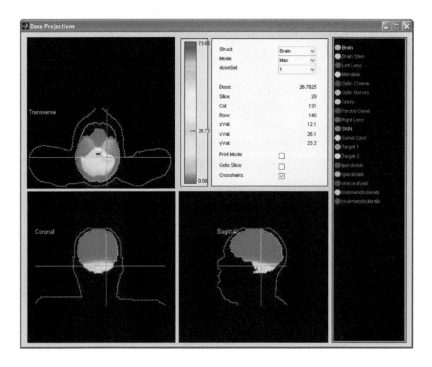

FIGURE 16.10 Dose projections. Dose projections are essentially maximum intensity projections applied to dose. They facilitate the rapid identification and localization of cold spots in a target volume or hot spots in a normal structure.

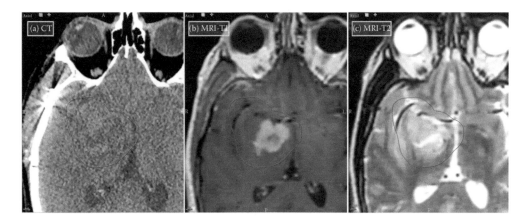

FIGURE 18.2 Example of CT and MRI imaging of a glioblastoma patient. Although the tumor is invisible in the CT data set, it can be easily visualized by the MRI. The red contour represents tumor boundaries, as defined on the MRI data sets and later transferred to the CT volume.

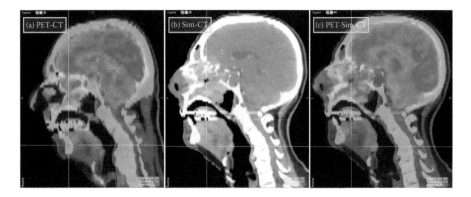

FIGURE 18.3 Example of integrating PET imaging into the treatment planning. The tumor was delineated on the PET/CT data set, of lower resolution, and transferred to the higher-resolution Sim-CT data set for planning. This case presented differences in posture and chin position that have to be considered when correlating these image sets.

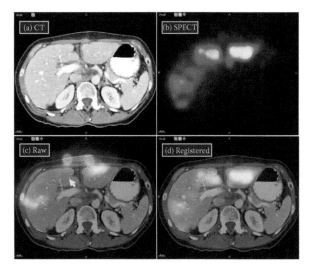

FIGURE 18.4 Typical imaging used for planning in targeted radiotherapy. In this approach, patient-specific data from CT (a) provides an anatomical model with resolutions on the order of one millimeter, whereas dose deposition estimation is based on the SPECT data set (b). The images are acquired on different scanners and appear displaced and unaligned as evident at arrow (c). An image registration procedure corrects this misalignment as illustrated in (d). After registration, the SPECT activity within the liver matches the liver shape as observed in the CT data set.

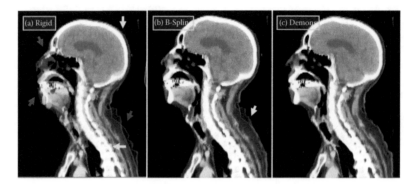

FIGURE 18.8 Comparison of rigid (a) and two popular deformable registration methods, B-spline (b), and demons (c). The result is shown as a red overlay on the fixed image, shown in background as a grayscale sagittal slice. In (a), the rigid registration is globally fairly good, but the areas marked with arrows are not well registered. In the brain, voxel values are fairly uniform and there will be very little detail with which to adjust the registration. A deformable registration algorithm should provide enough deformation at the arrows to correct the misregistrations and minimal warping in the brain. Notice that, visually, the quality of the two deformable registrations, (b) and (c), looks very similar, with a better result obtained by demons (c) at the yellow arrow.

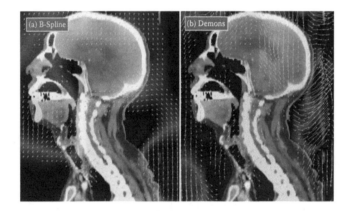

FIGURE 18.9 Display of the vector field magnitude for the solutions in Figure 18.8b and c illustrates the differences between these two deformable algorithms. For example, the B-spline has minimal warping in the brain, although there is significant warping for the demons algorithm. Such differences, which have to be assessed by on a case-by-case basis, may have significant implications for dose tracking and other voxel tracking applications.

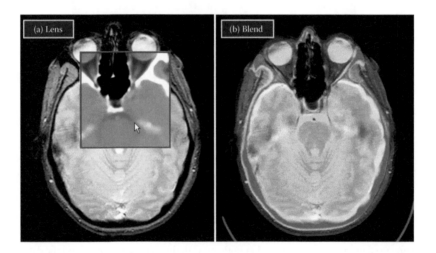

FIGURE 18.10 Examples of lens (a) and color blend (b) inspection of registration results. In the lens tool, the square region inside the red rectangle shows the CT data set and can be interactively moved over the MRI image to inspect the match at different locations. The color blend display allows a quick assessment of blurred regions where the match is suboptimal.

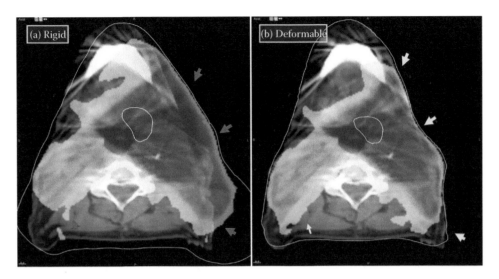

FIGURE 18.11 Usage of deformable registration for adaptive radiotherapy. The background shows the CBCT data set, whereas the overlays show the segmentations and doses computed in the planning process superimposed using rigid (a) and deformable (b) registrations. Deformable registration can be used to autosegment the CBCT data set by adapting the planning segmentation to the anatomical changes.

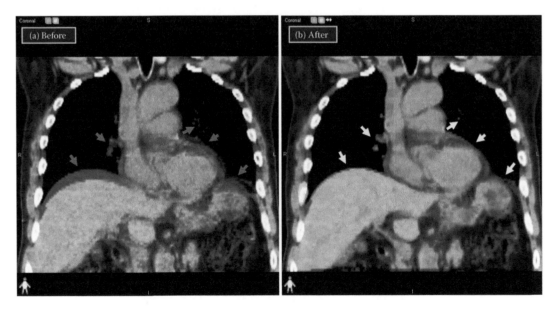

FIGURE 18.12 Sample result of deformable registration. In (a), two phases of a 4D CT are overlaid in a blend view, with arrows marking anatomical changes produced by the respiratory motion. Results of the deformable registration algorithm are presented in (b), showing liver and other organs deformed back to match their initial state.

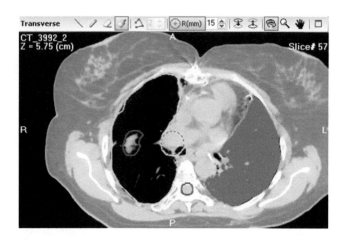

FIGURE 19.6 2D manual editing tools. The paint brush tool (selected) allows users to push a contour in or out using a cursor in the shape of a circle.

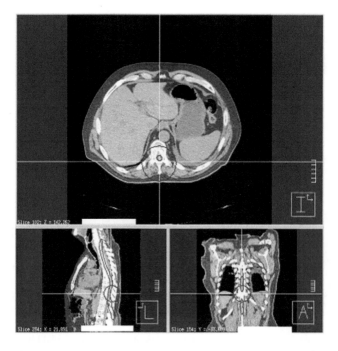

FIGURE 19.7 Manual 3D editing tool in Philips Pinnacle RTP system.

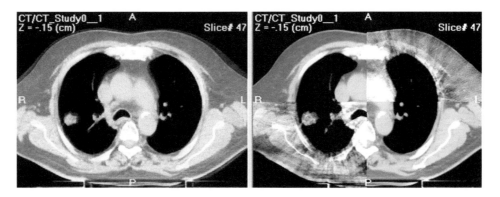

FIGURE 19.8 Fused image displays of a planning CT image and cone-beam CT image as a color overlay (left) and checkerboard (right).

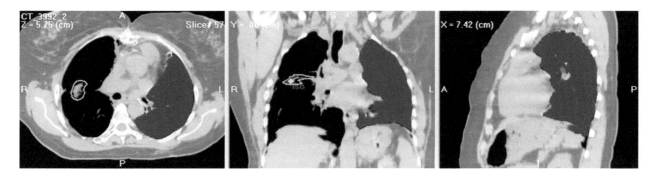

FIGURE 19.9 Example volume of interest displays on 2D cross-sectional images.

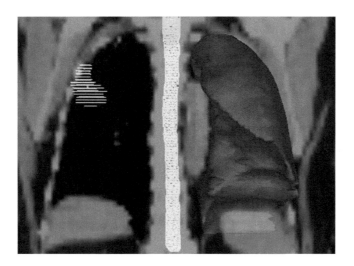

FIGURE 19.10 3D volume display. The anterior view shows a semitransparent coronal image midway through the patient, gross tumor volume (contour stack), cord (mesh), and left lung (tiled surface).

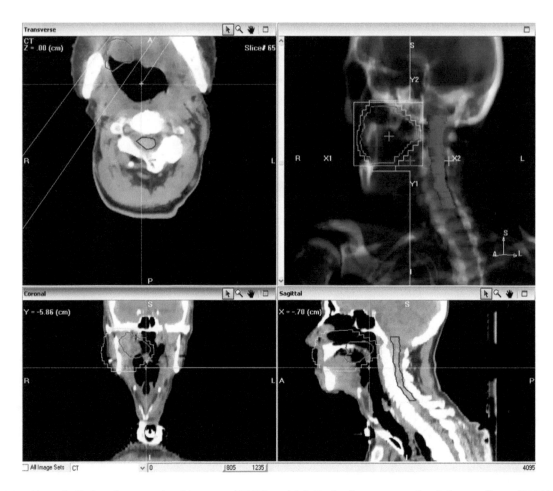

FIGURE 19.11 Example display of cross-sectional images and BEV (top right). On the three cross-sectional views, outlines of PTV (green), cord (blue), and MLC-shaped beam aperture (yellow) are displayed. BEV shows a DRR, PTV, beam aperture, and cord.

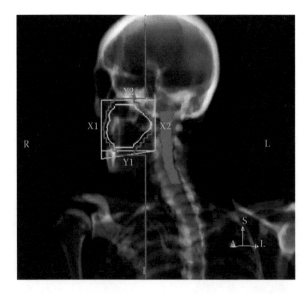

FIGURE 19.12 BEV display, in which the PTV is shown as a green outline, while cord is displayed as a solid blue surface. Radiation field information, such as MLC (pink), jaw, and wedge (orange), is also displayed. The volumes of interest, field information, and patient orientation are superimposed on a DRR.

Communication, Collaboration, IT, and Informatics Infrastructure for Clinical Trials

Ying Xiao
Jefferson Medical College

Mike Tilkin
American College of Radiology

14.1 Introduction

Information technology (IT) is the cornerstone for the successful conduct of modern clinical trials. A robust, efficient, secure and accessible IT system is essential for every component of the clinical trial process. From data preparation, submission, collection, transfer to archiving, and analysis, it serves the entire process of the clinical trial. To participate in a clinical trial, institutions must demonstrate their ability to conduct the trial and to provide data with the necessary quality. When data is submitted to a center involved in the information management of the trial. The data undergoes quality assurance (QA) testing to ensure that it meets the standards and protocols of the trial. At the end of the trial, the data is analyzed, archived, and shared.

This chapter describes, as an example, radiation oncology-related clinical trials and how IT serves both current and future needs. It captures the latest National Cancer Trial Network (NCTN) effort to consolidate the cancer clinical trial efforts.

14.2 Clinical Trial Process and Informatics

This section contains an overview of the radiotherapy (RT) clinical trial QA process in the context of the overall conduct

of the trials and of how informatics is the backbone of the processes.

14.2.1 Site Qualification and Credentialing

Credentialing is the process through which institutions or individuals are required to present evidence to demonstrate that they are able to provide data of the quality required for participating in multi-institutional clinical trials. General credentialing for RT, at present, involves institutions that are applying to become a member of a clinical trial group. Institutions fill out a comprehensive facility questionnaire documenting staffing and location information, treatment planning systems and delivery systems, basic QA documentation, and other relevant information. To maintain their membership status, institutions provide evidence of annual external audits of their machine calibrations by an expert center such as the Radiological Physics Center.

The dry run/dummy run/benchmark credentialing exercise applies to protocols that require a demonstration of an understanding of the protocol planning and data submission requirements. Institutions select similar cases to the one specified in the protocol or download prepared benchmark cases, define target or critical structures if needed, perform RT treatment planning as defined by the protocol, and submit the required data for review. In submitting digital data for centralized review, this process tests the data transfer/submission capability, target and critical structure delineation accuracy, and RT treatment planning appropriateness, in that they should meet protocol specified criteria. However, this process does not guarantee protocol-compliant RT delivery to the patients. This credentialing process applies to institutions, or individual physicians, and can be disease site or protocol specific.

The phantom irradiation QA assesses the complete treatment process for a specific treatment modality that might be common to many protocols (e.g., intensity-modulated RT). This end-to-end QA test of the treatment process includes imaging, treatment planning, setup for treatment, and actual delivery of the treatment. The institutions obtain phantoms from a QA center, with the delivered dose evaluated centrally by the QA center. The phantoms provide a consistent test to evaluate each institution's ability to deliver a specific RT treatment.

The individual case review process evaluates the RT treatment of a patient enrolled in the study. It includes an assessment of the target and critical structures defined by physician principal investigators. It also verifies dosimetric compliance.

These QA processes are essential for meaningful clinical trials (Peters et al. 2010; Abrams et al. 2012). However, they can sometimes be burdensome and hinder clinical trial accrual (Bekelman et al. 2012). It is essential to develop evidence-based and adaptive QA processes to ensure the quality and at the same time increase the efficiency of clinical trial conduct.

14.2.2 Trial Administration

Clinical trials are conducted through Network Group Operations Centers, which coordinate with associated statistics centers and other support groups for the operation. The develop and articulate an overall research strategy that reflects an integrated scientific approach both within and across specified disease areas, addresses important unmet clinical needs, and is consistent with national research goals. Scientific research committees are formed primarily to develop and oversee the conduct of clinical trials and studies within a defined research strategy (e.g., disease committee such as a breast committee that conducts trials in breast cancer, an experimental therapeutics committee, or a correlative science committee). These committees also play a key role in the development and conduct of correlative science studies. Administrative committees are formed to provide essential core service functions to support other aspects of research strategy (e.g., Patient Advocacy, Clinical Research Associates, Auditing, Pathology, and Surgery).

14.2.3 Data Acquisition, Transport, and Collection

All data, as well as any biospecimens collected, for an NCTN trial must be sent by the institutions or sites participating in the trial to the network group that is leading the trial, unless an exception is approved by the National Cancer Institute (NCI) Division of Cancer Treatment and Diagnosis to accommodate

the needs of a specific trial. The Network Group Operations Center is responsible for overseeing the timely collection and transmission of data and biospecimens from all its member institutions or sites to NCTN trials for patient accruals that are credited to the Network Group.

Network groups are required to use standard NCTN tools and services for all NCTN trials for data consistency. They include (a) Common Data Elements (CDE), the Common Data Management System for study design, study build, and data collection with case report reports designed by the Statistical Data Management Center that are compliant with the NCTN Program-approved sections of the data dictionary for CDEs in the NCI Cancer Data Standards Registry and Repository (caDSR); (b) NCTN information system for tracking biospecimen collection from NCTN trials (in development); (c) NCTN Oncology Patient Enrollment Network (OPEN) and Regulatory Support Services (RSS) via the Cancer Trials Support Unit (CTSU) for central registration and randomization of patients onto NCTN trials; (d) the NCI Common Terminology Criteria for Adverse Events; and (e) the Comprehensive Adverse Event and Potential Risks for agents, if available.

NCTN trials must be registered with not only the NCI Clinical Trials Reporting Program but also the U.S. National Library of Medicine clinical trials database (www.clinicaltrials.gov). Appropriate information updates, changes in the trial design and accrual, and results from NCTN trials must be reported in clinicaltrials.gov as required under the Food and Drug Administration (FDA) Amendments Act, Section 801. The Network Group Operations Center should work with its associated Network Group's Statistics and Data Management Center to coordinate activities to ensure information on Group NCTN trials is appropriately updated in these systems.

14.2.4 Data Processing, Cleansing, Review, and Analysis

Data are processed, cleansed, and prepared for central review and analysis. Committees are established for conducting central reviews of the following major data elements that affect the outcome of specific clinical trials, including the following:

- *Pathology:* cases in which known variability in the accuracy of histologic (or other) diagnosis is a potentially serious problem and in which pathology data are integral to appropriate study design and analysis.
- *RT:* review of treatment-planning studies and compliance with protocol-specified doses for individual patients may be required and should be provided via coordination with the Network Radiotherapy and Imaging Core Services Centers.
- *Imaging support, including diagnostic imaging:* When relevant, central review (either prospective or retrospective) of imaging in NCTN trials may be required for evaluating response, establishing a diagnosis, and/or screening of patients and should be provided via coordination with

the Network Radiotherapy and Imaging Core Services Centers.

- *Systemic therapies (chemotherapy, immune therapy, or other biological therapies):* Central review may be performed by the network group study team for the trial to determine protocol compliance with dose administration and dosage modification.
- *Surgery:* When relevant, adequacy of protocol-specified surgical procedures may be assessed (e.g., through review of operative notes, study-specific surgical forms, and pathology reports) by the network group study team for the trial.

14.2.5 Data Sharing for Research and Analysis

The network groups conduct clinical trials in cancer research. Each network group or NCTN study has a formal protocol document that includes a statement of the objectives of the study. Patient consent and authorization are obtained to collect the individual patient data required for addressing the study objectives. These data are sent from the treating or enrolling institution to the Network Group's Statistics and Data Management Center, where the data are reviewed, processed, and entered into an electronic database. The data may be submitted on paper or electronically. Not all information submitted on paper becomes part of the electronic database. The electronic database is used as the basis for the analysis of the Network Group's studies, with the analyses performed by the staff at the Network Group's Statistics and Data Management Center.

Each Network Group may have a more detailed set of procedures implementing the general data sharing policy. The general guiding policies from NCI are that individual-level de-identified data sets that would be sufficient to reproduce results provided

in a publication containing the primary study analysis would be available to individuals via a formal requesting procedure after a certain period post-publication. All research use of data collected on human subjects from NCTN studies led by the Network Group Operations Center with its associated Network Group's Statistics and Data Management Center is subject to applicable Office of Human Research Protections regulations and to applicable regulations of the Privacy Rule of the Health Insurance Portability and Accountability Act (HIPAA).

A simple, formal data use agreement that complies with regulatory considerations will usually be required. It specifies who will have access to the individual patient data, includes a statement that it will not be shared with others outside this specified set of individuals, and covers the release conditions. A fee may be requested for complex data preparation.

14.3 Clinical Trial Informatics Infrastructure—The IROC Cloud

A group is being formed to offer RT and imaging quality core services (Imaging Radiation Oncology Core; IROC) to the entire national cancer trial system (NCTN). It is designing the cloud infrastructure shown in Figure 14.1 and described in the following sections.

14.3.1 NCTN Technology Stack: OPEN/RSS/RAVE

OPEN is the Web-based registration system for patient enrolments into NCI-sponsored Cooperative Group clinical trials. The system is integrated with the CTSU Enterprise System for regulatory and roster data and with each of the Cooperative Groups'

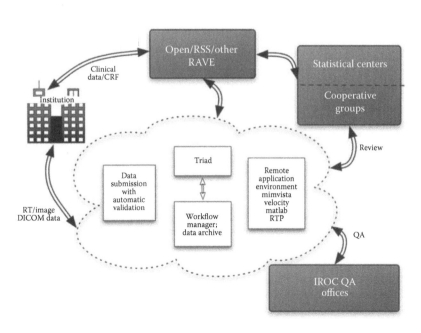

FIGURE 14.1 Imaging Radiation Oncology Core (IROC) cloud.

systems for patient registration and randomization. OPEN provides the ability to enroll patients on a 24/7 basis.

The OPEN portal application system will:

- enable patient registrars to record data and determine patient eligibility for inclusion on a given clinical trial;
- communicate checklist and related regulatory data (e.g., investigator status on a Group roster, etc.) to Group randomization processes using a data format that complies with industry standards (e.g., CDISC);
- communicate Group registration output data such as treatment assignment, patient ID, registration notes, etc., back to the OPEN portal user;
- reduce manual intervention in the registration process by providing a Web-based patient registration system for CTSU hosted clinical trials that offers near constant availability;
- reduce the training burden on site personnel by providing a system with a user interface that is standardized across all Groups and trials and that varies only as the content of the registration process requires;
- reduce the programming and administrative workload at participating Cooperative Groups by offering the OPEN portal as a complement to their existing Web registration systems (the OPEN portal is not expected to replace any existing Web registration system in its entirety);
- adhere as closely as possible to regulatory guidelines (e.g., GCP, 21 CFR 11, and FDA Guidelines) in the design and development of the system;
- make the OPEN portal available to other organizations as a service so that they do not need to use their resources to create a Web-based patient registration system;
- offer a standardized format for exchanging patient eligibility data among CTSU and the participating Cooperative Groups; and
- reduce the amount of paper involved in the patient enrollment process by offering digital equivalents, such as electronic signatures, to current practices that rely on paper and its distribution.

The CTSU RSS has a patient enrollment credentialing system (PEC) as one of its modules. The PEC is currently being used by the CTSU patient registrars to do the enrollments into the CTSU protocols. The PEC system has several checks that are performed to determine whether an institution/investigator can enroll a patient into a particular protocol through CTSU. The checks verify that:

- the enrolling institution has submitted institutional review board approval documents and satisfies any protocol-specific requirements;
- an investigator has an "active" investigator status in the database and is thus eligible to register a patient and, if applicable to the protocol, to receive drug shipments;
- the enrolling institution and/or investigator is affiliated with the particular Cooperative Group, which receives the accrual credit;

- the investigator is affiliated with the enrolling institution;
- the patient ID length and format meet the Cooperative Group requirements;
- an institution is a Community Clinical Oncology Program; and
- the site is eligible for payment from the CTSU.

The OPEN portal will interact with the PEC to apply all these rules to allow the site or the investigator to enroll patients into a protocol. The interaction will be through the backend database procedure calls to the CTSU database from the OPEN portal database.

Medidata Rave is a comprehensive system for capturing, managing, and reporting clinical research data in phase I to IV studies, streamlining the clinical trial process and helping life science organizations optimize their R&D investments. Throughout a clinical trial, Medidata Rave provides visibility to captured data. It is a single platform combining electronic data capture (EDC) and clinical data management capabilities. Medidata Rave's scalable, software-as-a-service architecture provides a cost-effective approach to quickly implement a single clinical trial or support an enterprise-wide deployment for multiple therapeutic areas, phases, and studies—including postmarketing observations and registries.

Medidata Rave's other capabilities include wide support of industry data standards, flexibility to implement any data management workflow with secure access for all study team members, and a set of on-demand data extraction and ad hoc reporting tools. It provides a platform to manage site, patient, and laboratory reported data (e.g., from EDC and other systems) and rapidly make them available for analysis and submission.

14.3.2 Electronic Image and Data Exchange

Medical imaging plays an important role in many NCI-sponsored trials, and making these data available throughout the life cycle of a trial is an important dimension of the clinical trial workflow. The process starts at the time of acquisition, with the need to transport imaging data from originating hospital systems to the QA center for review. This review process is designed to ensure the submitted data adheres to the study protocol and meets various quality standards. Upon confirmation of data quality, imaging studies are made available to reviewers per the study protocol. The review process may involve multiple phases. Ultimately, imaging studies will reside in an archive available to researchers for secondary analysis.

Given the continual increase in the size of imaging studies, there is an advantage to limiting the movement of data. Cloud-based technologies offer a number of solutions to expose individuals to data with minimal movement of large data sets. At the very least, however, studies must be moved from the facility responsible for acquisition to the QA center. Historically, physical media such as CDs or DVDs have been used to transport imaging studies and related data. Over the last several years, this mechanism has given way to Internet-based exchange. Several considerations need to be made when moving studies in this manner:

- Studies need to be de-identified to remove any patient-related information. This generally requires altering of Digital Imaging and Communications in Medicine (DICOM) tags when appropriate but also may require the blurring of "burned in" data that reside within the pixel data.
- Studies need to be associated with the appropriate clinical trial data. This includes ensuring that imaging studies submitted with a subject ID for a particular trial align with the clinical trial management systems at the receiving end. Minimizing the degree to which this is done manually is important for both efficiency and data integrity.
- Local validation, where possible, has the potential to avoid the transport of studies that do not adhere to protocol. This could reduce downstream iterations that may include both data cleaning and resending of studies.
- Given the size of imaging studies, efficiency of transport is important and should include features such as compression and the ability to manage large numbers of studies in an efficient manner.
- Flexibility to include DICOM and non-DICOM data is critical and needs to be done in a manner that ensures data integrity throughout the process.

There are a variety of commercial and open source tools designed to transport images, many of which address all or some of the considerations described above. One such system, the American College of Radiology (ACR) TRIAD system, employs a variety of web-based and premises-based technologies designed to address the full range of issues associated with clinical trial image submission including tight integration with the NCI NCTN systems. The flexibility of an image management platform is a critical feature that enables it to handle the variability within clinical trials.

For example, an integrated system such as TRIAD has several key components: First, there is a central service bus responsible for orchestrating the imaging workflow. Based on a service-oriented architecture (SOA), there are a variety of key services that allow a variety of tool-based and user-based access to resources. Key services include security and authentication through a role-based system designed so that users have the minimum level of access necessary to perform their tasks. Image management ensures the integrity of image and nonimage data as they flow through the system. Data exchange services allow the intake of data from external resources as well as the distribution of data to systems acting as data consumers. Workflow tools, including worklist management, help facilitate the processing of data at the various steps. Finally, there is a central portal that provides direct user access to central services such as the ability to manage users, configure anonymization profiles, and access worklists.

14.3.3 Internal Operations

An integrated system for communication is critical given the distributed nature of multi-center clinical trials. A central document repository should be used for SOPs and common work products. Chat, video, and web conferencing should be used to enhance collaboration. In the case of IROC, integration between systems that manage the distributed QA process and those that perform data management for cooperative groups become an important aspect of the end-to-end workflow.

14.3.4 External Access

Access to data via user-friendly-portals and system-friendly web services is critical in a distributed research environment. As seen in Figure 14.1, access to data in the IROC cloud is provided to individuals through a web portal while system-level access occurs through a variety of service-based interfaces. Increasingly, advances in thin-client visualization technology allows for access to imaging studies without needing to move large quantities of data. Since access to images occurs throughout the clinical trial workflow, minimizing movement of entire studies when possible is desirable.

14.3.5 Infrastructure

In a distributed environment, it's important to make services as location independent as possible. It's also important to enforce standards that cross institutional boundaries. The infrastructure required to support this environment includes a security fabric that is sufficiently flexible to accommodate the institutional constraints of the various entities involved. Attention to SOPs such as those for back-up and recovery and data handling is important and orchestration and enforcement must be done in a consistent manner. For IROC, a strong central hub provides the backbone, helps facilitate standards, and serves as an orchestrator for distributed data management activities.

14.4 Data and Systems Integrity: Challenges and Considerations

14.4.1 Promoting and Ensuring Standards Development

14.4.1.1 CDEs and NCI's caDSR

The NCI caDSR (Pathak et al. 2011) defines a comprehensive set of standardized metadata descriptors for cancer research data for use in information collection and analysis. It provides a database and a set of Application Programming Interfaces to create, edit, deploy, and find CDEs. It is based on the ISO/IEC 11179 model for metadata registration and uses this standard for representing information about names, definitions, permissible values, and semantic concepts for the CDEs. Various NCI offices and partner organizations have developed the content of the caDSR by registration of data elements based on data standards, data collection forms, databases, clinical applications, data exchange formats, UML models, and vocabularies. caDSR CDEs are leveraged for representing the data elements across the eMERGE and PhenX projects to enable interoperability across derived phenotypic data elements. It enables the utility of standards-based common metadata for the clinical research community. It is to be noted that independently, eMERGE and PhenX have been using caDSR to standardize their data elements, either by mapping to existing CDEs or curating new CDEs, where applicable, since the inception of both projects (Kerrien et al. 2007).

14.4.1.2 DICOM

DICOM provides standards for the transmission, encoding, and archiving of medical information. It is a widely accepted standard for the communication of data in clinical trials. An increasing number of manufacturers have complied with these standards, and the inclusion of RT objects has improved the interoperability within the RT community. DICOM is discussed in several other chapters in this book and we refer the reader to them for more details.

14.4.2 Privacy, Security, and System Robustness

The current threat landscape is always changing. Originally, the focus was on perimeter firewalls, Intrusion Detection Systems (IDS), and other edge devices. The next generation of security focused on defense in depth but still had a machine/technical tool focus. This evolved into focusing the protection around the data that companies had.

The new security focus is now on softer targets that have become the target of choice for hackers: the end users themselves. Concentrating on the end user presents a higher chance for success with the least amount of effort. Targeted attacks such as phishing, spear phishing, whaling, and other social engineering attacks have increased in volume and sophistication. This does not mean that we can lower our guard on the technical aspects of security such as firewalls, WAFs, IDS, and patching of systems. These are still part of the attack vector, and companies that fail to keep up with the industry pay the price.

Some recent public example of attacks are LinkedIn (6.46 million passwords reported stolen); CBS Music; E-Harmony (millions of passwords stolen); Sony (25 million accounts breached with full information); Apple IDs released (although this was not true, initial public perception can be damaging to reputation and public presence); and the Harvard MEEI Mass Eye and Ear Association's stolen unencrypted laptop that resulted in $1.5 million loss.

How many different passwords do you have? How many accounts use the same password? When Dropbox and Google were breached, they were breached by corporate accounts with the same password as their linked ones in (Dropbox) and Gmail (Google). Verizon's annual Data Breach Investigations Report (page 45) states that Web applications by companies remain a popular (fifty-four percent of breaches) and successful (thirty-nine percent of records) attack vector because it allows the expansion of the malicious surface area linked Web sites and applications.

Identity theft's primary objective is financial gain. People are the targets and the weakest link. Any large organization represents a lucrative source for hackers and a valuable target. How many have weak passwords or the same passwords for other accounts? Many of these systems have personally identifiable information (PII), which can then be used for identity theft.

CISOs and senior management are now focusing more on risk management, identifying the valuable resources and data and assigning risk values for protection. With this approach, proper allocation of funds and resources can be assigned for risk value protection. They focus on a reduction of potential future risk and not solely on existing threats and management of regulatory and compliance issues. Generally the goal is not only to meet all regulatory and compliance requirements now and the ones that are under review but also to reduce overall risk in a cost-effective manner.

Data need to be secured (1) "in-motion" (e.g., tunneling, SSL), (making sure that data are secured as it is transported and that there is a chain of custody providing verifiable user-time-date stamps on accessed or modified data), and (2) "at-rest," (making sure data are secured and have appropriate access restrictions). Encryption is desired to further protect data if other security measures are compromised while weighing performance considerations.

In clinical trials, the unique concern on the security front is largely about exposing patient-related information. Other concerns, such as the falsification of information or requests for restricted information, fall under more typical enterprise and application security. We must ensure that data that may have identifiable patient information be protected at the highest levels, including isolation, auditing, etc. We should treat data that may be de-identified as guilty until proven innocent, and we need to de-identify data locally before transfer. For example, TRIAD cleans DICOM headers and has utilities to clean burned-in data. Any data ingestion process should quarantine and track any suspicious data that may not be fully de-identified until appropriate measures have been taken to ensure that privacy concerns have been met.

Security programs should pay close attention to areas such as sufficient staffing, well-established policy and procedures, security awareness program and training, risk mitigation, third-party review of security and systems, and multiple solution systems (such as firewalls, WAF, IDS, Anti-Virus, Security Event Incident Management (SEIM), and Vulnerability Scanning).

14.4.3 Data Accessibility

14.4.3.1 Models and Technical Solutions for Distributed Querying

Finding an ideal ordering of the operations for a given query is the main role of the optimization layer. The main components of query optimization include the search space, the search strategy and the cost model (Ozsu & Valduriez 2011). The distributed queries are more complex than those in the centralized systems, with the major issue of dealing with join ordering. For the search space, with a complex query that involves many relations and many operators, the number of equivalent operator trees can be very high.

The most popular search strategy used by query optimizers is dynamic programming, which is deterministic. Dynamic programming builds all possible plans, breadth-first, before it chooses the "best" plan. Randomized strategies, such as simulated annealing and iterative improvement, concentrate on searching for the optimal solution around some particular points. An optimizer's cost model includes cost functions to predict the cost of operators, statistics and base data, and formulas to evaluate the sizes of intermediate results. The cost is in terms of execution time, so a cost function represents the execution time of a query.

14.4.3.2 Comprehensive Investigator Authentication Procedures

The access control is covered in three levels: system level, application level, and logging.

System-level access controls perform the following functions: (a) Internal Users on Premises: authenticate users to provide access; (b) Internal Users, Remote: authenticate users via VPN, provide a suite of predefined applications via Citrix Xen App, and allow application access for Web-based applications that are published and designed for Web-access; (c) External Resources, trusted users: provide a suite of predefined applications via Citrix Xen App and allow application access for Web-based applications that are published and designed for Web-access; and (d) External Resources, nontrusted users: prevent access to network resources and enforce application-level control for applications designed to be public-facing applications.

Application-level access means that each application is responsible for an authentication scheme that restricts users on a role basis. It can be divided into two categories: (a) Application Authentication—The authentication scheme varies by application. Some applications will use network credentials to log on, whereas others will have application-specific schemes. In many cases, the application-specific schemes interface with external security providers such as NCI CTEP credentialing (IAM) designed to facilitate single sign on. (b) Event-Based Authentication—Some applications will use credentials in a dynamic fashion in the course of specific data operations. For example, certain picture archiving and communication system systems that seek to ensure only specific users can execute queries for specific studies and may dynamically use network credentials within the query to establish access rights on the fly (i.e., query morphing).

Logging includes: (a) Access-based logging (which records who accesses systems and at what time, provides a record of activity, and helps establish a chain of custody for sensitive data activities, both at the system and application levels) and (b) Event-based logging (where some applications will log more details that capture specific types of activities). In the case of manipulating sensitive data and creating reproducibility, various applications will record whenever a user performs a specific

data operation. These data can also be used to establish performance metrics.

14.4.4 System Integrity and Robustness

System integrity and robustness includes disaster recovery, audit trails, reporting, redundancy, and business continuity. For software, we need to maintain its life cycle, integrity, guide lines for vendors, and upgrade procedures. For hardware, we are responsible for configuration management and routine maintenance.

14.5 Advancing Bioinformatics

Cancer research performed over the past decades shows that every tumor is different and every patient is different, and that a treatment should be tailored to an individual patient and his tumor. Although numerous diagnostic and therapeutic tools exist, it is at the moment extremely difficult for a physician and patient to decide on what is best in an individual case.

The availability of molecular tools now enables large-scale, parallel, quantitative, and inexpensive assessments of molecular states. The data from these tools is now also increasingly publicly available. The premier example of an internationally available data resource is GenBank, initially created in the early 1980s (Benson et al. 2007).

There is a culture of open sharing in molecular biology and bioinformatics that continues to grow: sharing of tools, data, findings, and publications. Important tools for bioinformatics are sometimes downloadable for free and, in many cases, have available source code.

The Radiation Oncology Therapy Group and the ACR have been continuously constructing and upgrading the data infrastructure to handle state of the art images, sophisticated dose and dose-volume evaluation for RT, new drugs for chemotherapy, quality of life, and the general electronic patient data.

The large amount of data generated from new blood tests, genomic data, and new biological modifiers, along with imaging, RT, and other clinical data, exist predominantly in separate databases. This current fragmented form is a major barrier to research integrating the biological, physical, and clinical sciences.

Improving the infrastructure for the development of outcomes data and creating a rapid learning healthcare system, where we can learn from each patient to guide practice, is increasingly viewed as crucial. With the recent explosion in the amount of data available, of data, it is essential to employ bioinformatics algorithms and strategies for analysis and modeling, incorporating biological, physical, and clinical data forms.

The American Medical Informatics Association (AMIA) recently augmented the scope of its activities to encompass translational bioinformatics as a third major domain of informatics. The AMIA has defined translational bioinformatics as "…the development of storage, analytic, and interpretive methods to optimize the transformation of increasingly voluminous biomedical data into proactive, predictive, preventative, and participatory health." The objective is to create an innovative framework that leverages data (medical, imaging, treatment, biological, lab, genomic, and psychosocial) that are already being collected in various databases and integrates the data without any demand on the type and way in which the data are collected. The data extraction system needs to be flexible enough to cope with all the differences in local data availability, structure, type, and meaning while at the same time present the data in a coherent model to facilitate analysis. A common metadata model can be developed for a certain type of cancer, specifically, to allow variation and integration in data collection of various forms.

The abundance of new medical information at the molecular, organ, and system levels and the trend towards record healthcare data electronically have opened up the way to using innovative bioinformatics approaches to gain new knowledge on diseases and their management (i.e., predictive models). This includes the development of predictive models that use data from multiple sources based on a combination of imaging, physical, clinical, and biological data, which potentially can improve prediction of survival as well as radiation-induced complications. Both classic statistical and biological models as well as more advanced machine learning techniques, such as support vector machines, will be used.

Whereas diagnostic models are typically used for classification, predictive models incorporate the dimension of time adding a stochastic element. Therefore, predictive models for a certain treatment should not only rely on *Discrimination* (C-statistic or AUC), the ability to separate those with various disease states, but also should assess *Calibration* (the Hosmer-Lemeshow goodness-of-fit test), the probability of correctly estimating a future non-existing event. *Risk classification* (the Net Reclassification Index) can aid in comparing the clinical impact of various models on risk for the individual. The models will not only be data driven but will also integrate available literature and local knowledge.

Other important tools for bioinformatics, such as Significance Analysis of Microarrays (Tusher et al. 2001), TM4 Multiple Expression Viewer (Saeed et al. 2006), GenePattern (Reich et al. 2006), GenMAPP (Dahlquist et al. 2002), and R and Bioconductor (Gentleman et al. 2004; Gentleman 2005), are downloadable for free and, in many cases, have available source code.

At the intersection of biology, medicine, mathematics, statistics, computer science, and IT, biomedical informatics involves the development and application of computational tools to the organization and understanding of biomedical information, so that new insight and knowledge can be discerned. There is now general consensus that biomedical research is transforming into an information science, and biomedical informatics is no longer an option but an integral component of all biomedical research.

Acknowledgment

The author thanks C. J. Myers for assistance with formatting.

References

Abrams, R. A. et al. (2012). Failure to adhere to protocol specified radiation therapy guidelines was associated with decreased survival in RTOG 9704—A phase III trial of adjuvant chemotherapy and chemoradiotherapy for patients with resected adenocarcinoma of the pancreas. *International Journal of Radiation Oncology Biology Physics, 82*(2), 809–816.

Bekelman, J. E. et al. (2012). Redesigning radiotherapy quality assurance: Opportunities to develop an efficient, evidence-based system to support clinical trials—Report of the National Cancer Institute Work Group on Radiotherapy Quality Assurance. *International Journal of Radiation Oncology Biology Physics, 83*(3), 782–790.

Benson, D. A. et al. (2007). GenBank. *Nucleic Acids Research, 35*(Database issue), D21–D25.

Cancer Data Standards Registry and Repository. Available at https://cabig.nci.nih.gov/community/concepts/caDSR/.

Cancer Trials Support Unit. Oncology Patient Enrollment Network Portal Development Plan. https://www.ctsu.org/open/Documents_For_OPEN/CTSU_OPEN_Plan.pdf.

Dahlquist, K. D. et al. (2002). GenMAPP, a new tool for viewing and analyzing microarray data on biological pathways. *Nature Genetics, 31*(1), 19–20.

Gentleman, R. (2005), *Bioinformatics and Computational Biology Solutions Using R and Bioconductor*, Springer Science + Business Media, New York, 2005.

Gentleman, R. C. et al. (2004). Bioconductor: Open software development for computational biology and bioinformatics. *Genome Biology, 5*(10), R80.

Kerrien, S. et al. (2007). Broadening the horizon—Level 2.5 of the HUPO-PSI format for molecular interactions. *BMC Biology, 5*, 44.

National Cancer Institute Clinical Trials Reporting Program. Available at http://www.cancer.gov/clinicaltrials/conducting/ncictrp/main.

Ozsu, M. T. & Valduriez, P., *Principles of Distributed Database Systems, Third Edition*, Springer Science + Business Media, New York, 2011.

Pathak, J. et al. (2011). Evaluating phenotypic data elements for genetics and epidemiological research: Experiences from the eMERGE and PhenX Network Projects. *AMIA Summits on Translational Science Protocols, 2011*, 41–45.

Peters, L. J. et al. (2010). Critical impact of radiotherapy protocol compliance and quality TROG 02.02. *Journal of Clinical Oncology, 28*(18), 2996–3001.

Reich, M. et al. (2006). GenePattern 2.0. *Nature Genetics, 38*(5), 500–501.

Saeed, A. I. et al. (2006). TM4 microarray software suite. *Methods in Enzymology, 411*, 134–193.

Tusher, V. G. et al. (2001). Significance analysis of microarrays applied to the ionizing radiation response. *Proceedings of the National Academy of Sciences, 98*(9), 5116–5121.

15

Research Data Management, Integration, and Security

Ashish Sharma
Emory University

Yusuf N. Saghar
Emory University

Joel H. Saltz
Emory University

15.1 Introduction

Information that is generated during clinical care is stored and managed in well-established systems. These data are accessed using established, predefined access patterns. For example, radiology images are stored in a Picture Archival and Communication System (PACS). The images are represented using the Digital Imaging and Communications in Medicine (DICOM) specification, which in turn defines an information model as well as a set of operations that a consumer uses to interact with the images. A radiologist will typically use a worklist to identify the images that they want to review, and this worklist contains all the necessary information to retrieve the images from the PACS. This predictable and consequently controlled flow of data makes it easier to manage information generated in clinical care. Consequently, there are numerous successful commercial solutions to address the issues around information sharing and data integration when applied in clinical care.

In biomedical, clinical, and translational research, the access patterns for data sharing and integration are quite different. Some of these changes stem from the fact that, in clinical operations, data tend to be patient centric and, in research, data are often study or trial centric. This change, although it may appear to be quite small, has major implications on how data are accessed and used. For example, in a patient-centric data management system, it can become very hard to share data that belong to a particular study if using clinical systems to manage the data. Because the data are rooted at the patient level, every patient in the database has to be searched to determine if the patient data set belongs to a particular research study. Contrast this with a study-centric data management system where all data are rooted at the study/trial and it is therefore trivial to identify all patients that belong to a particular

study. Beyond the obvious performance implications of such a shift, a study-rooted model also impacts vital access patterns, such as access control (who can access what components of a study).

Another difference between the access patterns of clinical data and research data lies in the inherent unpredictability of how data is discovered and used in research. Unlike a clinical operations, where access is very well defined, research data often have to be browsed in novel ways to devise a hypothesis. Such browsing of data requires the execution of queries that are novel and ideally require an on-demand integration of different data modalities. For example, a researcher, who is developing new algorithms to help craft tighter dose margins, will want to explore the planning computed tomography (CT), the radiotherapy (RT) dose and structures, patient diagnosis, and outcome data. To test their algorithms, the researcher may want to retrieve RT structures and possibly the associated planning CT, given a particular diagnosis, dose, and outcome. Most existing systems will have a hard time executing a query where some of the queryable attributes are in the electronic medical records (diagnosis and outcome) and a RT PACS (dose), whereas the objects of interest are in a RT PACS (structure) and radiology PACS (planning CT). From these examples, it becomes evident that existing clinical systems fall short when it comes to research use cases. Next, we explore some specific research data management systems, the benefits of a service-oriented architecture in facilitating data federation, and finally information security.

15.2 Research Image Data Management Systems

In biomedical research, imaging data as well as associated data sets, such as data derived from analysis or review of images, radiation therapy, and patient records, are often stored and managed in proprietary data management systems. Anecdotal evidence indicates that a common strategy is to store the data in folders and maintain the metadata in a spreadsheet. Some researchers create databases with Microsoft Access to manage the metadata. Both these solutions are valid solutions for individual researchers who are storing data for a single project and the data are usually in small quantity. However, if the researcher wanted to share these data with collaborators, then such a management solution will usually fall short. In other words, the system does not scale-out. Similarly, if the researcher wanted to increase the size of the data (i.e., store larger amounts of data or use the same system for multiple projects), then too the system will fall short (i.e., the system does not scale-up).

Recognizing these difficulties, numerous solutions have been developed to address the issue of medical image data management. All these solutions rely on an underlying database, with a generic schema and a user-facing component that allows researchers to explore and download data. Some of them are extensible and allow users to add their own data types and, in doing so, customize the system for their own specific use-case. Many provide an application programming interface (API) for programmatic access, which allows for the development of novel ways to access the data.

One of the most popular systems for data management, particularly in neuroimaging, is the Extensible Neuroimaging Archive Toolkit (XNAT) (Marcus et al. 2006). XNAT is a data management platform that is designed to store neuroimaging data sets in a secure repository, and allow users to browse the data, add new data, and download existing data, via a secure Web page or a REST API. In addition to data management, it also supports common data curation operations, such as project management and quality assurance. It supports a wide variety of image, data such as DICOM and NIFTI. XNAT associates all image data with projects that are defined by a user. Users have the ability to create new projects and add new data or reuse existing data. In other words, a single image data set can be associated with multiple projects. Users can browse the image metadata using a Web interface and visualize the images using the included Web viewer. XNAT also includes a DICOM browser and a gateway that allow clients that rely on DICOM messaging to communicate with XNAT. XNAT is highly configurable and extensible. Users can add new queryable attributes that are relevant to their study. This allows users to customize the way in which they search for their data by relying on data structures that are most relevant to their projects. The data can also be accessed programmatically, via a REST API, which opens the door for integration with other visualization clients, data integration frameworks, and analysis pipelines. Finally, XNAT also includes a pipeline engine that allows its users to create complex computational pipelines that can be easily executed on their data.

Another platform for data management is the National Biomedical Imaging Archive (NBIA) (National Biomedical Imaging Archive 2012). NBIA is a comprehensive image management system that is targeted toward the goal of collecting,

organizing, and sharing DICOM images. It is very well suited for use cases where images are submitted by multiple contributors, a strict QA process is required, and a mechanism is needed for researchers, with varying permissions, to access subsets of the images. It consists of a MySQL database that stores up to ninety DICOM metadata tags. These metadata tags constitute the attributes on which a user can formulate a query. Users can query the archive using a Web interface, which also allows them to browse the images, view thumbnails, and download images via a shopping-cart paradigm. New images are added using the Clinical Trial Processor (Perry 2012), which also allows the data to be de-identified. One unique feature of NBIA is its ability to communicate with other NBIA instances and create a federated archive. In such a scenario, users can connect to one archive and browse images at all other connected archives. NBIA also includes a caGrid (caGrid 2012) data service that provides programmatic access to the archive, allowing developers to create applications that can query and download images from a native client. A native client would be a client that can query and retrieve images directly from the archive without requiring a switch to the Web interface. Examples of native clients would be a radiology workstation or an image analysis program. NBIA is widely used to host multi-institutional or national imaging archives, such as the Cancer Imaging Archive, and the Osteoarthritis Initiative Image Archive.

Whereas NBIA and XNAT are directly targeted toward management of medical imaging and rely on metadata attributes that are either from or have direct correlates in DICOM, MIDAS (Jomier et al. 2010) is a more generic open-source toolkit that allows researchers to create Web-accessible data archives. MIDAS is an open source platform that is developed and maintained by Kitware. It is capable of storing images in the DICOM format as well as non-DICOM objects such as annotations, reports, and clinical records. It provides users with a familiar abstraction of data that is organized in collections of folders. However, the data and the metadata are stored and managed in an underlying database. Data can be uploaded using a programmatic API or via desktop applications that can connect and communicate using DICOM messaging. As data are added, certain operations such as anonymization, quality control, and metadata extraction can be automatically triggered. Data providers can restrict access to portions of data. MIDAS has been used to manage data in projects such as "Give a Scan," where patients can contribute their own medical scans for research. MIDAS is also being used to manage the journal archive of the *Insight Journal*.

The three systems described here are popular examples of free and open-source solutions to manage biomedical imaging data sets. Each of them provides a nice alternative to a PACS for image data management. However, they are usually geared toward managing a single data modality. With imaging becoming more popular in clinical and translational research, there is an increasing need for the ability to run integration queries, which include attributes from both imaging and nonimaging data sets.

15.3 Data Integration

Data integration is the ability to fuse data from multiple sources and create a single virtualized archive. The objective is not to replicate data from multiple sources into a single archive. Rather, it is to identify a subset of data using attributes from multiple data sources. For example, consider a hypothetical project where the PIs are developing strategies to predict outcome in low-grade glioma patients using diffusion-weighted magnetic resonance imaging. A possible integration query might be "Find DW-MR images of all patients whose Karnofsky Performance Score (KPS) is greater than 70 and who were less than 40 years when baseline imaging was acquired." To obtain these images, we have to query the clinical records from our IRB-approved cohort of low-grade glioma patients and query for only those patients whose KPS > 70. Next, for this set of patients, we query our image management system to identify a smaller subset whose age at time of baseline imaging was less than 40. Although it is conceivable that a user could execute this query manually, imagine a new query that also includes attributes from pathology reports, other laboratory reports, and gene expression data. Slowly, it becomes evident that a more automated strategy is needed for data integration.

Data integration is helpful when exploring data sets containing multiple modalities. Because data integration is directly related to the systems that manage the various data sources, it is very difficult to develop a generic data integration platform. These difficulties can be broken down into three categories. The first of these is data security. This is discussed in detail in the next section. The next set of obstacles center around the use of standards. Data are not always stored and represented with community-defined standards. Instead, researchers often use their own internal representations to describe the data or more importantly the metadata. Some of the issues related to standards can be mitigated with tools that ease data management and conversion. The tools described earlier achieve this goal for DICOM data.

Another major obstacle revolves around the availability of a programmatic interface that would allow users to develop systems that can interact with the various data sources. The tools described earlier all include APIs that are publicly available and well documented and thus allow users to develop data integration platforms. There have been a few community-driven efforts that help create public facing APIs for different data management systems. A recent and quite comprehensive effort was the caGrid toolkit (Oster et al. 2008; caGrid 2012) that was developed within the caBIG program. caGrid provides a services interface and allows developers to expose their data resources using a common vocabulary and a documented interface. It is possible to use caGrid and develop applications that can securely interact with a federation of data sources.

15.4 Security

The discussion so far has dealt with systems to manage imaging data sets in biomedical research and integrate them with other data types. However, the data may contain Protected Health Information (PHI) and access to such data must be regulated. Even if PHI is removed and data are anonymized and stripped of all identifiers, secure access maybe necessary to protect the intellectual property of the data provider. Therefore, a secured environment is necessary when dealing with biomedical or clinical research data. Data security is often achieved by a combination of policies, best practices, and technological solutions. The technical solutions cover a wide range of issues ranging from encryption (to secure data storage and communication) to ensuring that data are accessed by authorized users. Here, we discuss the latter, namely, how a user's identity is established and how systems can enforce policies to prevent unauthorized access.

15.4.1 Authentication

Authentication is the process of establishing an identity of an individual or an entity. It tries to answer the question "Is the person really who he/she claims to be?" Authentication is the first line of defense and generally the gateway to access most information systems. There is no one single or standard way to authenticate a user and establish identity. However, central to authentication is the concept of a "credential." A credential is like a unique key, which is tied to a user's identity, and must be presented to gain access to a resource. A common and well-understood form of a user credential is the combination of a username and password. Other examples of credentials include X-509, public key certificates, and Security Assertion Markup Language (SAML) tokens.

Another key aspect of authentication is the choice of authentication technique, which varies a lot depending on organizational policies, system complexity, economy of use, and factors such as level of assurance desired. What works in one setting may not work in another. It is important to understand the topology of the target organization to implement correct authentication technique. For example, if there is a single security domain (say a university where all consumers of data are employed by the university, and all data reside on systems that are secured by the university), authentication mechanisms are relatively easy to deploy. However, in a multi-institutional research project, consumers and data providers may reside in distinct organizational units and consequently rely on different authentication systems. In such an environment, it becomes much harder to communicate user credentials and establish a trust fabric among the different credential providers.

Single Institutional Setting. In a classic single institutional setting, there is typically one security domain. Most organizations use Lightweight Directory Access Protocol (LDAP) (Koutsonikola & Vakali 2004) for storing user identities along with their attributes: Full name, Phone number, E-mail, Group membership, etc. LDAP is also a protocol for authentication. Applications interface with LDAP at the programmatic level. When a user logs in, his credential is authenticated against the LDAP server. If authentication is successful, a session is established and remains active unless the user logs out or if he remains inactive for a long time.

Now if all applications within the organization fall under the same security domain and use the same LDAP server to source identities, a user can use the same credentials to access these applications. However, they may have to login multiple times, once for each application. Ideally, a user should be able to login once and be able to access all applications in the same session. This is addressed by Single Sign On (SSO) (The Open Group 2012) technology. With SSO, a user is required to login once. If successful, a security token, using the industry standard SAML (Cantor et al. 2005) format, is generated to attest the identity of the user. This token is a time-sensitive document that asserts the user identity and associated user attributes. Typically, a Certificate Authority (CA) is designated to digitally sign it. All applications involved in this transaction implicitly trust the CA. The token is associated with the user session and gets passed around as the user accesses different applications. All of this happens behind the scenes and is abstracted from the user. When a user logs out, the security token is revoked and all applications are notified. Shibboleth (Morgan et al. 2004; Cantor & Scavo 2005) is the most commonly used open sourced product for Web SSO.

Multi-institutional Setting. Things become interesting in a multi-institutional setting. Such a setup is desired for collaborative research and data sharing. Just as it is for a single institution, user identities are stored and managed in LDAP. However, because a credential from institution X would not be recognized by an application at institution Y, we need a mechanism to perform federated identity management and assertion. WS-Security and WS-Trust (Curbera et al. 2005; Nadalin et al. 2007) are open standards designed to address this issue. It describes an entity called Security Token Service (STS), a Web service that brokers authentication between a Service Provider (SP) and a Service Consumer (SC). The function of STS is threefold: issue, validate, and revoke security tokens. STS validates identities from an institution's identity provider (IDP) such as LDAP.

The sequence of transactions that occur when a user from institution X tries to access a resource at institution Y is as follows (Figure 15.1):

1. SC acting on behalf of User A requests a security token from the STS installed at institution X. This STS sources identities from a LDAP server hosted internally.
2. STS uses the credential supplied by SC to authenticate against the LDAP server.
3. Upon success, STS issues a security token.
4. SC embeds this security token in each request it makes to access resource at institution Y.
5. Institution Y cannot verify this token by itself because it originated at X. Thus, it takes the token and goes to the STS for verification.
6. STS verifies the token, because it was the one who issued it in the first place and sends the response back.
7. Institution Y now knows that the token confirms the identity of User A who belongs to Institution X. It grants access to User A.

Institutions avoid handling credentials from other institutions by having STS broker authentication between consumers and service providers across institutional boundaries. The interface provided by STS for issuing and validating a security token is standard, thus eliminating the need to change application code when configuring against a different IDP. The only change that happens when migrating from one IDP to another is reconfiguring STS, but that does not affect applications depending on it. There are other technologies like OpenID (Recordon & Reed 2006; O.I.D. Authentication 2011), OAuth (Hammer-Lahav 2010), and Shibboleth that are also very commonly used in a multi-institutional setting.

15.4.2 Authorization and Secure Data Sharing

Different authentication strategies help establish a users' identity but it is up to a data provider to consume that identity and determine if the authenticated user is allowed to access a particular data set that is stored on a given resource. Authorization systems try to answer "Who can do what?" by establishing and enforcing policies. There are different techniques that can be used to perform authorization. We will limit our discussion to a commonly

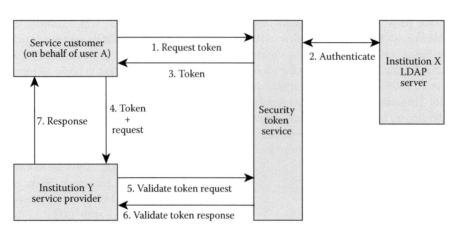

FIGURE 15.1 Security sequence for interinstitutional transactions.

used pattern, namely, Role-Based Access Control (RBAC), and an emerging standard in access control, namely, Extensible Access Control Markup Language (XACML).

At its core, every access-control policy tries to answer three questions, namely, the "who, what, and how." In other words, each policy typically contains:

"who" — the requester, formally known as the *subject*;
"what" — the *resource* that is being requested; and
"how" — the *action* of the request

In addition to these three elements, other information, such as time of request, request context, and target environment, may be relevant and necessary for crafting and enforcing a policy. It is imperative that all the subjects, resources, and actions have unique identifiers to avoid any ambiguity. A policy document may contain many rules. Each rule in a policy is evaluated to make an access-control decision. Because rules can be either affirmative or negative, it is possible to have conflicts in policy evaluation, that is, a subject may be allowed access by one policy and denied by another. When this happens, a conflict resolution strategy is needed. A commonly used conflict resolution strategy is "first-come first-served," where the policy that is applied first is the one that gets enforced. Other options such as "least restrictive" (apply the policy that grants most access) and "most restrictive" (apply the policy that grants least access) are examples of other conflict resolution strategies. Next we describe some widely used access-control techniques.

RBAC. As the name indicates, RBAC policies are written around the notion of a "role." Users are assigned roles depending on their type of work and responsibilities and roles are mapped to resources. For example:

ADMIN can read/write/delete on the database table PATIENT_INFO
RESEARCHER can read from the database table PATIENT_INFO

Here, ADMIN and RESEARCHER describe the roles, whereas PATIENT_INFO describes the resource. Roles are hierarchical in nature, meaning one role can inherit from multiple roles. For example, a Principal Investigator role inherits from a Researcher role. Also, a person may be associated with multiple roles. All these factors play an important role in policy evaluation.

RBAC is good when simplicity is desired. The RBAC model aligns well with the hierarchical organizational structure of most companies—roles are assigned based on separation of duty. Employees having different roles have different job functions and, by extension, different permissions. Moreover, roles can overlap with and/or inherit from other roles. It is straightforward to describe such a system using RBAC.

The simplicity of RBAC comes at a cost, though. There are situations where contextual information (e.g., the time of request, state of the system, and other environmental factors) needs to be taken into account when making an authorization decision. RBAC does not consider contextual information in making decisions. Another very important aspect that RBAC lacks is

delegation. The textbook definition of delegation is "the act of empowering to act for another." It is very common for someone to delegate responsibilities to someone else under special circumstances for a limited time period. A RBAC policy does not discriminate between a delegated credential and an actual one. Everyone in a certain role is treated the same way whether they are acting on their own or on someone else's behalf.

XACML. XACML (Moses 2005) is an Extensible Markup Language–based model for describing access-control policies. This specification was created to promote common terminology and interoperability between authorization implementations by multiple vendors. XACML is an example of Attribute-Based Access Control wherein subjects, actions, and resources are broken down into their respective attributes that are used to create fine-grain access control policies. For example, a subject may be associated with attributes such as name, organization, email address, role, etc. Thus, it is possible to describe a policy that reads "Allow all employees of Hospital A and role Physician access to Electronic Health Record of a patient X."

An XACML policy document is a collection of rules. Each rule in XACML is made of attributes from a subject, an action, and a resource. XACML employs a declarative functional language to describe how an attribute match occurs. More than one rule can match a given request; hence, a "rule combining algorithm" declared in the policy document determines the final decision. As mentioned earlier, it can be most restrictive, least restrictive, first matched, or a custom algorithm defined by the system administrator.

XACML is the most comprehensive standard for access-control. It is compatible with other standards such as RBAC and SAML. Policies written in XACML are portable across applications. Thus, a policy written for one system can be used in another. Delegation is also supported by XACML as of version 3.0 (Rissanen 2010). On the downside, XACML is complex and can be difficult to implement. For situations where RBAC suffices, XACML may be an over-engineered solution.

15.4.3 Data Authorization and Instance Based Authorization

The techniques we discussed so far are applicable for resources that are static in nature. Some examples include file, database, URL, API, etc. However, the data that reside in the database keep changing. When access-control decisions need to be made based on the state of the underlying database, these techniques fall short because there is no way of incorporating this dynamicity into authorization policies. The resource definition keeps changing as the data in the database change. Consider a policy that reads "Allow Jim (doctor) access to all those patients who have been diagnosed with diabetes in the last 3 months." Although it is possible to describe such a policy with XACML, it is extremely hard to evaluate it. This is due to the ever-changing definition of the resource—"patients diagnosed with diabetes in the last 3 months." The patients matching this description can never be predicted at the time of writing the policy. To evaluate the policy correctly, a database containing information about all the patients, their time of admission/diagnosis, disease, and other related information should be consulted. Assuming it is a relational database, the SQL query to fetch this information should look like:

```
"SELECT PATIENT_ID FROM PATIENT_INFO WHERE
DISEASE = 'DIABETES' AND DIFFERENCE (DATE_OF_
ADMISSION, TODAY) < 3"
```

This type of access-control is termed as Instance Level: each instance (row) in the database table is checked to see if it fits the definition of the resource and then the access-control policy is applied.

It is extremely useful in cases where binary decisions for authorization (allow or forbid) are insufficient. For example, if a patient tries to access the records of all the patients (including his own), clearly, it cannot be allowed and so access is denied. However, wouldn't it be great if instead of denying access, the patient were shown just his data, filtering out the results of the other patients?

Implementing systems that perform instance level authorization is not trivial. This is due to performance considerations. Two commonly used approaches are:

1. Rewrite query.
2. Filter query results.

Query rewriting involves modifying the original query according to user permissions. This involves fiddling with the syntax and grammar of SQL. Result filtering, on the contrary, is easy to implement but adversely affects the performance. There is yet another approach that is simple to implement and does not incur a performance penalty. In this approach, we leverage "Database Views" to define a custom view for each role. A view, created using a SQL query, is an overlay on top of a database table. It shows only that portion of data that the user is allowed to see. When a user requests some data, instead of hitting the actual database table, this view is summoned for execution of the query. It guarantees that the user sees only what he is allowed to see. Unfortunately, this technique is only good for read operations and limited to relational databases supporting the concept of "views."

15.5 Conclusion

This chapter focused on strategies for storing and managing imaging data and data related to, or derived from, imaging. Unlike clinical workflows, research workflows do not follow a consistent data access pattern; consequently, systems developed to manage data in clinical workflows do not meet the needs of research data management. These research data management systems also need to have the ability to support a variety of image

formats and be extensible. These systems often provide APIs for programmatic access, which in turn has enabled the development of various data integration systems. These data integration systems can create a unified view that brings together imaging with associated nonimaging data, such as clinical phenotypes. Finally, in order for these systems to work with data from human subjects, it is necessary that the systems provide a secure, regulated mechanism to reach the data. Security has various components, and the extent to which data are secured is dictated by the nature of the data, the requester, and the type of data access request.

References

caGrid (2012). Available at www.cagrid.org [cited 30 Sep 2012].

Cantor, S. & Scavo, T. (2005). Shibboleth architecture. Protocols and Profiles 10.

Cantor, S. et al. (2005). Assertions and protocols for the oasis security assertion markup language. OASIS Standard.

Curbera, F. et al., *Web Services Platform Architecture: SOAP, WSDL, WS-Policy, WS-Addressing, WS-BPEL, WS-Reliable Messaging and More*, Prentice-Hall PTR, 2005.

Hammer-Lahav, E. (2010). The oauth 1.0 protocol.

Jomier, J. et al. (2010). An open-source digital archiving system for medical and scientific research. Open Repositories 7.

Koutsonikola, V. & Vakali, A. (2004). LDAP: Framework, practices, and trends. *IEEE Internet Computing, 8*(5), 66–72.

Marcus, D. S. et al. (2006). XNAT: A software framework for managing neuroimaging laboratory data.

Morgan, R. L. et al. (2004). Federated Security: The Shibboleth Approach. *Educause Quarterly, 27*(4), 12–17.

Moses, T. (2005). Extensible Access Control Markup Language (XACML) version 2.0. OASIS Standard 200502.

Nadalin, A. et al. (2007). WS-Trust 1.3. *OASIS Standard 19*, 2007.

National Biomedical Imaging Archive (2012). [cited 30 Sep 2012]. Available at http://ncia.nci.nih.gov/ncia.

O.I.D. Authentication (2011). 2.0-Final.

The Open Group (2012). Single Sign-On. Available at http://www.opengroup.org/security/sso/ [cited 30 Sep 2012].

Oster, S. et al. (2008). caGrid 1.0: An enterprise grid infrastructure for biomedical research. *Journal of the American Medical Informatics Association, 15*(2), 138–149.

Perry, J. (2012). CTP-The RSNA Clinical Trial Processor. Available at http://mircwiki.rsna.org/index.php?title=CTP-The_RSNA_Clinical_Trial_Processor [cited 30 Sep 2012].

Recordon, D. & Reed, D. (2006). OpenID 2.0: a platform for user-centric identity management.

Rissanen, E. (2010). eXtensible Access Control Markup Language (XACML) version 3.0 (committe specification 01). Technical report, OASIS. Available at http://docs.oasisopen.org/xacml/3.0/xacml-3.0-core-spec-cd-03-en.pdf.

16

Open-Source Informatics Tools for Radiotherapy Research

Joseph O. Deasy
Memorial Sloan-Kettering Cancer Center

Aditya P. Apte
Memorial Sloan-Kettering Cancer Center

"The captain has said too much or he has said too little, and I'm bound to say that I require an explanation of his words." — *Treasure Island*, by Robert Louis Stevenson

16.1 Radiotherapy Research Informatics Needs in 2012

Radiotherapy is inherently a data-rich medical specialty, requiring imaging at many steps in the patient management process, including disease workup, treatment planning, treatment delivery, treatment follow-up, and possible retreatments. Data components produced include 3D images for diagnosis and simulation of treatment, markup of images for anatomic segmentation, target volume specification, volumetric (3D) dose calculations, treatment machine commands, treatment setup images, and patient management records. The relationships between all these data elements are also important. Newer data sources include intra-treatment monitoring of position via real-time fiducial marker location. Research and development require the intensive study of these data sources to better understand the relationship to outcomes. These factors create a need for adequate research informatics software tools whether for research or clinical development.

In this chapter, we provide a brief review of radiotherapy research informatics needs and relevant software. We emphasize the Computational Environment for Radiotherapy Research (CERR), both because of its widespread applications and our role in its development, but we also review other excellent tools. An associated Web page (opensource4rt.info) has been created, with links to tools mentioned in this review and to other tools for handling more general images.

16.2 Sharing Data in Radiation Oncology

Interest in comparing and sharing clinical radiotherapy data for research increased in parallel to the increasing complexity of digital data used for treatment planning and delivery. In the 1980s, it was recognized that conducting multicenter cooperative clinical trials based on volumetric treatment planning would require uniformity in target and normal tissue definitions (in addition to other issues). Achieving uniformity in these definitions required the development of systems to gather and review digital data to assure compliance with protocol parameters. Purdy et al. established the 3D Quality Assurance (QA) Center at Washington University in St. Louis, Missouri, to provide this service to the Radiotherapy Oncology Group (RTOG), which is funded by the National Cancer Institute (NCI) and conducts clinical trials, and later, to other clinical trial groups. The 3D QA Center, now named the Image Guided Therapy Center (as it is referred to hereafter), oversaw gradual improvements in the format of the American Association of Physicists in Medicine's (AAPM) Report 10 (Baxter et al. 1982), which is now referred to as the "RTOG format." Early 3D treatment planning systems incorporated the RTOG format as a standard output option, thereby facilitating the widespread participation of radiotherapy clinics in 3D RTOG (and other cooperative group) clinical trials. In the late 1980s and early 1990s, the Digital Image Communications in Medicine (DICOM) imaging standard to exchange radiological data was developed, providing the field of radiology, and eventually radiation oncology, with well-defined, machine-readable, data definitions that facilitated radiology data set communications. Although the RTOG format is still in use, DICOM has now become the dominant method for exchanging treatment planning data.

Radiation oncology is currently perhaps the most digitally complex field of medicine. Patient management is evolving away from a "snapshot" process and toward a more complicated process that involves the ongoing collection, analysis, and modification of treatment information. Figure 16.1 depicts this progression. It is clear that key standards and software systems (either commercial or open-source) that essentially tell the story of the patient via contextualized linkages of data are missing. CAISIS is an example of such an informatics system, developed at Memorial Sloan-Kettering Cancer Center (MSKCC), for surgical oncology outcomes research (Potters et al. 2003). CAISIS is patient-centric and chronologically organized, rather than organized by protocol, which is typical of research databases. However, CAISIS has not been extended to image-based medicine (although this is feasible in the future). Despite these major gaps, there are many tools that can provide help for radiotherapy researchers.

It must also be recognized that the field of radiation oncology, unlike, for instance, that of genomics, has yet to embrace data sharing or even anonymized data-pooling, and this slows progress considerably (Deasy et al. 2010).

16.3 Key Data Handling Challenges

A useful, although incomplete, list of key challenges in the handling of radiotherapy data includes the following:

1. Communications: The creation, validation, and transmission/reception of data in standard formats, especially the still-developing DICOM-Radiotherapy (RT) standard. The older RTOG format, which grew out of the AAPM

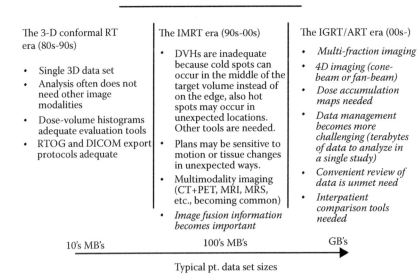

FIGURE 16.1 Evolving complexity of radiotherapy informatics needs. Data analysis is becoming more challenging as a wider variety of imaging, clinical, and patient management data are required for research analysis. Unmet challenges are italicized. Acronyms: ART, adaptive radiotherapy (e.g., therapy in which the treatment parameters are changed in response to patient/disease changes); DVH, dose-volume histogram, which plots the fraction of a defined tissue or organ receiving greater than the *x*-axis dose; IGRT, image-guided radiotherapy (e.g., therapy with frequent reimaging); IMRT, intensity-modulated radiation therapy.

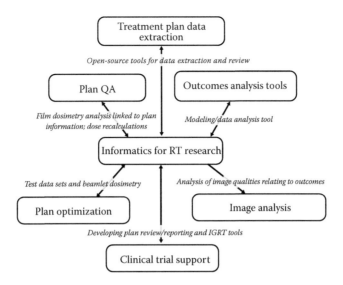

FIGURE 16.2 Schematic illustration of radiotherapy informatics needs. Outcomes analysis tools are the subject of a companion chapter by El Naqa.

Task Group 10 (Baxter et al. 1982) and a later report written by Michael Goitein (1985), is also still sometimes used but is fading in importance. Although non-RT DICOM tools (some hyperlinks are included on the Web page) can be of great value in radiotherapy, they are not the focus here.

2. Manipulation and analysis of radiotherapy and imaging data. Once data are in hand, tools to conveniently manipulate 2D, 3D, and now 4D data are needed. Applications for these tools include planning optimization research, QA analysis, image analysis, clinical trials support/plan review, and data extraction (Figure 16.2).

3. Archiving and query tools are needed. There are few open-source research archiving tools available that are radiotherapy enabled. A central unmet challenge is the need for defined archival storage schemes that store data in a way that preserves the proper context, story, and linkage (CSL) of the various data components. This again implies that chronological ordering may ultimately be indispensable.

4. Tools/Web sites to allow the sharing and comparison of image data. Another unmet need in data analysis is some kind of common space/Web site for radiotherapy researchers, where users could easily review data sets from other institutions and upload anonymized treatment data.

16.4 Why Open-Source Software?

Although there are many commercial packages that apply to various components of the radiotherapy informatics world, open-source software tools are needed for several reasons. First, research tools require unprecedented flexibility that would be difficult to support commercially. Such tools should be

modifiable and extensible to support cutting-edge research and development projects. Second, the need to deploy research tools is often unpredictable. Therefore, advanced budgeting for such tools can be difficult to justify. Last, any direct impact on institutional income is typically nonexistent. Commercial vendors are therefore unlikely to supply most research needs. At some point, the need for flexibility in research tools becomes much greater, and the potential audience small enough, that commercial motivation is inadequate.

Unfortunately, for similar reasons, funding agencies, such as the NCI, have seldom supported open-source software in radiation oncology, although there are important exceptions, such as the National Science Foundation–supported Slicer3D project and past NCI funding for the Plan University of North Carolina (PLUNC) system (further discussed below.) Thus, software projects typically must look for external funding "on the margins" for a range of projects where the software itself is not the main goal. Nonetheless, the motivation to create open-source software tools has been strong enough to result in the many available tools reviewed here.

Rather than being merely "money savers" (which they usually are not), successful open-source software projects can have several key benefits difficult to obtain from commercial products. In particular, they can (1) form a backbone of enabling technology for research, clinical prototype projects, or clinical components themselves; (2) provide a conduit for research and developmental collaborations; (3) support developments that incrementally build on previous work; (4) provide open tests of old or new standards [DICOM, dose-volume histogram (DVH) analyses, imaging tests, etc.]; and (5) encourage involvement and contributions.

Open-source projects usually suffer from varying degrees of defects in reliability, documentation, support, and maintenance (sometimes to a lesser extent than commercial software, however). Desirable features and specifications for a successful open-source research software project are briefly described below.

16.5 Questions to Consider Before Committing to an Open-Source Research Software Project

Before embarking on an open-source research software project, the scientist/clinician/developer should carefully consider the following questions. If the answer to most of these questions is "No," then the project is unlikely to be successful.

1. Are the needs to be addressed (use-cases) critical (i.e., "worth the effort")? Do many researchers have the same needs? Is this area of research or clinical need likely to continue as an area of focus for an extended period of time (more than five years)?
2. Is the system envisioned likely to be useful enough to the developers that continuous future maintenance seems assured?
3. Is future commercial competition that addresses key use-cases unlikely?
4. Will the software system be flexible, allowing for both incremental and large improvements in features?
5. Will the software be easy to program and maintain? Is it likely to be easy to recruit developers?
6. Will the software system work across the major computing platforms (Mac OS, Microsoft Windows, and Linux)?
7. If successful, does it seem likely that a new generation of developers and key supporters could be recruited to continue the development of the system? In other words, is it likely that there will be enough continued "payoff" to attract future collaborators/developers?

16.6 Requirements of Open-Source Research Software in Radiation Therapy

In addition to the preceding questions, open-source software should seek to fulfill certain software requirements, including the following:

1. Coding standards, including naming standards, should be adopted. A straightforward, human-readable, and human-comprehensible coding style should be the norm. Simplicity, yet generality, should be preferred. Most of the necessary information about a routine should be placed at the beginning of the relevant section of computer code in a completely standardized fashion. Code should be readable enough to avoid elaborate commentary (which can be wrong and/or obsolete).
2. Following good coding practice, functionality should be encapsulated in small, general-purpose routines/functions

whenever possible. A system library of such utility functions should be maintained.
3. The use of numerical constants, system preference values should be isolated to one point in the system. Any function-specific constants should be stated at the top of a routine.
4. QA methods should be used to minimize code defects, including independent code reviews.
5. The software architecture should be well planned and likely to age well. A fundamental characteristic of the system should be the capacity to add functions, modules, etc. Modularity of the system should allow some well-tested modules to be locked into place, whereas some modules are simultaneously modified and new modules are added.
6. Calculations performed by the system must be independently checked for accuracy, such as DVH reduction accuracy, tumor control probability calculations, and region-of-interest polygon conversion testing.
7. Documentation that is easy to use and read should be viewed as a necessity, not a luxury.
8. The tracking and (when needed) fixing of bugs should be an ongoing, high-priority activity.
9. Known bugs and their statuses should be listed on a publicly available Web page.

Open-source systems should all have a set of automated regression tests that are as comprehensive as possible. The purpose of these tests would be to ensure that changes to the system, for example, changes meant to speed up a routine, lead to results consistent with previous versions. To our knowledge, no systems fulfill this desirable feature.

16.7 Selection Criteria

Typical successful open-source software projects involve extensive collaborations within a community of developers (Scacchi 2002). The extent to which the software projects presented here incorporate feedback varies. Typically, user feedback is very important and there are only a small number of developers involved. This is perhaps not surprising given the expertise needed to effectively contribute. We selected open-source codes in this compilation based on a few simple requirements: there must be at least one publication describing the system, the focus must be primarily related to radiotherapy (codes useful in radiotherapy are listed on the Web page associated with this chapter), the code must be available for free download (possibly after course attendance, for example), the source code must be available, and the copyright must allow unrestricted research use (at least). Tools reviewed include the following: I Q Works QA software, IMAGE_RT nuclear medicine analysis software, plastimatch deformable image registration software, Histogram Analysis in Radiation Therapy (HART) dose-volume analysis software, Biological Evaluation of Radiotherapy Treatment Plans (BIOPLAN) outcome prediction software, EGSnrc dose calculation software, DPM dose calculation software, PLUNC treatment planning software, and CERR radiotherapy data analysis software.

The restriction to codes with a radiotherapy focus eliminates several important open-source tools that are nonetheless useful in radiotherapy, such as the Slicer3D visualization tool, the Insight Segmentation and Registration Toolkit, and various DICOM-focused tools. Links to those tools are nevertheless given on the Web site associated with this chapter: opensource 4rt.info.

16.8 QA: The I Q Works Software System for Automated Analysis of QA Images

I Q Works is a software program developed at the Edinburgh Cancer Center by Andrew Reilly et al., begun as part of Dr. Reilly's Ph.D. project. The idea for I Q Works came from a group of computed tomography (CT) users in the United Kingdom who recognized the need for software to perform QA on digital imaging equipment in radiotherapy (Reilly & Thwaites 2007). I Q Works is intended to provide a consistent, flexible, and convenient tool for the graphical and numerical analysis of DICOM test images. I Q Works has been designed to be useful for mammography, CT scan analysis, 2D image analysis (e.g., electronic portal image detector images), and magnetic resonance imaging (MRI). It includes a wide range of characteristics that are analyzed, including alignment, geometric distortion, the CT number/Hounsfield units calibration curve, noise power spectrum, modulation transfer function, and detector quantum efficiency.

I Q Works, according to the authors, has been extensively tested, has been found to agree with independent tests of the same quantities, and has been in routine use in the Edinburgh Cancer Center Radiotherapy Clinic for more than five years. I Q Works has been adopted by several professional groups in the United Kingdom as a basis for a new open-source QA system. I Q Works has a helpful, though incomplete Wiki page at iqworks. org. Technically, I Q Works is not yet open-source. However, the authors promise to release it as a SourceForge open-source project in the future.

16.9 Nuclear Medicine Image Analysis: The IMAGE_RT Package

IMAGE_RT is an open-source software tool developed by Graves and colleagues at Stanford (Graves et al. 2007). IMAGE_RT is based on the Interactive Data Language (IDL), which has roots in the analysis of planetary and space sciences data. The IDL language is a feature-rich, interpreted scripting, and data analysis vector language, similar to Matlab, and includes an integrated visualization and graphical user interface toolset. Like CERR and Matlab, IMAGE_RT has the advantage of rapid prototyping, but, like CERR, can be expected to be somewhat less computationally efficient than native C++ code. The issue of computational efficiency has gradually reduced over time with ever-increasing computational power.

Like CERR, IMAGE_RT can be modified and extended, either through adding IDL scripts or by linking to lower-level language programs. IDL's graphical user interface building toolset is extensive and easy to use.

IMAGE_RT is focused on the ability to read, register, and analyze images and structure sets relevant to radiotherapy. For example, positron emission tomography (PET) images can be overlaid and fused with CT images. Region-of-interest tools are extensive and include grouping, splitting, smoothing, morphing, etc. A full list of features can be found on the IMAGE_RT Web site hosted by SourceForge (http://rtimage.sourceforge.net/documentation.html).

16.10 Deformable Image Registration: The Plastimatch B-Spline Software System

Although many algorithms have been published for radiotherapy (CT/MRI/PET) image registration, few are open-source. Plastimatch (http://plastimatch.org/), developed at Massachusetts General Hospital/Harvard Medical School by Gregory Sharp and colleagues, is an open-source deformable image registration tool that implements efficient elastic deformable registration using B-splines (Wu et al. 2008).

Plastimatch has been shown to effectively and efficiently perform deformable image registration, for example, on the MIDRAS deformable image registration comparison (Brock and Deformable Registration Accuracy Consortium 2010). Plastimatch can effectively use multiple cores and graphical processor units (GPUs) to speed up the registration process (Sharp et al. 2007; Wu et al. 2008; Shackleford et al. 2010).

At MSKCC, we have used plastimatch for the challenging problem of matching the anatomy of different head-and-neck patients to potentially understand the effect of irradiating different muscle groups on the risk for radiation-induced trismus (inability of the patient to open his/her jaw). We found that plastimatch seemed to make the matches reliably according to collaborating radiotherapists who reviewed the results. In a (literally) head-to-head comparison, plastimatch outperformed another open-source registration tool, elastix.

Plastimatch was previously integrated with Slicer3D and now has been integrated with CERR. Plastimatch, available from plastimatch.org, is actively maintained and is under continuous development.

16.11 Histogram-Based DVH Analysis and Outcomes Predictions: HART

HART is an open-source Matlab-based program developed by Anil Pyakuryal and collaborators at Northwestern University Memorial Hospital to facilitate analysis of parameters derived from DVHs (Pyakuryal et al. 2010). HART was first released in 2007 and has been actively maintained. Like BIOPLAN and CERR, HART includes some common TCP and NTCP models, as well as many tools specifically implemented to analyze region-of-interest DVHs, including standard summary statistics, such as conformality and dose homogeneity index, generalized equivalent uniform dose, and tumor control probability. For example, HART implements a method for looking at partial DVHs as a function of relative lateral, vertical, or longitudinal planes. The system also includes some facility for analyzing longitudinal series of DVHs ("4D" analysis). HART, such as CERR, runs on top of Matlab and reads DICOM- or RTOG-formatted files as input to analyses. The developers maintain a list of active projects using HART, along with Wiki and FAQ documentation, at http://www2.uic.edu/~apyaku1/.

16.12 Predictive Models Based on Dose Distributions: The BIOPLAN Package to Facilitate Radiobiologically Driven Dose Prescriptions

BIOPLAN is a flexible software system designed, developed, and maintained by Alan Nahum and his colleagues (Sanchez-Nieto & Nahum 2000). More specifically than the other software considered in this chapter, it is designed to facilitate radiobiologically driven dose prescriptions based on plan-specific TCP and NTCP estimates. A range of TCP and NTCP models, including the Lyman-Kutcher-Burman (LKB) NTCP/gEUD model, seriality NTCP model proposed by Brahme and colleagues, Webb and Nahum TCP model 1993, and Poisson uniform radiosensitivity TCP model, are implemented. (For a broad discussion of NTCP and TCP models, see Moiseenko et al. 2005.) BIOPLAN includes some relatively novel information displays, such as the so-called delta-TCP method, which shows the relative impact of cold spots within a tumor on TCP. As is typical, only DVH data are needed. The philosophy of the system is to create a usable tool to understand the expected impact of potential changes in prescription dose or dose distribution on outcomes. A particularly useful feature is a plot of TCP and NTCP values as a function of a potentially varying prescription dose. BIOPLAN, which is under active maintenance, is freely available from Nahum and colleagues.

16.13 Caveats to Using Histogram Analysis Packages

In our view, there are some important caveats to applying radiobiological models to DVHs.

1. TCP models have many sources of uncertainty, including the basic formulation of the model. As pointed out elsewhere (Moiseenko et al. 2005), using a strict Poisson-based model of TCP that does not build in inter-tumor heterogeneity is problematic. Resulting model fits not only have

a radiobiologically unrealistic number of clonogens but also yield predictions that are too sensitive to cold spots (Webb & Nahum 1993). In fact, if tumors were of uniform radiosensitivity, there would be so few clonogens that the Poisson approximation itself would be inappropriate; voxels would often contain one or zero clonogens, clearly implying that only binomial, not Poisson, statistics would be appropriate. Tumor sequencing data makes it clear that most solid tissue tumors have a different range of genetic alterations, usually hundreds or more, supporting the idea that radiosensitivity is likely to vary between patients (The Cancer Genome Atlas Network 2012). Intratumor sampling of radiosensitivity parameters (Taghian et al. 1995) also supports the presence of significant heterogeneity (although intratumor heterogeneity is usually neglected in TCP models). Furthermore, there are currently little data to validate the dependency of any TCP model on cold spot characteristics. TCP model predictions, therefore, should be interpreted with care, as only one element in a decision-making process.

2. Frequently, the need arises to consider the effect of a dose distribution as if it were given with an alternative fractionation schedule, assuming the linear-quadratic equation. A common method used is to first transform all DVH bins to an effective dose, usually 2 Gy, thereafter making comparisons based on this transformed DVH. It should be noted that, although this method might seem intuitively correct, there is little validation that this is accurate when the volume effect is large (i.e., when mean doses correlate to the outcome). The relationship between the volume effect and fractionation corrections is currently only poorly understood.

3. The volume of DVHs reported by treatment planning systems may (unexpectedly) not be the total contoured volume. It is important to know the details of how commercial systems compute DVHs. In one commercial system (Pinnacle; we have not tested this in the latest version), if a portion of a contoured anatomic structure is outside the dose calculation matrix boundaries, then the DVH parameters only reflect the volume of the organ inside the dose matrix and neglect volumes outside the dose computation matrix. This can be an issue for esophageal volumes, for example.

16.14 Dose Calculations: The EGSnrc Monte Carlo Code System

The EGS systems of Monte Carlo codes were initially developed by Ralph Nelson and colleagues at the Stanford Linear Accelerator Center to simulate coupled electron-positron-gamma showers produced by high-energy collision experiments. A key motivation of this project was to design adequate shielding for high-energy experiments, and EGS was widely used for this purpose in the 1980s. Subsequently, at the National

Research Council of Canada, the Ionizing Radiation Standards (IRS) group, led by David W. O. Rogers (with key contributions from Alex Bielajew, Iwan Kawrakow, and others) developed "EGSnrc," which represents modifications and extensions to EGS that made the resulting codes highly accurate, flexible, and useful for the simulation of radiation cascades above a keV through the maximum energies used in radiotherapy. Key advances included transport algorithms that were more efficient, and extensive measurement, theoretical, and computational work to verify and fine-tune the interaction cross-sections involved. The over-arching twin goals of this effort are usefulness with very high accuracy.

The resulting code system was named EGSnrc and is now distributed by the IRS group at the Web site (http://irs.inms.nrc. ca/software/egsnrc/). This use of EGSnrc to simulate realistic radiotherapy accelerator geometries produced many insights and was a key development in the progress toward truly accurate machine-head models for clinical dose calculation algorithms and even for other types of algorithms.

In collaboration with Rock Mackie and colleagues at the University of Wisconsin-Madison, Rogers et al. developed extensions of EGS, specifically to simulate radiotherapy linear accelerators. This package of computer codes, called BEAM (now BEAMnrc), allows the user to compute dose to a CT phantom using either full incident-particle (phase-space) files (generated from previous head-model simulations) or head-model-based parameters. Software modules allow for the accurate geometrical modeling of accelerator head models (with visualization capabilities) and their resulting simulation. Graphical user interface modules (based on Tkl) are used to simplify and visualize the construction of accelerator components.

EGSnrc and associated codes have been used in hundreds of published studies, resulting in the most successful open-source software project in the field of radiotherapy. The EGS project has been open-source from an early phase. According to Rogers (2006), "Literally thousands of people got copies of the EGS4 code and applied it to a wide variety of problems resulting in extensive benchmarking that led to 'credibility.' It also meant that there was a constant stream of corrections and improvements being made as people worked with the code, and these corrections were fed back into the distributed version. This model of open software is common today in many contexts but at the time, it was not the norm."

At this point, the EGSnrc system is massive and has many capabilities beyond what we have space to describe. The level of quality of this code system is famously high, despite the massive technical complexity, and the diagnosis and correction of bugs takes a high priority. Support and maintenance is ongoing. Introductory courses are offered periodically and long-term user support (within reason) is provided to course attendees. The EGSnrc codes run across Windows, Linux, and (as stated) probably Mac OS. The IRS group maintains extensive documentation and offers download access on their Web site (irs.inms.nrc.ca/software/egsnrc/).

16.15 Dose Calculations: The Dose Planning Method Monte Carlo Code System

Whereas EGSnrc seeks to maintain impeccable accuracy over a wide range of materials and energy ranges, other codes have been developed to take advantage of algorithmic advances that dramatically decrease run-times (in particular, regarding boundary crossing and electron track segments) without significantly compromising accuracy. In particular, the Dose Planning Method (DPM) Monte Carlo software code is a completely open-source code for radiotherapy simulations that focuses on accurate simulations in materials with low atomic numbers (i.e., human tissue). The first DPM paper was published by the developers in 2000 (Sempau et al. 2000). DPM has an unusually open license and has been used for several research projects. For example, Jing Cui of our group developed a head model in DPM and the subsequent code was refined and validated by Davidson et al. (2008) for routine use at the Radiological Physics Center (RPC).

DPM is available for download (http://www-personal.umich. edu/~bielajew/). It is not clear what level of bug fixing is maintained. A major drawback is that there is no documentation in the usual sense; an effort to develop a useful Wiki for this code package would potentially be very helpful.

16.16 Clinical Treatment Planning: The PLUNC System

The open-source software system that has the highest clinical utility is probably the PLUNC treatment planning system. PLUNC has its origins in one of the first 3D beams-eye-view treatment planning systems, written by G. Sherouse. It has been subsequently developed and maintained by the UNC Department of Radiation Oncology (Tewell & Adams 2004). PLUNC has been supported by funding from the National Institutes of Health (NIH) and UNC.

PLUNC is a full-featured treatment planning system and has been used routinely in the clinic since 1996 by UNC (it was previously used by Duke University and the University of Chicago). PLUNC now has many advanced features, including an advanced convolution-superposition-based dose calculation algorithm; advanced autosegmentation methods; and intensity-modulated radiation therapy planning algorithms, including unusual fixed-field attenuator-based algorithms that do not require multileaf collimator field shaping. Standard treatment planning elements are supported in PLUNC, including DVH analysis and beams-eye-view.

Recognizing the complementary nature of PLUNC and CERR prompted us to write a PLUNC format import module within CERR. PLUNC has a unique role in the open-source radiotherapy world: it is currently the only true open-source, clinically usable treatment planning system. It provides a measure of flexibility for developmental projects that has not been matched by any commercial system. This alone is likely to provide a rationale for continued support of PLUNC, even if one day it is not the main clinical planning system used by any clinic.

PLUNC consists of C and C++ code and has been compiled for multiple platforms. A drawback is that much of the code and associated architecture is relatively old and difficult to correctly modify; there are apparently relatively few "PLUNC hackers." PLUNC is available for free for clinical or research use once a user has registered. The PLUNC system is well maintained by UNC.

16.17 General Data Handling: The CERR Software Package

CERR, pronounced "sir," was originally conceived in the early 1990s as a general, scriptable tool to provide convenient extraction, review, analysis, comparison, and manipulation of radiotherapy treatment planning data. CERR is basically an extensible research treatment plan analysis system see Figure 16.3. Before developing CERR, a review was undertaken of software technologies that could be used to implement CERR. Matlab was the obvious choice after data-type improvements were delivered in Matlab version 6.1 ("cell" and "struct" arrays) that made capturing and manipulating an entire patient's treatment plan, including metadata, in a single data object highly manageable, even convenient. In addition, Matlab provides graphical and numerical tools and is scriptable, a highly desirable feature of research informatics tools (Dubois & Yang 1999). CERR has been under continuous development in some way for more than ten years. The overall goal of the CERR project is to provide a general framework for technology that conveniently manipulates and analyzes heterogeneous radiation oncology data. Design goals included an easy-to-understand program organization, programming style, and transparent structure; a program architecture that allows us to easily maintain both forward and backward data compatibility; and an architecture that allows for the addition of arbitrary data elements and new functions.

Matlab features that support these design goals include an easy-to-learn syntax, vectorized matrix operations, cross-platform compatibility (Microsoft, Windows, Apple Macintosh, Linux, etc.), a compiler that makes stand-alone, freely redistributable executables, and a graphical debugger.

A primary goal of the project has always been to guarantee the accuracy of the data, and extensive QA work has been undertaken to ensure that results from CERR-based DVH analyses are correct. CERR has been used for several years by QA groups, including the QA Review Center (QARC), the Advanced Technology Consortium, and the RPC for various radiotherapy data handling applications. QARC has customized CERR extensively and integrated it into their protocol-compliant submission review process. An extensive independent validation test of seventy-three features of the customized system was performed

for multiple RTOG and DICOM data sets according to Food and Drug Administration standards (Ulin et al. 2010). Although several instances of disagreement between commercial systems and DVH results were found, these were considered easily explained, based on the slightly different approaches to computing DVH parameters, especially for small regions. CERR has been used to successfully import data from a wide range of commercial treatment planning systems (more than twelve), including gammaknife. CERR is also able to import PLUNC treatment plans.

For the last several years, CERR has been available free for download from cerr.info; successive versions of CERR have been downloaded many thousands of times. To date, CERR has been cited 170 times in the peer-reviewed literature. CERR was recently given a GNU (library) copyright. CERR has many features, of which we will highlight a few in this section. A CERR Wiki is maintained that contains extensive help information. A CERR Google group (http://groups.google.com/group/cerr-forum) has several hundred members and an active discussion rate.

16.18 Internal Structure of a CERR Archive

The internal structure of CERR treatment plan data follows the AAPM/RTOG data specification translated into Matlab high-level cell arrays and structures (keyword indexed arrays). This greatly facilitates structured access and organization of plan information. The CERR format is practically "self-describing." A single cell array holds all the planning data, as shown in Figure 16.4.

CERR also supports unloading CT scans and dose distributions onto local compressed files to keep the size of the archive manageable. New cell array data types can be introduced without breaking data compatibility. CERR itself is written to maintain data compatibility in two ways: either forward (archives created under older CERR versions, at least 3.0 or later, will continue to open under new versions) or backward (archives created under newer CERR versions will open without error under versions created in the last several years, at least back to version 3.0). Compact yet powerful Matlab script programs can be constructed using hundreds of built-in functions. As one example, the CERR command

```
[xDoseVals, yDoseVals, zDoseVals] =
getDoseXYZVals(planC{indexS.dose}(doseNum))
```

retrieves lists of the coordinates of dose voxel centers for the dose distribution indexed by the integer label doseNum.

A general tool is supplied to compute and compare plan metrics and biological models for selected dose distributions. For example, a built-in tool for the LKB/gEUD NTCP model interactively shows how parameter selection affects the model's results.

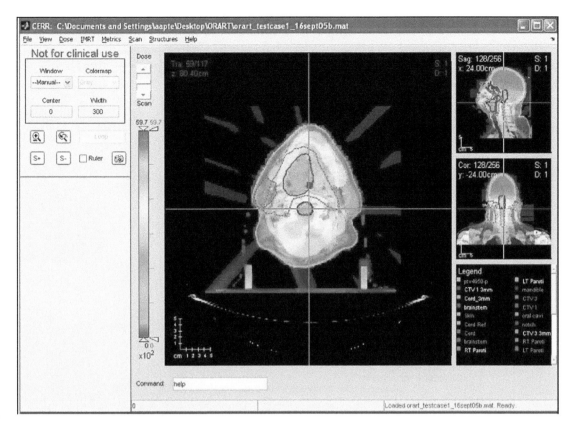

FIGURE 16.3 (See color insert.) Basic CERR view window shows transverse, sagittal, and coronal slice views along with the selected dose distribution. This plan was submitted to the authors by the University of Michigan and is distributed with CERR as part of the Operations Research Applications in Radiation Therapy Toolbox (Deasy et al. 2006). The view is highly customizable (see the Wiki link for more details). Different color bar color maps are available. The dose color-wash color bar itself can be interactively rescaled to cut-off cold dose regions. Isodose lines are also available. Window snapshots can be conveniently captured and automatically placed into a Web page, a feature the developers often use.

CERR provides an alternative, standard file format for storing radiotherapy treatment planning data. Relative to DICOM-RT, two key advantages of the CERR format are that the data are typically all held in a single file, thereby reducing effort and potential confusion, and perhaps more importantly, the data are accessible within a highly convenient system that facilitates customized interrogation or manipulation using Matlab and the entire CERR system.

CERR is a general, useful framework for a wide variety of analysis functions. Figure 16.5 shows a plan comparison window. Figure 16.6 shows an example of manual image registration. Figure 16.7 shows an image registration review tool unique to CERR: the "mirrorscope" (see caption for details). Figure 16.8 shows a novel "dose location histogram" (DLH) tool that indicates the location of cold spots relative to the edge of a target volume. Figure 16.9 shows a novel cohort review tool that can be used to compare, with a consistent color bar, dose distributions for different patients that surround the same structure type in up to sixteen different patients. Figure 16.10 shows a novel "dose projection" 2D display that, similar to maximum intensity projections, shows the location of cold spots within a target or hot spots in a normal structure.

16.19 Radiotherapy Research Informatics: An Ongoing Challenge

Because of the explosion of information in radiation oncology, progress in the field's research is inextricably linked to the quality of informatics tools. In the future, cutting-edge physics studies will probably use many sets of images, spread across modalities and time, which may be related by deformation maps. As discussed, the problem is made more complicated by the need to develop convenient and powerful computational tools and frameworks to analyze the data. Part of this challenge will be met by the ongoing increase in everyday computing power. In addition to the obstacle of a general lack of cooperation in sharing anonymized data sets within radiation oncology (Deasy et al. 2010), a key missing piece in the "informatics ecosphere"

```
Command Window
planC =
  Columns 1 through 5
    [1x1 struct]    [1x0 struct]    [1x1 struct]    [1x9 struct]    [1x14 struct]
  Columns 6 through 10
    [1x2 struct]    [1x7 struct]    [1x28 struct]   [1x1 struct]    [1x1 struct]
>> planC{1}
ans =
            tapeStandardNumber: 3.2000
         intercomparisonStandard: 3.2000
                    institution: 'WASHINGTON UNIVERSITY'
                    dateCreated: 2001100
                         writer: 'Lisa Westfall'
>> planC{4}(1)
ans =
                    imageNumber: 73
                      imageType: 'STRUCTURE'
                     caseNumber: 1
                    patientName: 'Patient, Prostate'
                  structureName: 'BLADDER'
           numberRepresentation: 'CHARACTER'
                structureFormat: 'SCAN-BASED'
                  numberOfScans: 72
              maximumNumberScans: 100
          maximumPointsPerSegment: 1024
           maximumSegmentsPerScan: 10
               structureEdition: 'set01'
                     unitNumber: ''
                         writer: ''
                    dateWritten: ''
                 structureColor: ''
           structureDescription: ''
             studyNumberOfOrigin: ''
                        contour: [1x72 struct]
>> planC{3}.scanInfo(1)
ans =
                    imageNumber: 1
                      imageType: 'CT SCAN'
                     caseNumber: 1
                    patientName: 'Patient, Prostate'
                       scanType: 'TRANSVERSE'
                       CTOffset: 1024
                     grid1Units: 0.0940
                     grid2Units: 0.0940
           numberRepresentation: 'TWO'S COMPLEMENT INTEGER'
                   bytesPerPixel: 2
              numberOfDimensions: 2
                sizeOfDimension1: 512
                sizeOfDimension2: 512
                         zValue: 127.7000
                        xOffset: 0
                        yOffset: 0
                          CTAir: 0
                        CTWater: 1024
                 sliceThickness: 0
                  siteOfInterest: 'Prostate'
```

FIGURE 16.4 A command-line example of accessing an underlying CERR structure, in this case associated with the CT scan. Keyword metadata follow previous standards (AAPM/RTOG or DICOM) as much as possible, although many unique keywords have been added to the system.

is an open-source Radiation Therapy Picture Archival and Communication System that supports sophisticated offline processing. Furthermore, tools and keyword ontologies that define data storage systems that preserve the clinical "storyline," or CSL, within a given patient's data archive need to be developed.

16.20 Continuing Need for Collaborative Efforts

Despite efforts to change the paradigm, it is unfortunately the case that the NIH, the primary source of funding for biomedical

research in the United States, provides only marginal funding to support important biomedical software advances. The limited ability to obtain resources makes collaboration in this area crucial to success. Thus, efforts to effectively organize consortia or other working groups are imperative. Moreover, in our opinion, it is naïve to believe that vendors will satisfy the software needs of academic radiation oncology research, many of which are considered neither low-hanging fruit nor the maximally profitable part of the commercial market. Therefore, research informatics will remain a key challenge for medical physicists, radiation oncologists, dosimetrists, and engineers.

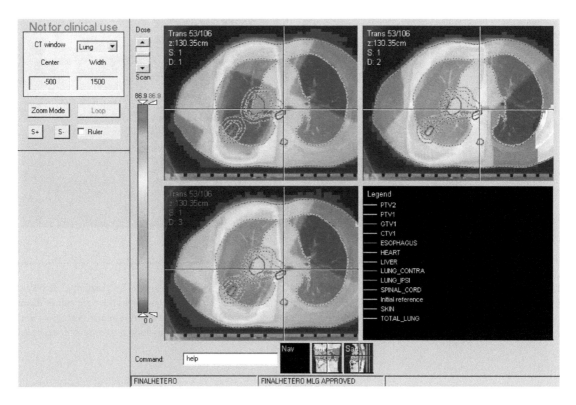

FIGURE 16.5 **(See color insert.)** CERR can conveniently compare multiple dose distributions registered to the same slices. An interactive, dragable, dose profile tool is available.

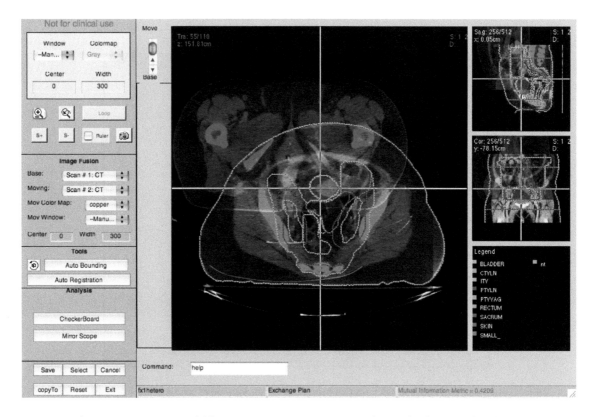

FIGURE 16.6 Manual image registration is available in CERR. Transparent images can be translated or rotated in all three views with the mouse.

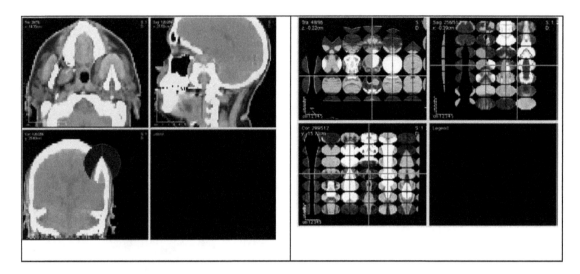

FIGURE 16.7 "Mirrorscope." As an example of a specialized data review tool in CERR, shown here are the "mirrorscope" displays of image registration. Within a moveable circle, the left side is the original image (shown outside the mirrorscope), whereas the right side of the mirrorscope is the mirror reflection of the registered image for the same region as the moving image. Thus, correct registration manifests as a clearly symmetric image. (left) A moveable mirrorscope applied to a single region; (right) the gridded mirrorscope gives a rapid overview of registration across the image.

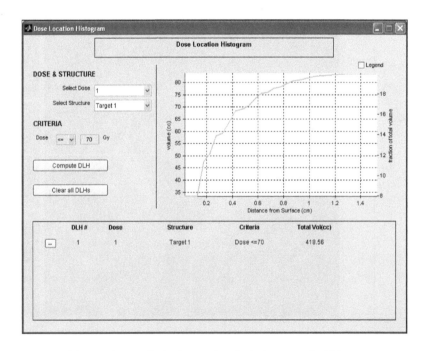

FIGURE 16.8 DLH. CERR has several dose analysis tools that better identify the location of cold/hot regions compared with DVHs alone. The DLH identifies the absolute volume of dose above or below a given dose within different maximum distances from the structure surface. In this case, the DLH shows the volume receiving less than 70 Gy for all voxels closer to Target1 than x cm.

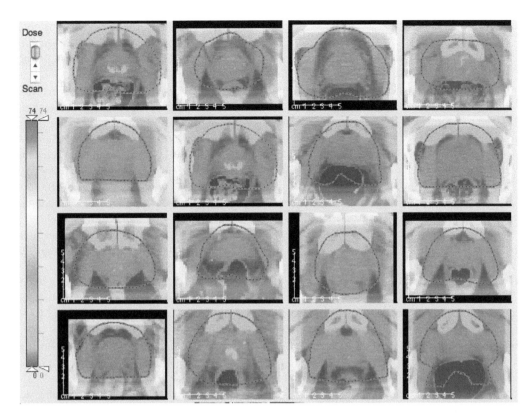

FIGURE 16.9 **(See color insert.)** Cohort review tools. This tool allows CERR to directly compare multiple plans for the same patient or different patients. Stepping through a slice moves all images in unison. Without such a tool, it is difficult to adequately appreciate changes in treatment plans within a population.

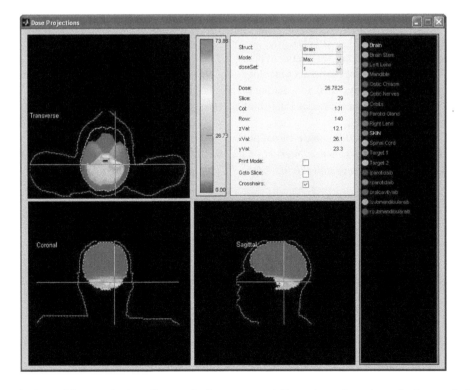

FIGURE 16.10 **(See color insert.)** Dose projections. Dose projections are essentially maximum intensity projections applied to dose. They facilitate the rapid identification and localization of cold spots in a target volume or hot spots in a normal structure.

References

Baxter, B. S. et al. (1982), *American Association of Physicists in Medicine (AAPM) Report No. 10: A Standard Format for Digital Image Exchange*, American Institute of Physics, New York, 1982.

Brock, K. K., and Deformable Registration Accuracy Consortium (2010). Results of a multi-institution deformable registration accuracy study (MIDRAS). *International Journal of Radiation Oncology Biology and Physics, 76*(2), 583–596.

Davidson, S. et al. (2008). A flexible Monte Carlo tool for patient or phantom specific calculations: comparison with preliminary validation measurements. *Journal of Physics: Conference Series, 102*, 012004.

Deasy, J. O. et al. (2006). A collaboratory for radiation therapy treatment planning optimization research. *Annals of Operations Research, 148*(1), 55–63.

Deasy, J. O. et al. (2010). Improving normal tissue complication probability models: The need to adopt a "data-pooling" culture. *International Journal of Radiation Oncology Biology and Physics, 76*(3 Suppl), S151–S154.

Dubois, P. F. & Yang, T. Y. (1999). Extending Python with Fortran. *Computing in Science & Engineering, 1*(5), 66–73.

Goitein, M., Specifications for tape format for exchange of planning information, version 2.2, in Evaluation of Treatment Planning for Particle Beam Radiotherapy, Ed. Goitein, M., et al., Sec. 5.11, National Cancer Institute, Bethesda, Maryland, 1985.

Graves, E. E. et al. (2007). RT_Image: An open-source tool for investigating PET in radiation oncology. *Technology in Cancer Research and Treatment, 6*(2), 111–121.

Moiseenko, V. et al., Radiobiological modeling for treatment planning, in The Modern Technology of Radiation Oncology: A Compendium for Medical Physicists and Radiation Oncologists, Ed. Van Dyk, J., Medical Physics Publishing, Madison, Wisconsin, 2005, 185–220.

Potters, L. et al. (2003). A chronological database to support outcomes research in prostate cancer. *International Journal of Radiation Oncology Biology and Physics, 56*(5), 1252–1258.

Pyakuryal, A. et al. (2010). A computational tool for the efficient analysis of dose-volume histograms from radiation therapy treatment plans. *Journal of Applied Clinical Medical Physics, 11*(1), 3013.

Reilly, A. J. & Thwaites, D. I. (2007). Automated analysis software for the objective assessment and optimisation of radiotherapy image quality. *International Journal of Radiation Oncology Biology and Physics, 69*(3 Suppl), S719.

Rogers, D. W. (2006). Fifty years of Monte Carlo simulations for medical physics. *Physics in Medicine and Biology, 51*(13), R287–R301.

Sanchez-Nieto, B. & Nahum, A. E. (2000). BIOPLAN: Software for the biological evaluation of radiotherapy treatment plans. *Medical Dosimetry, 25*(2), 71–76.

Scacchi, W. (2002). Understanding the requirements for developing open-source software systems. *IEEE Proceedings—Software, 149*(1), 24–39.

Sempau, J. et al. (2000). DPM, a fast, accurate Monte Carlo code optimized for photon and electron radiotherapy treatment planning dose calculations. *Physics in Medicine and Biology, 45*(8), 2263–2291.

Shackleford, J. A. et al. (2010). On developing B-spline registration algorithms for multi-core processors. *Physics in Medicine and Biology, 55*(21), 6329–6351.

Sharp, G. C. et al. (2007). GPU-based streaming architectures for fast cone-beam CT image reconstruction and demons deformable registration. *Physics in Medicine and Biology, 52*(19), 5771–5783.

Taghian, A. et al. (1995). In vivo radiation sensitivity of glioblastoma multiforme. *International Journal of Radiation Oncology Biology and Physics, 32*(1), 99–104.

Tewell, M. A. & Adams, R. (2004). The PLUNC 3D treatment planning system: A dynamic alternative to commercially available systems. *Medical Dosimetry, 29*(2), 134–138.

The Cancer Genome Atlas Network (2012). Comprehensive molecular portraits of human breast tumours. *Nature, 490*(7418), 61–70.

Ulin, K. et al. (2010). Validation of CERR for use as a digital data review tool at the Quality Assurance Review Center. *Medical Physics, 37*(6), 3245.

Wu, Z. et al. (2008). Evaluation of deformable registration of patient lung 4DCT with sub-anatomical region segmentations. *Medical Physics, 35*(2), 775–781.

IV

Imaging Informatics

17

NCI Cancer Imaging Program: Imaging Informatics and Radiotherapy Implications

John Freymann
SAIC-Frederick, Inc.

Justin S. Kirby
SAIC-Frederick, Inc.

C. Carl Jaffe
Boston University

17.1 Introduction

The Cancer Imaging Program (CIP) is one of the four programs in the Division of Cancer Treatment and Diagnosis of the National Cancer Institute (NCI). The mission of CIP is to promote and support the cancer-related basic, translational, and clinical research in imaging sciences and technology and the integration and application of these imaging discoveries and developments to the understanding of cancer biology and to the clinical management of cancer and cancer risk. An implemented informatics infrastructure to strengthen the value of cancer imaging data, especially that arising during clinical care, has long been seen as a critical building block to achieving those goals. A changing cultural dialogue is now occurring in science, which urges a transition to open research data. Within that new regime, larger, more integrated, data collections can be assembled from multiple sources offering greater statistical strength to scientific pursuits and minimizing investigator isolation and data redundancy. For example, the UK Royal Society recently issued a report urging researchers, funders, and institutions to shift away from a culture where data are viewed as a "private preserve." The report emphasizes that such data need to be not just accessible but also "intelligible" (i.e., accompanied by explanatory metadata). Consistent with that concept the NCI CIP, by early 2011, constructed The Cancer Imaging Archive (TCIA) with the purpose of providing a Web-accessible, dynamically growing repository of data sets that include both diagnostic images and clinical metadata.

17.2 Foundation of NCI Imaging Informatics—TCIA

Imaging has been increasingly acknowledged as possessing a useful endpoint in clinical trials, especially as either objective response or progression-free survival. Although imaging's use for treatment response assessment is becoming more accepted, many issues remain unresolved, particularly in the needs for image quantification and the challenge of standardizing imaging protocols across institutions participating as cooperative group members in clinical trials. For the past decade, NCI has supported the development of informatics tools and semantic standards, many of which support cancer imaging research. Several broad initiatives have addressed such issues:

1. The NCI/Association of American Cancer Institutes initiated the formation of Image Response Assessment Teams, now built upon by imaging efforts within Clinical and Translational Science Award (CTSA);
2. The workshop organized by the National Institute of Standards and Technology entitled "Imaging as a Biomarker: Standards for Change Measurements in Therapy" in 2006 (National Institute of Standards and Technology (NIST));
3. The Radiological Society of North America (RSNA) coordinated Quantitative Imaging Biomarkers Alliance between drug and equipment industries, imaging scientists, and professional societies with an aim to develop and

advance standards for using computed tomography (CT), 18-fluoro-deoxyglucose positron emission tomography (FDG-PET), and dynamic contrast enhanced (DCE) magnetic resonance imaging (MRI);

4. The Quantitative Imaging Network (QIN) (NCI Cancer Imaging Program) grown from the NCI program announcement "Quantitative Imaging for Evaluation of Responses to Cancer Therapies" and designed to promote research and development of quantitative imaging methods for the measurement of tumor response to therapies in clinical trial settings, with the overall goal of facilitating clinical decision-making.

5. As a means to supply data set repositories supporting research in those arenas, the NCI CIP funded TCIA (NCI Cancer Imaging Program) (Figure 17.1) to provide private health information (PHI)–protected deidentification and image data curation services to overcome traditional radiologic image data-sharing barriers. TCIA was conceived as a so-called "Big Data" concept of sufficient scale and functionality to acquire and distribute statistically robust large sample sets of images accompanied by clinical and other nonimaging metadata where possible. These tools arose to meet the various challenges to medical image sharing as a response to a science gap analysis as well as evolving policies by research sponsors that urged or required data sharing. The TCIA, which is operated as a subcontract to Washington University in St. Louis, has an operations and maintenance team consisting of project managers, image quality-control curators, subject

matter experts [cancer, imaging, Digital Imaging and Communication in Medicine (DICOM), and technology], systems and network administrators, software developers, and a customer support center. It is run on three 12 core servers, 2.66 GHz, mirrored 600 GB 15K RPM SAS drives, 48 GB memory (two additional 24 GB servers joining the "mini-private-cloud"), a three-head storage system with multiple RAID 6 FC-connected storage arrays with near-line SAS disks and redundant L7 (app level) load balancing switches with SSL offload.

17.2.1 Challenges to Data Sharing

Advances in network connectivity and infrastructure have opened an opportunity to overcome long-standing barriers that have impeded cross-disciplinary research. Despite widespread use in diagnosis, staging, and therapeutic response monitoring, clinical imaging has not achieved its full potential as a scientific resource, particularly in therapeutic clinical trials, because of long-standing cultural and institutional impediments. Clinical diagnostic images have benefited from a well-structured digital foundation and a decade or more of increased adoption and evolution of the DICOM image file structure, yet they are rarely shared in their original digital form because imaging's traditional purposes have been centered on individual patients and institutions. Thus, imaging data are usually available only as text reported results by subspecialty experts rather than shared with cross-disciplinary researchers who might wish to apply quantitative analysis and

FIGURE 17.1 TCIA provides free access to large imaging collections accompanied by metadata and information about ongoing research as well as a deidentification knowledge base.

image processing methods to those original images. Health Insurance Portability and Accountability Act (HIPAA) concerns are often cited as the rationale for rigorously restricting access to original clinical images for fear of releasing PHI. Research attempting to mine data from diagnostic medical record reports suffer from a lack of standard terminologies and ontologies and vary subjectively from expert to expert and day to day. A trusted resource that makes privacy-protected clinical images accessible in an organized repository with matching case-linked clinical metadata has been an acknowledged unmet science community aspiration for more than a decade.

17.2.2 The Cancer Genome Atlas

Intended to serve the cutting-edge molecular biology science community, a major public resource has recently emerged that offers comprehensive human cancer clinical data is the NIH Cancer Genome Atlas "The Cancer Genome Atlas (TCGA)" (National Institutes of Health). This multi-institutional NIH program obtains tissues from more than twenty cancer types and makes the results of its extensive genomic and proteomic analysis publicly available on the Web. Its original focus, to a large extent on molecular cellular pathways, bypassed inclusion of radiologic imaging such as MRI, PET, CT, and radiation therapy (RT) data. The CIP TCIA effort was planned in part to address that gap, as those technologies offer potentially untapped research value because they reflect the tissue phenotype expressed by intracellular genomic mechanisms. More usefully, imaging can noninvasively visualize the entire tumor and tumor changes over treatment times, whereas tissue genomic analysis is limited by subsampled biopsy material, usually obtained on a one-time or limited basis. Moreover, biopsies that only retrieve a portion of a tumor often undercharacterize a tumor's heterogeneous nature. Nonetheless, pairing imaging data with a research subjects' correlating genomic analysis has traditionally been challenging and is usually limited to small databases at individual institutions.

17.2.3 Confronting Barriers to Sharing

From a researcher's perspective, the benefits of sharing data have not always outweighed perceived costs and risks. Taking time to publish source data typically does not reward individual researchers with additional publications; often, there have been no serious efforts by sponsors to require researchers to do so. There is usually insufficient funding allotted to researchers to account for the time and energy spent on such activities. Therefore, it was absolutely critical for the TCIA mission to make the image archive submission process as trouble free as possible for data providers who might wish to contribute to archives like TCIA. Because DICOM (NEMA) is a near-universally adopted clinical imaging standard, CIP was able to encourage tools and processes to address specific curation and deidentification requirements of this data standard.

17.2.4 DICOM Image Data Deidentification

DICOM presents unique challenges to public data sharing that heretofore had not been addressed. Each DICOM image may contain hundreds of data elements describing acquisition parameters, patient demographics, etc., many of which are critical for downstream scientific research but which could contain PHI. In 2012, the DICOM standard incorporated detailed guidance for image deidentification (DICOM Supplement 142). NCI CIP collaborated with the National Electrical Manufacturers Association (NEMA) DICOM standards committee and developers of clinical imaging tools to implement standardized deidentification profiles into their software. TCIA adopted open-source tools and used them to implement new methodologies for preserving scientifically critical information, including that preserved in vendor-proprietary fields known as "private tags," while safely removing detectable PHI. A field-tested infrastructure was developed and is now available from TCIA, which uses software that allows image deidentification to occur inside the data contributor's firewall, before the data leave their institution. This has proven to ease concerns about potential misuses or leakages of PHI data that might leave the contributor's site. TCIA has adopted Clinical Trials Processor (CTP) for this task, which was developed by the RSNA. Substantial efforts are made by TCIA staff to preconfigure all software as much as possible and generally minimize any routine work on the part of the image data provider.

TCIA combines these major recent advances in standards and tooling, with a multidisciplinary team to significantly reduce the burden of deidentification on potential imaging data providers (Figure 17.2). TCIA maintains a full service team providing institutional review board (IRB)–approved curation and deidentification. It maintains a quality-control team, image submission experts for data providers, and a helpdesk for end-users. HIPAA-compliant deidentification and sharing of DICOM image data is now an achievable task that need not be a barrier to broader community research use of clinical image data. TCIA maintains

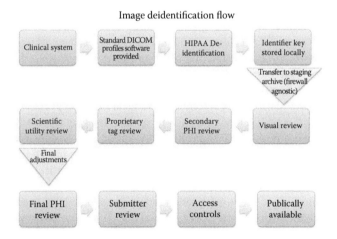

Image deidentification flow

FIGURE 17.2 TCIA provides sophisticated tools and processes to minimize the deidentification and submission burden on the image provider.

a public reference resource containing documentation and tools (TCIA). Data contributors can share clinical images in the original DICOM format without being locally burdened by onerous and costly deidentification efforts.

17.2.5 Clinical Metadata and Annotation

Human diagnostic imaging in DICOM format includes attribute tags that allow analysis of scan parameters such as noise properties and registration or change over time. However, it has its greatest research value when it can be assembled into groups based on some common clinical event such as survival or therapeutic course as is the case with progression-free survival. Such "metadata" are hosted on the TCIA Wiki site, contained in spreadsheets and XML files, or in external data repositories, which have been optimized for particular data types. Further research value has been achieved by providing linked expert-drawn poly-line boundaries of important anatomic structures that offer initiation points for image processors to match or enhance object recognition and quantitate tumor change. These resources can leverage other freely available software such as 3DSlicer, which is a sophisticated tool for constructing and visualizing volumes of interest. In like manner, TCIA is developing an additional analytic interface using the similarly freely available XNAT (http://xnat.org) software platform to expand the capability for storing and querying these types of associated image and clinical metadata.

17.2.6 Additional Imaging Informatics Resources and Activities

Ongoing efforts to ease contributor and user burdens include development of service-based architecture, thin client, and emerging network advances such as Cloud opportunities. Obtaining IRB approval from institutions planning to contribute data is much easier when detailed documentation and procedures are shared with them. TCIA's deidentification policies are all posted publicly so they can be easily referenced by IRBs and submitters as needed. Where there is a preplanned intent to share data in advance of deidentification among specific researchers who are known to each other, it may not be necessary to completely deidentify the image collection if the research team is defined to the IRB or allowed by the consent and the researcher participant signs a "data sharing agreement" (National Institutes of Health). TCIA is unique among large-scale medical imaging archives in that nearly all data that go into TCIA is available to anyone who registers. There is no cost associated with registration and no requirement for research proposals or any other potential barriers to entry that might discourage researchers from accessing the data. Beyond TCIA, information and links to additional biomedical imaging resources can be found at the CIP-conducted Survey of Biomedical Imaging Archives (Cancer Imaging Program). Of additional note to researchers, there is an independently maintained index of free medical imaging software, "IdoImaging" (Crabb).

17.2.7 Facilitating Research

The principal purpose motivating the TCIA effort has been an imperative to encourage cross-disciplinary research teams and allow them to easily form interest-based ad hoc dynamic multi-institutional teams. CIP first provided image collections with clinical metadata that was organized by specific diseases that presented a challenge for clinical image quantitative research. TCIA goes beyond just making the images and related data available. While funding is not provided, TCIA may be able to host regular teleconferences for ad hoc research teams, provide free use of the TCIA Wiki, and assist with leveraging open-source informatics software to analyze the data. TCIA demonstrates real-world evidence that production of significant scientific findings at low cost is possible by following an open-data, open science paradigm.

17.3 Successful Use-Cases

TCIA contains more than twenty purpose-built collections that are publicly accessible for download. Many include additional metadata and accompanying documentation. The Lung Image Database Consortium image collection consists of diagnostic and lung cancer screening thoracic CT scans with marked-up annotated lesions. It is a resource that can be used for development, training, and evaluation of computer-assisted diagnostic methods for lung cancer detection and diagnosis.

The six separate Reference Image Database to Evaluate Therapy Response collections are intended for generating an initial consensus on how to harmonize data collection and analysis of quantitative imaging methods as applied to measuring the response to drug or RT. The long-term goal is to provide a resource to permit harmonized methods for data collection and analysis across different commercial imaging platforms using imaging as a biomarker for therapy response such as would be required to support multisite clinical trials.

In 2004 in presenting to the NCI Executive Committee an American College of Radiology Imaging Network (ACRIN) proposal to conduct the ACRIN CT Colonography trial, justification was made that publicly accessible image data sharing would offer a valuable asset to a broad-based image processing research community. Adding to the many merits of that proposal, the data-sharing component was strongly endorsed by the Executive Committee. ACRIN completed the trial expeditiously and its results were published in the *New England Journal of Medicine* in fall 2008 to wide interest. ACRIN has allowed the wider research community access to a portion (825 cases) of the data from CT Colonography trial on the TCIA, including spreadsheets identifying the colonic location of polyp positive cases.

The Prostate-Diagnosis collection contains prostate, cancer-confirmed, MRIs obtained using endorectal and phased array surface coils at 1.5T (Philips Achieva). Pulse sequences include T1, T2, and DCE (designated as a "BLISS" sequence using Magnevist gadolinium contrast media injected intravenously).

Linked clinical information, including matched patient demographics, radiology, and pathology reports in XLS format, is also available.

TCIA also supports restricted-access collections for specific group purposes, as data sharing is often a requirement of ongoing research activities. The NCI QIN leverages the TCIA to facilitate sharing of data among its growing list of fourteen member institutions. Having the curated, deidentified data reside on the TCIA will facilitate creating a public resource when the opportunity to make the data public arises.

A growing model for new collections is the leveraging of TCGA Data Portal (http://cancergenome.nih.gov/) that provides rich genomic and clinical human cancer data sets and is now matched by TCIA radiologic data from the same cases. Presently, three cancer types are contained in TCIA with several more collections in development. The TCGA-GBM collection contains more than 150 imaging subjects and has stimulated more than twenty professional meeting abstracts by a self-assembled multi-institutional research group.

The TCGA Glioma Phenotype Research Group began as an ad hoc multi-institutional research team dedicated to discovering the value of applying controlled terminology to the MRI features of patients with gliomas. Research trials that incorporate imaging have unique challenges due to nonstandard use of terminologies, absence of uniform data collection, and validation. In the early stages of their work, the team used the REMBRANDT (NCI NINDS) image collection to develop a pilot glioma feature set visible on MRI but later addressed the TCGA-GBM data collection (https://wiki.cancerimagingarchive.net/display/Public/TCGA-GBM), which consists of images from subjects genetically analyzed as part of the NCI/National Human Genome Research Institute Cancer Genome Atlas. By combining the TCIA and TCGA data sets together, the group has launched research projects investigating the correlation of clinical imaging with genomics to help relate TCGA findings to a clinical setting.

Exploratory efforts have been made to augment these collections with available RT planning data associated with patients whose tissue was collected as part of the TCGA activity. Because common ontologies and data structures for RT data exchange are not as mature as those available to diagnostic imaging data, significant challenges remain. Numerous researchers have benefited from the availability of TCIA as evidenced by the list of publications, which have employed downloaded TCIA data (http://www.cancerimagingarchive.net/publications.html) and by that self-assembled research groups supported in the first year of operations (https://wiki.cancerimagingarchive.net/display/Public/Research+Projects).

17.4 Radiation Oncology in Collaborative Imaging Informatics

Radiation oncology continues to pursue progress to advance understanding of the interaction of radiation beams with target cancer tissues. A major focus pursues the need to create sufficient tumor boundary margins while minimizing exposure to adjacent normal tissues. Therapeutic plans rest considerably on advanced clinical imaging to provide a reliable tissue-delineated planning environment that accounts for tumor definition, energy deposition, protection of normal adjacent tissues, and beam path attenuation. These clinical needs become especially relevant as the more precise technologies of intensity-modulated RT (IMRT), Cyberknife, and proton beam become more broadly used. Specific challenges connected with quantitative imaging in radiation oncology include

- Minimization of uncertainties related to integrated imaging and radiation treatment procedures (e.g., patients imaged in the treatment position and motion compensation);
- Robustness of anatomic volumetric image analysis (e.g., deformable image registration);
- Appropriate 3D resolved response assessment (e.g., voxel-based endpoints for treatment outcome).

RT clinics have broadly adopted CT imaging as a critical element of that planning process. PET/CT and MRI are just beginning to establish an acceptable role but have yet to be incorporated in conventional workflow. Because dose fractionation over an extended time period implies an ongoing dynamic interplay between radiation beam and tumor control, it is likely that further progress in the field may incorporate more frequent time interval imaging to characterize this symbiotic dynamic. Moreover, greater understanding of adjuvant and neoadjuvant therapy that impacts RT might usefully explore closer engagement with a wider variety of diagnostic imaging tools, including MRI and perhaps even ultrasound. On this frontier, advanced imaging techniques, such as DCE MRI, which can define regional vascular perfusion behavior in the core and at the boundaries of large tumors, might provide potential yet to be exploited data. Many of these research opportunities will remain unexplored as long as present barriers persist between RT image-based planning structures and ongoing diagnostic imaging. Routine diagnostic imaging, after all, continues to occur post-RT based on clinical care needs as part of patient follow-up. However, to date, there has been only somewhat limited efforts to merge image structures between the two disciplines.

In fact, barriers that impede integration of RT plans with follow-up diagnostic imaging are not just attributable to the usual distinct IT infrastructure jurisdictions that exist between radiation oncology and diagnostic imaging picture archiving computer systems (PACS) but also at a more basic technical informatics level. RT planning use of imaging requires a more standardized acquisition and rigorous quantitative object-oriented record-keeping than is generally required for diagnostic purposes. More precise tumor boundary annotations and beam attenuation calculations usually impose a different approach to RT PACS, thus resulting in a different file structure for DICOM RT PACS compared with diagnostic DICOM. For the sake of advancing science in both disciplines, a capacity to crosswalk between those two types of images presents a challenge and motivates devising an integrated RT-DX infrastructure. Groundwork for the early effort benefitted by being able to capitalize on RT Advanced Technology Consortium (ATC) developed tools to capture digital treatment planning data for advanced-technology RT clinical trials. Those RT ATC efforts set the stage for development of digital data exchange formats—first RTOG Data Exchange and DICOM RT. Particularly important was ATC's participation in the NEMA DICOM Working Group 7 and the ASTROIHE-RO Initiative. Despite these advances, the need to robustly address the heterogeneity of various tumors remains. Any meaningful usable resource will require large patient case collections (preferably in the hundreds) that include relevant clinical metadata. Addressing this gap was a key aspiration for a special focus of the large curated, PHI-protected, image archive—TCIA. Providing a data-compatible join between RT planning documents and diagnostic DICOM image structures was the subject of some effort by the TCIA development team.

17.5 A Radiation Oncology Use-Case Pilot

As a use-case exercise to test the value of such an archive collection, the NCI CIP and Radiation Research Program programs identified an opportunity with an ongoing RTOG trial (0522). The 0522 trial was a randomized phase III trial of concurrent accelerated radiation and cisplatin versus concurrent accelerated radiation, cisplatin, and cetuximab for stage III and IV head-and-neck carcinomas. The trial, activated in 2005, had a target sample size of 945 patients. FDG-PET scans were optional but encouraged at baseline and eight to nine weeks after end of RT. More than 100 patients completed PET scans in addition to the RT protocol. Head-and-neck carcinomas were an attractive clinical model for a proposed imaging RT-DX analysis pilot because definitive chemoradiation therapy is the treatment of choice in locally advanced disease; despite significant therapeutic progress, long-term control remains challenging, rarely exceeding forty to fifty percent. This suggested a fifty percent event rate for correlating imaging surrogate endpoints with clinical outcome.

The main aim of the informatics effort was to combine dosimetric and diagnostic imaging information on the subset of patients receiving IMRT and the complete set of FDG-PET/CT scans (pretreatment and posttreatment). It offered the opportunity to combine the significant strengths of the imaging/data quality-control capabilities of both the RTOG and the ACRIN. Combining imaging and dosimetry data from the RTOG 0522 trial was particularly challenging because of the different paths and jurisdictions of data capture. Dosimetry and clinical outcome data were collected through the RTOG network, whereas the imaging data were collected by the ACRIN network. Each site used its own set of medical record numbers to identify cases, so a lookup table was generated and applied using the CTP lookup table pipeline during the submissions into the TCIA archive. The same exam date-offset value used for deidentification was applied to data from both sources. DICOM-RT internal reference tags (dose, structure) were maintained during the deidentification/submission process.

This pilot effort by the RT research investigators offered an opportunity to address controversies related to dose predictors of local/regional tumor response, perhaps resolving competing "optimal" dose parameters (e.g., minimum dose and generalized Equivalent Uniform Dose [gEUD]). It also provided data on the dose distribution within the PET-defined target (e.g., PET-weighted mean dose and PET-weighted gEUD), which might predict outcome. Using access to the common archive, investigators were able to assess the association between 3D dose distribution within the CT-defined target tissues (e.g., PTV, CTV, and GTV), PET-defined target tissues, and tumor-related outcomes (e.g., time to local/regional progression, survival times, and post-RT FDG-PET). As a result of these fundamental steps and such a process, which served exploratory investigational purposes, a potential for follow-on research projects have opened up. These include the association between therapy-associated short-term changes in CT measured tumor size, short-term changes in PET-defined tumor physiology, and longer-term tumor-related outcomes (e.g., local/regional progression and survival time).

The original effort, beginning with a few dozen cases, helped resolve a number of traditional roadblocks to RT-to-DX image data sharing and better definition of relevant anatomic tissues. Significant advances were made in the conversion of various RT data formats to DICOM-RT standard structures. Processes for data integrity verification and deidentification of DICOM-RT files were improved. Efforts continue to expand the value of this and similar data sets by improving online query capacity of RT-relevant data elements and efforts to expand the data set to include clinical information such as response and survival data.

Acknowledgments

This project has been funded in whole or in part with federal funds from the National Cancer Institute, National Institutes of Health, under Contract No. HHSN261200800001E. The content of this publication does not necessarily reflect the views or policies of the Department of Health and Human Services, nor does mention of trade names, commercial products, or organizations imply endorsement by the U.S. Government.

References

3DSlicer. Available at http://slicer.org.

ACRIN. ACRIN American College of Radiology. Available at www.acrin.org. Retrieved June 29, 2012.

Cancer Imaging Program. CIP Survey of Biomedical Imaging Archives. Available at https://wiki.nci.nih.gov/display/CIP/CIP+Survey+of+Biomedical+Imaging+Archives. Retrieved June 29, 2012.

Crabb, A. I Do Imaging. Available at http://www.idoimaging.com. Retrieved June 29, 2012.

CTSA. CTSA Clinical & Translational Science Awards. Available at https://www.ctsacentral.org/. Retrieved June 29, 2012.

National Institute of Standards and Technology (NIST). NIST Physical Measurement Laboratory. Available at http://www.nist.gov/pml/workshop_080306.cfm. Retrieved June 29, 2012.

National Institutes of Health. NIH Data Sharing Policy and Implementation Guidance. Available at http://grants.nih.gov/grants/policy/data_sharing/data_sharing_guidance.htm. Retrieved June 29, 2012.

National Institutes of Health. The Cancer Genome Atlas. Available at http://cancergenome.nih.gov/. Retrieved June 29, 2012.

NCI Cancer Imaging Program. Quantitative Imaging for Evaluation of Responses to Cancer Therapies. Available at http://imaging.cancer.gov/programsandresources/specializedinitiatives/qin. Retrieved June 29, 2012.

NCI Cancer Imaging Program. The Cancer Imaging Archive. Available at http://cancerimagingarchive.net. Retrieved June 29, 2012.

NCI NINDS. REMBRANDT. Available at http://caintegrator-info.nci.nih.gov/rembrandt. Retrieved June 29, 2012.

NEMA. Digital Imaging and Communications in Medicine. Available at http://medical.nema.org. Retrieved June 29, 2012.

TCIA. De-identification Knowledge Base. Available at https://wiki.cancerimagingarchive.net/x/ZwA2. Retrieved June 29, 2012.

Informatics for Multimodality Imaging

Eduard Schreibmann
*Emory University
School of Medicine*

Timothy H. Fox
*Emory University
School of Medicine*

18.1 Introduction

Combining imaging from different sources leads to an improved understanding of a patient's anatomical and physiologic processes that is essential in designing advanced treatment plans. A key piece of technology in this context is image registration that, when coupled with a rigid-type transform, is used to align image content and, when used with a deformable or elastic transform, can quantify changes in the patient's anatomy among image studies acquired at different time points. This chapter presents both the technical aspects of integrating various imaging modalities into a common system and its application to radiation oncology.

As illustrated in Figure 18.1, a typical process of delivering radiation treatment consists of simulation, planning, and delivery. During simulation, images of the patient's anatomy are acquired using a computed tomography (CT) simulator to produce a 3D representation of the patient's anatomy. These CT simulation images are further used in the planning process to predict how dose is absorbed and distributed within the patient. This imaging modality is the mainstay in the planning process because it can describe a patient's morphology with a resolution of typically less than one millimeter. However, it is limited in its ability to discern different tissues types as well as visualize physiology and function. New imaging modalities are increasingly used in addition to the CT simulation data set in the planning process to distinguish normal from diseased tissue by visualizing aspects other than anatomical structure. Magnetic resonance imaging (MRI) provides better contrast among soft-tissue

types by visualizing proton density, whereas positron emission tomography (PET) visualizes biological activity such as glucose consumption that is elevated in malignant cells.

To combine this complementary information available in different imaging modalities, a process commonly called image registration, image fusion, or image matching is frequently employed. Registering images is not specific to radiation oncology alone, having been developed mainly in the areas of computer vision and remote sensing applications. An example of image registration encountered in daily life is weather forecasting, in which color-enhanced imagery from satellite data showing weather activity is overlaid on geographic information (map) to show weather activity in a geographic area. Without this correlation, the satellite imaging will contain weather data, but it will lack location. Similarly, examining the geographic map alone will show location without any weather activity data. Incorporation of information from both sources into the radar weather maps commonly presented in weather forecasts enhances our understanding as opposed to interpreting the information separately.

Image registration is needed to ensure that the images to be combined represent the same location in the patient. Because each scanner has its own coordinate system and the patient may move, change posture, or undergo anatomical changes between the two scanning sessions, a procedure is needed to compensate for these unknown changes occurring between the scanning sessions. Technically speaking, the image registration procedure finds a transformation between the image sets, which brings one data set into the coordinate system of another by matching the

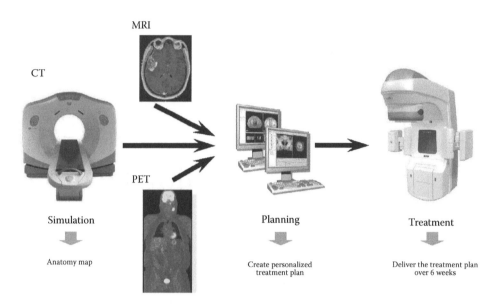

FIGURE 18.1 Multiple sources of imaging are used to better define the tumor and critical organs for external beam radiotherapy.

location of certain anatomical features within the images. The anatomical features that are matched may depend on the imaging modalities that are used.

The concept of using imaging information from different sources to improve quality can be extended outside the treatment planning process, because diagnostic imaging is increasingly used in radiotherapy to visualize changes occurring in time that can affect the accuracy of the delivery process. Although this chapter focuses on integrating imaging for the purpose of treatment planning, the concepts presented are quite general and can be applied as well to any process in oncology that involves tracking anatomy between two image sets.

18.2 Radiation Oncology Imaging Modalities

18.2.1 CT and MR Simulation

Modern CT simulators have been used on a routine basis in clinical departments to simplify acquisition of a patient's anatomical information in the radiotherapy planning process. The acquired CT study provides a 3D volumetric view of a patient's anatomy that can be easily integrated with the planning software to allow import, manipulation, display, and storage of images. CT simulation scanners are specially constructed for radiotherapy applications to simulate the treatment machine by having a flat treatment table and being equipped with a larger bore that allows imaging of the patient in the treatment position using various immobilization devices. Additionally, these scanners are also usually equipped with external patient marking/positioning lasers that define the scanner's local coordinate system and are used to reproduce patient positioning between simulator and therapy treatment machines.

MRI is added to the treatment planning process to better discern among different soft tissue types. Depending on the parameters used during the scanning session, this modality can image different properties of the tissue and not just density as provided by conventional CT. The technique finds major applications in radiotherapy of brain tumors, where the tissue is virtually invisible in CT imaging but can be distinguished from healthy brain tissue by inspecting specific sequences of a MRI plan as illustrated in Figure 18.2. The MRI also provides a significantly improved visualization of critical structures that have to be spared by the radiation such as the optical tract and brainstem.

Most of the sequences in an MRI data set can be registered to the CT images using mutual information (MI), a concept described later in this chapter, in a procedure that is implemented by most software vendors as an accepted clinical practice.

18.2.2 PET and PET/CT

PET is a molecular imaging modality capable of detecting small concentrations of positron-emitting radioisotopes. To date, most of PET imaging is performed on combined PET/CT scanners that were developed to provide an automated hardware solution to the need for coregistered anatomical and functional images. An accurate and precise matching of the CT and PET images is determined by carefully aligning the two scanners during installation and measuring their physical offset to providing a constant spatial transformation between PET and CT images that is known beforehand and is independent of patient positioning. The intrinsic registration accuracy of a combined PET/CT scanner is submillimeter, but the underlying assumption is that the patient does not move during the procedure.

For most radiation oncology departments, there are three scenarios for integrating the PET data into the simulation process.

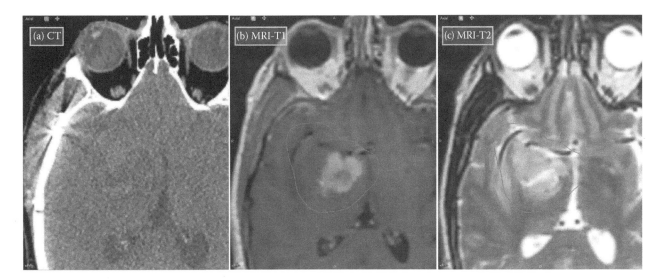

FIGURE 18.2 (See color insert.) Example of CT and MRI imaging of a glioblastoma patient. Although the tumor is invisible in the CT data set, it can be easily visualized by the MRI. The red contour represents tumor boundaries, as defined on the MRI data sets and later transferred to the CT volume.

The first scenario is the simplest where a PET/CT simulator is used for CT simulation of the treatment planning process. In this situation, there is no need to register another imaging data set to the CT simulation; the dual-scanner system uses the automatic physical registration for this process. This scenario ensures that the patient is in the treatment position for both the PET and CT scans. The second scenario is performing a CT simulation followed by the PET/CT procedure (Figure 18.3). The patient is scanned in the CT simulator to acquire the treatment planning CT data set. Following CT simulation, the patient is taken to the PET/CT scanner along with their custom-made immobilization system for the imaging procedure. After the scan, the PET/CT

data set is registered with the planning CT data set using various methods described in a later section. The registration is typically CT to CT to determine the coordinate system transformation. Once this is computed, the PET images are registered with the planning CT images for functional target delineation. The third scenario is performing the diagnostic PET/CT procedure followed by the CT simulation study. The patient receives a diagnostic PET/CT procedure before the CT simulation procedure. This is performed using diagnostic imaging protocols such as curved table tops and different patient positioning. After coming to radiation therapy department, a custom immobilization device is created, and the CT simulation procedure is performed

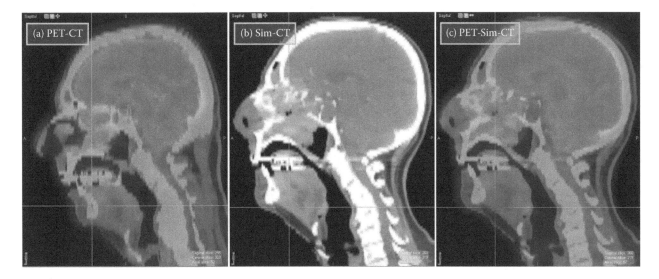

FIGURE 18.3 (See color insert.) Example of integrating PET imaging into the treatment planning. The tumor was delineated on the PET/CT data set, of lower resolution, and transferred to the higher-resolution Sim-CT data set for planning. This case presented differences in posture and chin position that have to be considered when correlating these image sets.

to acquire the treatment planning CT data set. The PET/CT images are then registered to the planning CT images. Accuracy is difficult to achieve because the patient was scanned in a different position between the two systems. This is not ideal because many rigid image registration methods fail to register the images correctly. Deformable image registration methods may be more appropriate to warp or stretch the PET/CT images to match the planning CT images.

From the above scenarios, there are challenges with the second and third scenarios that limit the accuracy of the image registration algorithms for multimodality images. One problem is the degree of similarity of the patient's position and shape during the imaging acquisitions. The other issue is the differences in time when acquiring the image data sets. Deformable or nonrigid image registration methods may be more appropriate to match the PET/CT images to the planning CT images. When performing scans in the thorax and abdomen region, motion artifacts will present a problem when registering PET/CT images with planning CT images. The use of a 4D PET/CT acquisition method, described later in this chapter, may address the motion artifacts encountered by respiration.

18.2.3 SPECT

Single positron emission computed tomography (SPECT) is a form of nuclear medicine tomographic imaging that uses a camera to detect gamma rays emitted from a radioisotope injected into the patient's bloodstream. The system operates similar to a CT scanner, with a detector rotating around the patient and capturing 2D images of the gamma rays emitted from within patient. These 2D images, or projection images, are further reconstructed into a 3D volume using a tomographic reconstruction algorithm.

SPECT imaging may be used similar to PET imaging in radiotherapy for visualizing anatomical function. Depending on the imaging application, various SPECT radionuclides can be used; for example, radiopharmaceuticals labeled with technetium-99m (^{99m}Tc) are preferred for bone and brain scans due to its physical characteristics, whereas the iodine isotopes ^{123}I or ^{131}I are preferred in neuroendocrine or neurologic imaging. One special application in radiotherapy where integration with standard anatomical imaging is needed is treating hepatic malignancies by radioembolization with microspheres containing yttrium-90 (^{90}Y) via the hepatic artery (1–4) (Figure 18.4). To verify that the microspheres are reaching the tumor and are not deposited by the blood flow in an undesired location, the patient is injected before treatment with a low dose of the radioactive-labeled drug. The deposition of the drug inside the patient is imaged using a SPECT camera system that provides a 3D representation of activity distributions with an approximately five millimeter spatial resolution. The approach improves accuracy over standard methods (5), but current image registration methods require significant user interaction (6, 7) due to the low tissue contrast in the CT data set and significantly different information in the SPECT data set as detailed in Figure 18.4.

Standard image registration algorithms developed for matching CT and MRI data sets match common information visible in both data sets and assume organs are imaged with uniform intensities, an assumption not fully met by these data sets. It is common to develop and tailor advanced registration algorithms for such nonstandard situations.

18.2.4 CBCT

Cone-beam computed tomography (CBCT) is an imaging technique producing volumetric images similar in quality and acquisition technique to the CT simulation data sets. In radiation therapy, CBCT images are acquired in the treatment delivery room using a hybrid X-ray source and radiation detector that is mounted on a radiation therapy linear accelerator. As opposed to conventional CT where a single narrow row of detectors is used, the detector for CBCT acquisition is a large flat panel of small amorphous silicon detectors. Both X-ray source and detector are attached to the treatment accelerator gantry that can rotate around the treatment couch. A volume of the patient's anatomy is reconstructed from projections acquired continuously, whereas the gantry-source-detector assembly rotates in an arc around the patient using tomographic reconstruction techniques similar to conventional CT.

With its ability to image the patient's anatomy just before treatment, the use of CBCT has found major applications in adaptive radiotherapy (ART). ART is treatment technique that aims to improve treatment accuracy by correcting for daily changes occurring during a typical radiotherapy treatment. At a minimum, the volumetric information is used to accurately position the patient with respect to the radiation field by translating the treatment couch until the online CBCT data set best matches the anatomy imaged before treatment. Current research uses the CBCT volumetric information to assess the geometry at the time of treatment (8), track the dose received by each voxel (9–14), or reoptimize remaining fractions in the plan (15–18) if large changes are observed. These calculations and corrections are performed either by compensating the remaining dose for the errors introduced in the previous fractions in an adaptive process or, ideally, by modifying the treatment plan to account for changes in the patient's anatomy just before delivering a fraction. Modifying the treatment plan before each fraction is technically harder to achieve due to time constraints, as all calculations and verifications have to be completed in the minimum amount of time, ideally just a few minutes. At the core of these plan adaptation technologies is the quantification of changes observed between the simulation CT and treatment delivery CBCT that can be achieved with the help of deformable image registration. These changes can be quantified in less than one minute using current algorithms and computing technology, allowing automated identification of tumor and critical structures in the pre-fraction CBCT data set.

One notes that, although the CBCT images are similar in terms of content to standard CT images, the modified acquisition CBCT technique using a large flat panel instead accounts

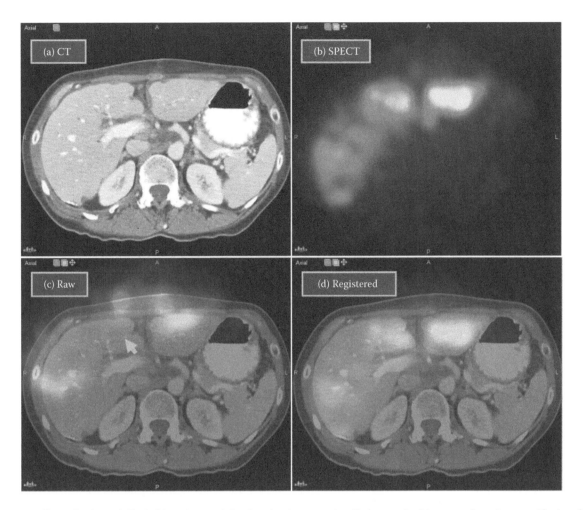

FIGURE 18.4 (**See color insert.**) Typical imaging used for planning in targeted radiotherapy. In this approach, patient-specific data from CT (a) provides an anatomical model with resolutions on the order of one millimeter, whereas dose deposition estimation is based on the SPECT data set (b). The images are acquired on different scanners and appear displaced and unaligned as evident at arrow (c). An image registration procedure corrects this misalignment as illustrated in (d). After registration, the SPECT activity within the liver matches the liver shape as observed in the CT data set.

for increased scatter at the detector, which in turn produces a higher inaccuracy in the reported Hounsfield units (HU) in the CBCT images, as well as a set of various artifacts when objects of high density are in the imaging field as illustrated in Figure 18.5. The image registration algorithm quantifying changes between these two image modalities should be able, technically, to handle these artifacts and differences.

18.2.5 Dynamic Imaging (4D)

Recently, state-of-the-art medical CT scanners allowing acquisition of dynamic or 4D images, in which a section of the body is repeatedly imaged to capture respiratory motion, are becoming increasingly used in medicine (19–27). The result of a 4D tomographic imaging session is a set of typically ten static 3D images sampling the patient's anatomy at different phases of the breathing cycle. This acquisition technique was primarily developed to visually assess individual tumor excursion during the breathing cycle and design treatment margins tailored to the motion based on the patient-specific breathing pattern for thoracic tumors that change position and shape with respiratory motion. The technique was developed initially to create 4D CT data sets but, due to its clinical advantages, was extended to other modalities such as CBCT (28, 29), MRI, and PET (30–32). The concept is similar in all these acquisition methods, where the patient's anatomy is sampled continuously at a table position during the breathing cycle and then reformatted according to the breathing phase information as obtained from an external marker or a camera-based patient monitoring system. One commercially available system (RPM Respiratory Gating System, Varian Medical Systems, Palo Alto, California) provides retrospective gating of the tomographic data set by taking 3D data sets at specific time intervals to create a time-dependent 4D CT imaging study. By incorporating 4D PET into the treatment planning process, it has been reported that tumor volumes have been reduced by as much as thirty-four percent (33).

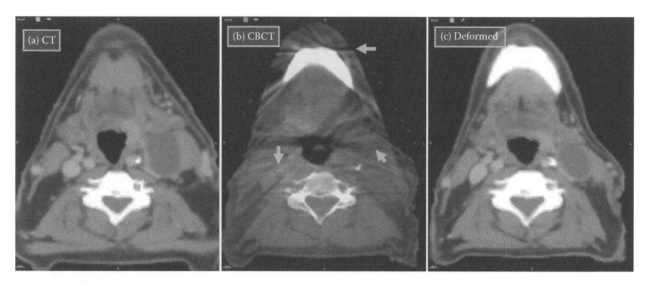

FIGURE 18.5 Adaptive radiotherapy uses deformable registration to compare the pretreatment anatomy (a) with the anatomy observed during treatment (b). Due to increased scatter reaching the detector, the CBCT acquisition technique is less accurate in reproducing correct HU units and is prone to a series of artifacts (marked as arrows). An image registration algorithm should be able to detect anatomical changes but ignore changes produced by differences in the imaging technique, as shown in (c).

18.2.6 Advanced MRI Modalities

Magnetic resonance spectroscopic imaging (MRSI) is an acquisition technique that can characterize regions of brain based on the presence of various metabolites and other substances. Metabolites that can be evaluated include choline (Cho) as a marker of cellular membranes, creatine (Cr) as a marker of energy metabolism, N-acetyl aspartate (NAA) as a marker of viable neurons, and lactate (Lac) as a marker of anaerobic respiration. Early investigators established that the spectra of certain brain tumors significantly differ from normal brain, with increased levels of Cho and Lac and decreased levels of NAA, suggesting that MRSI can add clinically useful information in the identification of these brain tumors (34, 35). MRSI has even been evaluated as a guide to radiotherapy planning when registered with the conventional MRI scans and treatment planning CT scans (36–39). Although MRSI was not widely used in clinical practice due to relatively low resolution, long acquisition time, and inefficient forms of data display, many of these limitations have now been overcome with state-of-the-art MRSI technology utilizing that acquisition of a high-resolution, 3D, whole-brain metabolite maps can now be obtained within twenty-minute scan times and in a form that can be registered with other MRI images.

18.2.7 Ultrasound Imaging

In ultrasound imaging, a sound wave is emitted in an arc-shape wave directed toward the anatomy to be imaged. The sound wave is partially reflected from tissue interfaces encountered along its path. The change in density along this path produces a partial reflection of the incoming sound wave, and the echo time, the time needed for the sound wave to bounce from different subcutaneous structures inside the patient, is recorded and decoded into an image of tissue densities. There is no dose received by the patient from an ultrasound study, but there are technical limitations to the anatomical site that can be imaged as the ultrasound wave cannot penetrate the bone. In radiotherapy, ultrasound is used for target localization and verification for certain pelvic cancers. Common applications are transrectal ultrasound for prostate radiotherapy, where this imaging modality is used to identify the prostate and guide real-time placement of radioactive seed implants. For gynecologic cancers, ultrasound can document changes such as uterus and cervix motion with respect to variations in bladder filling or rectal motion.

18.2.8 DICOM-RT Objects

Although integration of imaging was a challenge in radiation therapy planning more than fifteen years ago, this has been eliminated using the Digital Imaging and Communications in Medicine (DICOM), a vendor-independent standard for transmitting, medical imaging data between medical systems (40) that describe in detail methods of formatting and exchanging digital images such as CT, MR, and PET objects using standard network connections. Over the past decade, an extension to DICOM was developed for radiation therapy objects. Referred to as DICOM-Radiotherapy (RT), this extension handles the technical data objects in radiation oncology such as anatomical contours, digitally reconstructed radiograph images, treatment planning data, and dose distribution data. Many CT simulator and treatment planning vendors have adopted DICOM-RT for ensuring a cost-effective solution for sharing technical data in a radiation oncology department.

18.3 Multimodality Imaging Integration Methods

18.3.1 Multimodality Image Registration Methods

The aim of registration is to establish an exact point-to-point correspondence (coordinate system transformation) between the voxels of two image data sets acquired on different scanners, because the patient may lie differently, have a different posture during scan, or experience anatomical differences such as weight loss. Simply stated, an image registration procedure eliminates these differences by correlating the information within images to a common coordinate system.

Data sets obtained from different imaging systems are acquired independently, and these data sets would appear displaced if presented to the user without any correction. The image registration procedure recovers the unknown transformation between these independent coordinate systems by matching features or voxels intensities within the images. The transformation sought can be either rigid, to correct for changes in patient positioning and posture, or nonrigid (sometimes called curved, deformable, or elastic) to correct for complex anatomical changes. A simple rigid coordinate transformation allows only translations in three orthogonal directions and rotations in three directions. Advanced rigid registrations consider warping, shearing, and scaling in addition to rotations and translations. Nonrigid transformations are complex transformations in which the parameters are dependent of each voxel in the input data set but useful in modeling local anatomical changes in addition to global posture changes. The transform is recovered either manually, when the user adjust the transform interactively until the data sets match as judged subjectively by visual inspection, or the process can be automated in which a software algorithm automatically finds the optimal transform in an iterative search.

Input to this procedure are two images, one that is moved or transformed, called moving image, until it matches a second image, called fixed image. Initial implementations of image registration provided an interactive tool in which the physician could gain complete control over the registration process to create the coordinate transformation between the two image data sets by rotating and translating the moving image with respect to the fixed image data set (41) until the images are visually aligned. However, this is a tedious process of iteratively modifying and inspecting the subjective validation of the registration processes as the accuracy of registration depends on the user's judgment of the correlation between anatomical features. The main advantages of manual methods are intuitive handling, immediate display of results, and the fact that they do not need any time-consuming pre-processing. The disadvantages are poor reproducibility and no metric, indicating the goodness of the registration.

In a first attempt toward automation, landmark-based registration was introduced. The technique uses corresponding points located within different images to determine the spatial transformation between the paired points (42–45) that can be either external landmarks or internal landmarks. External landmarks are based on foreign objects introduced into the images such as a stereotactic frame attached to the patient for brain imaging. These external markers need to be visible on all image sets. This may be metal for CT images and ^{68}Ga for PET imaging. External methods rely on objects attached to the patient, which are visible with all of the imaging modalities. Because the markers can be easily seen in the imaging modalities, the image registration algorithms can be both automated and fast without the need for a complex optimization algorithm. One of the disadvantages of external landmarks is that the markers usually require an invasive procedure. Internal landmarks, referred to as anatomical markers, are points of internal anatomy that can be visualized or located within each image. A physician or clinical expert using an interactive software method of locating the points on both image data sets identifies these internal markers. After identifying the landmarks, the image registration algorithm calculates the geometric transformation by minimizing a cost function representing the mismatch between the image sets. This cost function may be the distance between the coordinates of these landmarks. Landmark-based methods are mainly used to find rigid transformations. In landmark-based registration, the set of identified points is sparse compared with the volume-based intensity methods, which makes for relatively fast optimization procedures. The identification of internal anatomical points is a segmentation procedure and involves locating four to ten points on corresponding sets of images.

Some of the most difficult cases of image registration are when multimodality imaging is used such as PET/CT, MR/CT, and PET/MR. Images acquired using these methods are difficult to register with landmark-based methods because it is difficult to identify or segment common structures in both image sets. Over the past fifteen years, automated image registration methods have increasingly used volume matching or intensity-based matching algorithms (46–48). Volume-based image registration methods are different from others because they operate directly with the image intensity (gray values) values without user interaction. The volume-based methods for 3D/3D image registration require large computational costs that have become available over the past decade to enable these methods in routine clinical practice. Automated versions of rigid or deformable image registration procedures are, in essence, optimization procedures customized to deal with medical images that recover unknown variables, the transform parameters, by minimizing discrepancies between the data sets as judged by a metric or cost function. As with any optimization procedure, the building blocks of an image registration process are as follows:

- A transform that defines the type of transformation or allowable mapping between the two data sets. The transform parameters are the variables to be optimized.

- A similarity metric (cost function) to measure how well two images are aligned; this measure characterizes in mathematical terms our concept of aligned images and can be constructed from anatomical features or voxel intensities within the images. During optimization, the cost function is used to judge the value of a match for the current transform parameters.
- An optimization procedure to find the optimal transformation parameters by making iterative changes to the transform parameters. After each iteration, match quality is evaluated through the metric and variations to the transform parameters are continued if the metric decreases, indicating a better match, or reversed if the metric increases indicating an erroneous change.

The similarity metric defines in mathematical terms a perfect match, with the transform parameters being iteratively modified by the optimizer until the metric value is minimized. The formulation of the cost function depends on the images to be matched. For simple matching problems such as images acquired under the same conditions using the same scanning modalities, a formulation of the metric is constructed using the differences between the voxel intensities in the two images. Indeed, when the images are aligned, the voxel intensities should be identical under these conditions and the difference between the two images should be zero. However, when the images are acquired with different modalities, the voxel intensities are proportional with different quantities; therefore, the same structure may be visualized with different intensity levels. A pure difference of intensities does not properly describe the optimal alignment, with other criteria needed to define the optimal alignment. Such criteria may consider locations of specific markers or statistical measures to describe an aligned image set.

In the single-modality images, it is assumed that a given structure has the same intensity in both images, an assumption frequently invalidated by artifacts or the acquisition protocol. A more general approach is to use multimodality registration, in which this constraint is relaxed to allow imaging of the same structure in the input images using different intensities; otherwise, the algorithm will generate unrealistic solutions.

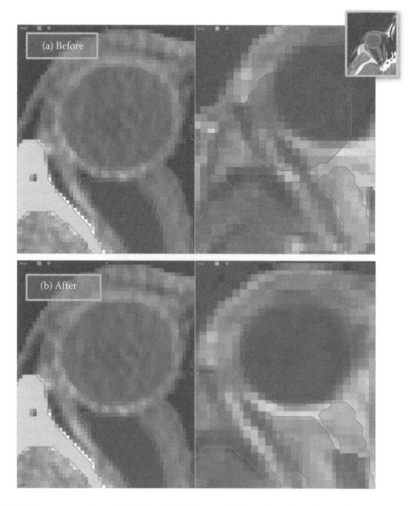

FIGURE 18.6 Concept of MI. The top row shows the CT and MRI data sets before the registration, where many pairs of intensities drawn from the two images can be identified inside a given structure. The bottom row shows the registration result that aims to minimize these pairs. See text for details.

A popular similarity measure using the voxel intensity histogram for multimodality image registration is the MI method. The use of MI for medical image registration applications was independently introduced in 1995 by both Viola and Wells and Collignon (47, 48). This method is based on the assumption that there is a correlation between groups of voxels that have similar values but does not explicitly model the correlation. The concept is illustrated on the display in Figure 18.6, showing a zoom-in on the left eye region for the CT-MRI data set presented in Figure 18.2. Various organs have been segmented to illustrate the concept on the CT data set and are shown as colored patches in the display also on the MRI volumes. Because the images come from different modalities, these various organs are imaged with different intensity levels. For example, the eye appears gray in the CT data set and dark in the MRI image. The MI metric counts the intensity pairs between the images for a given displacement and aims to minimize this number, the concept being illustrated in Figure 18.6. Before the registration (top row in the figure display), the MRI is displaced; therefore, voxels in the eye, shown as gray values, will map to different organs in the MRI. One can find voxels in the eye that map to darker voxels in the MRI, and voxels in the eye that map to brighter MRI regions, creating numerous pairs of intensities between the two data sets. When the images are aligned (bottom row), the same voxels in the eye map only to dark regions in the MRI. There is only one combination of image intensities in this match, excluding variations introduced by the inherent noise in the two data sets. The alignment can therefore be described in multimodality registration by minimizing the number of intensity pairs that can be constructed from the two input images.

18.3.2 Rigid Registration

As described earlier, the easiest form of image matching is rigid registration that uses translations and rotations to align two image sets. This procedure is exemplified in Figure 18.4. Before registration, the images appear displaced, as the activity within the liver in the SPECT images is displayed outside the patient when superimposed on the CT data set. After applying the rigid transformation found by the registration, the SPECT activity is displayed in the liver, as expected.

The iterations of the algorithm in finding these displacements are shown in Figure 18.7, in which the first panel shows the MI metric values as minimized in the optimization procedure for a rigid translation (blue line) and a deformable registration (yellow line). As only the transformation model differs between these graphs, the figure illustrates the concept that rigid and deformable registration differs only by the transformation between the two image sets. The registrations have in common both a registration metric (that judges the match) and an optimization procedure (that varies the transformation parameters to minimize the metric). For the rigid registration, the algorithms starts from higher values, denoting a worse match, and progressively decreases to lower values as the images are matched by the algorithm. The gray markings in the figure are the translations along

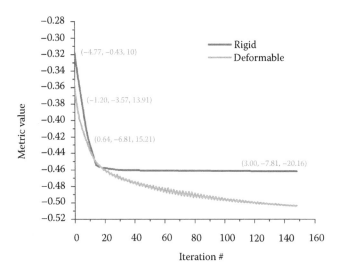

FIGURE 18.7 Registration metric value as a function of iteration for registering the images in Figure 18.5 using a rigid and a deformable transform. The lower the metric, the better the matching, with the deformable registration providing a better values as its model can describe local anatomical changes.

the x, y, and z axes at the corresponding iterations. One notes that this mimics the manual procedure in which the physician would modify the translation progressively and visually judges the match; only here the whole procedure is automated with the cost function. The metric value improves when switching from a rigid to a deformable transform, suggesting that a better match can be obtained when considering a nonrigid match.

18.3.3 Deformable Registration

For many applications in radiotherapy, a rigid matching is insufficient as, in reality, organs in a patient's body will deform under forces exercised between them and from the surrounding medium. Although these forces cannot be measured in vivo and are unknown, their effect, such as organ compression, inflation, and displacement, can be visualized using different imaging techniques. A deformable registration algorithm is in essence an optimization procedure that estimates the magnitude of these forces by comparing images acquired before and after the deformation.

Indeed, deformable registration extends the concept of rigid registration by improving the transform to allow increased freedom in describing organ changes. This is achieved as the translation of each voxel is considered independently rather than viewing the whole data set as a block. An example is given in Figure 18.8, where rigid registration is compared against two popular versions of deformable registrations for matching CT and CBCT data sets of the head-and-neck case. The rigid registration was successful in matching nonwarping tissue such as the bony anatomy or external contour (green arrows) but could not model soft-tissue changes caused by differences in the head-and-neck posture. In contrast, both deformable registrations were able to take account for these differences by creating

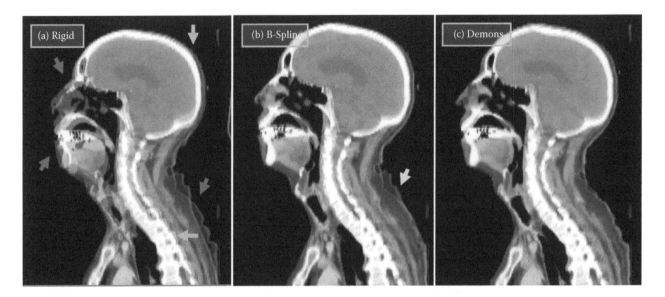

FIGURE 18.8 **(See color insert.)** Comparison of rigid (a) and two popular deformable registration methods, B-spline (b), and demons (c). The result is shown as a red overlay on the fixed image, shown in background as a grayscale sagittal slice. In (a), the rigid registration is globally fairly good, but the areas marked with arrows are not well registered. In the brain, voxel values are fairly uniform and there will be very little detail with which to adjust the registration. A deformable registration algorithm should provide enough deformation at the arrows to correct the misregistrations and minimal warping in the brain. Notice that, visually, the quality of the two deformable registrations, (b) and (c), looks very similar, with a better result obtained by demons (c) at the yellow arrow.

a displacement map that better correlates the two data sets. The first algorithm, generating a sparse deformation field defined only on a lattice of locations, did not model small deformation, as marked with the yellow arrow. The second algorithm, generating a deformation vector for each voxel, did an improved matching of small anatomy at the expense of some restrictive assumptions on the type and quality of the input images.

Technically, deformable registration is much more complex computationally when compared with rigid registration due to the increased number of variables. Different deformable registration methods have been proposed to account for changes in modern radiotherapy, with each algorithm performance restricted by its underlying assumptions. Many deformable registration algorithms have been proposed in recent years, each one making specific assumptions in its mathematical model to simplify the problem at hand. To date, we feel there is no clear winner over a broad range of problems. A class of algorithms may perform better than average on a specific problem, but it may be outperformed by a different type of algorithm on another problem. Three main classes of algorithms have emerged:

- Variational Techniques (49–52): Also known as demons or optical flows, the deformation is defined for each voxel in the input data set and is evolved under the influence of a partial differential equation derived from the voxel intensities in the two data sets. Common critiques are that, in its standard form, it allows only for monomodality registration and thus cannot be used to deform images acquired from different modalities. A more general approach is to use multimodality registration, where this constraint

is relaxed to allow imaging of the same structure in the input images using different intensities. Variants of this algorithm that can match multimodality data sets have been recently proposed (53).

- Finite Element Method (54–57): The algorithm uses physical equations to model organ deformation. It starts by segmenting organs from one image and warping the segmentation using finite element models in an equation that incorporate anatomical information of structure stiffness and elasticity (55). To date, the values of some physical coefficients are unknown and cannot be measured in vivo.

- B-Spline Technique (58–60): This is an optimization approach where the deformation is defined only on a grid of nodes that is superimposed on the moving image. Deformation at any other location is obtained from the grid using B-spline interpolation. The optimization iteratively modifies the node deformation until differences between the fixed and warped moving image are minimized. By using the B-spline grid, the number of variables to be optimized is kept to a reasonable number, whereas the B-spline interpolation provides the flexibility to define small local deformations.

In general, a deformable registration algorithm will try to match the information on a set of images by modifying the parameters of a motion model, until the images to be matched are alike. The description of the deformation field is the key aspect of a given deformable registration model, as it defines the trade-off between mathematical complexity and accuracy based on the expected smoothness of the deformation field. The complexity is generated by the dimensionality of the problem at hand, as the deformable registration aims to account for the displacement of every voxel in the input images. This issue results in a multifaceted mathematical problem, with typical deformable models involving thousand of parameters to describe local tissue changes. The optimization procedure has to be adapted to the complexity of the deformation model. The strategy to find the displacement field between two images is adapted to the representation of the transformation, being either parametric or nonparametric. In the case of a parametric representation, a finite number of parameters, representing the deformation field with reasonable complexity, and an optimization procedure is used to find the optimal parameter values. One classic example is the B-spline transform that approximates the deformation field on sparse matrix of control points, with interpolation between control points used to obtain displacement values in any location. The B-spline model was initially proposed to model local and smooth deformations over small volumes. However, the model becomes impractical when large data sets are used or when the problem tries to model local, unsmooth deformations because a large number of nodes are needed to provide the necessary accuracy in the interpolation.

When it comes to estimate a dense deformation field in which the displacement vectors at each voxel are considered separately, the dimension of the optimization space tends to be so big that a variational method is often the preferred approach. The most popular variational registration technique is the optical flow algorithm (61), a model where the deformation is defined for each voxel in the input data set and is evolved under the influence of a partial differential equation derived from the voxel intensities in the two data sets. The classic implementation was initially designed for monomodality registrations as well as small displacement fields, but recent research reported on the integration of multimodality metrics measuring local changes into these types of algorithms.

Independent of the algorithm used, output of a registration algorithm is a set of vectors describing its displacement as all voxel-based algorithms (58, 62) assign a vector to each voxel in the input images and iteratively vary magnitude using different approaches until the pre-deformed image—warped with the deformation field—matches the postdeformation image. An example is given in Figure 18.9, in which the computed displacements are visualized using arrows whose magnitude and direction are proportional to the displacement intensity and direction. One notes that although the resulting images are very similar as illustrated in Figure 18.8, the actual mapping is significantly different. This aspect should be carefully checked in clinical practice as detailed in the next paragraph.

18.3.4 Validation of Image Registration Methods

Registration, either rigid or deformable, is currently validated clinically by visual comparison of the fixed image against the moving data set deformed with the registration result. Some typical evaluation tools used for this task are presented in Figure 18.10. The lens tool displays parts of registration result in a small lens-like window that can be interactively moved over the fixed image, with evaluation performed in this case by

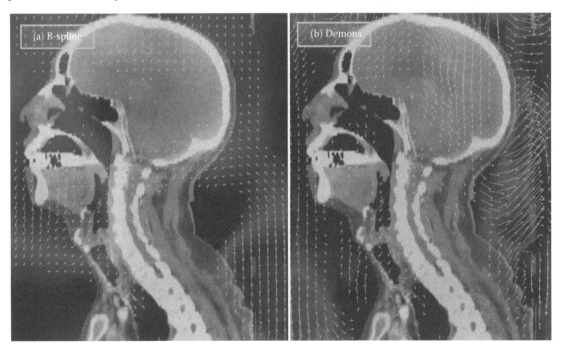

FIGURE 18.9 (**See color insert.**) Display of the vector field magnitude for the solutions in Figure 18.8b and c illustrates the differences between these two deformable algorithms. For example, the B-spline has minimal warping in the brain, although there is significant warping for the demons algorithm. Such differences, which have to be assessed by on a case-by-case basis, may have significant implications for dose tracking and other voxel tracking applications.

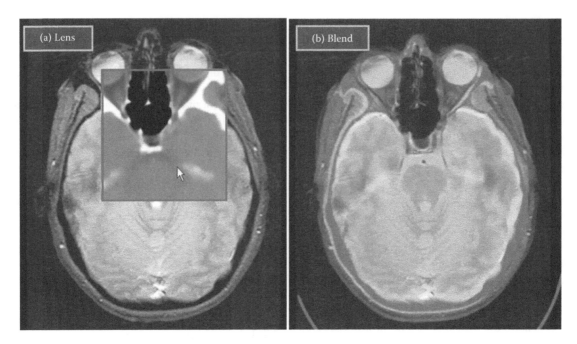

FIGURE 18.10 **(See color insert.)** Examples of lens (a) and color blend (b) inspection of registration results. In the lens tool, the square region inside the red rectangle shows the CT data set and can be interactively moved over the MRI image to inspect the match at different locations. The color blend display allows a quick assessment of blurred regions where the match is suboptimal.

visually comparing prominent anatomical features at the lens-background interface. An alternative is the blend view having the advantage of showing and enabling result evaluation in the whole slice at a glance. In this type of display, the fixed image is shown in the background, whereas the deformed moving image is shown as a color-coded semitransparent overlay superimposed on the fixed image. Regions of misalignment are visualized as blurred regions in this type of display.

Rigid registration accuracy can be increased by using a region of interest on the images to eliminate clinically irrelevant regions, as most common sources of registration failures in day-to-day clinical operations are differences in the field of view, objects present in one image but not in the other one, or image artifacts. With a region of interest, the metric is computed only on a rectangular region defined by the user ignoring irrelevant regions such as background and noise. The approach can be further refined by using only voxels within an organ or at a given distance from organ borders (63, 64).

Although, for rigid registration, the accuracy can be relatively easily assessed visually by comparing the locations of anatomical structures or implanted markers, in the case of deformable registration, validation becomes a practically difficult problem. This is because deformable registration is technically an underconstrained problem, with many similar solutions that would deform the images in a similar fashion. For example, consider the case presented in Figure 18.9, in which voxels inside the brain are imaged with uniform intensities, with any solution swapping the locations of these voxels being mathematically equivalent as these voxels are indiscernible one from the other. Voxels in such regions of constant intensity are indiscernible to

most deformable registration algorithms (65); thus, there is little information for the algorithm to build a displacement field that accurately mimics anatomical changes. Indeed, when visualizing voxels, as in Figure 18.8, the solutions found by the B-spline and demons algorithms are very similar by visual analysis of the resulting image. However, visualizing the deformation field rather than the deformed moving image, as illustrated in Figure 18.9 for the two solutions, may provide additional information to judge the clinical plausibility of the deformation field by comparing its smoothness and other properties with the expected anatomical behavior of the tissue or organ imaged. The tissue inside the brain for the B-spline solution is smooth in this case, as expected, whereas the solution found by the demons has large displacements that seem unnatural. As deformable registration models are in essence synthetic models not necessarily based on underlying anatomy, careful evaluation as exemplified in Figure 18.9 is needed to ensure that the solution found by the registration conforms to expected anatomical deformations. The use of rigid or deformable registration may improve the solution, but the physician should always realize that image registration is an approximation that has to be visually reviewed on a case-by-case basis before usage for clinical purposes.

It is important to note that the technique to measure accuracy depends also on the clinical endpoint. If the deformation algorithm is used for segmenting new images by warping a presegmented template to the image to be segmented through deformable registration, an academic approach is to mathematically quantify registration errors by distances between the automated and user-delineated contours through Hausdorff or Dice measures (66–77) as the characteristics and behavior inside the field are irrelevant. However, surface comparisons are not sufficient for other applications. When the resulting displacement field from a deformable registration based on CT data sets is used to warp supplementary information such as a dose matrix, it is crucial that the displacement field inside a structure of constant intensity follow real anatomical changes (65). The reason for this issue is that the CT data set voxels inside a structure cannot be discerned, whereas the dose data sets associated with these voxels differentiate them. For such applications, tools have to be developed and validated clinically to allow inspection of the displacement field. Measures derived from vector analysis can be employed here to detect regions where the deformation field displays unnatural anatomical motion or to identify regions of contraction or expansion. These can be correlated with tumor shrinkage or expansion for response assessment or are used between extreme phases of 4D CT data sets of lung patients to deduce ventilation (78, 79) and incorporate this information into the treatment planning process (80).

18.4 Multimodality Clinical Applications for Radiation Oncology

18.4.1 Registration of CT-CBCT for Adaptive Radiotherapy

The addition of information available in the CBCT data set to the treatment planning data set is expected to have a major impact upon our ability to accurately define the extent of the target and shape the dose accordingly with major applications in the head-and-neck, prostate, rectal, or bladder cancers. In this concept

(12, 81, 82), the plan can be adapted to changes observed in the CBCT imaging with many compelling advantages in defining the tumor volume and designing treatment plans that better spare the critical structures. With the information within the CBCT data set, the dose can be tracked after the fraction was delivered and the plan can be readapted if needed. To adapt the plan, a fast and accurate segmentation method is needed to track changes and segment target and critical organs (83, 84).

The method relies on transferring target and critical structures segmentations already defined in the planning CT images to the CBCT data set to provide insight into changes in shape of the tumor and critical organs. A point-to-point mapping obtained by a deformable image registration is used to match the CT with the CBCT image for the case illustrated in Figure 18.5. Once the point-wise transformation is obtained, the same transformation is applied on the structures defined on the CT data set to warp them onto the data set to be segmented. Figure 18.11 illustrates this process, where, for simplification purposes only, the tumor and external body contour are shown, but the displacement field can be applied on any structure to warp it from the CT to the CBCT data set. The dose matrix can also be warped for an estimation of the delivered dose in the presence of anatomical changes.

18.4.2 Registration of 4D CT for Dose Mapping

One of the first applications of deformable registration was tracking dose across breathing phases for a better understanding of how the respiratory motion affects deposited dose (85, 86). Deformable registration is an essential tool as it quantifies trajectories of individual voxels through the breathing cycle.

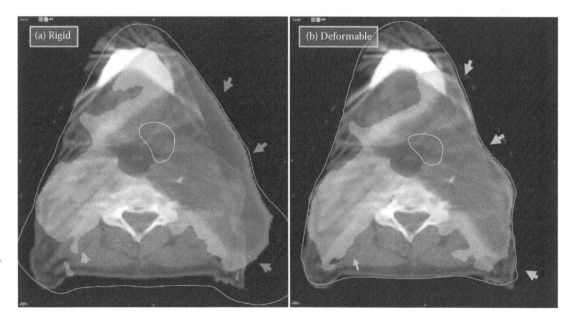

FIGURE 18.11 (**See color insert.**) Usage of deformable registration for adaptive radiotherapy. The background shows the CBCT data set, whereas the overlays show the segmentations and doses computed in the planning process superimposed using rigid (a) and deformable (b) registrations. Deformable registration can be used to autosegment the CBCT data set by adapting the planning segmentation to the anatomical changes.

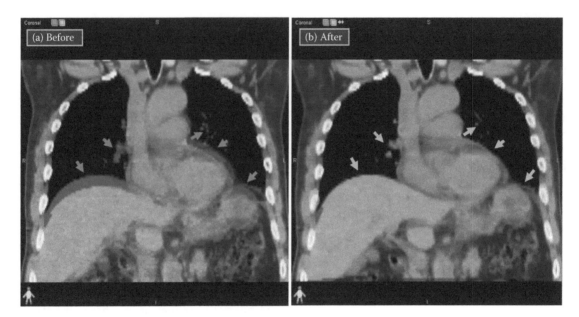

FIGURE 18.12   (**See color insert.**) Sample result of deformable registration. In (a), two phases of a 4D CT are overlaid in a blend view, with arrows marking anatomical changes produced by the respiratory motion. Results of the deformable registration algorithm are presented in (b), showing liver and other organs deformed back to match their initial state.

The method starts by quantifying anatomical motion using a 4D CT data set that samples patient's anatomy at different phase of the breathing cycle. This is exemplified in Figure 18.12, where for a lung tumor patient deformable registration was used to recover the respiratory motion. The left image shows the expiration phase overlaid on the expiration data set before registration. Anatomical changes produced by respiration are marked with arrows. The deformable registration is able to eliminate these changes by warping each voxel in the second data set to its original location, as observed in the right image in which the differences between the two phases are removed. To obtain cumulative dose volume histograms, the dose matrix at the second phase can be warped with the deformation field to the reference phase, and the dose is added by simple summation to deduce dose deposition from all phases (87, 88).

18.5 Conclusion

Development of complementary imaging techniques provides effective means for clinicians to understand and identify the dependence of the involved structures and help design better strategies for targeting tumors. At the same time, it necessitates a robust rigid or deformable image registration algorithm to maximally use the image data derived from different modalities or acquired under different conditions. Voxel-based automated deformable registration is capable of providing an accurate and robust spatial match between two data sets acquired with different modalities. Although rigid registration corrects for shifts and posture changes, when using deformable registration, a direct correspondence for each voxel is obtained that included possible warping and local deformations of soft-tissue organs.

References

1. Khalaf, H. et al. (2010). Use of yttrium-90 microsphere radioembolization of hepatocellular carcinoma as downstaging and bridge before liver transplantation: A case report. *Transplant Proceedings, 42*(3), 994–998.

2. Bult, W. et al. (2009). Microsphere radioembolization of liver malignancies: Current developments. *Quarterly Journal of Nuclear Medicine and Molecular Imaging, 53*(3), 325–335.

3. Vente, M. A. et al. (2009). Yttrium-90 microsphere radioembolization for the treatment of liver malignancies: A structured meta-analysis. *European Radiology, 19*(4), 951–959.

4. Stuart, J. E. et al. (2008). Salvage radioembolization of liver-dominant metastases with a resin-based microsphere: Initial outcomes. *Journal of Vascular and Interventional Radiology, 19*(10), 1427–1433.

5. Campbell, J. M. et al. (2009). Early dose response to yttrium-90 microsphere treatment of metastatic liver cancer by a patient-specific method using single photon emission computed tomography and positron emission tomography. *International Journal of Radiation Oncology Biology Physics, 74*(1), 313–320.

6. Glatting, G. et al. (2005). Internal radionuclide therapy: The ULMDOS software for treatment planning. *Medical Physics, 32*(7), 2399–2405.

7. Giap, H. B. et al. (1995). Development of a SPECT-based three-dimensional treatment planning system for radioimmunotherapy. *Journal of Nuclear Medicine, 36*(10), 1885–1894.

8. Ghilezan, M. et al. (2004). Online image-guided intensity-modulated radiotherapy for prostate cancer: How much improvement can we expect? A theoretical assessment of clinical benefits and potential dose escalation by improving precision and accuracy of radiation delivery. *International Journal of Radiation Oncology Biology Physics, 60*(5), 1602–1610.

9. Yan, D. et al. (1999). A model to accumulate fractionated dose in a deforming organ. *International Journal of Radiation Oncology Biology Physics, 44*(3), 665–675.

10. Yan, D. et al. (2000). An off-line strategy for constructing a patient-specific planning target volume in adaptive treatment process for prostate cancer. *International Journal of Radiation Oncology Biology Physics, 48*(1), 289–302.

11. Schaly, B. et al. (2004). Tracking the dose distribution in radiation therapy by accounting for variable anatomy. *Physics in Medicine and Biology, 49*(5), 791–805.

12. Yan, D. et al. (1997). Adaptive modification of treatment planning to minimize the deleterious effects of treatment setup errors. *International Journal of Radiation Oncology Biology Physics, 38*(1), 197–206.

13. Pavel-Mititean, L. M. et al. (2004). A geometric model for evaluating the effects of inter-fraction rectal motion during prostate radiotherapy. *Physics in Medicine and Biology, 49*(12), 2613–2629.

14. Hoogeman, M. S. et al. (2002). A model to simulate day-to-day variations in rectum shape. *International Journal of Radiation Oncology Biology Physics, 54*(2), 615–625.

15. Wu, C. et al. (2002). Re-optimization in adaptive radiotherapy. *Physics in Medicine and Biology, 47*(17), 3181–3195.

16. Wu, Q. et al. (2006). Application of dose compensation in image-guided radiotherapy of prostate cancer. *Physics in Medicine and Biology, 51*(6), 1405–1419.

17. Unkelbach, J. & Oelfke, U. (2005). Incorporating organ movements in IMRT treatment planning for prostate cancer: Minimizing uncertainties in the inverse planning process. *Medical Physics, 32*(8), 2471–2483.

18. Unkelbach, J. & Oelfke, U. (2004). Inclusion of organ movements in IMRT treatment planning via inverse planning based on probability distributions. *Physics in Medicine and Biology, 49*(17), 4005–4029.

19. Vedam, S. S. et al. (2003). Acquiring a four-dimensional computed tomography dataset using an external respiratory signal. *Physics in Medicine and Biology, 48*(1), 45–62.

20. Ford, E. C. et al. (2003). Respiration-correlated spiral CT: A method of measuring respiratory-induced anatomic motion for radiation treatment planning. *Medical Physics, 30*(1), 88–97.

21. Low, D. A. et al. (2003). A method for the reconstruction of four-dimensional synchronized CT scans acquired during free breathing. *Medical Physics, 30*(6), 1254–1263.

22. Rietzel, E. & Chen, G. T. (2006). Improving retrospective sorting of 4D computed tomography data. *Medical Physics, 33*(2), 377–379.

23. Rietzel, E. et al. (2005). Four-dimensional image-based treatment planning: Target volume segmentation and dose calculation in the presence of respiratory motion. *International Journal of Radiation Oncology Biology Physics, 61*(5), 1535–1550.

24. Pan, T. et al. (2004). 4D-CT imaging of a volume influenced by respiratory motion on multi-slice CT. *Medical Physics, 31*(2), 333–340.

25. Pan, T. (2005). Comparison of helical and cine acquisitions for 4D-CT imaging with multislice CT. *Medical Physics, 32*(2), 627–634.

26. Wink, N. et al. (2006). Phase versus amplitude sorting of 4D-CT data. *Journal of Applied Clinical Medical Physics, 7*(1), 77–85.

27. Zeng, R. et al. (2005). Respiratory motion estimation from slowly rotating X-ray projections: Theory and simulation. *Medical Physics, 32*(4), 984–991.

28. Li, T. et al. 4D cone-beam CT (CBCT) using an on-board imager. 48th AAPM Annual Meeting, 2006.

29. Sonke, J. J. et al. (2005). Respiratory correlated cone beam CT. *Medical Physics, 32*(4), 1176–1186.

30. Nehmeh, S. A. et al. (2004). Four-dimensional (4D) PET/CT imaging of the thorax. *Medical Physics, 31*(12), 3179–3186.

31. Nehmeh, S. A. et al. (2004). Quantitation of respiratory motion during 4D-PET/CT acquisition. *Medical Physics, 31*(6), 1333–1338.

32. Thorndyke, B. et al. (2006). Reducing respiratory motion artifacts in positron emission tomograph through retrospective stacking. *Medical Physics, 33*(7), 2632.

33. Nehmeh, S. A. et al. (2002). Effect of respiratory gating on reducing lung motion artifacts in PET imaging of lung cancer. *Medical Physics, 29*(3), 366–371.

34. Go, K. G. et al. (1995). Localised proton spectroscopy and spectroscopic imaging in cerebral gliomas, with comparison to positron emission tomography. *Neuroradiology, 37*(3), 198–206.

35. Nelson, S. J. et al. (1999). Serial evaluation of patients with brain tumors using volume MRI and 3D 1H MRSI. *NMR in Biomedicine, 12*(3), 123–138.

36. Narayana, A. et al. (2007). Use of MR spectroscopy and functional imaging in the treatment planning of gliomas. *British Journal of Radiology, 80*(953), 347–354.

37. Chang, J. et al. (2006). Image-fusion of MR spectroscopic images for treatment planning of gliomas. *Medical Physics, 33*(1), 32–40.

38. Graves, E. E. et al. (2001). Registration of magnetic resonance spectroscopic imaging to computed tomography for radiotherapy treatment planning. *Medical Physics, 28*(12), 2489–2496.

39. Pirzkall, A. et al. (2004). 3D MRSI for resected high-grade gliomas before RT: Tumor extent according to metabolic activity in relation to MRI. *International Journal of Radiation Oncology Biology Physics, 59*(1), 126–137.

40. Law, M. & Liu, B. (2009). DICOM-RT and its utilization in radiation therapy. *Radiographics, 29*(3), 655.

41. Rosenman, Ph. D. et al. (1998). Image registration: An essential part of radiation therapy treatment planning. *International Journal of Radiation Oncology Biology Physics, 40*(1), 197–205.

42. Fox, P. et al. (1985). A stereotactic method of anatomical localization for positron emission tomography. *Journal of Computer Assisted Tomography, 9*(1), 141.

43. Maurer, Jr., C. et al. (2002). Registration of head volume images using implantable fiducial markers. *IEEE Transactions on Medical Imaging, 16*(4), 447–462.

44. Schad, L. et al. (1987). Three dimensional image correlation of CT, MR, and PET studies in radiotherapy treatment planning of brain tumors. *Journal of Computer Assisted Tomography, 11*(6), 948.

45. Strother, S. et al. (1994). Quantitative comparisons of image registration techniques based on high-resolution MRI of the brain. *Journal of Computer Assisted Tomography, 18*(6), 954.

46. Hutton, B. et al. (2002). Image registration: An essential tool for nuclear medicine. *European Journal of Nuclear Medicine and Molecular Imaging, 29*(4), 559–577.

47. Maes, F. et al. (2002). Multimodality image registration by maximization of mutual information. *IEEE Transactions on Medical Imaging, 16*(2), 187–198.

48. Wells, III, W. et al. (1996). Multi-modal volume registration by maximization of mutual information. *Medical Image Analysis, 1*(1), 35–51.

49. Wang, H. et al. (2005). Implementation and validation of a three-dimensional deformable registration algorithm for targeted prostate cancer radiotherapy. *International Journal of Radiation Oncology Biology Physics, 61*(3), 725–735.

50. Lu, W. et al. (2004). Fast free-form deformable registration via calculus of variations. *Physics in Medicine and Biology, 49*(14), 3067–3087.

51. Noe, K. O. et al. (2008). GPU accelerated viscous-fluid deformable registration for radiotherapy. *Studies in Health Technology and Informatics, 132*, 327–332.

52. Sharp, G. C. et al. (2007). GPU-based streaming architectures for fast cone-beam CT image reconstruction and demons deformable registration. *Physics in Medicine and Biology, 52*(19), 5771–5783.

53. Janssens, G. et al. (2011). Diffeomorphic registration of images with variable contrast enhancement. *International Journal of Biomedical Imaging, 2011*, 891585.

54. Bharatha, A. et al. (2001). Evaluation of three-dimensional finite element-based deformable registration of pre- and intraoperative prostate imaging. *Medical Physics, 28*(12), 2551–2560.

55. Brock, K. K. et al. (2005). Accuracy of finite element model-based multi-organ deformable image registration. *Medical Physics, 32*(6), 1647–1659.

56. Brock, K. K. et al. (2008). Accuracy and sensitivity of finite element model-based deformable registration of the prostate. *Medical Physics, 35*(9), 4019–4025.

57. Brock, K. K. et al. (2006). Feasibility of a novel deformable image registration technique to facilitate classification, targeting, and monitoring of tumor and normal tissue. *International Journal of Radiation Oncology Biology Physics, 64*(4), 1245–1254.

58. Mattes, D. et al. (2003). PET-CT image registration in the chest using free-form deformations. *IEEE Transactions on Medical Imaging, 22*(1), 120–128.

59. Shackleford, J. A. et al. (2010). On developing B-spline registration algorithms for multi-core processors. *Physics in Medicine and Biology, 55*(21), 6329–6351.

60. Schreibmann, E. et al. (2006). Image interpolation in 4D CT using a BSpline deformable registration model. *International Journal of Radiation Oncology Biology Physics, 64*(5), 1537–1550.

61. Thirion, J. P. (1998). Image matching as a diffusion process: An analogy with Maxwell's demons. *Medical Image Analysis, 2*(3), 243–260.

62. Wang, H. et al. (2005). Validation of an accelerated 'demons' algorithm for deformable image registration in radiation therapy. *Physics in Medicine and Biology, 50*(12), 2887–2905.

63. Ng, L. & Ibanez, L. (2003). Narrow band to image registration in the Insight Toolkit. *WBIR 2003, LNCS 2717*, 271–280.

64. Schreibmann, E. & Xing, L. (2005). Narrow band deformable registration of prostate magnetic resonance imaging, magnetic resonance spectroscopic imaging, and computed tomography studies. *International Journal of Radiation Oncology Biology Physics, 62*(2), 595–605.

65. Lawson, J. D. et al. (2007). Quantitative evaluation of a cone beam computed tomography (CBCT)-CT deformable image registration method for adaptive radiation therapy. *Journal of Applied Clinical Medical Physics, 8*(4), 96–113.

66. Makni, N. et al. (2009). Combining a deformable model and a probabilistic framework for an automatic 3D segmentation of prostate on MRI. *International Journal of Computer Assisted Radiology and Surgery, 4*(2), 181–188.

67. Suh, J. W. & Wyatt, C. L. (2009). Deformable registration of supine and prone colons for computed tomographic colonography. *Journal of Computer Assisted Tomography, 33*(6), 902–911.

68. Bender, E. T. et al. (2009). On the estimation of the location of the hippocampus in the context of hippocampal avoidance whole brain radiotherapy treatment planning. *Technology in Cancer Research and Treatment, 8*(6), 425–432.

69. Hwang, A. B. et al. (2009). Can positron emission tomography (PET) or PET/computed tomography (CT) acquired in a nontreatment position be accurately registered to a head-and-neck radiotherapy planning CT? *International Journal of Radiation Oncology Biology Physics, 73*(2), 578–584.

70. Chao, M. et al. (2008). Auto-propagation of contours for adaptive prostate radiation therapy. *Physics in Medicine and Biology, 53*(17), 4533–4542.

71. Wijesooriya, K. et al. (2008). Quantifying the accuracy of automated structure segmentation in 4D CT images using a deformable image registration algorithm. *Medical Physics, 35*(4), 1251–1260.

72. Chao, M. et al. (2008). Automated contour mapping with a regional deformable model. *International Journal of Radiation Oncology Biology Physics, 70*(2), 599–608.

73. Orban de Xivry, J. et al. (2007). Tumour delineation and cumulative dose computation in radiotherapy based on deformable registration of respiratory correlated CT images of lung cancer patients. *Radiotherapy & Oncology, 85*(2), 232–238.

74. Bender, E. T. & Tome, W. A. (2009). The utilization of consistency metrics for error analysis in deformable image registration. *Physics in Medicine and Biology, 54*(18), 5561–5577.

75. Ostergaard Noe, K. et al. (2008). Acceleration and validation of optical flow based deformable registration for image-guided radiotherapy. *Acta Oncologica, 47*(7), 1286–1293.

76. Castadot, P. et al. (2008). Comparison of 12 deformable registration strategies in adaptive radiation therapy for the treatment of head and neck tumors. *Radiotherapy & Oncology, 89*(1), 1–12.

77. Lawson, J. D. et al. (2007). Quantitative evaluation of a cone-beam computed tomography-planning computed tomography deformable image registration method for adaptive radiation therapy. *Journal of Applied Clinical Medical Physics, 8*(4), 2432.

78. Zhang, G. et al. (2008). Functional lung imaging in thoracic cancer radiotherapy. *Cancer Control, 15*(2), 112–119.

79. Guerrero, T. et al. (2006). Dynamic ventilation imaging from four-dimensional computed tomography. *Physics in Medicine and Biology, 51*(4), 777–791.

80. Yamamoto, T. et al. (2011). Impact of four-dimensional computed tomography pulmonary ventilation imaging-based functional avoidance for lung cancer radiotherapy. *International Journal of Radiation Oncology Biology Physics, 79*(1), 279–288.

81. Yan, D. et al. (1997). Adaptive radiation therapy. *Physics in Medicine and Biology, 42*(1), 123–132.

82. Yan, D. et al. (1998). The use of adaptive radiation therapy to reduce setup error: A prospective clinical study. *International Journal of Radiation Oncology Biology Physics, 41*(3), 715–720.

83. Wu, Q. et al. (2009). Adaptive replanning strategies accounting for shrinkage in head and neck IMRT. *International Journal of Radiation Oncology Biology Physics, 75*(3), 924–932.

84. Ahunbay, E. E. et al. (2009). An on-line replanning method for head and neck adaptive radiotherapy. *Medical Physics, 36*(10), 4776–4790.

85. Velec, M. et al. (2011). Effect of breathing motion on radiotherapy dose accumulation in the abdomen using deformable registration. *International Journal of Radiation Oncology Biology Physics, 80*(1), 265–272.

86. Flampouri, S. et al. (2006). Estimation of the delivered patient dose in lung IMRT treatment based on deformable registration of 4D-CT data and Monte Carlo simulations. *Physics in Medicine and Biology, 51*(11), 2763–2779.

87. Zhang, G. et al. (2010). Generation of composite dose and biological effective dose (BED) over multiple treatment modalities and multistage planning using deformable image registration. *Medical Dosimetry, 35*(2), 143–150.

88. Wu, Q. J. et al. (2008). The impact of respiratory motion and treatment technique on stereotactic body radiation therapy for liver cancer. *Medical Physics, 35*(4), 1440–1451.

19

Imaging for Radiation Treatment Planning

Gig S. Mageras
Memorial Sloan-Kettering Cancer Center

Yu-Chi Hu
Memorial Sloan-Kettering Cancer Center

Stephen McNamara
Memorial Sloan-Kettering Cancer Center

Hai Pham
Memorial Sloan-Kettering Cancer Center

Jian-Ping Xiong
Memorial Sloan-Kettering Cancer Center

19.1 Introduction

This chapter discusses medical images, their associated information, and how these images are used in 3D radiation treatment planning (RTP) systems. In the context of informatics, emphasis will be placed on communication of imaging information between imaging and RTP systems, design of the RTP system, and a description of some algorithms for information processing. We will focus on external beam treatment, although many of the concepts also apply to brachytherapy planning systems.

In current common practice, basic steps in the workflow of the RTP process are (1) acquisition of volumetric images, (2) transfer of images and data to the RTP system, (3) definition of volumes of interest, (4) design of treatment machine parameters, (5) design of reference images for localization, (6) calculation of radiation dose, (7) review and approval of the treatment plan, and (8) transfer of the radiation treatment plan and reference images to the system that controls the localization imaging equipment and treatment machine. In this chapter, we discuss Steps 1 to 5 and the infrastructure for accomplishing these steps. Other aspects of treatment planning and delivery are discussed elsewhere in this book.

19.2 Acquisition of Volumetric Images

Computed tomography (CT) is the primary imaging modality for RTP. CT provides cross-sectional images of patient anatomy, which are well suited for RTP, particularly for defining volumes of interest and for calculating radiation doses. Each element of the image represents a small area or pixel; the product of a pixel area and the slice thickness constitutes a volume element or voxel. The digital value reconstructed for each pixel (or voxel) is expressed as a CT number and represents the linear attenuation coefficient of that volume. CT images are usually acquired with a 120- to 140-kV peak x-ray beam; therefore, a nonlinear conversion table is used to obtain electron densities for dose calculations relevant to the much higher energy x-rays used for treatment. The relationship between CT numbers and electron densities is derived empirically and expressed as two or three linear regions and stored in a "lookup" table (Seco & Evans 2006).

Software systems for performing CT simulation have become increasingly widespread and have supplanted the traditional radiographic simulator in radiation therapy clinics. These systems provide treatment simulation functions with volume representation from a CT image set. CT simulation functions include the following:

- Patient alignment based on scout view images;
- Target and normal organ localization performed by defining contours on axial CT images;
- Virtual fluoroscopy, which mimics a conventional fluoroscopic simulator and allows definition of a treatment portal without requiring target contours on all CT slices;
- Definition of reference isocenter points from the volumetric CT data, followed by CT table positioning, such that alignment lasers indicate triangulation points for marking the patient surface;

- Treatment portal design including entry of block or multileaf collimator (MLC) outlines;
- Generation of digitally reconstructed radiographs (DRRs); and
- Transfer of CT images and associated simulation data to an RTP system for dose calculation and plan evaluation.

Other imaging modalities provide complementary information. Magnetic resonance images (MRI) often provide better soft-tissue contrast than do CT images, thus improving discrimination between tumor and normal tissues in some disease sites. Positron emission tomography (PET) imaging allows tumor visualization based on the elevation of glucose metabolism. However, several characteristics have limited the applicability of other image modalities alone for treatment planning. MR pixel intensities do not correlate with electron density as they do in CT; thus, they cannot be used directly in dose calculations. The images may be distorted by variations in local magnetic fields caused by imperfections in the machine itself and by the presence of metal objects in the environment and within the patient. PET alone does not provide anatomical information and has poorer spatial resolution than CT. Because of these limitations, MR and PET images are usually registered to CT images, and the information they provide is transferred to the CT study for treatment planning purposes. The registration of CT and other volumetric images is described later in this chapter.

19.3 Infrastructure and Data Transfer Between Systems

Some general concepts in creating a network for transferring data for external beam RTP are discussed here. Henceforth,

the term "network" will indicate a local area network (LAN) that supports, at a minimum, 100 Mbit/s connectivity to all equipment.

Building an infrastructure for RTP requires the expertise of staff from many areas: medical physics, networking, computer support, and radiation therapy. The medical physics and radiation therapy staff are most knowledgeable in clinical workflow and how information needs to be transferred from imaging systems (e.g., CT, MR, and PET) through simulation, treatment planning, and finally to treatment delivery (aka "record and verify") systems. The roles of the networking and computer support staff are to lay the groundwork and put in place the mechanisms necessary for the transfer of information as defined by the physics and therapy staff.

Figure 19.1 is a clinical workflow example. Patient images are acquired in one or more modalities. The images are then transmitted over the network to one or more systems. In the example, CT images are sent directly to a CT simulation system. The patient may also have had a MRI or PET scan that the physician wishes to fuse with the CT data in either the CT simulation or external beam RTP system. In this example, the MRI study might be sent directly from the MR console or via a picture archiving and communication system (PACS). Likewise, PET images may be transferred via the PACS to the CT simulation system or the RTP system. Radiation therapists or physicians may then add points, contours, and other information, which are then transferred to the RTP system. Note that the thickness of the arrows in the figure illustrates the increasing amount and complexity of the information being transferred.

In the next step, physicists and dosimetrists, together with physicians, create the patient's treatment plan on the RTP system (in some institutions, the CT simulation and RTP systems

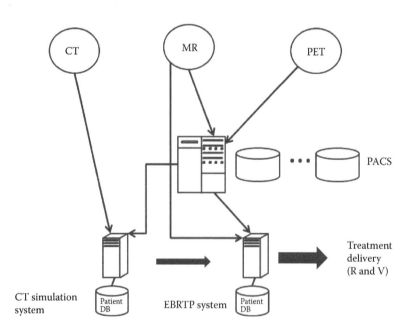

FIGURE 19.1 Diagram of a clinical workflow involving patient imaging via CT, MRI, or PET for radiation treatment simulation. The arrow thickness indicates the amount and complexity of the data being transferred. DB, database; EBRTP, external beam RTP; R and V, record and verify.

may be one and the same). The treatment plan is then transferred to the treatment delivery system. The prevailing method for transferring this information over a network uses the Digital Imaging and Communications in Medicine (DICOM) standard, discussed in greater detail in the next section. To configure two systems to communicate using DICOM, one needs three pieces of information:

- *Application Entity (AE) Title.* This is used to define the specific DICOM entity that will either send or receive the data. When receiving DICOM data, a system can be configured to receive from anywhere (also known as "promiscuous" mode).
- *Port Number.* This is the TCP/IP port on which the DICOM entity "listens." The default is 104, but in many instances it can be changed.
- *IP Address.* This is the network address of the system (i.e., imaging modality, computer, and film printer).

To successfully transfer data, these three pieces of information must match exactly. The following paragraphs discuss these DICOM parameters in more detail and offer insight into planning an RTP infrastructure.

AE Titles. AE titles are case sensitive. Most vendor systems come with a predefined AE title, which in many instances is fine; however, in larger networks, this can lead to confusion. Some commercial systems do not allow multiple destinations with the same AE title. Furthermore, having multiple DICOM systems on the network with the same generic AE title is difficult to manage and troubleshooting can be problematic. Changing the AE title to reflect location, vendor, modality, or some other feature will help in the long run.

Port Numbers. As mentioned above, the default port number for DICOM communication is 104. However sometimes it is necessary to change this. For example, if multiple DICOM receivers are running on a given computer, each receiver must be given a unique port number. Most vendor systems allow the port number to be defined.

IP Addresses. It is vital that the IP addresses used for DICOM communication are permanent or "static." IP addresses are often provided automatically to computers over the network; however, these "dynamic" addresses may change over time. If that happens, then DICOM communication breaks down. The networking or computer support staff should be able to provide a static IP address.

19.3.1 Sample DICOM Receiver Configurations

A DICOM "receiver" is a system to which data are sent using DICOM. There are many possible configurations for a DICOM receiver system. In the simplest configuration, a system (i.e., a CT console or an RTP computer) runs a single process that receives DICOM data from the network. Figure 19.2 illustrates this configuration. The system runs a single DICOM "client" process to receive data from the network. To add this system to the list of DICOM destinations on another system, for example the PACS, one would provide the following information to the PACS Administrator: AE Title = RTP_PLAN; Port Number = 104; IP Address = 100.100.100.101. In this example, all DICOM data received are written to a folder on the local C: drive. Many institutions may prefer (or even mandate) that patient data be written to a network drive. In most DICOM applications, this should be a straightforward change to the configuration.

In some institutions, this configuration may not suffice. For example, many hospitals have multiple sites or campuses. In these settings, sending all patients to a single system may prove unwieldy and inefficient. The configuration in Figure 19.3 illustrates a single computer capable of receiving DICOM data from five different locations or campuses. Say, for example, that a vendor is installing a new imaging modality at Campus #3. To receive images from that modality requires providing the vendor with the following information: AE Title = CAMPUS3, Port Number = 106, IP Address = 100.100.100.101. Note that each DICOM client process writes the data it receives into different folders. As mentioned previously, these folders can be local or

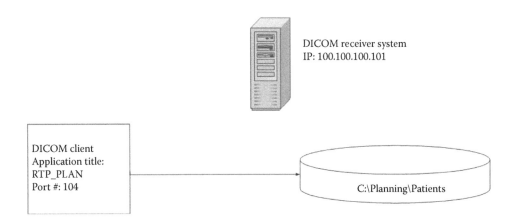

FIGURE 19.2 Diagram illustrating a simple system that runs a single DICOM "client" process to receive data from the network.

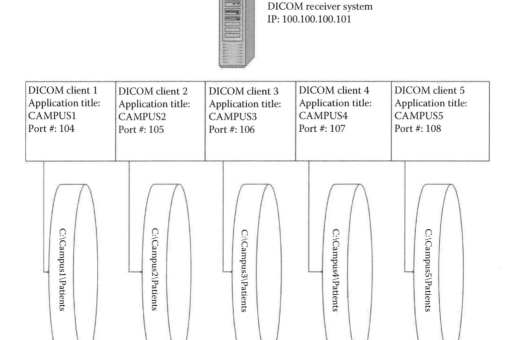

FIGURE 19.3   Diagram illustrating a single computer receiving DICOM data from five different campuses or sites.

on a network. A similar configuration could be used to separate DICOM data by modality, department, or any other criteria.

19.3.2 Data Storage Considerations

The amount of disk storage required for imaging is significant. A single CT slice uses approximately 0.5 MB of space. A typical CT study consists of roughly 200 slices, which translates to 100 MB. Larger CT studies are not uncommon. For 4D studies, the storage requirements grow by a factor of 10. Also, some patients undergo periodic cone-beam CT scans, which increase the storage requirements by approximately 35 MB per patient per scan.

When designing an infrastructure for RTP, it is important to include disk space that is reliable and extensible. Writing data to a network drive has several advantages. First, multiple systems can share the data, eliminating the need to transfer duplicate data to multiple destinations. Second, if the network storage is in a data center or other secure location, it may help protect patient information (as required in the United States by the Health Insurance Portability and Accountability Act). It may also allow the information to be backed up to tape or other offline media. Third, placing the data on a network disk provides more fault tolerance. If the DICOM receiver system should go down or become unavailable, the DICOM data are still accessible (although new data cannot be sent).

Although DICOM facilitates the transfer of patient information over a network, there is one unfortunate byproduct: information duplication. Referring back to Figure 19.1, it is apparent

that a patient's image study can easily exist in three different places: the image modality's console computer, the CT simulation system, and the RTP system. Although providing significant amounts of disk storage for all these various systems is important, thought must also be given to archiving and, as appropriate, deleting these data. Radiation therapy archiving systems are described in more detail in Chapter 9.

19.3.3 DICOM

The American College of Radiology and the National Electrical Manufacturers Association (NEMA) developed the DICOM standard to ensure the interoperability of systems used to produce, store, display, process, send, retrieve, query, or print medical images (NEMA 2009a). The DICOM standard defines the general format of information (Parts 3–6), standardized workflow of network communication among different vendor systems (Parts 4, 7, and 8), and media interchange (Parts 10–12). Modules and data elements related to radiation therapy are included in DICOM, known as DICOM-RT. Some important concepts in DICOM that are relevant to RTP are reviewed briefly below.

19.3.4 Information Object Definitions

Data are organized based on information object definitions (IODs), which provide an abstract definition (Part 3) of appropriate real-world objects (e.g., CT and MR images). Each IOD consists of modules, and each module includes a set of data elements. For example, a CT image IOD contains up to 17 modules. Among these modules, a patient module has data elements for patient information and an image pixel module includes data elements for image dimension and image pixels. Elements are identified by a tag composed of a group ID and an element ID. For example, (0028, 0010) is the tag for the number of rows in the image.

19.3.5 Network Communication

The most import component in the DICOM standard is the network communication between systems to take advantage of the high bandwidth of modern networks. The two parties on each side of the network connection can be considered AEs (defined in Section 19.3). An AE is a computer program that provides the DICOM services for service classes defined in Part 4. Service classes are defined based on service (such as storage or print) object (such as CT or MR) pairs (SOPs). An AE can play multiple roles: it can be a service class provider (SCP), a service class user (SCU), or both. When an AE initiates a network connection to another AE, the peer AEs first negotiate what service classes are supported and what roles each sides will play. Other information, such as transfer syntax (byte order) and maximum length for a data unit, is also specified in the association. Within the established association, various commands can be issued by the SCU for the agreed-upon services provided by the SCP. The standard defines the behaviors of SCU and SCP as to how the commands should proceed. Some commonly used services are described below.

19.3.5.1 Verification

C-ECHO command verifies the communication between peer AEs. This is mostly used to test if the DICOM settings such as AE titles are properly configured. Usually, a list of all the remote AE titles allowed for connection is set up on the local AE. The calling AE title of the remote client should be on the list; in addition, the called AE title specified by the remote client should match the local AE title.

19.3.5.2 Storage

Once the association is established, the SCU issues the C-STORE command for each image sent to the SCP. Based on the SOP class specified in the command, an SOP instance is created for each image transferred and should have a unique universal identifier (UID). The UIDs are important for storage purposes as well as for connecting associated data. A study instance UID can be referred to by other data such as contours associated with the study. The data in the instance are encoded based on the corresponding IOD. To control the data flow, a data block can be broken up into smaller blocks according to the agreed-upon maximum length when the association is established.

19.3.5.3 Query/Retrieve

The C-FIND command is used by the SCU to query data stored at a remote site. The query can target any of three levels: patient, study, and series (a fourth, at the individual image level, is not often used in common practice). The SCP responds with level-related attributes. For example, for querying at the patient level, the SCP returns a list of patients that match the query criteria. Based on the query result, the SCU can issue either C-MOVE or C-GET commands to retrieve patient data. Both will initiate C-STORE suboperations for transferring data. C-MOVE is used to ask the SCP to send data to a specified storage destination in a separate association for the storage SCP specified, whereas C-GET causes the SCU to accept C-STORE within the same association; thus, the SCP/SCU roles are reversed for C-STORE suboperations.

19.3.5.4 Print

Print management service classes are used for printing image and image-related data on a hard-copy medium, usually a film printer. The SCU issues various commands for (1) retrieving the printing device's configuration (N-GET), (2) creating a film box to accommodate the images being printed (N-CREATE), (3) setting image box parameters to format how an image should be printed (N-SET), (4) initiating printing (N-ACTION), and (5) completing the job (N-DELETE).

19.3.6 Media Interchange

When the network is not available or when patient data come from other institutions, the images can also be transferred via portable media such as CDs and DVDs. Similar to data transferred through a network, each SOP instance (e.g., an image) is a file. The files are collected into file sets and a unique and mandatory file with the file name DICOMDIR is included in each file set. The DICOMDIR file itself is a DICOM SOP class instance containing the information about the SOP instances stored in the file set. Each SOP instance has a File ID specified in this file, which is the full path of the SOP instance under the directory where DICOMDIR resides. For example, File ID PA1\ST1\SE3\IM114 can read as "image file IM114 is under folder series 3 (SE3) of folder study 1 (ST1) of folder patient 1 (PA1)." The valid characters for the file ID (path) are A to Z (uppercase), 0 to 9, and "_" (underscore). The file in the file set not only has an SOP instance encoded that is the same as the one transferred via network, but it also has metadata for necessary information such as transfer syntax, which is specified in association if data are transferred via network, allowing the applications to read the data correctly.

19.3.7 DICOM-RT

As noted in Section 19.3.3, DICOM-RT is an extension of DICOM to radiation therapy. The RT extension allows transfer of radiation therapy–related data between different devices, treatment planning systems, departments, and institutions. New DICOM-RT IOD objects have been introduced, namely, RT Image, RT Structure Set, RT Plan, RT Dose, and RT Treatment Record. The RT Treatment Record further includes the RT Beam Treatment Record, RT Brachy Treatment Record, and RT

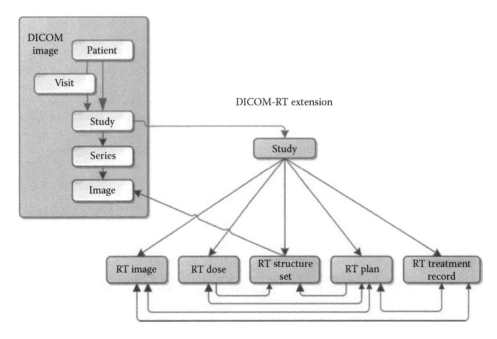

FIGURE 19.4 Diagram illustrating the radiation therapy extension to DICOM-RT objects as an extension of the DICOM standard and their relation to one another and to the DICOM image object. (From Law, M. Y. Y. & Liu, B., *Radiographics*, 29, 655, 2009. With permission.)

Treatment Summary Record. Within each patient study, these RT objects are related to each other by references (Figure 19.4). These DICOM-RT objects are briefly described in the subsections that follow.

19.3.7.1 RT Image IOD

The RT Image IOD contains attributes that describe RT-specific characteristics of a projection image, which is acquired or calculated by using a conical imaging geometry. The image can be a portal radiograph, simulator image, or DRR. Information associated with image generation, such as gantry angle and field shape, can be stored directly as an attribute or by referencing the RT Plan object. RT Image also defines the image coordinate system by means of the RT Image Orientation (3002, 0010), RT Image Position (3002, 0012), and Image Plane Pixel Spacing (3002, 0011).

19.3.7.2 RT Structure Set IOD

The RT Structure Set IOD addresses the requirements for transfer of patient structures defined on CT scanners, virtual simulation workstations, treatment planning systems, and similar devices. One can specify 3D point structures, open/closed and coplanar/noncoplanar 3D contour structures, as well as their display attributes such as color and contour slab thickness.

19.3.7.3 RT Plan IOD

The RT Plan IOD addresses the requirements for transfer of treatment plans from a virtual simulation or treatment planning system to a treatment delivery system. The RT Plan IOD contains the RT Fraction Scheme module (containing referenced beams, meter settings, and beam doses), RT Prescription module, RT

Tolerance Table module (machine tolerances), RT Patient Setup module (patient setup information), and RT Beams module (complete beam information).

19.3.7.4 RT Dose IOD

The RT Dose IOD contains radiation dose for a treatment plan. It may contain a 2D or 3D dose array, isodose lines, and dose-volume histograms. The dose summation type may be a single radiation beam, a fraction group, or a complete treatment plan (potentially the sum of multiple fraction groups). RT Structure Set and RT Plan are referenced in RT Dose for relating information on regions of interest with treatment plan information.

19.3.7.5 RT Treatment Record IOD

The RT Treatment Record is further divided into RT Beams Treatment Record, RT Brachy Treatment Record, and RT Treatment Summary Record. The RT Beams Treatment Record addresses the requirements for transfer of treatment session reports generated by a treatment verification system during a course of external beam treatment or gathered during treatment delivery, with optional cumulative summary information. Similarly, the RT Brachy Treatment Record addresses brachytherapy session reports. The RT Treatment Summary Record contains the cumulative radiation treatment summary information generated after completion a course of treatment (NEMA 2008).

19.4 RTP

The primary goal of radiation therapy is to deliver a tumoricidal dose to the malignant site while keeping radiation exposure to surrounding organs to acceptably low levels. Imaging the location of tumor and nearby organs at risk is essential to achieving this goal. Application of images in RTP requires accurate spatial information about a patient's tissues as well as tissue density and composition for dose calculation. To be able to do so, volumetric image sets must meet certain requirements. For example, an image set must contain information that defines a positional relation between images that form the 3D volume. Pixel size must be uniform throughout the images. Images should have sufficient contrast for discerning boundaries between anatomical structures. Also important is the ability to assign attenuation coefficients (CT number) to voxels for conversion to tissue densities used in dose calculations.

Images are used in RTP primarily to define the target for treatment [planning target volume (PTV)] and organs at risk, define the radiation beams constituting the treatment plan, and evaluate the resultant dose distribution from the treatment plan. An RTP system is a computer simulation system that provides the capabilities to perform these tasks. A report of the National Cancer Institute Photon Treatment Planning Collaborative Working Group (1991) outlined requirements for volumetric image-based RTP systems that were still largely relevant when this chapter was written (see also reviews by Kijewski 1994 and Kalet & Austin-Seymour 1997). The basic requirements of such a system are to (1) delineate the volumes of interest in the patient, including the volume enclosed by the patient surface, the PTV, and the volumes of the organs-at-risk; (2) model the geometry and dosimetric properties of the radiation beams; and (3) display anatomical, beam geometry, and dosimetric information. The functions of an RTP system are discussed further in the sections below.

19.4.1 Display of Volumetric Images

For radiation therapy, a correct spatial relationship is essential in the display of volumetric images and the superposition of treatment planning information. To appropriately relate these two sets of information, an RTP system uses information contained in the image set that defines a positional relation between images. For example, each image in a DICOM image set contains an Image Position tag (3D coordinate of the center of the first pixel of the image) and an Image Orientation tag (direction cosines of first row and first column with respect to patient). To correctly reconstruct a 3D image volume, the RTP system checks that all images have the same Image Orientation. It then sorts individual images to form a stack by using values stored as Image Position tags. Because 3D image volumes are usually stored as a 3D array, the pixel sizes (stored in Pixel Spacing tag) must be the same for all axial images. Slice thickness is not important for the reconstruction of the 3D image. The CT number of each pixel, defined in Hounsfield units, is calculated from the 12-bit unsigned integer Pixel Value stored for each pixel, and from the Rescale Slope and Rescale Intercept values in the DICOM image header, according to the formula:

CT number = Pixel Value × Rescale Slope + Rescale Intercept.

RTP systems display images in two and three dimensions to help planners and physicians visualize 3D image volumes. 2D image displays usually consist of axial, coronal, and sagittal cross-sectional images, which can be scrolled through to navigate a 3D image volume. Some systems, such as GE Advantage Windows, also include a 2D oblique image view. The oblique image planes are useful for displaying radioactive seed implants (Kalet & Austin-Seymour 1997). Because a CT image set contains only a stack of axial images, coronal and sagittal images must be rapidly interpolated and displayed as needed from the axial images. The spatial resolution of the reconstructed coronal and sagittal images greatly depends on the axial image slice spacing. The basic steps of image display are to (1) extract CT values from the 3D data set, (2) preprocess CT values, (3) map CT values, and (4) display the images. One such implementation, used in an RTP system developed at the Memorial Sloan-Kettering Cancer Center, is described below.

19.4.1.1 Extract CT Values from the 3D Data Set

The CT values are extracted either directly from the original images (axial) or through interpolation (coronal and sagittal). Coronal and sagittal image planes are chosen so that they align with the rows and columns of pixels in axial images, respectively, which reduces the image computation time. Thus, calculation of the CT number of a coronal or sagittal image pixel requires interpolation from only two neighboring pixels, those from the axial images above and below.

19.4.1.2 Preprocess CT Values

Preprocessing functions serve to enhance anatomical boundaries in the images. Common preprocessing functions are unsharp masking, Sobel edge detection, and histogram equalization (Gonzalez & Woods 1992). Unsharp masking enhances fine detail by subtracting a blurred copy of the image from the original. Sobel edge detection is an operator that displays the gradient of the image intensity at each pixel. Histogram equalization uses a transformation function to produce an image whose gray levels have a uniform density; the resultant image gray levels are spread out between black and white, thus increasing image contrast.

19.4.1.3 Map CT Values

The use of window/level settings is the most common way to improve image readability. Based on user-specified window/level values, CT numbers are mapped to displayed pixel intensities. The window width specifies the range of CT numbers to be mapped to the monitor display range. The level is at the center of the window width. The window width is inversely proportional to the displayed image contrast. In Figure 19.5, the window width is 500 and the level value is 950. CT numbers less than 700 are displayed as black and CT numbers greater than 1200 are displayed as white; intermediate values are linearly mapped to grayscale intensities between 0 and 255 (this example assumes that the graphics card and monitor are capable of 256 grayscale levels).

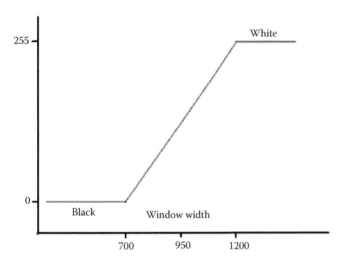

FIGURE 19.5 Concept of window/level definition for image display.

19.4.1.4 Display Images

It is often desirable to magnify images on the monitor to view finer details. Digital image magnification changes the spatial sampling of the image and uses an interpolation algorithm such as nearest neighbor, bilinear, or bicubic to calculate pixel gray levels. These algorithms differ in the tradeoff between quality of resampled image and speed. Nearest-neighbor interpolation is the fastest and lowest-quality algorithm, in which each destination pixel takes the value of the nearest source pixel. Bilinear interpolation uses the four neighboring source pixels, whereas bicubic interpolation uses the 16 nearest neighbors. Bicubic interpolation is the slowest but can produce smoother images and preserve fine details better than other algorithms. Bilinear interpolation provides a balance between quality and speed.

19.4.1.5 3D Image Display

Image data are usually rendered by using one of two methods: surface-based rendering or direct volume rendering such as ray casting, splatting (Westover 1991), and shear warp (Lacrout & Levoy 1994). In surface-based rendering, image data are represented as polygonal meshes created by extracting surfaces of equal CT numbers from the image volume. Marching cubes is a commonly used algorithm for this purpose (Lorensen & Cline 1987). Direct volume rendering is a technique that uses transfer functions to map voxel values to opacities and colors. This technique can produce very high-quality images but is also very computationally intensive.

19.4.2 Coordinate Systems

All images and data for modeling the treatment plan must be related by a common coordinate system or by transformations between coordinate systems. The DICOM standard defines a patient coordinate system in which the x-axis direction is toward the patient's left, y points toward the posterior, and z is toward the patient's head. For each image in the 3D set, two elements

determine the image plane's position and orientation: the Image Position (0020, 0032), which specifies the x, y, and z coordinates of the upper left-hand corner of the image, and the Orientation (0020, 0037), which specifies the direction cosines of the first row and first column with respect to the patient. For example, direction cosines of (1, 0, 0) for the first row indicate that the right side of the image is the patient's left-hand side. Note that the image plane is not necessarily orthogonal to the coordinate system axes. Modeling of the patient and treatment machine makes use of different coordinate systems and transformations between them. The International Electrotechnical Commission 61217 standard defines the coordinate systems and scales to be used for radiation therapy equipment and includes matrices for transformation to and from the DICOM patient coordinate system (International Electrotechnical Commission 2007).

19.4.3 Definition of Volumes of Interest: Image Segmentation

The relevant anatomy for treatment planning, such as target volumes and organs at risk, must be defined and delineated in the volumetric image set. Volumes of interest are represented by a set of contours outlined on axial images, with each contour containing three or more ordered points or vertices. The point coordinates are defined in the image coordinate system relative to the upper left corner of each image. Combined with the known positional relation between axial images, the outlined contours define a 3D volume. When all contours for a volume are defined, a polygonal mesh is generated by using a contour triangulation algorithm. In some RTP systems (e.g., Varian's Eclipse and Philips Medical System's Pinnacle), a user can start with a mesh (chosen from a database) to represent the volume. The user modifies the mesh by moving vertices on the axial, sagittal, or coronal images. The mesh is then resliced to generate contours on the axial image slices.

Bit encoding is another useful means of representing volumes of interest, by assigning each voxel in an image a value such that the ith bit is 1 if the voxel is inside the ith volume and 0 otherwise. Bit encoding allows rapid identification of voxels inside a volume, for use by surface reconstruction algorithms such as marching cubes (Lorensen & Cline 1987). Another application is to deform structures defined on a planning CT to other breathing phases of a respiration-correlated CT set (Yang et al. 2008). The bit-encoded volume is defined on the planning CT and deformed to the image sets at the other breathing phases using the predicted displacement fields derived from deformable image registration (Section 19.4.4.1).

In current practice, delineation of volumes of interest, or image segmentation, in RTP systems continues to be done by using manual or semiautomatic tools in most cases, with manual drawing of contours on axial images being the most common method (Mundt & Roeske 2005). Figure 19.6 is a transverse CT scan that has been marked with common manual tools. A line drawing tool allows users to specify contour points by clicking on the axial images. The points are then connected by using

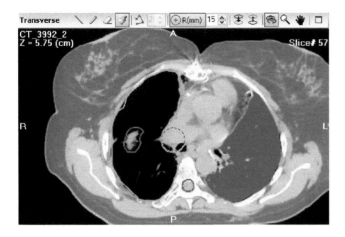

FIGURE 19.6 (**See color insert.**) 2D manual editing tools. The paint brush tool (selected) allows users to push a contour in or out using a cursor in the shape of a circle.

straight lines. In pencil drawing mode, users can draw freeform lines by tracing the mouse cursor while holding down the left mouse button or the shift key. A paintbrush tool allows users to push a contour inward or outward using a cursor in the shape of a circle. This tool is very effective for editing contours. An eraser tool is used to remove contour points. As users move the eraser (the mouse cursor), any points within the eraser tip are removed.

Semiautomatic methods accelerate the volume delineation process by means of algorithms that are initiated with less user input than manual segmentation. These techniques can be broadly categorized into region-based, boundary-based, and hybrid methods. Typically, region-based methods examine within-region similarities, and boundary-based methods search for anatomical boundaries based on image edge features.

Region-based methods partition the image voxels into separate regions based on the homogeneity criteria of pixel intensities. Because of noise and other artifacts present in the images, constraints on piecewise continuity are important to prevent oversegmentation (i.e., splintering into many small regions). Region-based methods need an additional step to extract contours from the segmented regions, usually by using isoline and isosurface algorithms, such as marching squares (for 2D) and marching cubes (for 3D). Boundary-based methods start with some initial contour, which is usually drawn manually and then changed in shape by minimizing some cost/energy function that finds the edge of the target structure. Therefore, piecewise continuity is guaranteed. However, for structures where the boundary is not clear or where surrounding structures have higher contrast boundaries, there can be "leakage" of the contour such that it includes tissue outside the structure of interest. Hybrid methods attempt to exploit the advantages of both methods, for example, by using a boundary-based method for evolving a contour while maintaining the homogeneity of the region within the contour, therefore avoiding oversegmentation and reducing the chance of leakage. Some of the most commonly used semiautomatic methods are discussed below. A more complete review of segmentation methods is given by Hu et al. (2009).

19.4.3.1 Thresholding

Thresholding is the simplest and fastest region-based segmentation method. A voxel is examined to determine if it belongs to the target structure by comparing its gray-level value to one or more thresholds. The thresholds can be chosen manually or automatically via analyzing the shape of the intensity distribution of the image (histogram) (Rosenfeld & Torre 1983; Sezan 1990) or some statistical measurement such as minimizing variance within the tissue class (Otsu 1979). Thresholding works best in segmenting patients' outer and bony structures, but it is less suitable for soft tissues in CT because the intensity distributions for some tissue classes are not well separated.

19.4.3.2 "Growing" a Region of Interest

Expanding a region of interest, referred to as region growing (Adams & Bischof 1994) starts with a seed representing the target region, usually manually specified on the image. The region is then "grown" successively by adding neighboring voxels of the region if the voxels satisfy the region's homogeneity predicate. A simple predicate can be defined as $P(x, r) = |f(x) - \mu_r| < T$, where x is a neighboring voxel to region r, $f(x)$ is the gray value of x, μ_r is the mean gray value of the region, and T is a threshold. If P is true, then x is added to the region and μ_r is recalculated. This is repeated until no more neighboring voxels to the region can be added. Because μ_r can be computed very rapidly in each iteration, one can choose T interactively and the result can be updated in real time. The use of a region's mean value and growing with only the region's neighboring voxels provides a constraint on piecewise continuity. Region growing can be done in two or three dimensions.

19.4.3.3 Live Wire

Live wire (Mortensen & Barrett 1995) is an interactive 2D segmentation method allowing the user to rapidly delineate a contour in the same way as manual drawing but with fewer mouse clicks. The method finds the minimum-cost path between the pixel at the previous mouse click and the pixel at the current cursor position. A subimage formed from the two anchor pixels is modeled as a rectangular matrix or graph. Each pixel of the image is a vertex of the graph and has edges connecting to the eight neighboring pixels. The edge costs are defined based on a cost function such that higher the contrast, the lower the cost. Each time the cursor is moved, the path is updated and drawn on the image in real time; the user confirms it as a desired contour segment by a mouse click. Then, a new graph is formed for the next segment. During the process, the original edge cost function can be substituted by a statistical cost function that estimates the probability that the edge is a boundary from the training of previous paths in the process. Higher probability reduces the cost, thereby improving the robustness of the desired organ segmentation while avoiding nearby high-contrast organs.

Some RTP systems, such as Varian Eclipse and Philips Pinnacle, allow manual editing of a 3D polygonal mesh. Figure 19.7 shows a 3D mesh of the spinal cord edited with Pinnacle's

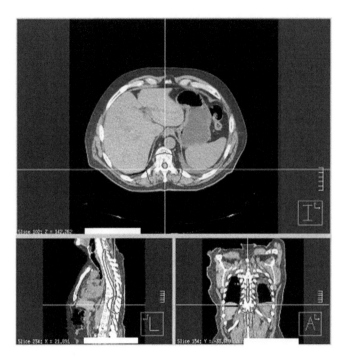

FIGURE 19.7 **(See color insert.)** Manual 3D editing tool in Philips Pinnacle RTP system.

3D contouring tool. The user imports a model from an atlas that is a mean shape for a specific anatomical structure and manually positions it to the corresponding site on the image. The surface of the model is then deformed by minimizing an energy function consisting of an external energy term that attracts the surface to image edge features and internal energy term that penalizes shape deviations from the model (Pekar et al. 2004; Ragan et al. 2005). The atlas is derived from a library of prior segmentations. The user can modify the result by moving the mesh vertices. When modifications are complete, the mesh is resliced to create contours on the axial images.

19.4.4 Image Registration

Different image modalities provide complementary information (discussed in Section 19.2) that is often desirable or necessary in a treatment plan. Thus, methods are required to form a composite view of patient anatomy from different image sources. These processes are called image registration and image fusion. Image registration is the determination of a geometrical transformation that aligns (maps) points of an object in one image with corresponding points of that object in another image. Image fusion involves combining the two registered images to form a single image One common form of the former process is rigid registration, which determines a 3D translation and rotation such that the two image sets spatially coincide. Scaling along each axis may also be required. Nonrigid or deformable registration methods attempt to account for tissue deformations caused by respiration or changes in the patient between imaging sessions.

Automated image registration methods minimize a cost function that measures the similarity between the two image sets for a given spatial alignment. Often, a volume of interest is defined within the reference image for computing the registration. Commonly used similarity measures, mutual information (MI) and cross-correlation, are discussed briefly below.

MI is suitable for registration of images from different modalities (Zitova & Flusser 2003). The MI between two image sets can be calculated as

$$MI = H_A + H_B - H_{AB}$$

$$MI = \sum_i^n (-P_A(i)\text{Log}(P_A(i))) + \sum_j^m (-P_B(j)\text{Log}(P_B(j)))$$

$$+ \sum_{i,j}^{n,m} (-P_{AB}(i,j)\text{Log}(P_{AB}(i,j))) \quad .$$

In the above equations, A is the reference image; B is the target image; i and j are the indices of the pixel value in images A and B; H_A and H_B are the entropies of image A and B, respectively; and H_{AB} is the joint entropy. Entropies are calculated based on the probability distribution of pixel values, $P_A(i)$, $P_B(i)$, and the joint distribution $P_{AB}(i,j)$. The best match between two images is when MI is maximized.

Normalized cross-correlation (NCC) is a fast and robust registration method for images with the same modality and is given by

$$NCC = \sum_{p=0}^{n-1} (G_A(p) - \bar{G}_A)(G_B(p) - \bar{G}_B) \Bigg/ \left[\sqrt{\sum_{p=0}^{n-1}(G_A(p) - \bar{G}_A)^2} \sqrt{\sum_{p=0}^{n-1}(G_B(p) - \bar{G}_B)^2} \right] ,$$

where $G_A(p)$ and $G_B(p)$ are the grayscale values at pixel p and $\bar{G}_A$, $\bar{G}_B$ are the means of grayscale values of the reference and target images (Maintz & Viergever 1998).

19.4.4.1 Deformable Registration

Unlike rigid registration, where the transformation is global and voxel independent, nonrigid or deformable registration finds a voxel-dependent vector displacement field u, such that, for a voxel at position x in image A, its corresponding position in image B is $x + u(x)$. Its computation is commonly done by minimizing an energy function $E(u) = Sim(A,B,u) + \lambda Reg(u)$, where $Sim(A,B,u)$ is the similarity measurement of the image A and the deformed image B, and $Reg(u)$ is a regulation term (described below), weighted by a constant λ. For same-modality freeform nonrigid registration (Lu et al. 2004), $Sim(\cdot)$ is simply the intensity difference between the two images: $Sim(A,B,u) = \sum_x |A(x) - B(x+u)|^2$. The regulation term is the smoothness of the displacement field, defined as $Reg(u) = \|\nabla u\|^2 = \sum_x \sum_i \|\nabla u_i\|^2$, which controls the variation in displacement between a voxel and its nearest neighbors. Other methods replace the regulation term by a more sophisticated smoothness model such as B-spline (Rueckert et al. 1999) or viscous fluid flow (Christensen et al. 1996). The registration uses a multiresolution scheme such that large deformations are determined at lower resolution and fine-tuned at higher resolution. Kashani et al. (2008) reported a comparison of the performance of various nonrigid registration methods.

Nonrigid registration is an active area of research. One application is in respiratory-correlated 4D CT, in which one segments the anatomical structures in the image set of one respiration phase and then propagates the contours to the images at other phases via the registration, thereby providing quantification of tumor and other anatomical motion (Pevsner et al. 2006; Zhang et al. 2008). Another application is 4D planning, which calculates the summed dose distribution over the whole or respiration-gated breathing cycle and uses this information in the design of treatment plans (Keall 2004; Rietzel et al. 2006).

19.4.4.2 Display of Registration Information

The two most commonly used methods of image fusion display are color overlay and checkerboard, both shown in Figure 19.8. Steps to create a fused image on a 2D plane are as follows.

1. Create a primary image on the plane (described in Section 19.4.1).
2. Create a secondary image on the plane. Because a forward transformation transforms the secondary image to the primary image's coordinate system, an inverse transformation is needed to compute the secondary image's CT numbers. For each point on the plane (in the primary image coordinate system), an inverse transformation is applied to transform the point into the secondary image's coordinate system. The transformed point is used to interpolate the CT number from the secondary image.
3. Fuse primary and secondary images. Techniques used to combine the two images are different depending of the type of views. For checkerboard displays with four tiles (Figure 19.8, right), a simple copy of parts of the secondary image onto the primary image is sufficient. For a checkerboard with many small square tiles, a mask can be used for faster painting. A common technique in color overlays is the use of red-green-blue components of the display. For example, one can assign the primary image to the R component, and the secondary image to the G and B components. The pixels in the fused image will display in shades of gray where the CT numbers of the two images match and in red or cyan where the images do not match (Figure 19.8, left). Other methods to combine images use hue, saturation, and value components (i.e., by assigning the primary image to the V component and the secondary image to the H or S components) or use averaging or multiplication of the color components of the primary and secondary images (Stokking et al. 2003). In the DICOM standard, the blending softcopy presentation state IOD is used to specify how to fuse two sets of images. RTP systems can specify (1) the output color space, (2) the grayscale contrast transformation, (3) a color palette to map secondary image grayscales, (4) area of the blended images to display, and (5) graphics and text annotation (NEMA 2008).

19.4.5 Display of Volumes of Interest and Beam Information

An advantage of defining volumes of interest is that they can be displayed and manipulated on a computer screen. This begins with constructing a surface from the stack of 2D contours (Rosenman et al. 1989). The simplest approach is to assign a width to each contour equal to the CT slice spacing, thus defining a

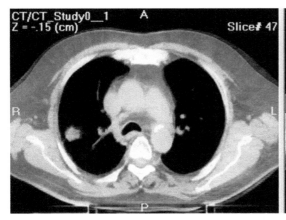
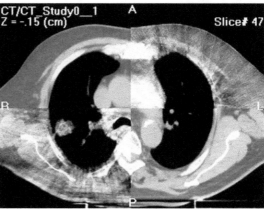

FIGURE 19.8 **(See color insert.)** Fused image displays of a planning CT image and cone-beam CT image as a color overlay (left) and checkerboard (right).

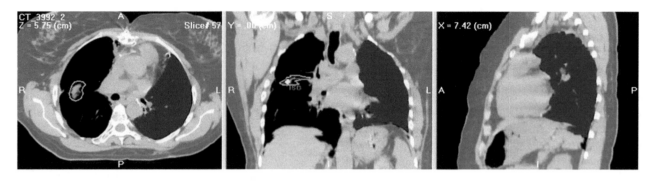

FIGURE 19.9   (**See color insert.**) Example volume of interest displays on 2D cross-sectional images.

stack of ribbons. A better approach is to define a polygonal mesh, made up of polygons (usually triangles) with vertices on adjacent contours, using a contour triangulation algorithm. Figure 19.9 shows the gross tumor volume (shown as an outlined contour) and left lung (shown as a semitransparent region) superposed on a 2D image plane. These outlined contours or regions are constructed by intersecting the currently displayed image plane with the 3D polygonal meshes. Figure 19.10 shows an anterior view that contains a semitransparent coronal image approximately midway through the patient, depicting the gross tumor volume (blue, contour stack), spinal cord (green, mesh), and left lung (red, tiled surface). In the current DICOM-RT framework (RT Structure Set), volumes of interest are defined as stacks of 2D contours, but alternative representations such as surface meshes are not handled.

RTP systems enable one to superimpose radiation fields and view their relationship to the delineated target and organ-at-risk volumes. The information can be displayed on 2D cross-sectional images (referred to as back projection) (Goitein et al. 1983), on a projected 2D image viewed from the source of a treatment field [referred to as "beam's eye view" (BEV)], or in a 3D view.

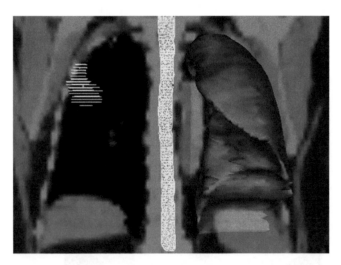

FIGURE 19.10   (**See color insert.**) 3D volume display. The anterior view shows a semitransparent coronal image midway through the patient, gross tumor volume (contour stack), cord (mesh), and left lung (tiled surface).

Displayed beam information on 2D cross-sectional images includes beam central axis (CAX), wedge, and radiation field edge. The CAX is drawn on the image plane by projecting the beam source and isocenter points onto the image plane and connecting the two points by a line. Different line styles can be used to indicate if the CAX is on the image plane. To display a wedge, the intersecting segments between image plane and wedge surfaces can be used. The position of the displayed wedge usually requires adjustment to ensure that it lies within the image. Radiation field edges can be computed in the following way. First, the jaw positions, MLC settings, and any other apertures (e.g., blocks) are used to compute the shape of the beam. One or more contours of the beam shape is obtained, depending how individual apertures are drawn. Next, a ray is cast from the beam source point to each beam shape vertex and extended to a distant position. A cone shape of the beam is obtained by forming triangles from two adjacent rays. By using the image plane to cut the beam cone surface, a projection of the beam onto the image plane is obtained (Figure 19.11).

In BEV, volumes of interest are viewed from the source of the treatment beam and projected onto a plane perpendicular to the beam CAX. In addition to the anatomical structures, the apertures, blocks, MLC, jaw settings, and wedge are displayed and interactively modified in the BEV display. In the case of intensity-modulated radiation therapy or volume-modulated arc therapy consisting of a sequence of MLC positions, some BEV displays provide scrolling capability through the MLC sequence. A principal use of the BEV display is the definition of apertures, blocks, MLC, and jaw settings, either manually or automatically, in relation to volumes of interest. Automatic aperture design (Brewster et al. 1993) typically requires the user to select a volume of interest and a margin between the projected volume at the BEV plane and the aperture (Figure 19.12). Integration of DRRs, described further in the next section, into the BEV display aid in treatment field design. Outlines of volumes of interest and radiation fields can be superimposed onto DRRs.

19.4.6 DRRs

Verification of treatment requires localization of anatomical volumes of interest and radiation fields. DRRs have been the predominant form of reference image for this purpose, although

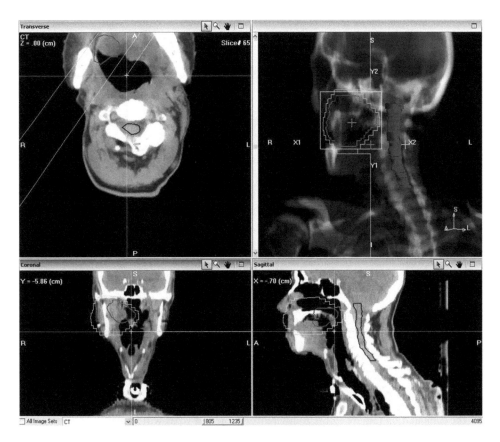

FIGURE 19.11 (**See color insert.**) Example display of cross-sectional images and BEV (top right). On the three cross-sectional views, outlines of PTV (green), cord (blue), and MLC-shaped beam aperture (yellow) are displayed. BEV shows a DRR, PTV, beam aperture, and cord.

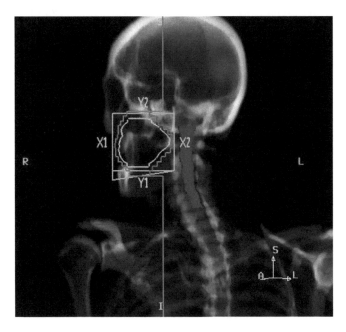

FIGURE 19.12 (**See color insert.**) BEV display, in which the PTV is shown as a green outline, while cord is displayed as a solid blue surface. Radiation field information, such as MLC (pink), jaw, and wedge (orange), is also displayed. The volumes of interest, field information, and patient orientation are superimposed on a DRR.

CT-based image guidance has led to increased usage of the planning CT as reference.

A DRR is the digital equivalent of a conventional simulation film. It is generated by projecting the 3D image volume onto a 2D image, in which image brightness corresponds to a density-dependent attenuation (Goitein et al. 1983). Ray casting is commonly used to project the 3D volume data. Grid point coordinates on the DRR plane are defined by the pixel size, the x- and y-unit vectors of the projection plane. A ray is constructed from the imaging source to each grid point, and the pixel intensity is computed by integrating the attenuation coefficients along the ray. This algorithm is also referred to as "image order" rendering because a ray is cast to the position of each projected image pixel. Interpolation is needed to compute the attenuation coefficients for the sampling points along the ray. Interpolation methods affect the accuracy of the generated DRR. Because the rays must visit every 3D voxel and some voxels are visited repeatedly by different rays, the algorithm is computationally intensive. The selection of interpolation methods affects DRR quality. Trilinear interpolation gives the best results but takes longer than other methods, such as Z-slice interpolation. One way to reduce DRR computation time is to use ray casting onto a coarse grid on the projection plane and then interpolate from these grid points to obtain intensities for the remaining grid points. This method trades off DRR accuracy for speed of computation. Another

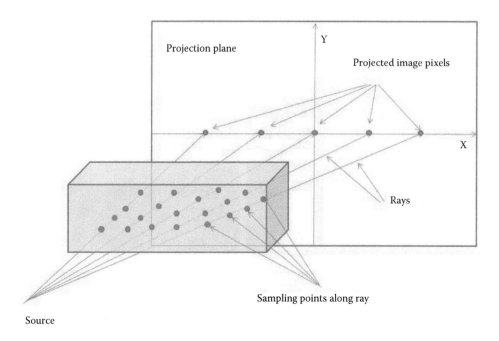

FIGURE 19.13 Schematic of ray casting. Rays are cast from the image source to pixels on the projection image plane. Attenuation coefficients are accumulated along the ray for each sampling point.

method is to choose a subset of the 3D image set, within a volume of interest, to do the computation (Figure 19.13). A different approach is voxel projection, also called "object order" rendering, in which each 3D voxel is sequentially projected onto the DRR plane. Its intensity is accumulated to the image pixel on which the voxel projection falls. The voxels are processed in storage order, which allows the 3D data set to be traversed quickly, and no interpolation of voxel attenuation coefficient is needed;

thus, this algorithm is faster than ray casting. In general, a 3D projected voxel will not fall exactly on a projection image pixel, so a spread filter is needed to distribute a fraction of its intensity to several neighboring pixels of the projected position (Westover 1990) (Figure 19.14).

Another novel DRR algorithm uses a shear-warp factorization method (Lacroute & Levoy 1994; Weese et al. 1999), which combines the advantages of image order and object order. A 2D

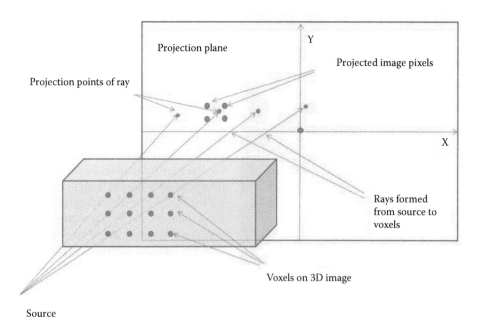

FIGURE 19.14 Schematic of voxel projection. Rays are cast from the image source to voxels in the 3D image. Attenuation coefficients are computed for each voxel and distributed to neighboring pixels on projection image plane.

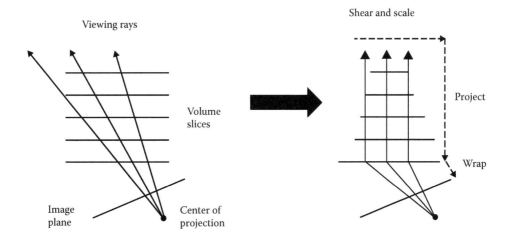

FIGURE 19.15 Schematic of shear-warp factorization method. A volume is transformed to the sheared object space for a perspective projection by translating and scaling each slice. The projection in the sheared object space is thereby simple and efficient. (From Lacrout, P. & Levoy, M., *Proc. SIGGRAPH '94, Orlando, Florida*, 451, 1994. With permission.)

intermediate image is computed whose coordinate axes parallel those of the 3D image data set. The viewing transformation is decomposed into a shear matrix and warp matrix. The shear matrix transforms the 3D volume data into sheared 3D volume voxels such that the rays are parallel to each other. The pixel intensities of the intermediate image are defined by the integral of the 3D voxel intensity along rays from the source to the pixels; however, the integration is not done voxel by voxel along the ray as in ray casting. Instead, one slice after another of the sheared 3D image is added to the intermediate image. The final DRR can then be obtained from intermediate image by a warp transformation. This algorithm uses scan-line-order volume rendering, which is more efficient than ray casting (Figure 19.15).

With the advent of CT technology in the treatment room, CT-guided localization and correction of patient position (referred to as image-guided radiation therapy) has become more widespread (discussed further in Chapter 23). Usually, the planning CT is used as the reference image for registration to the CT image set acquired at treatment. The associated information in the planning CT, such as isocenter and volumes of interest, are transferred to the image-guided radiation therapy registration application by using the DICOM and DICOM-RT objects described in an earlier section.

19.5 Summary and Conclusions

The infrastructure required for external beam RTP is complex and evolves over time. Staff from various disciplines must work together to create a reliable infrastructure that is easy to support, maintain, and expand. The DICOM standard is an integral piece of this infrastructure. Documenting DICOM destinations (IP addresses, AE titles, and port numbers) can help in managing day-to-day support and maintenance tasks and in planning for new imaging modalities, systems, and applications.

The goals of radiation therapy require quantification of the radiation dose received by target and normal tissues and association of that dose to the treatment plan. RTP systems are specifically designed to provide these capabilities. Images play a critical role in this process: they must provide accurate information on spatial and tissue density characteristics of the patient's anatomy. RTP systems require the definition of information in addition to that used in radiology, hence the need for the radiation therapy extension to DICOM when communicating between systems.

The evolving development of RTP systems is strongly coupled to advances in imaging and computer technology. Multimodality imaging provides complementary information and is increasingly incorporated into the treatment planning process. Further, tomographic imaging in the treatment room via image guidance systems provides a wealth of new information about patient changes during treatment and more accurate estimates of delivered dose. The developments in imaging have introduced new considerations to the treatment planning process, including more rational assessment and reduction of treatment margins and modification of treatment plans according to information gathered as treatment progresses. Current developments in RTP systems have been motivated by the general interest in exploring these capabilities.

Advancements in medical imaging reveal shortcomings in the DICOM-RT standard and the need to address them. One example mentioned earlier is that volumes of interest within RT structure sets are limited to stacks of 2D contours. Reliably interpolating between adjacent contours is technically difficult, particularly in cases of bifurcated structures. More reliable representations of segmentation such as 3D bit maps or triangular meshes are needed. The development of new-generation objects to support technological advances is being addressed by a DICOM working group for radiation therapy objects (NEMA 2009b).

References

Adams, R. & Bischof, L. (1994). Seeded region growing. *IEEE Trans Pattern Anal Mach Intell, 16*, 641–647.

Brewster, L. et al. (1993). Automatic generation of beam apertures. *Med Phys, 20*, 1337–1342.

Christensen, G. E. et al. (1996). Deformable templates using large deformation kinematics. *IEEE Trans Image Proc, 5*, 1435–1447.

Goitein, M. et al. (1983). Multi-dimensional treatment planning: II. Beam's eye-view, back projection, and projection through CT sections. *Int J Radiat Oncol Biol Phys, 9*, 789–797.

Gonzalez, R. C. & Woods, R. E. (1992). *Digital Image Processing.* Reading, MA: Addison-Wesley, 1992, 161–231.

Hu, Y.-C. et al. Survey of recent medical image segmentation techniques, in *Biomedical Engineering.* Rijeka, Croatia: InTech, 2009, 321–346.

International Electrotechnical Commission. *Radiotherapy Equipment—Co-ordinates, Movements and Scales.* Rep. IEC 61217, ed. 1.2. Geneva: IEC, 2007.

Kalet, I. J. & Austin-Seymour, M. M. (1997). The use of medical images in planning and delivery of radiation therapy. *J Am Med Inform Assoc, 4*, 327–339.

Kashani, R. et al. (2008). Objective assessment of deformable image registration in radiotherapy: A multi-institution study. *Med Phys, 35*, 5944–5953.

Keall, P. (2004). 4-Dimensional computed tomography imaging and treatment planning. *Semin Radiat Oncol, 14*, 81–90.

Kijewski, P. Three-dimensional treatment planning, in *Radiation Oncology: Technology and Biology*, Mauch, P. M. & Loeffler, J. S., Eds. Philadelphia: WB Saunders, 1994, 10–33.

Lacrout, P. & Levoy, M. Fast volume rendering using a shear-warp factorization of the viewing transformation, in *Proc. SIGGRAPH '94, Orlando, Florida*, July 1994, 451–458.

Law, M. Y. Y. & Liu, B. (2009). DICOM-RT and its utilization in radiation therapy. *Radiographics, 29*, 655–667.

Lorensen, W. E. & Cline, H. E. (1987). Marching cubes: A high resolution 3D surface construction algorithm. *ACM SIGGRAPH Comput Graph, 21*, 163–169.

Lu, W. et al. (2004). Fast free-form deformable registration via calculus of variations. *Phys Med Biol, 49*, 3067–3087.

Maintz, J. B. A. & Viergever, M. A. (1998). A survey of medical image registration. *Med Image Anal, 2*, 1–37.

Mortensen, E. N. & Barrett, W. A. Intelligent scissors for image composition, in *SIGGRAPH '95: Proceedings of the 22nd Annual Conference on Computer Graphics and Interactive Techniques.* New York, NY: ACM Press, 1995, 191–198.

Mundt, A. & Roeske, J. C. *Intensity Modulated Radiation Therapy: A Clinical Perspective.* Hamilton, ON: BC Decker, Inc., 2005, 110.

National Electrical Manufacturers Association (NEMA) (2008). *Digital Imaging and Communications in Medicine (DICOM), PS3 Part 3, 2008.* Available at ftp://medical.nema.org/medical/DICOM/2008/08_03pu.pdf. Accessed 02 November 2009.

National Electrical Manufacturers Association (NEMA) (2009a). *DICOM Brochure.* Available at http://medical.nema.org/DICOM/geninfo/Brochure.pdf. Accessed 18 August 2009.

National Electrical Manufacturers Association (NEMA) (2009b). *Strategic Document.* Available at http://medical.nema.org/DICOM/geninfo/Strategy.pdf. Accessed 9 November 2009.

Otsu, N. (1979). A threshold selection method from grey-level histograms. *IEEE Trans Syst Man Cybern, 9*, 62–66.

Pekar, V. et al. (2004). Automated model-based organ delineation for radiotherapy planning in prostatic region. *Int J Radiat Oncol Biol Phys, 60*, 973–980.

Pevsner, A. et al. (2006). Evaluation of an automated deformable image matching method for quantifying lung motion in respiration-correlated CT images. *Med Phys, 33*, 369–376.

Photon Treatment Planning Collaborative Working Group (1991). Evaluation of high energy photon external beam treatment planning: project summary. *Int J Radiat Oncol Biol Phys, 21*, 9–23.

Ragan, D. et al. (2005). Semiautomated four-dimensional computed tomography segmentation using deformable models. *Med Phys, 32*, 2254–2261.

Rietzel, E. et al. (2006). Design of 4D treatment planning target volumes. *Int J Radiat Oncol Biol Phys, 66*, 287–295.

Rosenfeld, A. & Torre, P. D. L. (1983). Histogram concavity analysis as an aid in threshold selection. *IEEE Trans Syst Man Cybern, 13*, 231–235.

Rosenman, J. et al. (1989). Three-dimensional display techniques in radiation therapy treatment planning. *Int J Radiat Oncol Biol Phys, 16*, 263–269.

Rueckert, D. et al. (1999). Nonrigid registration using free-form deformations: application to breast MR images. *IEEE Trans Med Imaging, 18*, 712–721.

Seco, J. & Evans, P. M. (2006). Assessing the effect of electron density in photon dose calculations. *Med Phys, 33*, 540–552.

Sezan, M. I. (1990). A peak detection algorithm and its application to histogram-based image data reduction. *Comput Vision Graph Image Process, 49*, 36–51.

Stokking, R. et al. (2003). Display of fused images: Methods, interpretation, and diagnostic improvements. *Semin Nucl Med, 33*, 219–227.

Weese, J. et al. Fast voxel-based on shear-warp factorization, in *Medical Imaging 1999: Image Processing*, Hanson, K. M., Ed. *Proc SPIE 3661*, 1999, 802–810.

Westover, L. (1990). Footprint evaluation for volume rendering. *Proc SIGGRAPH 90, Comput Graph, 24*, 367–376.

Westover, L. A. (1991). Splatting: A parallel, feed-forward volume rendering algorithm. PhD thesis. Chapel Hill: Department of Computer Science, University of North Carolina.

Yang, D. et al. (2008). 4D-CT motion estimation using deformation image registration and 5D respiratory motion modeling. *Med Phys, 35*, 4577–4590.

Zhang, G. G. et al. (2008). The use of 3D optical flow method in mapping of 3D anatomical structure and tumor contours across 4D CT data. *J Appl Clin Med Phys, 9*, 2738.

Zitova, B. & Flusser, J. (2003). Image registration methods: A survey. *Image Vision Comput, 21*, 977–1000.

V

Informatics for Treatment Planning, Delivery, and Assessment

Evaluation of Treatment Plans

Kevin L. Moore
University of California, San Diego

Deshan Yang
Washington University in St. Louis

20.1 Introduction

Now into the third decade of computerized 3D treatment planning, nearly all radiotherapy patients' intended treatments will be reviewed on a computer screen in a commercial treatment planning system (TPS). The purpose of treatment plan evaluation might appear self-evident, but it is worthwhile to explicitly state that the goal is to ensure that the course of radiotherapy simulated in the TPS has fulfilled the physician's therapeutic intent while minimizing harm to healthy tissues. In simple palliative treatments (e.g., spinal cord compressions or whole brain) (Halperin et al. 2008), this evaluation can be as straightforward as verifying (a) the patient's setup (b) the shape of a collimating block relative to the patient's anatomy, (c) confirming that the dose prescription and fractionation are appropriate to the patient's diagnosis, and (d) that the beam energy and monitor units correspond to that prescription. A review of 3D dose distributions and dose-volume histograms (DVHs), should any exist, would appear unnecessary as they add little value to the evaluation process.

In contrast, the clinical review of intensity-modulated radiation therapy (IMRT) delivering a curative course of therapy for an advanced head-and-neck cancer involves several orders of magnitude greater complexity than simple palliative treatments. There are usually multiple planning target volumes (PTVs) receiving concurrent but differing cumulative doses (simultaneous integrated boost or SIB), multiple organs-at-risk (OARs) of various radiosensitivities, and an intricate sequence of manual and automated tasks involved in the design of modulated radiation fields. For this level of complexity, review of the dose distribution and DVHs is but the beginning of the evaluation process. Some, but not all, of the other elements in a

computerized treatment plan that should be reviewed for IMRT head-and-neck cases include the following:

- Fusion of secondary imaging
- Definition of clinical target volumes and appropriate dose/fractionation to each target
- Margin expansion of PTVs
- Normal tissue contouring, including the presence of all relevant OARs
- Inhomogeneity corrections, including any density overrides in TPS software
- Beam selection (e.g., arrangement of static fields or arc trajectories)
- Consistency of beam delivery elements
- Avoidance of proximal fluence through shoulders
- Appropriate dose gradients near the most critical OARs (e.g., spinal cord and parotid glands)

This is by no means an exhaustive list, and the above items are iterated merely to convey the myriad elements clinicians must consider when evaluating a modern radiotherapy treatment plan. The amount of information that must be consumed and interpreted has achieved a level of complexity that stresses a human's ability to process everything in a single sitting. And, like any human process, there will be a range of ability and experience levels in the individuals generating a treatment plan and the people conducting its pretreatment evaluation. Any process that hinges so strongly on human subjectivity and relative experience will be susceptible to errors, such as misdetection or misdiagnosis of problems, especially when the presented information is as elaborate and multifaceted as an IMRT treatment plan.

It need not be thus. The nature of the treatment planning information is digitized and highly quantitative, meaning that most elements of a computerized treatment plan evaluation would be receptive to intelligent distillation of the clinical information needed to evaluate the clinical aspects of a treatment plan. Beyond replacing verification activities currently performed by humans, advanced analytics can facilitate the confirmation of plan quality elements not typically incorporated into on-screen reviews (e.g., identifying logical inconsistencies between IMRT optimization parameters) (Yang & Moore 2012).

Furthermore, informatics enables both (a) the analysis of multiple prior treatment plans and (b) the incorporation of any quantitative distillation of prior experience into ongoing quality control elements in the treatment planning process. The literature is replete with large-scale analyses of treatment planning information embodying the former, although nearly all involved substantial manual data collection and analysis. Far less common is the latter instance where analysis tools incorporate prior studies into clinical tools for future treatment plan evaluations. It is this informatics application that holds substantial promise for dramatic clinical improvements in quality and efficiency for the plan evaluation process.

This chapter will review the current state of the clinical treatment planning process and how informatics can be brought to bear in making the evaluation of treatment plans more quantitative and less dependent on human elements. Beyond a review of the current state of affairs, a primary goal will be to consider the use of informatics tools to improve the pretreatment plan evaluation process. Insofar as adaptive radiotherapy (ART) stretches the treatment plan review process into an evolving course of treatment, such plan evaluation tools can extend their usefulness into the ongoing clinical management of patients' care.

As future TPS will surely advance from the present state, this chapter will further describe a vision for how informatics tools should be designed to retain maximal flexibility in the face of inevitable software evolution. As there is no commercial "one-size-fits-all" informatics system for treatment plan evaluation, software designed to assist clinicians in this task is typically developed and customized by individual institutions to suit their own particular needs. The development and maintenance of "homegrown" ancillary software elements represent a non-trivial expenditure of clinical resources, and whenever possible, the associated costs will be considered aside the clinical benefits. As the authors strongly champion the use of informatics tools in the plan evaluation process, we would be remiss if we did not note that administrators must weigh these costs not only against potential clinical benefits but also against the costs of doing nothing (i.e., maintaining the current system of human-based verifications). It is possible, perhaps even likely that plan evaluation informatics systems are cost-effective based purely on efficiency gains, long-term cost savings, and quality improvements.

20.2 Treatment Plan Evaluation Strategies

Figure 20.1 depicts the flow of treatment planning data through the various databases that will contain the information as well as the flow through "time" as the data progress from simulation to treatment. A patient's treatment plan might be reviewed multiple times by multiple individuals at different time points and in different software systems. It is thus important to denote the context of any act of reviewing treatment planning information by both where and when the evaluation occurred. Prior to initiation of treatment, one can distinguish in situ from ex situ plan evaluation, respectively defined as the act of pretreatment plan evaluation either within the active TPS infrastructure or outside of it. Ex post facto evaluations encompass all studies on treatment planning data that occur after the patient has begun treatment. In this section, the advantages and disadvantages of each strategy are discussed, and these considerations will shape the following sections that describe the use of informatics tools in each regime.

20.2.1 In situ Plan Evaluation

In situ plan evaluation, the most common form of review, is defined as any review that is conducted within the framework of the commercial TPS used to develop the treatment plan. In most cases, this evaluation will occur directly in the TPS software interface itself, although the definition covers instances of direct analysis of a vendor's proprietary database structure. As outlined in Table 20.1, the benefits of this approach are many. Perhaps the most important advantages of in situ plan evaluation is that errors are caught at the earliest possible time point and can be corrected in the TPS before any erroneous data have metastasized to other databases such as the treatment management system (TMS) or a picture archiving and communication system (PACS).

From an informatics perspective, however, there are some drawbacks to relying on in situ plan evaluation exclusively. As the data will likely be exported as Digital Imaging and Communications in Medicine-Radiation Therapy (DICOM-RT) objects to the TMS and/or PACS, there is the potential for mismatch between the TPS data at the time of plan review and the exported DICOM data. Data discrepancies can be lessened with either manual checks by, for example, trained medical physicists, or by independently developed informatics software designed to reconcile the TPS and TMS databases (Siochi et al. 2009; Yang et al. 2012).

20.2.2 Ex situ Plan Evaluation

Ex situ plan evaluation is defined as any review that is conducted outside of the TPS infrastructure yet still prior to the initiation of treatment. Ex situ plan evaluation necessarily involves the transfer of data from the TPS to some other system, be it digital data transfer to another database or even a physical printout of plan information. By far the most common format for such TPS data export is DICOM-RT, though the term "data export" encompasses both the use of non-DICOM data formats and/or treatment plan summary printouts, the latter of which is a common means for a pretreatment plan check conducted by a

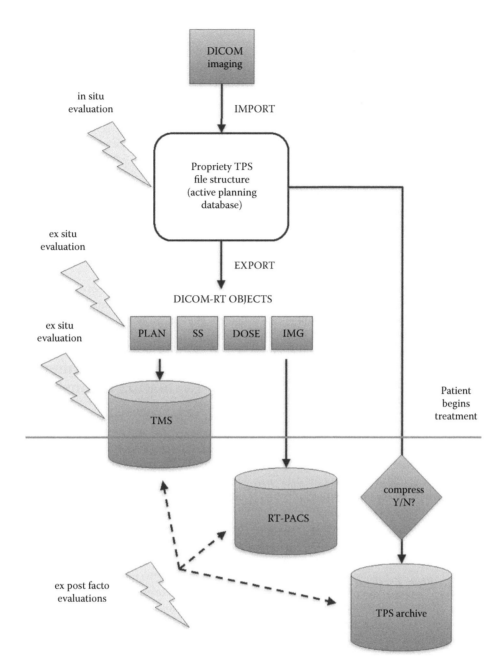

FIGURE 20.1 Block diagram outlining data flow from TPS input to data export and archiving after planning. In situ, ex situ, and ex post facto analyses are represented in the time point and action upon the associated databases. Note: Configurations that unify databases between TPS and TMS (e.g., Varian's Eclipse and Aria) would require modification of this diagram by eliminating the TMS as a separate database to which the TPS exports DICOM-RT objects: plan, structure set (SS), dose, and images (IMG).

medical physicist. One common example of ex situ plan evaluation is the use of ancillary software to analyze the DICOM-RT plan data object for secondary verification of the TPS-calculated monitor units. This evaluation step analyzes only one aspect of the treatment plan yet, for this particular application, can be very robust and efficient. For most radiotherapy clinics, such secondary MU verifications (including phantom-based quality assurance tests) may represent the only ex situ evaluations that make use of computerized systems to establish the technical integrity of a plan, with the rest of the pretreatment evaluation

conducted by medical physicists confined to comparisons between data in the TMS and a printed summary of the TPS planning data.

Ex situ plan evaluations have the potential to be quite comprehensive if all DICOM-RT data objects are incorporated into the review. A further advantage in the case of evaluating DICOM information exported to a TMS is that these objects correspond to the data that will in fact be used for patient treatment. The disadvantages of ex situ plan evaluations lie in the conversion of the TPS data to another format and the transfer of that data to

TABLE 20.1   Advantages and Disadvantages of Different Plan Evaluation Strategies

	Advantages	Disadvantages
In situ plan evaluation	• All planning information is available • No time lost in data export • Any problems detected at time of plan evaluation have lowest correction costs • Visual display of information	• Limited to operations on proprietary data formats of TPS vendor • Potential mismatch between TPS data and exported data • Requires access to TPS for reviewing party to access data
Ex situ plan evaluation	• Analyzes the DICOM data ultimately used for patient treatment • Any DICOM-based analyses are TPS independent • Platform independence could enable remote review	• Some planning data lost in export from TPS • Potential mismatch between TPS data and exported data • Detected problems have higher correction costs if caught at later time point in process
Ex post facto plan evaluation	• Naturally suited for aggregate and/or batch analyses of multiple planning data sets • Can incorporate on-treatment data into analyses	• Maximal planning data loss at the end of the data chain • Potential mismatch among TPS data, treatment data, and archived data • Reduced or eliminated opportunities for corrective action

another database. Currently this conversion process necessarily involves some loss of information (NEMA 2012). Resources will have to be expended in the export of the data and the confirmation of its concordance to the original TPS data (Siochi et al. 2009, 2011). Should any problems be detected, it will require more time to resolve issues as the data have been duplicated onto multiple systems and the error detection will almost certainly have occurred at a later time point in the treatment planning workflow.

Thus far, the discussion of ex situ evaluations has assumed a separation between TPS and TMS. This is not always the case, for example, Varian Medical System's integrated Eclipse and Aria systems (assuming properly updated caching for database and application state consistency). In such installations, any pretreatment analyses on TMS data should be categorized as in situ evaluations, except for any treatment plan summary printouts that better fit the definition of ex situ.

20.2.3 Ex Post Facto Plan Evaluation

Ex post facto evaluations are defined as analyses conducted on planning data after treatment has begun. Ex post facto analyses are characterized not by how or where they are conducted but when. This broad definition includes on-treatment reviews at chart rounds, weekly physics chart checks that review treatment planning data, and planning studies for quality improvement initiatives or academic research. Depending on the purpose or time point of the evaluation, such ex post facto reviews might be conducted in the TPS, TMS, a RT-PACS, or TPS archive. The TPS archive might compress the information for storage efficiency, complicating further the retrieval of data for ex post facto analyses.

Any disadvantages are tautological in that, by definition, ex post facto analyses do not apply to pretreatment evaluations. Any planning errors detected after the start of treatment incur the maximum system costs and, depending on where the patient is in the treatment process, could entirely preclude any corrective action. Nevertheless, ex post facto plan evaluations are a necessary part of any comprehensive quality assurance program in

radiation oncology (Kutcher et al. 1994) and hold many opportunities for treatment planning quality improvement if they can be brought to bear for future cases.

20.3 Improving Treatment Planning Evaluation through Informatics

Chapter 16 focused on informatics tools for research in radiation oncology, and any research tools that are dedicated to processing treatment planning data (e.g., the open-source software system CERR) (Deasy et al. 2003), fit nicely into the ex post facto definition. As this chapter focuses on informatics tools that specifically aim to influence pretreatment planning evaluations, we begin by exploring how ex post facto studies could and should be incorporated into real-time decision support tools for the plan evaluation process. While the benefits of carefully reviewing prior data to inform future treatment planning decisions are perhaps obvious, the impediments to efficiently conducting such investigations are numerous.

As an example, suppose Physician A is conducting a pretreatment review of a head-and-neck IMRT plan with an apparently high mean dose to an ipsilateral parotid gland, a known dosimetric correlate to radiation-induced salivary injury (Houweling et al. 2010). One reasonable course of action would be to direct the planner to reoptimize the plan and reduce the excess dose to the parotid, but how should Physician A proceed if the planner comes back several hours later having been unable to significantly reduce the parotid dose without compromising other aspects of the plan? Dr. A could be forgiven for simply approving the treatment plan at this point, though it should be vexing that after so many years of clinically delivering IMRT that the expected range of parotid dose variations is unknown. Perhaps Dr. A's curiosity is piqued and she wants to know what this range has been for all of her previous 100 head-and-neck patients. She has several options at this point, none particularly desirable. First, she could try a review of the paper or electronic charts, looking up the DVH point of interest from plan printouts. Second, she could try reloading each patient in the TPS and measuring mean dose, a process that can range from very to exceedingly time-intensive depending on the data archiving infrastructure. Third, if the data have been exported to a RT-PACS, the dosimetric information could be manually accessed there. If she is successful, Dr. A might find that the patient's treatment plan is on the high end but within two standard deviations of a normally distributed set of parotid mean doses. But wait, why is there such a wide distribution for her patients? And how should one compare her results to her colleague, Dr. B, who practices at another institution with identical equipment and a larger proportion of head-and-neck patients?

The inefficiency of this process should be apparent, yet this is typically how such investigations must be conducted. Or, perhaps more accurately, such studies rarely occur due to the enormous resource expenditures required just to answer a single clinical question. In this case, Dr. A simply wanted to quantitatively understand the performance of her clinical practice, both in terms of internal consistency and with respect to an external benchmark. A reasonable strategy in a resource-constrained clinic would be to abandon the retrospective review strategy and just begin tracking all parotid mean doses going forward. This will eventually yield benefits, unless it is discovered at a later time that other variables that might explain the dose variations were not tracked.

Contrast this contrived (but realistic) example of attempting to control the quality of IMRT planning with ex post facto studies that employed informatics tools to carry out such retrospective planning studies on precisely the problem that plagued

Physician A. Several investigators have attempted to solve this problem by analyzing prior planning data to establish expected values for OAR sparing in IMRT (Vineberg et al. 2002; Hunt et al. 2006; Wu et al. 2009, 2011; Moore et al. 2011; Zhu et al. 2011; Appenzoller et al. 2012; Petit et al. 2012). Notably for this problem, the various approaches investigated all employed quantitative analysis of the geometric relationship between OARs and PTVs, with the more recent studies involving very sophisticated case-based (Wu et al. 2009, 2011) or model-based (Zhu et al. 2011; Appenzoller et al. 2012) techniques that would be impossible to develop and implement without ancillary software systems conducting ex post facto analyses on digital DICOM data. Investigators at Washington University in St. Louis (Moore et al. 2011) and Johns Hopkins University (Wu et al. 2011) have reported converting these ex post facto analyses into in situ clinical quality control tools, with the Washington University group reporting significant clinical gains in organ sparing after implementation (Figure 20.2).

The particular example of IMRT quality control is but one of many applications for which ex post facto informatics systems play an indispensible role. At this point in time, it is all but impossible to envision any studies involving the treatment planning process that would not need to make use of ex post facto analyses on digital data. Less clear is the manner by which the results of these studies can be incorporated back into the active treatment planning process via in situ or ex situ informatics tools, as particular TPS installations can have different capabilities.

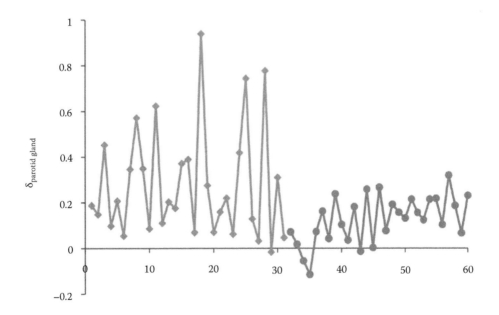

FIGURE 20.2 Washington University implemented an in situ quality-control tool using PINNACLE scripts, observing significant reductions in excess parotid gland dose and reduced variability between treatment planners. A predictive model for parotid mean dose, developed through an ex post facto analysis of prior clinical treatment plans, gave a prediction for the mean dose (D_{pred}), which could be compared with the observed mean dose (D_{mean}). The parameter $\delta = (D_{mean} - D_{pred})/D_{pred}$ quantified the percentage by which the clinical treatment plans exceeded the model predictions. Red closed circles and blue open circles represent head-and-neck treatment plans before and after the clinical implementation of the QC tool, respectively. (Reprinted from Moore, K. L. et al., *Int. J. Radiat. Oncol. Biol. Phys.*, *81*(2), 545, 2011. With permission.)

20.3.1 Aiding Plan Evaluation with In situ Informatics Tools

Direct incorporation of customized analysis routines into the clinical treatment planning process has been a very limited enterprise in radiation therapy due to the restrictions on most TPS platforms to any kind of application programming interface (API). The planning data are technically accessible in the TPS database, although at the plan development stage the information is stored in proprietary data formats and file structures that would have to be reverse engineered when developing analytic programs, raising immediate safety concerns for any responsible clinician.

The PINNACLE (Philips Radiation Oncology Systems, Fitchburg, WI) TPS is a well-known exception, with an interface that gives users the ability to record assemblies of internal command messages in the form of text files. These "scripts" facilitate complicated sequences of commands to be recorded and re-invoked for future uses. Because the scripts are stored as text files on the TPS database, the files may be edited and modified, allowing further customization of the command sequences outside of the original recorded button clicks (Geoghegan 2007). As PINNACLE is mounted on a Unix platform, other informatics tools are available to users, for example, PERL (Sheppard 2000), a general purpose programming language standard on all UNIX systems. With the ability to directly write TPS command sequences and to parse and analyze data files with PERL,

the in situ informatics capabilities of PINNACLE are currently unmatched in the commercial TPS space with several researchers exploiting these capabilities for numerous applications (Zhang et al. 2007; Holdsworth et al. 2011; Kim et al. 2012; Yang & Moore 2012; Janssen et al. 2013).

PINNACLE scripting can be categorized as "read-write" capable, in that scripts that are invoked inside the TPS can modify the state of the active treatment plan by sequencing PINNACLE command messages. In contrast, the Eclipse TPS recently released an API that does not allow modification of the state of the TPS system. Such "read-only" APIs can facilitate data queries but preclude the active creation/modification/deletion of system elements such as beams, contours, and reference points.

Both read-write and read-only APIs are immensely powerful informatics tools that can assist the plan evaluation process. One application accessible to both is dosimetric reporting, which summarizes numerous aspects of a treatment plan. The DVH that is represented in an active treatment plan can be parsed, queried, and directly displayed while the TPS software is still running. An example of this can be seen in Figure 20.3, where the Quantitative Analysis of Normal Tissue Effects in the Clinic (QUANTEC) dosimetric guidelines (Bentzen et al. 2010) were applied to each OAR present in a head-and-neck IMRT plan. In the QUANTEC article that described the use of normal tissue complication probabilities in the clinic (Marks et al. 2010), the authors stated that they "hope that at least some of the summary tables, graphs, and models presented will be reproduced

---

QUANTEC Report
Normal Tissue Complication Probability

NAME:
PHYSICIAN:

Plan ID: HEAD_NECK
Rev: R03.P01.D02

Total Dose: 7000 cGy
ROC: ███████
PLAN DATE: 2011-07-29 15:22:15

Trial ID: MD_APPROVED

Per Fraction Dose: 200 cGy/Fx

BRAINSTEM	Permanent cranial neuropathy or necrosis	Dmax = 4779.2 cGy	< 5% rate
CORD	Myelopathy	Dmax = 4229.4 cGy	0.2% rate
PAROTID_L	Long term parotid salivary function reduced to <25% of pre-RT level	Mean dose = 2630.1 cGy	> 20% rate
PAROTID_R	Long term parotid salivary function reduced to <25% of pre-RT level	Mean dose = 3355.1 cGy	> 20% rate
LARYNX	Vocal dysfunction	Dmax = 6509.2 cGy	< 20% rate
LARYNX	Aspiration	Mean dose = 2634.3 cGy	< 30% rate
LARYNX	Edema	Mean dose = 2634.3 cGy	< 20% rate
LARYNX	Edema	V50 = 6.0%	< 20% rate
ESOPHAGUS	Grade >= 3 acute esophagitis	Mean dose = 1936.4 cGy	5-20% rate
ESOPHAGUS	Grade >= 2 acute esophagitis	V35 = 1.6%	< 30% rate
ESOPHAGUS	Grade >= 2 acute esophagitis	V50 = 0.0%	< 30% rate
ESOPHAGUS	Grade >= 2 acute esophagitis	V70 = 0.0%	< 30% rate

FIGURE 20.3 Example of in situ plan evaluation based on analysis within the PINNACLE TPS scripting environment, that is, computing operations are performed on transient PINNACLE data, not DICOM. Report summarizes NTCPs based on QUANTEC values by analyzing the state of a PINNACLE treatment plan amidst the planning process.

and posted in resident workrooms, dosimetry planning areas, and physician offices, as is currently done with the Emami et al. (1991) tables." While the benefits of physically posting common planning guidelines and standards cannot be disputed, also having a software tool that applies those same endpoints directly to the dosimetric values observed in an on-screen plan can greatly reduce the probability of inadvertently exceeding organ tolerances.

Beyond DVH reporting, in situ plan analysis can facilitate a comprehensive review of all the technical planning parameters, ensuring the physical and dosimetric integrity of plans at the pretreatment stage. Figure 20.4a outlines a PINNACLE scripted report that made use of PERL subroutines to systematically verify numerous technical elements of a treatment plan (Figures 20.4b–d). A single analysis routine and subsequent report can verify any and all elements of the patient's setup, reference points, contours, beams, prescriptions, dose calculation parameters, and, if applicable, IMRT optimization parameters. The latter item is of particular interest as these are currently not exported via DICOM-RT and are only accessible from within the TPS. In addition to dramatically decreasing the time required to check the many elements that comprise a radiotherapy treatment plan, such routines can conduct analyses that would be impractical for humans to carry out. For example, it can be quite difficult to check all IMRT optimization parameters to ensure that there are no mutually exclusive objectives (Figure 20.4d). Such conflicts can hinder the IMRT planning process and, while manually checking all parameter permutations would be time-consuming and error-prone, a subroutine designed to compare all overlapping structures' objectives efficiently detects and flags any conflicts.

In any in situ plan analysis tool, there can be significant initial costs to gain experience with the particular data elements of the TPS. There is also the risk that the TPS vendor will, in future releases, modify software elements upon which these tools rely. And of course, should a particular institution choose to move from one TPS software vendor to another, all in situ informatics tools will require a complete reinvention in the new TPS infrastructure.

Undoubtedly, the cost to develop in situ informatics software is substantial. However, these expenditures can be legitimately weighed against the long-term costs, in both financial resources as well as likelihood of error detection, of the time spent by physicians and medical physicists comparing numerical entries between separate databases. Another important advantage of employing informatics tools for treatment plan quality control is that all of the information is fully represented, i.e. no data have been lost in transferring data to other systems. Errors caught at this stage are more easily and cheaply remedied because the data have not yet left the TPS.

As any reliance on a vendor's proprietary data structure leaves ancillary in situ informatics tools vulnerable to changes in the TPS or TMS architectures, the specific design of such software is critical to long-term functionality and minimizing maintenance costs. Any subroutines that can be written in general programming languages as opposed to the TPS scripting language itself will decrease any replacement costs incurred by changes in the TPS.

20.3.2 Aiding Plan Evaluation with Ex situ Informatics Tools

There have been several published studies regarding the use of informatics tools for ex situ plan evaluations in the TMS. Azmandian et al. described an error checker designed to mine multiple patient data in a TMS and use clustering techniques (Forgy 1965; MacQueen 1965) to detect outliers with respect to inferred normative behavior (Azmandian et al. 2007). Furhang

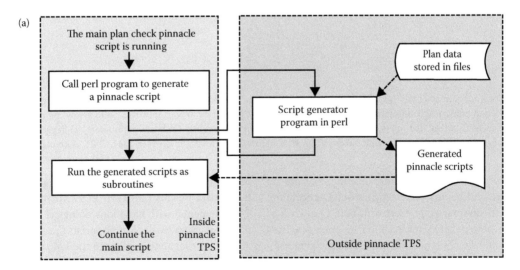

FIGURE 20.4 In situ IMRT plan integrity verification report. (a) Workflow of dynamic PINNACLE script generation that facilitates logical comparisons and numerical computations of the state of a treatment plan prior to plan approval. Solid arrows represent program flow, and dashed arrows represent data flow. This method outputs detailed report modules such as (b) beam parameters, (c) flagged logical violations, and (d) mutually exclusive IMRT objectives. (Reprinted from Yang, D. & Moore, K. L., *Med. Phys.*, 39(3), 1542, 2012. With permission.)

(b)

			Beams for prescription - "LT NECK" Type - IMRT		
#	1	2	3	4	5
Name	1A RPO 185 HN	1B RPO 210 HN	1C LAO 40 HN	1D LAO 70 HN	1E LPO 100 HN
Field ID	1A	1B	1C	1D	1E
Is Setup Beam	No	No	No	No	No
Type	SMLC	SMLC	SMLC	SMLC	SMLC
Isocenter	PREVIEW	PREVIEW	PREVIEW	PREVIEW	PREVIEW
Machine	TR3_TRILOGY2	TR3_TRILOGY2	TR3_TRILOGY2	TR3_TRILOGY2	TR3_TRILOGY2
Modality	Photons	Photons	Photons	Photons	Photons
Energy	6MV	6MV	6MV	6MV	6MV
Gantry	185	210	40	70	100
Couch	0.0	0.0	0.0	0.0	0.0
Collimator	0.0	0.0	0.0	0.0	0.0
X1	6.5	7.0	3.0	4.0	3.8
X2	3.0	3.0	7.0	7.5	7.0
X1+X2	9.5	10.0	10.0	11.5	10.8
Y1	5.0	5.0	5.0	5.0	5.0
Y2	6.0	6.0	6.5	6.5	6.5
Y1+Y2	11.0	11.0	11.5	11.5	11.5
Dose Engine	AC	AC	AC	AC	AC
MU	75.0	96.0	69.0	98.0	79.0
SSD	88.55	83.71	94.05	94.69	94.52
Optimization Type	DMPO	DMPO	DMPO	DMPO	DMPO
Segments	7	5	6	8	10
MU per segment	1 - 5.0 2 - 10.9 3 - 22.8 4 - 8.5 5 - 7.5 6 - 8.7 7 - 11.6	1 - 13.5 2 - 11.9 3 - 39.2 4 - 17.6 5 - 13.8	1 - 14.8 2 - 9.5 3 - 7.7 4 - 12.8 5 - 11.2 6 - 13.0	1 - 5.0 2 - 5.0 3 - 9.8 4 - 11.3 5 - 27.2 6 - 21.9 7 - 5.9 8 - 12.0	1 - 8.2 2 - 5.0 3 - 8.0 4 - 11.0 5 - 19.2 6 - 5.0 7 - 5.0 8 - 5.0 9 - 7.2 10 - 5.5
Bolus	- no bolus -	- no bolus -	- no bolus -	- no bolus -	- no bolus -
Total MU	515.0				
Total Segments	48				
Summary	• All beams use the same delivery machine • All beams use the same isocenter • All beams are co-planar.				

FIGURE 20.4 (Continued) In situ IMRT plan integrity verification report. (a) Workflow of dynamic PINNACLE script generation that facilitates logical comparisons and numerical computations of the state of a treatment plan prior to plan approval. Solid arrows represent program flow, and dashed arrows represent data flow. This method outputs detailed report modules such as (b) beam parameters, (c) flagged logical violations, and (d) mutually exclusive IMRT objectives. (Reprinted from Yang, D. & Moore, K. L., *Med. Phys.*, 39(3), 1542, 2012. With permission.)

et al. used a combination of in-house programming, commercial reporting software, and Microsoft Excel (Microsoft Corporation, Redmond, WA) to summarize pretreatment data in a TMS as well as compare new patient's information to prior practices (Furhang et al. 2009). Siochi et al. designed and implemented an electronic radiotherapy quality assurance system that compared planning data in multiple TMS to multiple TPS, demonstrating a very robust method of verification between numerous software permutations (Siochi et al. 2009). Yang et al. also described a decision support system that reconciles the electronic record between the TMS and the TPS, including analysis of many different data objects (DICOM, PDF, Microsoft Word, TPS, and TMS elements) with violations of logical comparisons flagging potential issues to medical physicists (Yang et al. 2012).

Systems that analyze plans in the TMS prior to treatment can greatly improve the standard pretreatment physics checks as well as provide detailed quality assurance of the data that will be transferred to the delivery devices. Besides a necessary reliance on the TMS architecture, there are really no downsides to such implementations beyond its late stage in the pretreatment

(c)

#	1	2	3	4	5	6
Name	1A RPO 175 HN[7]	1B RAO 210 HN[13(It's RPO)]	1C LAO 40 HN[6,10]	1D 70 HN[13(It's LAO)]	1F LPO HN[7]	LPO 130 HN[4,9]
Field ID	1A	1B		1D	1E	1F
Summary	<ul><li>All beams use the same delivery machine</li><li>All beams use the same isocenter</li><li>All beams are co-planar.</li><li>[4] Beam name does not start with digits.</li><li>[5] Beams are out off sequence</li><li>[6] Beam does not have field ID.</li><li>[7] Beam has no gantry angle in their names</li><li>[9] Beam name does not start with the beam ID.</li><li>[10] Beam ID does not start with digits.</li><li>[13] Incorrect beam orientation in the beam name.</li></ul>					

(d)

Optimization objectives overlapping / conflicting analysis

ROI1	ROI2	Overlapping (cm^3)
uninvolved colon Max Dose = 3950	PTV 45 Min Dose = 4600 Max Dose = 4950	9.4871
PTV 45 Min Dose = 4600 Max Dose = 4950	PTV 50 Min Dose = 5100 Max Dose = 5300	593.633
	Small Bowel Max Dose = 3950	4.333
PTV 50 Min Dose = 5100 Max Dose = 5300	Small Bowel Max Dose = 3950	0.00572202

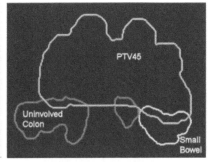

FIGURE 20.4 (Continued) In situ IMRT plan integrity verification report. (a) Workflow of dynamic PINNACLE script generation that facilitates logical comparisons and numerical computations of the state of a treatment plan prior to plan approval. Solid arrows represent program flow, and dashed arrows represent data flow. This method outputs detailed report modules such as (b) beam parameters, (c) flagged logical violations, and (d) mutually exclusive IMRT objectives. (Reprinted from Yang, D. & Moore, K. L., *Med. Phys.*, 39(3), 1542, 2012. With permission.)

process. If the errors originated in the TPS, it will require significant resources to correct problems detected in the TMS. The costs of error mitigation are greatly reduced by detecting problems prior to physician and export to the TMS.

On the face of it, ex situ analytics would not appear to be a very desirable strategy for pretreatment plan evaluation that is used directly in the planning stage. The added step of exporting DICOM data from a TPS will, at present, incur some loss of treatment planning information as well as duplicate potentially erroneous data on yet another database (Siochi et al. 2011). Nevertheless, these costs can be outweighed by the independence from TPS proprietary data formats that DICOM provides. Informatics tools developed for DICOM-RT can be assured of much greater operational longevity, and any changes to the DICOM standard will be discussed well in advance of any implementation (NEMA 2012).

An example of a DICOM-RT driven ex situ plan evaluation tool that has been used in clinical practice was developed at Washington University for keeping track of dosimetric values for combined IMRT and multifraction high-dose rate (HDR) brachytherapy treatments (Figure 20.5) (Richardson et al. 2012). The stand-alone software tool functions by processing DICOM-RT objects, including associated structure DVHs, exported from Varian's BrachyVision and Eclipse software.

Similar to GEC ESTRO guidelines (Pötter et al. 2006), the tool allows rapid review of Biological Effective Doses (BEDs) and Equivalent Dose in 2 Gy fractions (EQD$_2$) as well as tracking running dose parameters for each contoured structure and providing projections for expected final dose parameters. The advantage of this tool for both real-time planning and ex post facto analyses are hard to overstate. Ongoing brachytherapy treatments are immediately and efficiently compared to previous fractions, and thresholds can be set and adjusted based on prior experience. The final treatment summaries themselves become an object for ex post facto analyses of a very efficient and controlled distillation of clinical treatment parameters. Further, as the DICOM-RT objects for each patient's treatment are already exported to the database the tool references, any ex post facto analyses can add new dosimetric analyses with great efficiency of effort.

Consider as well the case of deciding between multiple competing treatment plans that have made different clinical trade-offs, a problem for which a decision analytic tool was proposed 20 years ago (Jain & Kahn 1991; Jain et al. 1993) and several variants have since been proposed and prototyped (Yu 1997; Yu et al. 1997; Jones & Hoban 2000; Craft et al. 2005; Monz et al. 2008). Currently, on-screen DVH comparison is available in all commercial TPS, but only with advanced programming in a

GTV-CERVIX

(a)

	BTFX1	BTFX2	BTFX3	BTFX4	BTFX5	BTFX6	BTX AVG	BTX TOTAL	BTX+EBRT (MTV CERVIX)	STDV
Vol (cc)	1.70	1.64	1.42	0.05	1.46	0.82	1.18			0.64
V100 (%)*	100.00	100.00	100.00	100.00	100.00	100.00	100.00			0.00
D100 (Gy)	6.15	2.52	2.11	2.48	2.35	3.83	3.24	19.44	60.35	1.55
D90 (Gy)	6.90	3.01	2.55	2.77	2.86	4.53	3.77	22.62	68.71	1.69
D_Mean (Gy)	9.48	3.71	3.91	3.44	5.84	5.36	5.29	31.73	82.85	2.27
D_Max (Gy)	23.63	5.31	12.77	4.24	45.22	8.35	16.59	99.52	153.69	15.69
EQD$_2$(Gy),D100	8.28	2.63	2.13	2.58	2.42	4.41	3.74	22.45	61.52	2.36
EQD$_2$ (Gy),D90	9.72	3.26	2.67	2.95	3.06	5.49	4.52	27.15	71.88	2.74
EQD$_2$ (Gy),D_Mean	15.38	4.23	4.53	3.85	7.71	6.86	7.09	42.57	92.94	4.34
EQD$_2$(Gy),D_Max	66.22	6.77	24.23	5.03	208.09	12.77	53.85	323.12	376.99	78.88

OAR-Bladder

(b)

	BTFX1	BTFX2	BTFX3	BTFX4	BTFX5	BTFX6	BTX AVG	BTX+Mean EB (Bladder)
Vol (cc)	70.09	71.84	62.76	64.10	99.24	60.27	71.38	
D100(Gy)	1.33	0.47	0.64	0.63	0.58	0.69	0.72	49.33
D90(Gy)	1.73	0.62	0.81	0.87	0.78	0.89	0.95	50.69
D_Mean (Gy)	3.18	1.03	1.26	1.34	1.31	1.38	1.58	54.49
D_Max (Gy)	6.82	2.19	2.50	2.60	2.72	2.73	3.26	64.55
D2cc (Gy)	6.20	1.99	2.27	2.35	2.57	2.45	2.97	62.82
EQD$_2$(Gy),D100	1.15	0.33	0.46	0.46	0.41	0.51	0.55	44.77
EQD$_2$(Gy),D90	1.63	0.45	0.62	0.67	0.59	0.69	0.77	46.10
EQD$_2$(Gy),D_Mean	3.92	0.83	1.07	1.16	1.13	1.21	1.55	50.77
EQD$_2$(Gy),D_Max	13.37	2.27	2.74	2.91	3.11	3.12	4.59	68.97
EQD$_2$(Gy),D2cc	11.39	1.98	2.39	2.51	2.86	2.67	3.96	65.24
Ratio(D2cc/A_AVG Dose)	1.07	0.92	1.06	1.16	1.22	1.10	1.09	
Ratio:EQD(D2cc)/EQD(A_AVG)	1.49	0.91	1.10	1.23	1.35	1.17	1.21	

FIGURE 20.5 Example of ex situ plan evaluation of multifraction HDR brachytherapy treatments based on analysis of DICOM data exported from Varian's BrachyVision and Eclipse TPS. Portions of report summary dose statistics on (a) target volumes and (b) OARs alike, including both BED calculations, integration of brachytherapy and external beam plans, and color-coded alerts for thresholds that exceed expected historical values.

TPS-specific scripting language could such decision analytic systems be developed into an in situ evaluation tool. Implementing such a framework in a software tool that processes DICOM-RT would be a nontrivial enterprise yet, once developed, would be able to make comparisons between plans from any TPS.

20.3.3 Plan Evaluation Tools for Adaptive Radiotherapy

ART is an area of clinical investigation that is difficult to envision without incorporating automated plan evaluation into the workflow. For example, in online ART (Yan, Wong, et al. 1997; Yan, Vicini, et al. 1997), new "plans-of-the-day" are developed based on on-treatment imaging such as cone beam computed tomography (CBCT). Such adapted plans require pretreatment review, yet it does not seem feasible to expect physicians or physicists to be stationed at every treatment machine reviewing every patient's adapted plan. In this regime, automated plan evaluation tools are a practical necessity.

Needless to say, both offline and online ART blur the distinction of a "pretreatment" plan review. The expectation of adaptive planning means that the entire software chain from simulation to treatment needs to be designed around the expected adjustment of an initial treatment. Certainly focusing on in situ plan evaluations exclusively would appear unwise, as their ART benefit will be limited only to on-treatment data that is imported back into the TPS, a potentially time-consuming step in a process that demands efficiency. Unless there is an integrated TPS, TMS, image-guided radiation therapy, and adaptive planning software system, the use of the general DICOM-RT data format for ART plan analysis software would be almost a necessity.

20.4 Building a Robust Informatics Framework for Treatment Plan Evaluation

Institutions considering the development of an informatics system to aid in treatment plan evaluation should begin the design process by outlining the goals of the process and the capabilities of the software systems at hand. Different TPS and TMS combinations have unique and evolving capabilities, and while it would be impractical to describe all permutations, here it is not difficult for clinicians to evaluate the capacity of their

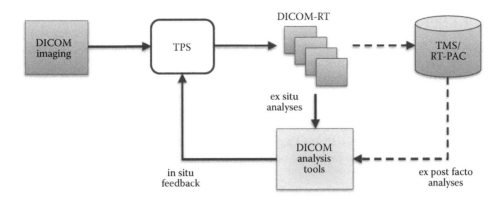

FIGURE 20.6 Diagram representing an idealized informatics infrastructure around the treatment planning process, with DICOM-RT as the core data format and feedback from prior experience is integrated directly into the in situ plan evaluation process.

institution's software systems. As for the goals of the process, this will be a largely universal desire to build a versatile informatics suite capable of multifaceted analyses while reducing the upfront and ongoing costs to the analytic software, for example,

Sample Goals for Plan Evaluation Software

- Maximize computational approaches to data analysis
- Minimize costs to error correction
- Maximize robustness to changes in vendor-specific architecture
- Minimize costs to retrospective analyses
- Minimize maintenance costs

It is the authors' opinion that the ideal plan evaluation infrastructure should rely primarily on analysis of DICOM-RT data while still exploiting any in situ capabilities of one's particular TPS and TMS. Figure 20.6 depicts a simplified block diagram of such a system, with the resemblance of this architecture to a feedback loop being no accident. The primary engine of this system is a software system designed to process DICOM-RT data. The main goal as depicted is to detect errors during planning and while interim treatment planning data will have to be exported to an ancillary system that can be programmed in any architecture. Should the TPS, TMS, or both be replaced, the main ex situ analysis engine remains intact. If a TPS has either read-write or read-only API functionality, these can be used to digest any analyses back into the TPS to increase automation. Ex post facto analyses—conducted on extracted data from the TMS, a RT-PACS, or another DICOM-RT database—can be conducted in the same analytics software and synthesized into new pretreatment evaluation tools. While no architecture can ever be perfect, given the current and expected state of radiotherapy treatment planning software such a system would best satisfy the five design goals.

20.5 Quality Assurance of Plan Evaluation Tools

While using informatics tools for plan evaluation can typically be characterized as a quality control mechanism, any software that becomes part of the clinical workflow must itself be included in a radiation oncology quality management program. When considering the QA issues related to using advanced informatics for treatment plan evaluations, it is important to consider that, once incorporated into clinical use, these tools are fully part of the treatment planning process. Depending on intended use, decision support tools can be as harmful as an erroneous dose calculation should they fail to operate properly. A treatment plan summary like the simple QUANTEC report shown in Figure 20.3 has numerous potential failures:

- Mismatch between OAR and toxicity
- Mismatch between OAR and complication rate
- False positives—OAR dose threshold not exceeded but flagged to user as overtolerance
- False negatives—OAR dose threshold exceeded but not flagged to user as overtolerance
- Incorrect calculation of OAR dosimetric values
- OAR(s) erroneously not included in report

The latter three failures would be particularly troubling, as these could mislead a physician or physicist to approve a treatment when an unacceptable dosimetric deviation has occurred. Like all critical data entries in radiation therapy, these values should be routinely second checked. If exhaustive testing of all reported elements is warranted, employing benchmark cases that efficiently verify functionality can reduce what would otherwise be a very laborious task.

Strict revision control, where revisions to source code are indexed and introduced in a controlled manner, is a standard quality practice in software configuration management and is appropriate for such informatics tools. The source code should be maintained by knowledgeable personnel and well documented, also following standard industry practices.

Should the plan evaluation tools involve data transfer, the quality assurance of this aspect is within the scope of AAPM Task Group 201 regarding the Quality Assurance of External Beam Radiotherapy Data Transfer (Siochi et al. 2011), to which the reader is referred.

20.6 Future of Plan Evaluation

The recent history of radiotherapy suggests that, with ever-greater computing power and the steady incorporation of new delivery technologies, the complexity of the average treatment will continue to increase. Coupled with expected increases in patient populations and the near-certain incorporation of adaptive treatment management, this increased complexity will stress severely the current treatment planning workflow that is designed around manual software operation and human-performed verification checks.

One technological advance on the horizon that will likely play a role in reducing the human element in the plan design stage is automated treatment planning, a technique where the sequence of planning tasks are accomplished by an automated routine as opposed to manual button clicks (Purdie et al. 2011; Zhang et al. 2011). Patient-specific IMRT optimization objective tuning is a particularly important step in current commercial TPS, and new IMRT optimization techniques (Alber & Nusslin 2001; Halabi et al. 2006; Thieke et al. 2007; Clark et al. 2008; Monz et al. 2008; Qi et al. 2009; Voet et al. 2013) hold the potential of reducing or eliminating the manual process of constructing a cost function. Automated planning will likely rely heavily on both in situ scripting routines and ex post facto informatics studies, as evidenced by recent work from Wu et al. (2011). Needless to say, as automated planning develops from research endeavors to clinical implementation, new quality control elements must be brought to bear to ensure that these advanced software elements do not open up new possibilities for errors that could harm patients.

In the face of these trends, treatment planning decision support tools are almost certain to move from a convenience afforded by research academic centers to a necessity for all functioning radiotherapy centers. Luckily, the costs of computational resources—computational power, storage, networking architecture, and electronic medical records—needed to implement informatics-based plan evaluation tools are being lessened by the same technological forces that are driving the clinical necessity. Thus, it has never been easier or more cost-effective to develop and put into practice decision support tools for treatment plan evaluation. Online consortia of software users, such as that found at www.medphysfiles.com (Geoghegan 2007) for PINNACLE scripting, can also facilitate information sharing and dispersion of best practices. Future offerings from commercial vendors might fill some of these clinical needs, but until such time as they do clinicians interested in developing such plan evaluation tools should pay close attention to the literature and abstracts at scientific meetings presented to find approaches that fit their infrastructure best.

References

Alber, M., and Nusslin, F. (2001). Optimization of intensity modulated radiotherapy under constraints for static and dynamic MLC delivery. *Phys Med Biol, 46*(12), 3229–3239.

Appenzoller, L. M., Michalski, J. M., Thorstad, W. L., Mutic, S, and Moore, K. L. (2012). Predicting dose-volume histograms for organs-at-risk in IMRT planning. *Med Phys, 39*(12), 7446–7461.

Azmandian, F., Kaeli, D., Dy, J. G., Hutchinson, E., Ancukiewicz, M., Niemierko, A., and Jiang, S. B. (2007). Towards the development of an error checker for radiotherapy treatment plans: A preliminary study. *Phys Med Biol, 52*(21), 6511–6524.

Bentzen, S. M., Constine, L. S., Deasy, J. O., Eisbruch, A., Jackson, A., Marks, L. B., Ten Haken, R. K., and Yorke, E. D. (2010). Quantitative analyses of normal tissue effects in the clinic (QUANTEC): An introduction to the scientific issues. *Int J Radiat Oncol Biol Phys, 76*(3 Suppl), S3–S9.

Clark, V. H., Chen, Y., Wilkens, J., Alaly, J. R., Zakaryan, K., and Deasy, J. O. (2008). IMRT treatment planning for prostate cancer using prioritized prescription optimization and mean-tail-dose functions. *Linear Algebra Appl, 428*(5–6), 1345–1364.

Craft, D., Halabi, T., and Bortfeld, T. (2005). Exploration of tradeoffs in intensity-modulated radiotherapy. *Physics in Medicine and Biology, 50*(24), 5857.

Deasy, J. O., Blanco, A. I., and Clark, V. H. (2003). CERR: A computational environment for radiotherapy research. *Med Phys, 30*(5), 979–985.

Emami, B., Lyman, J., Brown, A., Coia, L., Goitein, M., Munzenrider, J. E., Shank, B., Solin, L. J., and Wesson, M. (1991). Tolerance of normal tissue to therapeutic irradiation. *Int J Radiat Oncol Biol Phys, 21*(1), 109–122.

Forgy, E. (1965). Cluster analysis of multivariate data: Efficiency versus interpretability of classifications. *Biometrics, 21*, 768–780.

Furhang, E. E., Dolan, J., Sillanpaa, J. K., and Harrison, L. B. (2009). Automating the initial physics chart checking process. *J Appl Clin Med Phys, 10*(1), 2855.

Geoghegan, S. (2007). *Scripting in the Pinnacle-3 Treatment Planning System.* Available at www.medphysfiles.com.

Halabi, T., Craft, D., and Bortfeld, T. (2006). Dose-volume objectives in multi-criteria optimization. *Phys Med Biol, 51*(15), 3809–3818.

Halperin, E. C., Perez, C. A., and Brady, L. W. (2008). *Perez and Brady's Principles and Practice of Radiation Oncology.* 5th ed. Philadelphia: Wolters Kluwer Health/Lippincott Williams & Wilkins.

Holdsworth, C., Hummel-Kramer, S. M., and Phillips, M. (2011). Scripting in radiation therapy: An automatic 3D beam-naming system. *Med Dosim, 36*(3), 272–275.

Houweling, A. C., Philippens, M. E. P., Dijkema, T., Roesink, J. M., Terhaard, C. H. J., Schilstra, C., Ten Haken, R. K., Eisbruch, A., and Raaijmakers, C. P. J. (2010). A comparison of dose-response models for the parotid gland in a large group of head-and-neck cancer patients. *Int J Radiat Oncol Biol Phys, 76*(4), 1259–1265.

Hunt, M. A., Jackson, A., Narayana, A., and Lee, N. (2006). Geometric factors influencing dosimetric sparing of the parotid glands using IMRT. *Int J Radiat Oncol Biol Phys, 66*(1), 296–304.

Jain, N. L. & Kahn, M. G. (1991). Ranking radiotherapy treatment plans using decision-analytic and heuristic techniques. *Proc Annu Symp Comput Appl Med Care*, 1000–1004.

Jain, N. L., Kahn, M. G., Drzymala, R. E., Emami, B. E., and Purdy, J. A. (1993). Objective evaluation of 3-D radiation treatment plans: a decision-analytic tool incorporating treatment preferences of radiation oncologists. *Int J Radiat Oncol Biol Phys, 26*(2), 321–333.

Janssen, T., van Kesteren, Z., Franssen, G., Damen, E., and van Vliet, C. (2013). Pareto fronts in clinical practice for Pinnacle. *Int J Radiat Oncol Biol Phys, 85*(3), 873–880.

Jones, L. C. & Hoban, P. W. (2000). Treatment plan comparison using equivalent uniform biologically effective dose (EUBED). *Phys Med Biol, 45*(1), 159.

Kim, B., Park, H. C., Oh, D., Shin, E. H., Ahn, Y. C., Kim, J., and Han, Y. (2012). Development of the DVH management software for the biologically-guided evaluation of radiotherapy plan. *Radiat Oncol J, 30*(1), 43–48.

Kutcher, G. J., Coia, L., Gillin, M., Hanson, W. F., Leibel, S., Morton, R. J., Palta, J. R., Purdy, J. A., Reinstein, L. E., and Svensson, G. K. (1994). Comprehensive QA for radiation oncology: Report of AAPM Radiation Therapy Committee Task Group 40. *Med Phys, 21*(4), 581–618.

MacQueen, J. B. (1965). Some methods for classification and analysis of multivariate observations. In *Proceedings of the 5th Berkeley Symposium on Mathematical Statistics and Probability*.

Marks, L. B., Yorke, E. D., Jackson, A., Ten Haken, R. K., Constine, L. S., Eisbruch, A., Bentzen, S. M., Nam, J., and Deasy, J. O. (2010). Use of normal tissue complication probability models in the clinic. *Int J Radiat Oncol Biol Phys, 76*(3 Suppl), S10–S9.

Monz, M., Kufer, K. H., Bortfeld, T. R., and Thieke, C. (2008). Pareto navigation: Algorithmic foundation of interactive multi-criteria IMRT planning. *Phys Med Biol, 53*(4), 985–998.

Moore, K. L., Brame, R. S., Low, D. A., and Mutic, S. (2011). Experience-based quality control of clinical IMRT planning. *Int J Radiat Oncol Biol Phys, 81*(2), 545–551.

NEMA (2012). *The DICOM Standard*. Available at http://medical.nema.org/standard.html. Cited October 9, 2012.

Petit, S. F., Wu, B., Kazhdan, M., Dekker, A., Simari, P., Kumar, R., Taylor, R., Herman, J. M., and McNutt, T. (2012). Increased organ sparing using shape-based treatment plan optimization for intensity modulated radiation therapy of pancreatic adenocarcinoma. *Radiother Oncol, 102*(1), 38–44.

Purdie, T. G., Dinniwell, R. E., Letourneau, D., Hill, C., and Sharpe, M. B. (2011). Automated planning of tangential breast intensity-modulated radiotherapy using heuristic optimization. *Int J Radiat Oncol Biol Phys, 81*(2), 575–583.

Pötter, R., Haie-Meder, C., Van Limbergen, E., Barillot, I., De Brabandere, M., Dimopoulos, J., Dumas, I., Erickson, B., Lang, S., Nulens, A., et al.; GEC ESTRO Working Group (2006). Recommendations from gynaecological (GYN) GEC ESTRO working group (II): Concepts and terms in 3D image-based treatment planning in cervix cancer brachytherapy-3D dose volume parameters and aspects of 3D image-based anatomy, radiation physics, radiobiology. *Radiother Oncol, 78*(1), 67–77.

Qi, X. S., Semenenko, V. A., and Li, X. A. (2009). Improved critical structure sparing with biologically based IMRT optimization. *Med Phys, 36*(5), 1790–1799.

Richardson, S., Sun, B., Garcia-Ramirez, J., and Grigsby, P. (2012). PO-276 real-time dosimetric quality assurance for high-dose-rate gynecological brachytherapy. *Radiother Oncol, 103*(Supplement 2), S111.

Sheppard, D. (2000). *Beginner's Introduction to Perl*. Available at http://www.perl.com/pub/2000/10/begperl1.html.

Siochi, R. A., Pennington, E. C., Waldron, T. J., and Bayouth, J. E. (2009). Radiation therapy plan checks in a paperless clinic. *J Appl Clin Med Phys, 10*(1), 2905.

Siochi, R. A., Balter, P., Bloch, C. D., Santanam, L., Blodgett, K., Curran, B. H., Engelsman, M., Feng, W., Mechalakos, J., Pavord, D., et al. (2011). A rapid communication from the AAPM Task Group 201: Recommendations for the QA of external beam radiotherapy data transfer. AAPM TG 201: Quality assurance of external beam radiotherapy data transfer. *J Appl Clin Med Phys, 12*(1), 3479.

Thieke, C., Kufer, K. H., Monz, M., Scherrer, A., Alonso, F., Oelfke, U., Huber, P. E., Debus, J., and Bortfeld, T. (2007). A new concept for interactive radiotherapy planning with multi-criteria optimization: First clinical evaluation. *Radiother Oncol, 85*(2), 292–298.

Vineberg, K. A., Eisbruch, A., Coselmon, M. M., McShan, D. L., Kessler, M. L., and Fraass, B. A. (2002). Is uniform target dose possible in IMRT plans in the head and neck? *Int J Radiat Oncol Biol Phys, 52*(5), 1159–1172.

Voet, P. W., Dirkx, M. L., Breedveld, S., Fransen, D., Levendag, P. C., and Heijmen, B. J. (2013). Toward fully automated multicriterial plan generation: A prospective clinical study. *Int J Radiat Oncol Biol Phys, 1*(85), 866–872.

Wu, B., Ricchetti, F., Sanguineti, G., Kazhdan, M., Simari, P., Chuang, M., Taylor, R., Jacques, R., and McNutt, T. (2009). Patient geometry-driven information retrieval for IMRT treatment plan quality control. *Med Phys, 36*(12), 5497–5505.

Wu, B., Ricchetti, F., Sanguineti, G., Kazhdan, M., Simari, P., Jacques, R., Taylor, R., and McNutt, T. (2011). Data-driven approach to generating achievable dose–volume histogram objectives in intensity-modulated radiotherapy planning. *Int J Radiat Oncol Biol Phys, 79*(4), 1241–1247.

Yan, D., Vicini, F., Wong, J., and Martinez, A. (1997). Adaptive radiation therapy. *Phys Med Biol, 42*(1), 123–132.

Yan, D., Wong, J., Vicini, F., Michalski, J., Pan, C., Frazier, A., Horwitz, E., and Martinez, A. (1997). Adaptive modification of treatment planning to minimize the deleterious effects of treatment setup errors. *Int J Radiat Oncol Biol Phys, 38*(1), 197–206.

Yang, D. & Moore, K. L. (2012). Automated radiotherapy treatment plan integrity verification. *Med Phys, 39*(3), 1542–1551.

Yang, D., Wu, Y., Brame, R. S., Yaddanapudi, S., Rangaraj, D., Li, H. H., Goddu, S. M., and Mutic, S. (2012). Technical Note: Electronic chart checks in a paperless radiation therapy clinic. *Med Phys, 39*(8), 4726–4732.

Yu, Y. (1997). Multiobjective decision theory for computational optimization in radiation therapy. *Med Phys, 24*(9), 1445–1454.

Yu, Y., Schell, M. C., and Zhang, J. Y. (1997). Decision theoretic steering and genetic algorithm optimization: Application to stereotactic radiosurgery treatment planning. *Med Phys, 24*(11), 1742–1750.

Zhang, T., Chi, Y., Meldolesi, E., and Yan, D. (2007). Automatic delineation of on-line head-and-neck computed tomography images: Toward on-line adaptive radiotherapy. *Int J Radiat Oncol Biol Phys, 68*(2), 522–530.

Zhang, X., Li, X., Quan, E. M., Pan, X., and Li, Y. (2011). A methodology for automatic intensity-modulated radiation treatment planning for lung cancer. *Phys Med Biol, 56*(13), 3873–3893.

Zhu, X., Ge, Y., Li, T., Thongphiew, D., Yin, F. F., and Wu, Q. J. (2011). A planning quality evaluation tool for prostate adaptive IMRT based on machine learning. *Med Phys, 38*(2), 719–726.

Human–Computer Interaction in Radiation Therapy

Junyi Xia
*University of Iowa
Hospitals and Clinics*

John E. Bayouth
*University of Iowa
Hospitals and Clinics*

21.1 Introduction

Human–computer interaction (HCI) is an inescapable part of everyday life. For example, a modern car has as many as 50 microprocessors that constantly process signals from hundreds of sensors on board to obtain and maintain optimal performance. When driving, we seldom notice these interactions because car manufacturers hide the complexity from drivers. Interpreting extraneous details, such as intake air flow rate, is an overwhelming distraction to drivers. The car dashboard is the interface where the human (driver) interacts with the computer (microprocessors and control systems); a well-designed dashboard will not only show the necessary information but also provide the best way for the driver to interact and respond to changes in normal car operating conditions. For example, when the low fuel warning light illuminates, most drivers are not interested in the number of fuel sensors and their locations in the gasoline tank or the algorithms for determining low fuel volume, only the distance to the next fueling station.

With the widespread development and implementation of advanced technologies in radiation therapy centers, they have evolved into an inherently heterogeneous environment: hardware and software from different vendors must be interfaced to provide safe, accurate, and optimal treatments. The operators of radiation therapy centers face the challenge of efficiently interacting with the different systems, while the vendors struggle to design user-friendly interfaces. Because each vendor has different design models and design philosophies, some are more intuitive to use, while others need substantial training to be effectively used. Complicated interfaces may easily confuse or frustrate the users, and more importantly impose increasing risks of treatment errors. The purpose of this chapter is to identify general HCI design principles that facilitate the effec-

tive interaction between the users and computer/control system, used in radiation therapy.

Figure 21.1 shows a contemporary radiation therapy delivery system, which includes a medical linear accelerator [1], record and verify (R&V) system, patient monitoring system, couch control system, surrogate for respiratory monitoring, and treatment delivery interface. Although the linear accelerator is the main component, each peripheral component has its unique interface to the linear accelerator and to the end user. First, we will provide a brief description of the function of each component.

A *linear accelerator* generates high-energy x-ray or electron beams and delivers them to the patient based on the planned treatment parameters.

The *R&V system* receives treatment parameters from the treatment planning system and transfers them to linear accelerator. At the same time, actual delivered treatment parameters, such as leaf positions, couch position, dose rate, and monitor units (MU), are recorded by the R&V system.

The *patient monitoring systems* include not only the microphone and the closed circuit cameras but also optical-based intrafractional monitoring systems, such as the surface-based optical system (VisionRT) or the marker-based optical system (RadioCam).

The *couch control system* refers to the traditional four degrees of freedom (DOF) couch or more advanced six DOF couch in which pitch and roll adjustments can made.

The *surrogate for respiratory gating* refers to the system that turns on the radiation beam when the respiratory surrogates are in the correct predetermined position, for example, the Anzai gating system (Anzai Medical Co., Tokyo, Japan) uses a strain gauge as a surrogate to measure the respiratory traces by positioning a strain gauge belt around a patient's abdomen; the Varian Real-time Position Management (RPM) system [2]

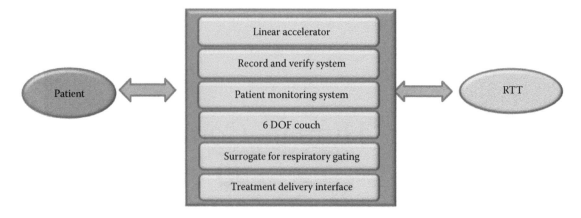

FIGURE 21.1 Overview of the radiation therapy treatment system. A radiation therapy treatment system includes a linear accelerator, R&V system, patient motion monitoring systems, couch control system, surrogate for respiratory gating system, and treatment delivery interface.

monitors respiratory cycles by tracking reflective markers using an infrared camera to monitor the range of motion.

The *treatment delivery interface* acts like a gateway to the accelerator. For example, the Elekta MOSAIQ Sequencer interfaces with the Siemens linear accelerator using a software application called the Coherence RTT 4.2 Workspace.

Figure 21.2 shows a treatment console area at the University of Iowa Hospitals and Clinics, where the Radiation Therapy Technologists (RTT) face the challenge of simultaneously viewing up to seven different monitors, namely,

1. Accelerator console;
2. MOSAIQ R&V computer;
3. RTT workstation 4.2 computer, which acts as middleware between the accelerator control console and R&V system;
4. Radiocam computer, which uses an optical array for stereotactic radiosurgery or stereotactic radiation therapy;
5. VisionRT computer to track intrafractional patient motion by real-time surface information;
6. Protura six DOF couch computer for patient positioning; and
7. Closed circuit television for visual observation of the patient.

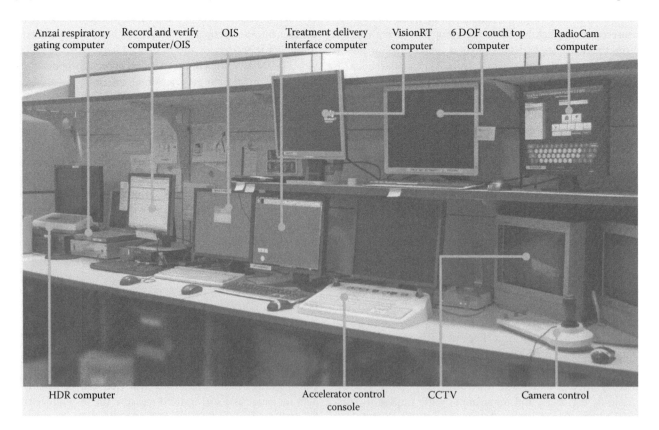

FIGURE 21.2 A snapshot of a radiation therapy treatment control area (Department of Radiation Oncology at the University of Iowa Hospitals and Clinics, Iowa City, IA). For a complex radiotherapy delivery, the radiation therapists have to monitor up to seven different computer displays.

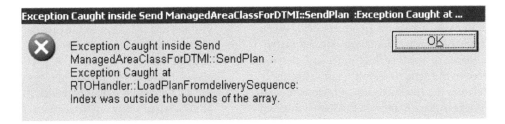

FIGURE 21.3 An error message of MOSAIQ 2.3 R&V system when trying to download a large amount of treatment fields.

Interpreting error messages from any software application may not be a simple task. For example, Figure 21.3 shows a MOSAIQ 2.3 Sequencer error message when too many fields are downloaded to the linear accelerator in one session. This error message is best written for software developers or people with programming background to understand the source of the error, not those operating the radiation therapy system. In addition to the challenge of interpreting messages describing error sources, system operators (RTTs or medical physicists) are faced with an overwhelming amount of information to be processed.

21.1.1 Information Overload in Radiation Therapy

The term "information overload" was popularized in the 1970 bestselling book *Future Shock* by Alvin Toffler; it refers to the difficulty of interpretation and cognition when too much information is presented. During radiation therapy treatment, the therapists must watch several displays to observe the radiation treatment delivery. Designing a system to enable them to process the information to gain an appropriate view of a treatment without becoming overwhelmed is a daunting task. Human nature may lead therapists to ignore some displays that are perceived as an "always OK" system. This can be dangerous; abnormalities of the system can have serious effects on patient treatment. Another way to overwhelm the therapists is to set the warning or alarm sensitivity too high. If therapists receive frequent false alarms, they tend to ignore them when there is a real emergency condition. Ultimately, information overload can result in either delaying decision-making or promoting making wrong decisions. HCI should be designed to address and minimize information overload, providing intuitive controls and appropriate feedback to the system operators.

Compared with computers that can handle many calculations at the same time, the human brain is only capable of handling a limited number of processes simultaneously. Systems must be designed to match machine information with the human's ability to process the information. Analog and digital data need to be processed, filtered, and presented to the RTT in an understandable format and at the appropriate time.

Radiation therapy is one of many industries struggling to address information overload. The aeronautical industry faced similar challenges in providing pilots succinct information in the cockpit. Guidelines to minimize pilot information overload [3] include the following:

- Do not overload with too much information.
- Do not provide unnecessary information.
- Do not ask the pilot to perform a task that can be performed faster or more accurately by a machine.
- Do not give the pilot tedious, repetitive tasks if they can be automated.
- Do not ask the pilot to monitor instruments to detect rare events; instead, provide alarm systems.

Although these HCI guidelines were proposed for the aeronautical industry, they may be applied to the design of user interfaces in radiation therapy. It is imperative that information should be kept succinct and relevant and grouped to reflect relationships and the hierarchy, with critical information highlighted.

A well-designed interface can greatly assist users in their ability to accomplish tasks, while a poorly designed user interface frustrates users and may put human lives at risk. Unfortunately, history provides many examples of poorly designed interfaces contributing to errors: the medical device industry, the aviation industry, and the nuclear industry provide three obvious examples of bad interface designs that have caused serious injuries and death. One such example is the Therac-25 accident.

21.1.2 Therac-25 Accident [4]

The Therac-25 was a dual-mode medical linear accelerator produced by the Atomic Energy of Canada Limited. It could delivery either 25 MV photon beams or 5 to 25 MeV electron beams. Between June 1985 and January 1987, the Therac-25 was involved in accidents that massively overdosed six patients. These accidents are described as the worst in the history of medical accelerators.

A significant contributor to the accident was introduced by the wrong energy mode of the accelerator being executed for treatment in contrast to the energy mode selected by the RTT, as the data entry by the RTT was faster than the control system could register. In a linear accelerator, x-ray photons are produced by focusing high-energy electron beams to a target. Part of the electron energy is converted into photons and most of the electron energy is dissipated as heat. The resulting x-ray beam is highly peaked in the forward direction, so a flattening filter is used to selectively attenuate the x-ray beam to provide a beam of uniform intensity across the treatment field. When these accidents happened, neither the target nor the flattening filter were in place, so massive high-intensity electron beams were delivered to the patients, resulting in deliveries of approximately 100 times the intended dose. Previous models had hardware interlocks in place to prevent this happening, but the Therac-25 had removed them, depending solely on software interlocks for safety. These software interlocks were susceptible to failure.

From the HCI point of view, the Therac-25 did not communicate optimally with the user: (1) When the system noticed something was wrong, it only displayed the error code by the word "MALFUNCTION" followed by a number from 1 to 64. The error code did not provide any details of the error, not even in the user manual. (2) The users were made insensitive to the error message because it was frequently occurring and users could simply override the error message to continue treating.

21.2 Human–Computer Interaction

Human–computer interaction, as the name suggests, is the study of the interaction occurring between users and computers. It is the design, implementation, and evaluation of user interfaces. A user interface relates to the device where the user interacts with the system to complete specific tasks. The mode of interaction may include a graphical user interface (GUI) design, such as the layout and arrangement of the textboxes or dropdown lists, but also includes the hardware, such as mouse, keyboard, or touch screen. HCI permeates modern existence. From setting the temperature on our thermostats, clicking the key fob to unlock our car or heating food in a microwave, the quality of our experience is dependent on the successful design of these seemingly trivial elements.

21.2.1 Principles of HCI

Donald A. Norman wrote an excellent introductory design book called "The Design of Everyday Things." In the book, he illustrated some important principles for the successful design of hardware for better human interaction [5]:

1. **Visibility.** The correct components must be visible, and they must convey the correct message. It should be easy for the user to tell what the state of the device is and know what alternatives for action are available. In radiation therapy delivery, important treatment parameters include (1) MU, (2) dose rate, (3) field size and the position of the multi-leaf collimator (MLC), (4) gantry angle, (5) couch position, and (6) energy. Further descriptions of these parameters are provided below. The status of these parameters should be optimally arranged and easily visible to the users. Key information such as the number of delivered MUs and total MUs should be prominently placed. Using a graphical representation instead of numbers, to represent the gantry position and MLC size would ease the task. Remember: "a picture is worth a thousand words."

2. **A good conceptual model.** A good conceptual model allows the users to predict the results of their action. Without a good conceptual model, the users will operate blindly, based solely on instructions. If errors occur or an unanticipated situation arises, a system with a good conceptual model will not leave the users in the dark and will help to predict the results of their possible actions. Any system should provide a good conceptual model for the user, with consistency in the presentation of operations and results and a coherent, unified system image.

3. **Good mapping.** Mapping refers to the relationship between two steps in a procedure. For example, the action of a user and the natural expectation of the result. To turn to the left, we move the steering wheel to the left; it is very intuitive, easy to learn, and always remembered.

4. **Feedback.** Provide users information about what has been done and the result of their action. Imaging trying to paint a picture using brushes but no paint; you cannot see where you have painted. In good accelerator design, it could be as simple as showing "BeamON" when the "BeamON" button is pressed, or producing beeping sounds when the accelerators are generating x-rays. A progress bar is another good example of feedback user interface. Imagine a dosimetrist (the person responsible for creating the treatment plan) trying to register a large data set using accurate software, but without a feedback element: the user interface only responses to user input when registration is complete. After beginning the registration process, the UI seems frozen, and there is no visual cue indicating progress. Imagine the dosimetrist's consternation when after 5 minutes nothing appears to have happened. Frustration follows: Is there a hardware or software problem? Should I restart the computer or relaunch the program? Why is it taking forever? In the ideal scenario, the software developer adds a progress bar to the same software, showing the dosimetrist real-time progress when the registration process is started. The dosimetrist can say to him/herself, "Since everything is working well and it will take 5 minutes, I will go out and get a cup of coffee." There is a clear difference made by a feedback interface in changing the users' experience in the execution of the desired action.

21.2.2 Components of the Treatment Delivery System

During radiation therapy treatment delivery, the RTT has to observe and manage an array of monitors, as shown in Figure 21.2. These monitors show information regarding the patient, the treatment delivery system, and the interaction of the patient with the system. Although the number of RTTs involved in treatment delivery is commonly greater than one [6], the number of monitors continues to increase to accommodate the growing number of software applications needed to treat patients. Below we will describe some of the information provided by these monitors and their interrelationship between the RTT, patient, and treatment delivery system.

21.2.2.1 Audio and Video

Radiation therapy treatment rooms are designed to treat the patient remotely. Although the radiation beam is collimated so the majority of the radiation (~98%) is directed toward the intended treatment volume, radiation scatter occurs, making it unsafe for RTTs to remain in the room during therapy. Consequently, the RTTs control the delivery from an adjacent space, separated by concrete walls approximately 2 meters thick.

During treatment delivery, the patient can be seen and heard at all times by the RTT. A two-way intercom system enables audio communication and a series of closed-circuit video cameras provides video monitoring of the patient. Usually, the patient is observed from three distinct directions, enabling continuous observation from at least two directions even when one of the cameras is obscured by moving components of the treatment machine. This allows the RTT to pause or terminate treatment due to patient request, distress, or general noncompliance. For example, patients with head-and-neck cancer are positioned lying flat on their back with a mask covering their head, jaw, neck, and shoulders. Many of these patients have difficulty swallowing and may begin to choke during treatment; this can pose a serious threat to the patient if not addressed immediately by the RTT.

21.2.2.2 Radiographic Imaging for Patient Positioning

Although the patient is initially aligned relative to the treatment beam by using marks on the patient's skin and laser lights within the room directed toward a known location, the internal anatomy moves relative to the skin marks on the surface. Consequently, x-ray imaging is frequently used to verify the positioning of the patient based on the internal anatomy prior to treatment delivery. The x-ray imaging approach may be an orthogonal pair of planar x-ray images or 3D image acquired using computed tomography. Common among these approaches, daily images of the patient are compared with reference images taken prior to treatment to determine the required adjustments to the patient's position. The software application used to perform this task typically interfaces with the image acquisition hardware, a Picture Archiving and Communication System containing the reference images, and the patient positioning table. The RTT must evaluate the application to determine (i) the adequacy of the acquired images, (ii) the acceptability of automated image registration or required adjustments, and (iii) the transfer of vector displacement commands to the patient positioning table.

Figure 21.4 shows the interface of an ImageGrid RT viewer, which is used to verify patient position by physicians. The interface is divided into four panels: (1) Panel A: A digitally reconstructed radiograph (DRR) from the treatment planning system. A DRR represents the ideal patient treatment position; the goal of patient position verification is to make sure daily portal image match the DRR. The green crosshair indicates the image center and serves as a reference line for patient shift evaluation. (2) Panel B: A portal image. Portal images are taken when a patient is set up on the treatment table. Physicians can then shift the portal image to match the corresponding feature in the DRR image. The shifts are shown on the panel D. (3) Panel C: Fused images. Different color schemes can be chosen to check the image alignment by overlaying the portal image on top of

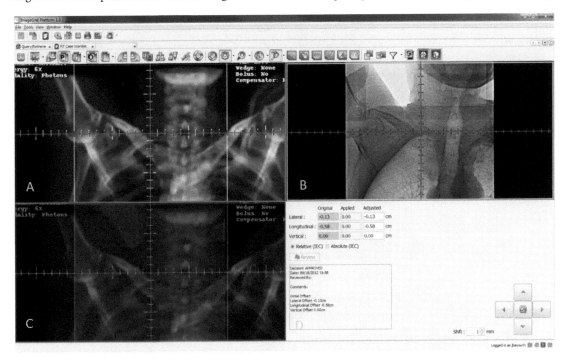

FIGURE 21.4 Patient position verification using portal images on ImageGrid RT Viewer (Siemens Medical Systems, Inc., Issaquah, Washington).

the DRR image. (4) Panel D: Shift reporting and portal image approval. The physician can approve or reject the shifts made by the therapists. The utility at the bottom right corner is to move the portal image a selected step size. Overall, this interface is intuitive; it makes the key elements (DRR and portal image) visible to the users. The layout of the DRR image and the portal image panel is consistent with the software interface physicians usually use at the treatment console, thus reducing the learning curve for the physician. The four-arrow adjustment panel is a very intuitive map: clicking the right arrow moves the portal image to the right, clicking the left arrow moves the portal image to the left. One important aspect of the interface is to provide feedback to the physician about the quality of the image alignment: is it a good fit? The current interface uses both crosshairs and color image fusion as the feedback mechanisms to evaluate image alignment. However, in some cases, applying a color scheme may not be the best method; the user experience may be improved if additional validation methods, such as a checkerboard image or mouse motion synchronization between the portal and DRR images, are available. Other ways to improve registration feedback are (1) displaying residual errors if auto image registration algorithms are available and (2) availability of statistical models to indicate the confidence level of shifts, which are compared with the historical data for a specific site.

21.2.2.3 Patient Positioning Table

During pretreatment imaging and treatment delivery, the patient is immobilized on a patient positioning table. Historically, these tables have a top parallel to the floor and are extended by a pedestal that accommodates movement in all three planes and rotations around the axis perpendicular to the floor (yaw). The range of motion for the patient positioning table is typically ±15 cm laterally, ±35 cm longitudinally, ±40 cm from the floor to the ceiling, and ±90° yaw rotation. Because the movement of the table is capable of causing a collision with the treatment delivery unit, great care must be taken to assure automated re-positioning of the patient is executed safely.

The current generation of patient positioning tables allow automated patient positioning with six DOF, adding rotations of pitch and roll. These tables may be designed to directly interface with the radiographic imaging system and apply the computed patient positioning shifts. Alternatively, the RTT may be required to manually enter the patient positioning table offsets into the software application, either from within the treatment room or remotely from the treatment control console. One of the many complicating factors for the RTT is the source of the shift information; this may be provided by one of several imaging systems, each with its own coordinate system. Consequently,

FIGURE 21.5 Control interface for Protura six DOF couch top.

the RTT may have to manually select the imaging system from which the shift information is being provided.

Figure 21.5 shows the control interface for a Protura, 6 DOF couch top. This touch screen–enabled user interface is clearly divided into four zones: (1) zone A shows patient-specific settings and parameters, including site name, field name, and coordinate system; (2) zone B is used for data entry, including current pedestal position and the desired shifts for the couch top; (3) zone C contains the "action" buttons, which send commands to the couch top to start or stop couch motion; and (4) zone D shows the status of the couch top: Is it moving? What is the current image-guided radiation therapy coordinate system? Does the control computer communicate with the couch top hardware? Good visibility makes it straightforward to operate the couch top. However, due to its need to interface with different coordinate systems from different imaging systems, understanding the direction of motion is somewhat challenging. Improvements could be made to optimize the mapping visibility by using a "preview" window, in which graphics or animations could be used to demonstrate the proposed motion with respect to current couch top position. This is similar to the "print preview" feature available seen in word processing software.

21.2.2.4 Respiratory Gating

Radiation therapy treatments from a linear accelerator usually take between 10 and 20 minutes. During this time interval, tumors of the thorax and abdomen move because of respiratory

motion. Several strategies exist to account for this motion during treatment, one of which is to turn the radiation beam on and off based on the patient's breathing pattern [7]. The treatment beam remains off until the patient reaches the end of the breathing cycle (end exhale), and then the x-ray treatment is delivered before inhalation begins.

The patient's breathing pattern is measured indirectly by a surrogate; one example is a pressure sensor within a strain gauge belt wrapped around the patient's thorax. As the patient breathes in, the tension on the belt increases and the amplitude of the signal measured by the pressure sensor also increases. This information is processed by a software application operated by the RTT. If the amplitude of the breathing falls within the window preset by the RTT, a signal allowing the radiation beam to turn on is sent to the linear accelerator. During treatment delivery, the software application monitors the patient's breathing pattern and starts or stops the radiation beam. The RTT monitors this process to verify the proper integration of the treatment delivery system with the patient's breathing. If this process is executed incorrectly, the tumor will not be treated correctly and regions of healthy tissue will receive damaging radiation doses.

Figure 21.6 shows the interface of the Anzai respiratory gating software. It includes all the necessary information for the gating application: (1) Respiratory traces. This is critical to respiratory gating. The wave chart makes it easy to distinguish between normal breathing waves (0%–100% amplitude) and abnormal ones. (2) Preset respiratory windows. These can also be visualized in a

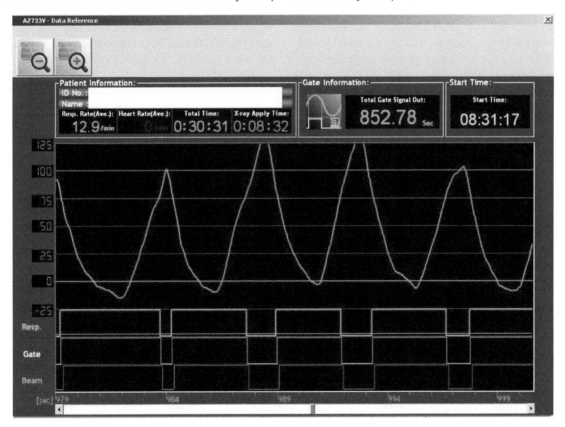

FIGURE 21.6 Interface to the Anzai respiratory gating system.

rectangular waveform, together with gating-on signal and beam-on signal. The real-time respiratory rate in the patient information panel provides an alternative way to check the consistency between the preset respiratory rate and the actual respiratory rate. The interface would benefit if the average respiratory rate and the trend of the respiratory rate were also displayed. Given this information, the RTT may be able to distinguish between a random respiratory rate change and a systemic one. One way to improve visualization is to directly display beam-on signal in the wave chart and provide users the feedback necessary to help them tell which part of the respiratory trace triggers the radiation beam-on signal. Better visualization will help users spot an unusual gating pattern and take corrective action.

21.2.2.5 Continuous Localization System

In addition to respiratory motion, other types of patient motion may occur during radiation therapy. For example, the internal anatomy may move due to motion within the digestive tract (peristalsis), beating of the heart, or patients may simply reposition their anatomy while lying on the patient positioning table. The RTT may monitor patient motion during treatment delivery using a continuous localization system, one that localizes the patient, tumor, or tumor surrogate at a frequency of ≥5 Hz and above. Failure to localize appropriately may result in the tumor being missed.

Many types of continuous localization systems are commercially available, relying on different technologies. One approach is to use light patterns to map the patient's surface and compare this with the expected or desired surface map. An example of a system that we use has two imaging pods along with their analysis software running on a standard personal computer with a two-channel frame-grabber card mounted in the treatment room. Each pod consists of two charge-coupled device (CCD) stereo cameras, a CCD texture camera, speckle flash with lens, a clear light flash without a lens, and a speckle projector. The two camera pods are mounted in the ceiling of the treatment room and each covers approximately 120 degrees of axial body surface

from the midline to posterior flank of the body. The stereo cameras acquire topological data of interest on the patient's surface. The system can be used in two principal modes: a single-frame mode and a continuous mode. In the single-frame surface acquisition mode, two flashes are illuminated, one with the speckle pattern and one without. The latter flash enables acquisition of the textured images. In the continuous mode, the speckle pattern is projected continuously while acquiring image data at a rate of 6.5 frames per second. The continuous localization interface displays both the patient's current surface map and the desired surface map to the RTT. These images are reduced to a single vector, providing guidance to the RTT as to what the required translation and rotation of the patient needs to be to realign them to the desired position.

Figure 21.7a shows the interface of an infrared marker localization system, in which infrared markers are used as a surrogate of patient motion. The most essential information, the translation and rotation of motion, is displayed using a large font size and is clearly visible from a long distance. However, because the infrared maker localization system does not communicate with the linear accelerators, the RTTs have to manually pause the beam if the motion is out of tolerance. The time delay in observing the abnormal motion to making a decision and pressing the pause button could be as long as 1 second. Assuming a high-dose delivery rate of 2000 MU per minute, the radiation delivered due to a human delayed response could be as large as 33 MU or approximately 6% of the daily dose for stereotactic body radiation therapy patients. Therefore, auto beam-hold is a highly desirable feature for a future continuous localization system. Our clinic also lacks a mechanism for exporting historical records from the localization system, making retrospective analysis difficult.

Another second example of a frameless continuous localization system is shown in Figure 21.7b. Unlike the infrared marker-based localization system, the AlignRT system does not require any markers, relying instead on patient surface information. Communication between the AlignRT system and the

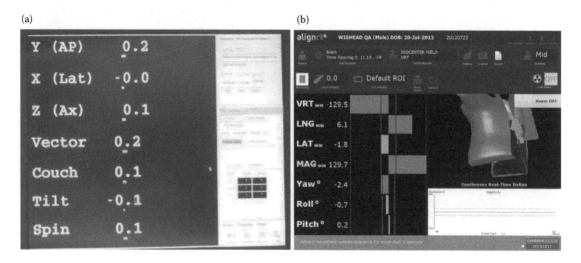

FIGURE 21.7 (a) An interface of an infrared marker localization system. (b) An interface to a surfaced based localization system.

linear accelerators enables it to display beam-on status and stop delivery of the radiation beam if the computed shift is outside of the tolerance. The UI is clear and legible; the color bars help to identify any out-of-tolerance motion. Because the accuracy of the surface-based localization system is strongly affected by the quality of the surface and registration convergence, this information could be displayed to further assist users in identifying the confidence level of the computed shifts.

21.2.2.6 Oncology Information System

Oncology Information Systems (OIS) contain the electronic radiation therapy record of the patient. This software application is also referred to as the "R&V" system. Although these software applications are capable of containing a complete medical record, we will limit our discussion to the elements being sent to and received from the linear accelerator for radiation treatment. Most radiation therapy treatments are executed using a computer-based treatment plan. Separate software tools are used to generate these treatment plans, which are then exported from the treatment planning system and imported into the OIS. The specific data elements of the treatment plan that are imported into the OIS include details regarding the dose, machine parameters, couch parameters, or other particular information for a particular treatment. These parameters are received from the

treatment planning system and stored in the OIS. At the time of treatment delivery, these parameters are transferred to the linear accelerator for execution of the treatment plan. Patients are usually treated once daily but for multiple sessions, with the same parameters; 10 to 35 sessions are common for many types of treatment. In addition to containing the intended treatment plan, the OIS stores or records the actual treatment parameters as delivered. Differences in these values may be the result of random variations in the way the patient is positioned on the treatment couch or minor differences in mechanical/directional settings for the beam. Tolerances for these differences are defined prior to treatment, and they too are stored in the OIS and verified for each beam prior to execution. For this reason, the OIS is often referred to as the "R&V" system.

An example of the OIS interface from Elekta is shown in Figure 21.8. The RTT selects the beams to be sent to the linear accelerator for treatment delivery from the patient's treatment calendar. Ease of reading this OIS interface is affected by the large number of parameters to be sent and monitored, but consistency in intuitive mapping helps the RTT to efficiently review parameters. In some instances, if the OIS identifies a deviation in the actual parameter setting greater than the tolerance value allows; the specific parameter will be highlighted with a bright yellow background and treatment suspended. Feedback for this application is clear

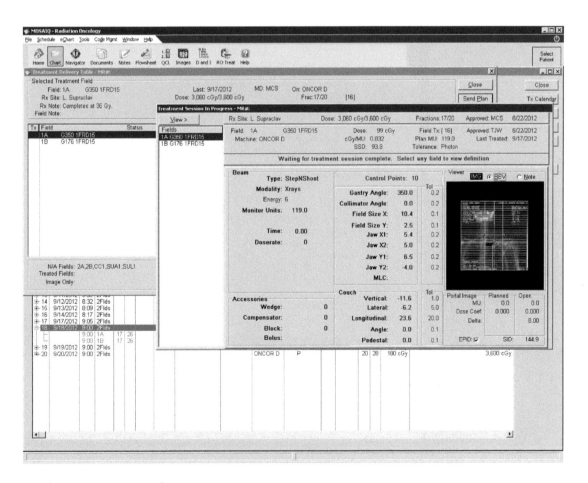

FIGURE 21.8 Elekta MOSAIQ OIS interface.

and definitive, but it does not happen in real-time. Information for all treatment fields within a plan are transferred from the OIS to the linear accelerator, and return information only arrives after all treatment fields have been delivered.

21.2.2.7 GUI for Treatment Delivery

The GUI for treatment delivery is a software application that serves as an interface between the OIS and the linear accelerator control console (described below). This system receives the treatment parameters from the OIS and displays them graphically for the RTT. Unlike the OIS, the GUI for treatment delivery is in continuous communication with the linear accelerator control console; if discrepancies are identified, they are displayed via this application to the RTT. An example GUI for treatment delivery is shown in Figure 21.9. This example shows a list of nine radiation beams (area A) that have been received from the OIS for the treatment of head-and-neck cancer. The graphic in area B shows the arrangement of each beam, with beams already delivered for treatment having a checkmark next to their label in

area A and displayed in blue in area B. The beam displayed in orange is the beam currently being delivered, while the beam in yellow is the beam yet to be delivered. Because this treatment is an intensity-modulated radiation therapy (IMRT) treatment plan, each beam has multiple "segments" or "control points." Each segment has a unique amount of radiation to be delivered determined by MUs, which are shown in area C. This example has seven segments within the current beam (1B, g250) and the third segment is being treated. This segment requires delivery of 69 MU, and at the time of image capture, 21.7 MU of the 69 MU had been delivered. This is also shown graphically by the orange progress bar next to the label "Beam MUs." In addition to a segment being unique in MU, the field shape of the segment is also unique and is created by the 100+ leaves of the MLC. The graphical display of the arrangement of those leaves is shown in area D.

The GUI for treatment delivery is designed to improve visibility of treatment parameters as well as serve as a good conceptual model that can clearly inform the RTT of deviations using both graphics and text. The layout of information is well arranged and allows the

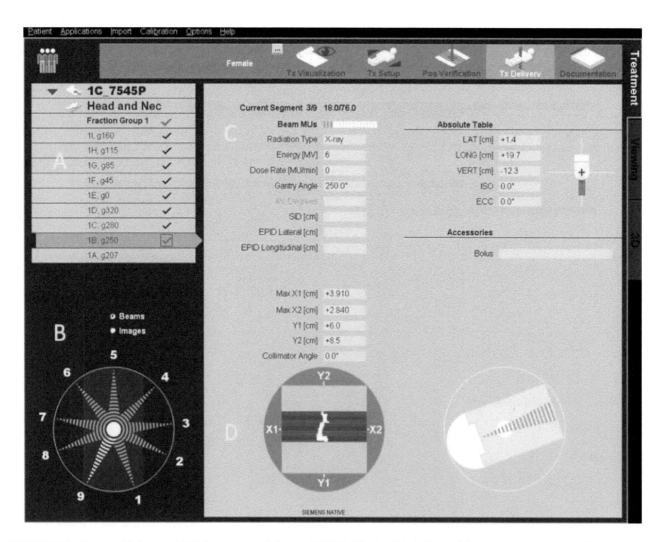

FIGURE 21.9 Siemens "Coherence" GUI for treatment delivery—RTT 4.2. Displayed is the beam delivery sequence for a nine-beam treatment delivery of IMRT.

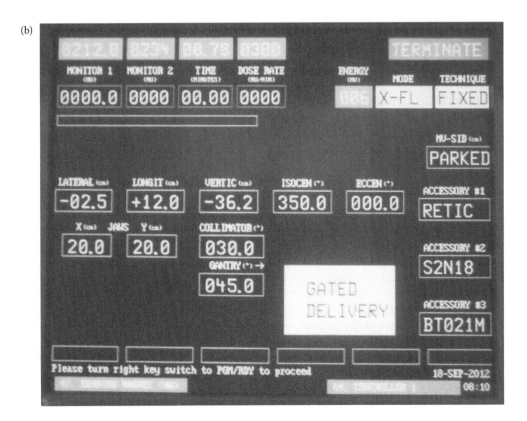

FIGURE 21.10 (a) Elekta linear accelerator treatment console. (b) Siemens linear accelerator treatment console. (c) Varian linear accelerator treatment console.

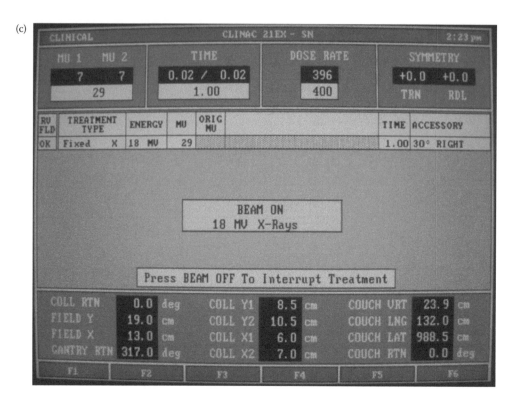

FIGURE 21.10 (Continued) (a) Elekta linear accelerator treatment console. (b) Siemens linear accelerator treatment console. (c) Varian linear accelerator treatment console.

RTT to display all required treatment parameters for each segment of each beam. Feedback is in real time. This application can also extend the functionality of the OIS to enable additional treatment parameters to be included for a treatment plan. For example, the required use of gating the x-rays on or off, based on surrogates from the patient due to respiratory motion or changes in positioning, can be stored in the GUI for treatment delivery application.

21.2.2.8 Linear Accelerator Treatment Console

Linear accelerators manufactured today cannot operate without computer control. Scores of commands are executed to prepare a linear accelerator for beam generation and delivery, requiring precise integration and simultaneous monitoring of multiple systems. Some examples of the major components include the source of the microwave power, the electron gun that provides electrons to be accelerated, the bending magnet that directs the beam, the water cooling system, and the primary electrical power distribution system. Each of these has multiple subsystems, all of which are being monitored and automatically adjusted by the control system. The interface between this control system and the RTT is called the linear accelerator treatment control console. This console has several modes of operation, each with a unique interface, depending on the intent of the RTT. Routine operation assumes treatment parameters are downloaded from the OIS to the linear accelerator via the GUI for treatment delivery (described above). A manual mode of operation also exists, enabling direct data entry from a keyboard for a small subset of parameters. Finally, a service mode is available to allow

technical service personal and medical physicists to investigate and, if necessary, adjust operational parameters.

Figures 21.10a–c show examples of linear accelerator treatment consoles for three separate manufacturers. These systems show substantial variation in size of text and graphics, contrast, information layout, and the amount of information being displayed. These systems have varying degrees of legibility. The conceptual model lends itself to consistency in operation and is effective in communicating to the RTT that the system has aborted operation due to the detection of inconsistencies. These systems often struggle to provide an adequate description of the inconsistencies to identify the root cause. This is a common frustration for the RTTs that imply the need for an improvement to the conceptual model. The control consoles tend to demonstrate good intuitive mappings, where intervention by the RTTs results in anticipated changes. Feedback from these systems is generally excellent under normal operation, limiting the volume of information to only that which is required for the RTT to verify treatment delivery is consistent with the intent. Feedback is limited, however, when the linear accelerator halts operation due to a detected fault. An example is seen in Figure 21.10b. Here, the system clearly indicates the radiation beam has stopped by displaying the word "TERMINATE" in the blue box in the upper right-hand corner. At the bottom of the screen are two red boxes that indicate the cause for, or are an additional consequence of, a system failure. The two messages, "BENDING MAGNENT" and "CONTROLLER 1," lack any useful information to the inexperienced RTT.

21.3 Conclusions

Radiation therapy is a complex, technically driven subspecialty of medicine that requires effective interaction between human operators and multiple integrated computer-controlled systems. In this chapter, we have introduced several example components of the treatment delivery system and explained how they interact with one another, the patient, and the human operator. Just as with other industries that require humans to manage information overload, a well-designed HCI having good visibility, conceptual model, mapping, and feedback is essential for safe, effective, and accurate radiation therapy.

References

1. Greene, D. & Williams, P. C., *Linear Accelerators for Radiation Therapy*. 2nd ed. Medical Science Series 1997, Bristol, UK; Philadelphia: Institute of Physics Pub. xvii, 268 pp.
2. Chang, Z. et al. (2011). Evaluation of integrated respiratory gating systems on a Novalis Tx system. *Journal of Applied Clinical Medical Physics, 12*(3), 3495.
3. Lovesey, E. J., Information overload in the cockpit, in *IEEE Colloquium on Information Overload*, 1995, 5/1–5/5.
4. Leveson, N. G. & Turner, C. S. (1993). An investigation of the Therac-25 accidents. *Computer, 26*(7), 18–41.
5. Norman, D. A., *The Design of Everyday Things*. 1st Basic paperback. ed. 2002, New York: Basic Books. xxi, 257 pp.
6. Zietman, A. L. et al., *Safety Is No Accident: A Framework for Quality Radiation Oncology and Care*, American Society for Radiation Oncology, 2012.
7. Jiang, S. B. (2006). Technical aspects of image-guided respiration-gated radiation therapy. *Medical Dosimetry, 31*(2), 141–151.

Informatics for Image-Guided Radiation Therapy

Peter Balter
University of Texas MD Anderson Cancer Center

George Starkschall
University of Texas MD Anderson Cancer Center

22.1 What Is IGRT?

Image-guided radiation therapy (IGRT) is a broad field (Verellen et al. 2008; Chen et al. 2009) but will be simplified for this discussion. For the purpose of this work, we will define IGRT as the use of radiographic imaging, either projection or volumetric, to align the patient with the radiation isocenter prior to each treatment. This definition neglects some important developmental technologies such as 3D surface optical imaging (Vision RT Ltd., London, UK) and radiofrequency beacon fiducial tracking (Calypso Medical Technologies, Inc., Seattle, Washington), but many of the informatics issues encountered in radiographic IGRT are also applicable to nonradiographic technologies.

Treatment setups in IGRT are based on two types of imaging, projection imaging and volumetric imaging. Projection images are 2D images. These images can be real images, in which a source of x-rays passes through the patient to a detector, or they can be virtual images, in which 2D images are calculated based on attenuation of a photon beam through a 3D computed tomography (CT) data set. Volumetric images are 3D images, usually CT images, although they could be reconstructed images from other modalities such as positron emission tomography (PET) or magnetic resonance imaging (MRI).

Projection imaging has been used for verification of patient setup since the advent of radiation therapy (Munro 1995; Dong & Boyer 1996; McGee et al. 1997). Projection imaging was typically done by acquiring a megavoltage portal image, in which the beam portal and the surrounding anatomy were imaged. Based on the image, the radiation oncologist could verify that the correct anatomy was contained within the portal and that critical structures were excluded. The acquired image was displayed on film and was compared with a reference image, also on film, acquired on a radiation therapy simulator that mimicked the geometry of

the radiation therapy delivery machine. With the advent of CT simulators, the film reference image was replaced by a digitally reconstructed radiograph (DRR) (Sherouse et al. 1990; Galvin et al. 1995) that was numerically produced from the simulation CT image set. Although these DRRs could be printed on film and used as a comparison for the acquired images, they were intrinsically digital images. The next step forward was to digitize the acquired portal radiographs and compare them directly with the DRRs in a patient management system.

In recent years, 3D conformal radiation therapy (3DCRT) (Armstrong et al. 1993; Purdy 1999) and intensity-modulated radiation therapy (IMRT) have made the concept of portal images less useful, as the oblique angles and small multileaf collimator (MLC) openings make it difficult or impossible to verify and correct setup based on the beam portal. For these cases, orthogonal images of the isocenter location are used rather than portal images. Simultaneous with the widespread use of digitizers was the adoption of electronic portal image detectors (EPID) (Munro 1995) and photostimulable phosphor plates (PSP) (Shalev et al. 1989) as the image receptors, allowing direct digitization of the acquired radiographs.

A major issue in communicating image information is obtaining and transferring the scaling of the image. The method for obtaining this scaling information depends on the method by which the projection images are acquired. Planar radiographic images can be acquired digitally using either computed radiography (CR) plates placed at arbitrary locations with respect to the beam or with EPIDs mounted on a stable arm at known orientations with respect to the treatment beam. If an EPID is used, its location with respect to the beam, the location of the isocenters with respect to the array of image values, and the magnification of the image with respect to the plane of the isocenters should be reported to the patient management system.

It is the responsibility of the physicist at each site to verify that these values are reported properly and give the proper orientation and scaling. For CR-based detectors, the scale and location are typically embedded into the image, as was done with film, using a mechanical graticule placed in the treatment beam prior to imaging. If this approach is used, a system should be in place to then localize and scale these images in the patient management system. This is often done by placing and then scaling an electronic graticule on top of the one embedded in the image.

A special case of projection x-ray imaging uses a room-mounted orthogonal x-ray system both to set up the patient initially and then to monitor the target during treatment (Chen et al. 2009). An example of such a unit is illustrated in Figure 22.1. These systems, as currently implemented, operate outside of the patient management system. They receive a separate copy of the patient setup data from the TPS and either give instructions to the therapists as to shifting the patient or they may control an add-on to the treatment couch that can perform small corrections from translation and rotation. These systems use pairs of 2D x-ray images for set up but use 3D volumetric data as the reference, with an iterative fast system for generating DRRs that can generate new reference images on the fly to find the best match for translation and rotations. These systems require that the reference data be downloaded and prepared prior to the treatment. Care must be taken with these systems to ensure the patient and plan data used for localization matches the data used for treatment.

Projection imaging has many limitations and, in general, is not sufficient to perform high-precision setup, with the exception of situations when radio-opaque targets, such as the spine, are imaged. Because of this limitation, volumetric imaging is also available from most manufactures for daily setup. Volumetric imaging has become common in treatment planning and target definition for radiation therapy.

Volumetric imaging information is currently available mainly based on CT. These can be acquired in a traditional CT, slit-scan, geometry using either a CT-on-rails system (Uematsu et al. 1996) or the treatment beam in a helical tomotherapy system (Mackie et al. 1993). CT scans also be acquired in a cone-beam geometry using a conventional accelerator gantry to rotate an x-ray source and detector around the patient images (Jaffray & Siewerdsen 2000). The x-ray source can be a kilovoltage, diagnostic, x-ray tube or can be the megavoltage treatment beam.

As was the case with projection imaging, location information from the volumetric imaging needs to be associated with the image data set. If the CT data set is acquired using the accelerator gantry, then the localization of the data set with respect to the treatment isocenters should be well known. The position of the isocenters, the orientation of the patient, and the pixel sizes must be transferred to the patient management system along

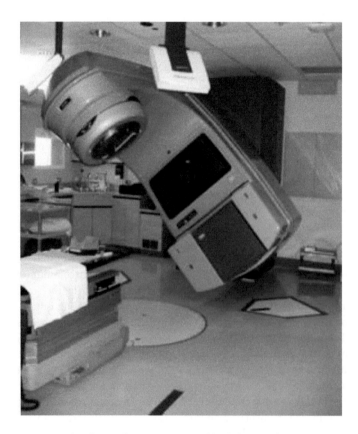

FIGURE 22.1 An example of a room-mounted orthogonal x-ray system used both for initial patient setup and for on-treatment monitoring of patient positioning.

with the acquired image data set. This can be done with a spatial registration object (SRO). If the CT data set is acquired using an independent gantry, then some system must be in place to perform this localization; this is often done with imageable fiducials placed at known locations with respect to the accelerator isocenter. As with projection images, it is the responsibility of the physicist to ensure that the relationship between acquired CT images and the treatment isocenter is correct.

Volumetric imaging has also been done using B-mode ultrasound units linked to the machine isocenters. These units currently require a separate transfer of the plan data to the local system and do not interact with the patient management system. These units are largely being replaced by the integrated imaging capabilities of the linear accelerators.

22.2 How Is IGRT Informatics Different from Non-IGRT Informatics?

One way in which IGRT informatics is different from non-IGRT informatics is the amount of data required. A traditional record of a treatment fraction involves storing the linear accelerator settings such as monitor units, field sizes, couch positions, etc. (Figure 22.2). These data constitute less than 50 values in addition to space for comments from the radiation therapists; this amounts to less than 1 kB per treatment per day. If IGRT is used, resulting in a pair of medium resolution orthogonal x-ray images added to these data ($1024 \times 768 \times 2$ bytes per pixel), this is increased by 3 MB. In practice, images acquired both before and after isocenter shifts are often acquired, resulting in an additional 6 MB per treatment. If cone-beam CT (CBCT) is used, the data requirements are larger still at 32 MB per treatment. If we assume 40 patients per day, each linear accelerator will generate approximately 61 GB per year of imaging data if orthogonal x-rays are used for set-up on all patients or 325 GB per year if CBCT is used for setup on all patients. In practice, a mix of CBCT and orthogonal x-ray images is often used with the data storage requirements somewhat between these values. With high-quality, enterprise-level storage costs currently between $2

and $5 per GB, with the price decreasing rapidly, the cost of storage is negligible. The only limitation becomes the network bandwidth and the stability of databases containing large objects.

For IGRT to be used, the reference data must be present at the control console for the radiation therapists performing the treatment. Ideally, acquired images, as well as the reference data, should also be available for real-time review by physicians in their offices or at remote locations. For pairs of 2D planar images, adequate infrastructure currently exists to accomplish these goals. However, movement and display of volumetric imaging drains resources. The short-term solution to the long download time of reference data from a data server to the viewing console is to cache a local copy of the reference data at the treatment workstation. If, due to scheduling or maintenance issues, the patient is treated on another machine, a second copy must be cached at another machine. Unfortunately, current patient management systems do not cache copies of the reference images at review workstations so, for the review to occur, both the reference data and the acquired data must be loaded prior to review; the time required for loading makes remote review of volumetric data while the patient is in the treatment position difficult. At least one vendor is currently developing a system to preprocess the images on a back-end server and send reconstructed planar images on-the-fly to review workstations; this type of implementation allows optimization of hardware graphic processors, without forcing excessive requirements on individual workstations as well as reduces the data bandwidth from the server to the review workstations.

Another challenge in the implementation of IGRT is to develop and deploy tools that will allow the efficient comparison of acquired images to reference images. Currently, each vendor restricts review tools to those that they supply. If an institution is using a patient management system from a different vendor, then any remote review tools will have a different look and feel from those on the console. In addition for both 2D and 3D images, the simultaneous use of remote and local review tools is often restricted by a data flow model that assumes remote review will only be done after local review is accomplished. In practice, this may or may not be true. One workaround is to use remote

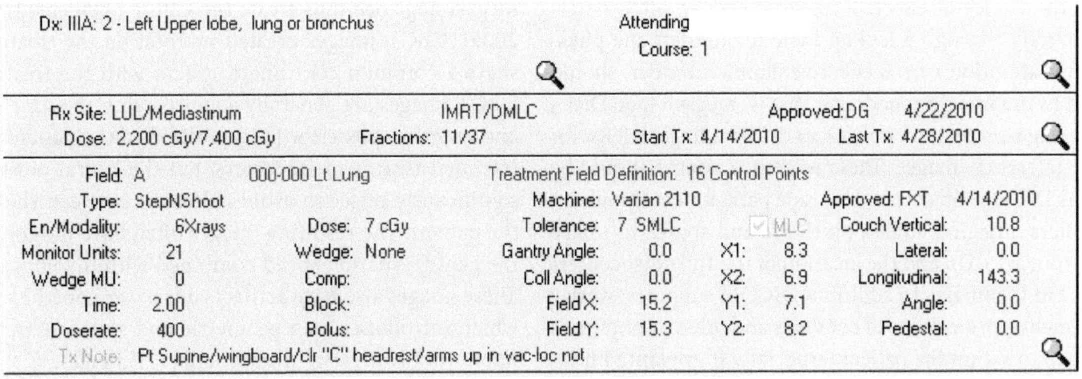

Dx: IIIA: 2 - Left Upper lobe, lung or bronchus						Attending Course: 1			
Rx Site: LUL/Mediastinum			IMRT/DMLC				Approved: DG	4/22/2010	
Dose: 2,200 cGy/7,400 cGy		Fractions: 11/37			Start Tx: 4/14/2010		Last Tx: 4/28/2010		
Field: A	000-000 Lt Lung		Treatment Field Definition: 16 Control Points						
Type: StepNShoot			Machine: Varian 2110				Approved: FXT	4/14/2010	
En/Modality: 6Xrays		Dose: 7 cGy	Tolerance: 7 SMLC		☑ MLC		Couch Vertical:	10.8	
Monitor Units: 21		Wedge: None	Gantry Angle: 0.0	X1:	8.3		Lateral:	0.0	
Wedge MU: 0		Comp:	Coll Angle: 0.0	X2:	6.9		Longitudinal:	143.3	
Time: 2.00		Block:	Field X: 15.2	Y1:	7.1		Angle:	0.0	
Doserate: 400		Bolus:	Field Y: 15.3	Y2:	8.2		Pedestal:	0.0	
Tx Note: Pt Supine/wingboard/clr "C" headrest/arms up in vac-loc not									

FIGURE 22.2 An example of information recorded for a typical radiotherapy fraction.

review tools at the console so that the radiation therapists and the physician can simultaneously look at the images.

One potential source for error in image review is communicating desired shifts to the treatment machine. When using tools at the treatment console, this is not a concern as the couch corrections are automatically computed and applied. Remote review is more challenging as shifts are often transmitted back to the radiation therapists in terms of the isocenter, which is opposite from the patient (couch) shift. This also requires manual entry of the shifts, which can always be a potential source of error. Ideally, remote review tools would allow the operators to shift images of isocenters and report the results as new target couch coordinates, and these would be automatically transmitted back to the treatment machine.

22.3 Details of IGRT Information Flow

The majority of IGRT is based on CT data sets acquired for the purpose of radiation therapy planning. These data sets are used by a treatment planning system (TPS) to develop a model of the patient that can be used to position beams and determine required settings on the linear accelerator to achieve the desired dose distributions. Each radiation therapy plan is based on a primary data set with a unique Digital Imaging and Communications in Medicine (DICOM) identifier. Structures and points of interest such as isocenters are created in the 3D coordinate system defined by the primary data set. IGRT is based on moving either the primary data set or the associated objects or derived (simplified) versions of these data to the treatment machine to serve as a reference to align the patient.

The most common method of generating images to support IGRT is still megavoltage portal imaging, usually done on a weekly basis. However, this is not generally what is referred to as IGRT; the standard method of IGRT is based on the acquisition and review of orthogonal pairs of kilovoltage images. In either case, these acquired images are compared to reference images, typically DRRs derived from the planning CT data set by the TPS. Care should have been taken to generate DRRs of sufficient spatial resolution and bit depth to enable comparison with daily images at the precision required to verify the boundaries of the planning target volume (PTV) (ICRU 1993). It should also be noted that the choice of effective imaging energy, often determined indirectly through a lookup table to simulate the physics of photon attention versus effective atomic number, should be matched to the imaging modality, that is, megavoltage DRRs for megavoltage projection images and kilovoltage DRRs for kilovoltage reference images. These reference images should be generated as DICOM objects and include patient demographics, field identifiers, imaging angles, pixel size and source to image receptor distance (SID), and the location of treatment isocenters with respect to the image. In addition, DICOM supports overlay structures such as projections of contours and other annotations that may help to set up the patient, especially if implanted fiducials are used. As an alternative, these overlays can be burned into the pixel matrix, but this is less desirable as the annotation

cannot be turned on and off by the operator and can interfere with automatic matching systems.

A special case of projection imaging exists in which room-based orthogonal oblique images acquired during treatment. These types of systems are available from two manufacturers (BrainLAB AG, Feldkirchen, Germany, and Varian Medical Systems, Palo Alto, California) and include integrated x-ray systems and analysis software. A 2D–3D matching algorithm is used where DRRs are generated on the fly to best match the acquired images to projections through the patient. The calculated corrections include both translations and rotations. If these systems are used with a six degrees-of-freedom couch, these corrections can be quickly applied and verified. The oblique images generated from these systems are often difficult to interpret; thus, corrections cannot be determined without computer assistance. These systems also tend to have small fields of view because the detectors are far from the isocenters, and because the images are projection images, they often cannot resolve soft-tissue targets. Current implementations of these systems do not integrate with patient management systems and the setup information and reference images must be transferred separately to the accelerator control system and the imaging system. Care must be taken to ensure the data sent to each system is consistent.

Daily volumetric matching is becoming a common method of IGRT. Volumetric imaging offers a few distinct advantages over projection imaging. Soft-tissue targets that are not visible on projection images, such as many tumors, can be used for direct setup, and volumetric images are often easier and faster to interpret than projection images. Downsides to the use of volumetric imaging include the fact that imaging dose is much higher than that of projection kilovoltage imaging (Murphy et al. 2007), although it may be lower than that of projection megavoltage imaging, and the acquisition and processing times are longer than those for projection images, but this may be offset by the lower time required to interpret the images.

The general process of daily volumetric imaging is to use the gantry of the linear accelerator as an imaging gantry to rotate an x-ray source and a flat-panel detector around the patient to create a CBCT data set. The most common x-ray sources are kilovoltage sources mounted orthogonal to the treatment beam (Varian, Elekta), but at least one manufacturer uses the megavoltage beam to create the CBCT (Siemens) (Pouliot et al. 2005). CBCT images created by rotating the treatment gantry share a common coordinate system with the treatment beam. CBCT images are generally acquired with the superior–inferior and anterior–posterior (AP) position of the patient close to the intended treatment isocenters, but the lateral position is often significantly offset to avoid collisions between the gantry and the patient. The resulting images often have motion blur due to the gantry rotation speed combined with involuntary motions. These images also have artifacts due to the cone-beam geometry, which introduces both geometric and scatter artifacts that are not normally present in traditional CT images.

Another currently available, but not widespread, method for in-room volumetric imaging is the CT-on-rails (Barker et al.

2004). This system consists of a diagnostic CT unit mounted in the treatment room on a rail system that translates the CT gantry rather than the patient to accomplish the scan (Figure 22.3). This configuration allows the patient to be scanned on the treatment table in the treatment position. Generally, the treatment couch is rotated 90 or 180 degrees from the treatment position and the CT is then moved over the patient. These systems provide diagnostic quality images with large fields of view and scanning times of 10 to 15 seconds for the entire treatment areas with little or no motion artifacts. Unlike the CBCT systems, they do not share a gantry with the treatment beam; thus, some system must be created to transfer the isocenter coordinates into the CT imaging geometry. The lack of integration of the CT-on-rails to the linear accelerator also extends to the setup software, and in-house third-party systems must be used to calculate couch shifts. This is being addressed by one manufacturer but is just becoming available.

In radiation treatment planning, the patient is represented by a single volumetric CT data set. This CT data set has objects associated with it such as radiation plans and structures (contours). Additional volumetric data sets belong to the plan that belongs to the primary CT.

The majority of radiation treatment planning is CT based. Multiple CT data sets may be imported into the TPS for planning and target delineation, but all radiation treatment plans will have a unique planning data set on which the geometry is defined and the dose is calculated. It is easier to understand the linked data objects by considering that the CT has a plan rather than the plan contains the CT. The TPS does not modify this data set but does create other DICOM objects that are referenced to this data set. These objects include contours (RT-STRUCT) and plans (RT-PLAN). Other DICOM objects can also be derived from these data sets such as DRRs (RT-IMAGE). These objects are transferred from the TPS to the record and verify (R&V) system to be used for positioning the patient in the correct location with respect to the treatment beam and to set the parameters on the linear accelerator for each field. If volumetric setup methods are to be used, the original CT data set should also be sent to the R&V system for transfer to the volumetric 3D–3D matching system.

R&V systems typically have a method to import the images and structures and associate them to both a patient and a treatment site. Special care should be used be taken when a patient has multiple treatment sites or multiple versions of a plan to ensure the proper isocenter information and the proper CT and/ or DRRs are present and associated to the correct site. This is further complicated by the fact that not all planning systems use DICOM coordinates when specifying isocenter locations; thus, the isocenter point in the TPS may be different from that displayed by the TPS. One of the TPS systems has the coordinates matched if the patient is head-first prone, but coordinate conventions are different if the patient is in other orientations making verification difficult (Figure 22.4). A newer release of this planning system now displays the DICOM coordinates but only in one of the many viewing windows.

Volumetric image-based setup is still not the most common method of daily patient alignment; skin mark-based setup

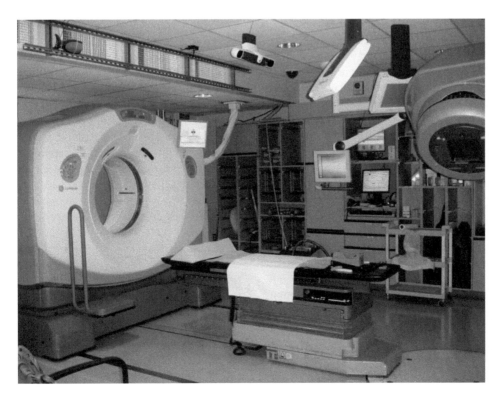

FIGURE 22.3 A CT-on-rails mounted in the treatment room.

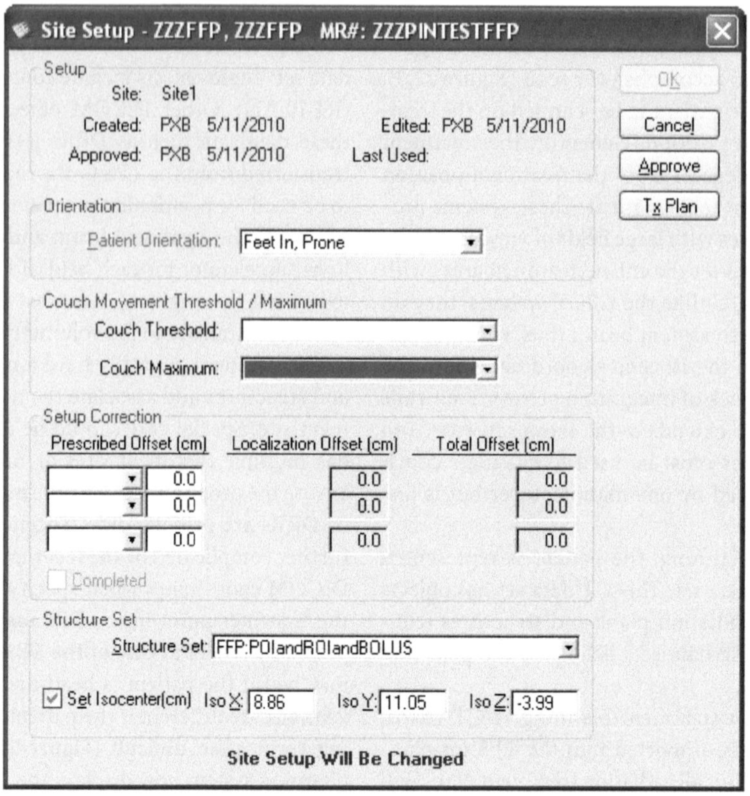

FIGURE 22.4 Defining coordinate conventions in an R&V system.

verified by projection imaging is. This model originated from traditional planning techniques in which a simulation film was acquired and was used for planning and then as a reference for setup verification. CT-based planning and 3DCRT techniques, including IMRT, do not have planning films or easily verifiable field shapes; for these cases, reference fields have replaced the beam ports and DRRs have replaced the simulation films. The reference fields are generally orthogonal projections, AP and lateral, through the isocenters that easily allow shifts to be computed to correct setup errors.

DRRs are, in general, created by the TPS. A true DRR would mimic the physics of x-ray projection imaging including the atomic number dependence on x-ray attenuation and the large scatter component in large field imaging. To simplify the problem, scatter is ignored and the energy and atomic number dependence of the x-ray attenuation is replaced by a lookup table that maps attenuation to electron density to achieve the desired look of the imaging x-ray (Figure 22.5). At least one manufacture allows the lookup table to be created on-the-fly to emphasize ranges of values (Figure 22.6). The end user of the DRR can adjust brightness and contrast (window and level) but cannot adjust the energy. The DRRs should be generated at a bit depth of 12 or higher (4096 gray levels) to allow the end user the ability to adjust the display to match the acquired images and/or to allow automated matching systems the greatest amount of data. The DRR should be generated as a DICOM RT-IMAGE object

including scaling and a referenced to the treatment isocenter. Ideally, annotation should not be burned into the pixel data but rather included as DICOM tags and should be turned on and off by the user as needed. At this time, not all systems include tools to include annotations as overlays or to display them so, for the time being, we must accept them burned into the pixel data. Each DRR is created based on either a treatment or a reference field, and in the ideal world, the DRRs and fields would automatically associate with each other when imported into the R&V system; this is currently not supported by all systems and can introduce a significant error if done wrong by the require manual association. An important part of pretreatment quality assurance is assuring these associations are done properly.

We have discussed that reference images and objects are required for IGRT. The primary object is the planning CT, but the planning CT alone is not useful without other objects generated by the TPS. The simplest, but most important, object is the isocenter coordinate in imaging space. This along with the patient orientation is required for 3D–3D matching and can be automatically transferred from the TPS to the R&V system and then onto the treatment console. These coordinates are not stored as part of the treatment fields but rather as part of the treatment site, which then contains the fields. This assignment of isocenter coordinates creates the possibility of fields and treatment sites being mismatched. Useful, but not required, objects for 3D–3D matching are contours of critical and target structures. These

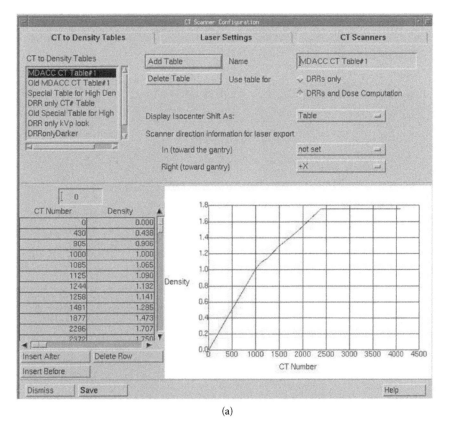

(a)

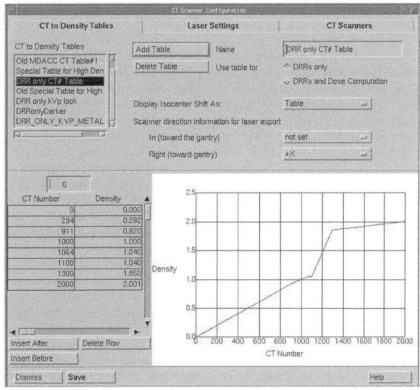

(b)

FIGURE 22.5 (a) CT number to density conversion table for dose calculations. (b) CT number to density conversion table for DRR generation.

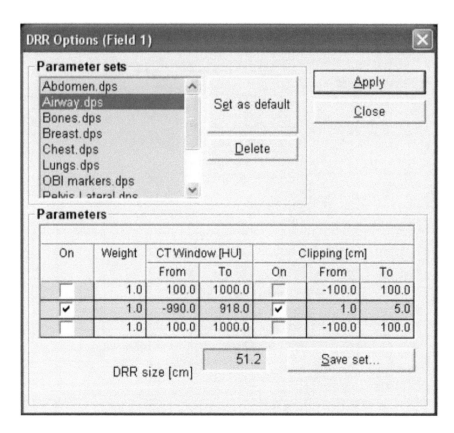

FIGURE 22.6 CT number to density conversion table that can be changed on the fly.

can be used to create a region of interest for automated matching algorithms, can be used to perform or verify an alignment by the operator, and can be used in evaluating changes in the patient anatomy over the course of treatment. 2D–2D matching requires that the TPS generate DRRs; these DRRs should be DICOM objects with appropriate bit depth, scaling, and localization information. They should be, but usually are not, automatically associated to treatment or reference fields. They should also be of sufficient quality and have the appropriate effective photon energy to reduce errors in the 2D–2D matching process.

Tools are required at the treatment console for comparing acquired images to reference images. These tools may be incorporated into the treatment console, may be incorporated into the R&V system, or may be third-party tools that provide functionality not available in either. Whatever tool is chosen should share several characteristics: (1) It should allow review immediately after image acquisition, (2) it should receive all data in DICOM format, (3) it should report shifts in couch coordinates to avoid possibilities of misinterpretation of directions, and (4) it should allow for remote, real-time, review as radiation oncologists are often not physically present at the treatment console but are often required to approve setup images prior to treatment.

2D–2D matching tools are available from multiple vendors; the simplest version displays a scale and isocenter indicator, that is, a graticule, on both the reference and acquired images. The operator uses these scales and the underlying anatomical

information to determine patient shifts required to align the patient in the planned position. More sophisticated tools have additional functions, such as "magic window," split screen, and/or color blend functions that allow the operator to view various mixes of the reference and acquired images to accomplish the match. In addition, most systems offer some type of auto-alignment system using mutual information (Viola & Wells 1997) or cross-correlation algorithm (Murphy 1997). The limited information and poor quality in 2D projection images generally limit the usefulness of automatic matching algorithms for 2D–2D matching. 2D–2D systems can, in general, be used to calculate shifts in five dimensions: three translational and two rotational. Patient rotation in the directions perpendicular to the incident x-ray beams can be determined but only the one along the direction of the couch rotation can be easily corrected. In practice, only the three translation directions in the patient are generally used. 2D–2D matching is generally available both at the console and on the R&V system for immediate review locally and for distant review.

3D–3D matching tools are also available from all vendors who supply in-room volumetric imaging. These are similar to the 2D–2D matching tools, with the exception of structure overlay being an import feature in 3D–3D matching, as it allows the operator both the ability to set up the patient based on previous contoured structures and the ability to estimate changes in the patient anatomy subsequent to the time of simulation. 3D–3D matching is superior to 2D–2D in that soft-tissue targets can be directly visualized without fiducials, six degrees-of-freedom can easily be calculated, and there is less ambiguity in the matching as much more information is available. 3D–3D tools, however, do require more time to acquire and reconstruct images than do 2D–2D tools, require more time to load the patient's reference data to the treatment system, and require a higher imaging dose to the patient. The first three of these concerns will likely be solved by technological improvements, but the higher imaging dose is fundamental to the imaging process and will be needed until nonionizing radiographic imaging becomes available in the treatment room. Remote review is currently a challenge for 3D–3D matching as the acquired CT data (30 MB) must be moved from the console to the patient management system and then to the remote view station, which must also receive the much larger (100+ MB) reference data and contours. A solution to this problem will be backend processing of the 3D data and just sending views on the fly to the remote workstation, but this is not available at this time. An additional problem, which is not an informatics problem, is that the time to review a 3D data set is greater than that to review 2D images, and a radiation oncologist with 40 patients under treatment may not be able to review 3D data sets on all the patients each day.

22.4 IGRT Data Storage Requirements

One consideration in the implementation of IGRT is the requirement for data storage. Reference CT image data sets require approximately 100 MB for the images plus an additional 5 to 15 MB for the associated structures. Acquired daily CT images range in size from 15 to 30 MB depending on the reconstruction options. If these images are acquired daily for 35 fractions, the data storage requirement will be 1 GB; a linear accelerator treating 40 patients per day will generate approximately 300 GB per year of CT data plus 30 GB of reference data. Projection images used in radiation therapy are usually of lower resolution than those for diagnostic imaging ($1024 \times 768 \times 2$ bytes = 1.5 MB per image), and two to four of these images are acquired per patient per day (if CBCT is not acquired); this will result in 60 MB per machine per year. In reality, a mix of CBCT and projection imaging will occur on most machines, resulting in a storage requirement between 60 and 360 GB per year. Fortunately, because the surge of IGRT began approximately 5 years ago, these data storage requirement have become somewhat trivial and the limitations become managing the TPS database with these large objects. In addition, better database architecture in the next few years should help address this data set size issue. In the meantime, at least one manufacturer (TeraMedica, Inc., Milwaukee, Wisconsin) has created a system to archive the large objects into a traditional Picture Archiving and Communication System (PACS) at the conclusion of patient treatment; the ideal architecture would be to have all the objects in the PACS and only store references in the R&V system. This, however, does not facilitate rapid review tools as it limits preprocessing by the R&V system.

TCP/IP communication over 100 Mbit/s Ethernet has become the standard for data communication in most hospitals. A 1.5 MB (1.3×10^7 bits) image should be transmitted in 0.12 seconds; in reality, this is not the case as some of the bandwidth is spent on overhead for the physical and software layers. In addition, the data path often requires multiple sends of the same image from the acquisition system to the local R&V workstation to the R&V server and then to the remote review client. Also, TCP/IP networks rarely operate near the rated speeds. In addition, the real need for remote review is often for off-site review over consumer Internet feeds that rarely achieve 10 Mbit/s or even mobile feeds (0.3 Mbit/s for 3G). Luckily, as with storage, these problems seem to be correcting themselves as technology improves (1 Mbit/s over Ethernet has been available for a number of years and 4G wireless offers a theoretical 70 Mbit/s). In addition, server-side processing, or the use of remote servers with the client only serving as a terminal, further reduces the required bandwidth for remote review. However, it should be noted that, in the IGRT world, treatment machines cannot be used effectively outside of an R&V environment, and the bandwidth from the R&V system the treatment machine must be maintained. The only effective way to accomplish this is through proper network architecture that separates or prioritizes data packets between key devices such as the R&V system and the treatment console.

22.5 QA for IGRT Informatics

Informatics systems for IGRT enable efficient and precise treatments but can also introduce large errors into the treatment process. Centers performing IGRT need to include informatics testing of their IGRT systems into the acceptance testing and routine QA of their TPS, R&V, and treatment consoles. These tests should concentrate on the orientation and localization of the treatment isocenters and the correctness of the calculated shifts required to align the patient.

Acceptance testing of new machines or new software versions should minimally consist of an end-to-end test in which a phantom is imaged and marked (Bissonette et al. 2012). The phantom image is then transferred to the TPS and planned. The plan is transferred to the R&V system and finally localized with all forms of IGRT in normal use with the position verified by visual indicators or a radio-opaque object visible in a true portal image. The testing should be done with all normal patient orientations, head-first supine, feet-first supine, head-first prone, and feet-first prone. The phantom should be set up using a marked isocenters and then shifted to a final isocenter. The final isocenter should be offset by at least 5 cm in all three directions. The shifts should be calculated at the treatment machine using all normally used setup methods, including skin marks and couch shifts, 2D–2D matching based on DRRs, and 3D–3D matching based on CT data.

Daily QA should be done for IGRT to ensure treatment and imaging isocenters are in agreement, but tests should also be arranged in such as way as to ensure the information systems are working correctly. One method is to create a QA patient in the R&V system that contains imaging fields for a daily imaging QA device. Varian, for example, provides a cube with external scribe marks and an internal target. If daily isocenter tests are done using the R&V system to serve the reference images of the QA device and record the acquired images, it also ensures that all communication channels are working that morning and gives the therapists a chance to call service before a problem is discovered with a patient on the table.

Plan checks should also be expanded to ensure IGRT-related data transfer has occurred correctly for each patient. Part of a routine plan check should be to verify that the isocenters and scaling of each DRR are correct and that the qualities of the DRRs are sufficient for the matching tasks. If CBCT is to be used, the reference CT data set and structures should also be checked as well as the locations of the treatment isocenters. This is especially important if multiple isocenters exist on the same patient. The only method of checking the integrity of the CT transfer at this time is to ensure the number of CT slices is correct and that the data visually look correct and that the transferred contours match the CT image.

22.6 Conclusions

IGRT is rapidly becoming the standard of care for external beam radiotherapy. IGRT requires that images and supporting DICOM objects such as structures and isocenter locations flow smoothly and correctly from the virtual simulation system to the TPS through the R&V system to the treatment system and back again. Errors in any of these transmission steps could cause the images not to be available for initial setup or post-treatment review or, more importantly, could cause the patient to be treated in an incorrect location.

In order for IGRT to be effectively performed, the network infrastructure needs to support the transport of large data sets in a highly reliable system. The network and R&V system are as important to the treatment delivery as the electricity to run the accelerator because without them the treatment cannot be performed. Reliable robust networks designed and maintained by a team that includes the medical physicist can help to achieve this.

Quality assurance needs to extend beyond the radiation dose calculations and accelerators setting and move into the IGRT world as well. The initial acceptance testing of any imaging, planning, data transfer device, or software version should include testing of the correct flow of imaging data including the images, the associated DICOM objects, and the isocenter locations. These should be tested for all patient orientations normally used. Pretreatment plan checks should include verification of the images and isocenter locations for that treatment, and extra care needs to be taken for patients with multiple isocenters.

References

Armstrong, J. G. et al. (1993). Three-dimensional conformal radiation therapy may improve the therapeutic ratio of high dose radiation therapy for lung cancer. *Int J Radiat Oncol Biol Phys, 26*(4), 685–689.

Barker, J. L. et al. (2004). Quantification of volumetric and geometric changes occurring during fractionate radiotherapy for head-and-neck cancer using an integrated CT/linear accelerator system. *Int J Radiat Oncol Biol Phys, 59*(4), 960–970.

Bissonette, J.-P. et al. (2012). Quality assurance for image-guided radiation therapy utilizing CT-based technologies: A report of the AAPM TG-179. *Med Phys, 39*(4), 1946–1963.

Chen, G. T. et al. (2009). A review of image-guided radiotherapy. *Radiol Phys Technol, 2*(1), 1–12.

Dong, L. & Boyer, A. L. (1996). A portal image alignment and patient setup verification procedure using moments and correlation techniques. *Phys Med Biol, 41*(4), 697–722.

Galvin, J. M. et al. (1995). The use of digitally reconstructed radiographs for three-dimensional treatment planning and CT-simulation. *Int J Radiat Oncol Biol Phys, 31*(4), 935–942.

ICRU, *Prescribing, Recording, and Reporting Photon Beam Therapy. ICRU Report 50.* Bethesda: International Commission on Radiation Units and Measurements, 1993.

Jaffray, D. A. & Siewerdsen, J. H. (2000). Cone-beam computed tomography with a flat-panel imager: Initial performance characterization. *Med Phys, 27*(6), 1311–1322.

McGee, K. P. et al. (1997). The value of setup portal films as an estimate of a patient's position throughout fractionated tangential breast irradiation: An on-line study. *Int J Radiat Oncol Biol Phys, 37*(1), 223–228.

Mackie, T. R. et al. (1993). Tomotherapy: A new concept in the delivery of dynamic conformal radiotherapy. *Med Phys, 20*(6), 1709–1719.

Munro, P. (1995). Portal imaging technology: Past, present, and future. *Semin Radiat Oncol, 5*(2), 115–133.

Murphy, M. J. (1997). An automatic six-degree-of-freedom image registration algorithm for image-guided frameless stereotaxic radiosurgery. *Med Phys, 24*(6), 857–866.

Murphy, M. J. et al. (2007). The management of imaging dose during image-guided radiotherapy: Report of the AAPM Task Group 75. *Med Phys, 34*(10), 4041–4063.

Pouliot, J. et al. (2005). Low-dose megavoltage cone-beam CT for radiation therapy. *Int J Radiat Oncol Biol Phys, 61*(2), 552–560.

Purdy, J. A. (1999). 3D treatment planning and intensity-modulated radiation therapy. *Oncology, 13*(10 Suppl 5), 155–168.

Shalev, S. et al. (1989). Video techniques for on-line portal imaging. *Comput Med Imaging Graphics, 13*(3), 217–226.

Sherouse, G. W. et al. (1990). Computation of digitally reconstructed radiographs for use in radiotherapy treatment design. *Int J Radiat Oncol Biol Phys, 18*(3), 651–658.

Uematsu, M. et al. (1996). A dual computed tomography linear accelerator unit for stereotactic radiation therapy: A new approach without cranially fixated stereotactic frames. *Int J Radiat Oncol Biol Phys, 35*(3), 587–592.

Verellen, D. et al. (2008). A (short) history of image-guided radiotherapy. *Radiother Oncol, 86*(1), 4–13.

Viola, P. and Wells, W. M. (1997). Alignment by maximization of mutual information. *Int J Computer Vision, 24*(2), 137–154.

Patient Assessment Tools

Todd McNutt
The Johns Hopkins University

Joseph M. Herman
The Johns Hopkins University

23.1 Clinical Assessment of Toxicity

In general, the goal of radiation therapy is to eradicate the tumor while sparing adjacent normal tissues. Patients tolerate radiation treatment differently, and the incidence and severity of complications associated with that treatment vary as well. For this reason, a process by which patient toxicity information is prospectively evaluated and recorded through weekly assessments, and late toxicity through subsequent follow-up visits, has been developed. A standardized way of collecting this information across disease sites allows physicians to evaluate treatment-related toxicity for all patients receiving radiation therapy.

The assessment of a patient's clinical status is a key component to managing care during radiation therapy. Early side effects may depend on the disease site and treatment technique; however, some effects such as fatigue and nausea are common for all patients. The prospective collection and monitoring of data on side effects allows interventions to be introduced that may decrease or eliminate common side effects across disease sites. Further, patients are not all in the same physical condition at the beginning of treatment. The initial assessment of patients enables physicians to establish a baseline status for each patient that can be used to identify changes and evaluate trends in side effects during the course of treatment.

The description of a clinical assessment can be challenging, as it must accurately reflect the status of the patient, and, ideally, should be capable of being compared against assessments of other patients (Chen et al. 2006; Davidson et al. 2007). Historically, medical information has been recorded in textual dictations that are descriptive. These dictations can convey physician impressions and convey critical medical information among caregivers. However, such dictations are difficult to quantify and hence make comparing assessments of different patients problematic (DeHart & Holbrook 1992). Laboratory tests and prescription doses are typically quantifiable and can be compared easily, but

patient assessments and physician impressions can be subjective and are much more challenging to quantify.

Quantifying toxicities can also be challenging, as each toxicity may be multifaceted and multisymptomatic. The basic approach in many areas of oncology is one in which sets of symptoms or characteristics are grouped into organ system categories (e.g., hematologic, cardiac, and gastrointestinal) and ranked in terms of grade or severity (Williams et al. 2003). The National Cancer Institute's Common Terminology Criteria for Adverse Events (NCI-CTCAE) (Trotti et al. 2003; National Cancer Institute 2009) is one such system created to enable quantification and grading of patient assessments and physician impressions. Grading scales are provided for various toxicities, pain, and quality-of-life measures. These grading scales are used extensively in clinical trials, and their adoption for the general population of cancer patients is becoming more prevalent with the increased use of information technology (IT) (Palazzi et al. 2008). Indeed, consistent use of grading systems such as the CTCAE allows reliable prospective collection of data that can be used for statistical analyses, thus allowing us to gain broader knowledge from individual medical records.

The basic principle of the NCI-CTCAE scale is to rank or grade various types of toxicity, expressed as clinical symptoms or signs, according to the following general definitions. Grade 1 effects are minimal, are usually asymptomatic, and do not interfere with functional endpoints; interventions or medications are generally not indicated for these minor effects. Grade 2 effects are considered moderate, are usually symptomatic, and interventions such as local treatment or medications may be indicated. Grade 2 effects may or may not interfere with specific functions but not enough to impair activities of daily living. Grade 3 effects are considered severe and very undesirable and may include multiple, disruptive symptoms. More serious interventions, including surgery or hospitalization, may be indicated. Grade 4 effects are potentially life-threatening, catastrophic, disabling,

TABLE 23.1 Samples of the NCI-CTCAE Grading Scales

Adverse Event	Grade				
	1	2	3	4	5
Dysphagia (difficulty swallowing)	Symptomatic, able to eat regular diet	Symptomatic and altered eating/swallowing (e.g., altered dietary habits and oral supplements); IV fluids indicated <24 hours	Symptomatic and severely altered eating/swallowing (e.g., inadequate oral caloric or fluid intake); IV fluids, tube feedings, or TPN indicated ≥24 hours	Life-threatening consequences (e.g., obstruction and perforation)	Death
Anorexia	Loss of appetite without alteration in eating habits	Oral intake altered without significant weight loss or malnutrition; oral nutritional supplements indicated	Associated with significant weight loss or malnutrition (e.g., inadequate oral caloric and/or fluid intake); IV fluids, tube feedings or TPN indicated	Life-threatening consequences	Death
Diarrhea	Increase of <4 stools per day over baseline; mild increase in ostomy output compared with baseline	Increase of 4–6 stools per day over baseline; IV fluids indicated <24 hours; moderate increase in ostomy output compared with baseline; not interfering with ADL	Increase of ≥7 stools per day over baseline; incontinence; IV fluids ≥24 hours; hospitalization; severe increase in ostomy output compared with baseline; interfering with ADL	Life-threatening consequences (e.g., hemodynamic collapse)	Death
Fatigue (asthenia, lethargy, and malaise)	Mild fatigue over baseline	Moderate or causing difficulty performing some ADL	Severe fatigue interfering with ADL	Disabling	
Nausea	Loss of appetite without alteration in eating habits	Oral intake decreased without significant weight loss, dehydration, or malnutrition; IV fluids indicated <24 hours	Inadequate oral caloric or fluid intake; IV fluids, tube feedings, or TPN indicated ≥24 hours	Life-threatening consequences	Death

Note: ADL, activities of daily living; IV, intravenous; TPN, total parenteral nutrition.

or may result in loss of organ, organ function, or limb. Grade 5 events refer to death related to the adverse effect. Table 23.1 gives examples of the types of toxicity often associated with radiation therapy and describes grade rankings for each type.

The vocabulary of medicine is very complex, and consistency in the meaning of the various terminologies is crucial when comparing toxicities across groups of patients. Faria et al. (2009) showed that reports of radiation-induced lung toxicity varied greatly depending on which scoring criteria were used for the assessment. Several initiatives are under way that strive to harmonize the terminology used for clinical assessments. One such initiative, Medical Dictionary for Regulatory Activities (MedDRA; http://www.meddramsso.com), is a dictionary that provides consistent definitions. The latest version of the NCI-CTCAE, version 4.0, was designed to be compatible with the MedDRA. Another effort, the Vocabularies & Common Data Elements Workspace (Cimino et al. 2009) of the Cancer Bio-Informatics Grid (caBIG; https://cabig.nci.nih.gov) (Fenstermacher et al. 2005; caBIG Strategic Planning Workspace 2007) is responsible for developing standards for the representation of ontologies and vocabularies used throughout the caBIG environment. The NCI-CTCAE terminology has also been included in the caBIG environment.

23.2 Outcome and Quality of Life

The outcome of therapy is often challenging both to assess and to define. "Cure" is typically defined as being free of disease at 5 years after completion of treatment; however, "cure" may not always be the intent of treatment. Many radiation therapy treatments are given with palliative rather than curative intent and enhance the quality of the patient's life as well as extending their survival time (Velikova et al. 2008). Assessments of outcome and quality-of-life issues again require quantification of the assessment to compare results in meaningful ways.

The outcome of various treatment regimens has been compared through the use of quality-of-life questionnaires that have been standardized for patients receiving treatment for various disease sites. For example, the standard of care for localized rectal cancer is conventional three-field radiation therapy and concurrent 5-fluorouracil–based chemotherapy. This treatment treats the rectal tumor but also results in irradiation of normal adjacent structures such as the bladder, femoral heads, and sexual organs. A modified form of radiation therapy known as intensity-modulated radiation therapy (IMRT) delivers focused radiation to the rectal tumor and at-risk lymph nodes while sparing radiation dose to these normal structures. Presumably, IMRT results in less toxicity and improved quality of life; however, it is difficult to conduct randomized clinical trials comparing the two treatment techniques because of lack of funding and it can take years to complete. Alternatively, physician assessments of toxicity and patient assessments of quality of life via standardized questionnaires can be administered at the point of service for all patients and can quantitatively determine if IMRT is indeed superior to conventional radiation therapy techniques.

In general, it is thought that having a sphincter sparing surgery for rectal cancer (lower anterior resection) results in a better quality of life than undergoing a surgery where your sphincter is removed (abdominal perineal resection). Surprisingly, little is known whether patient quality of life is better when patients undergo a lower anterior resection following neoadjuvant chemoradiation (Gervaz et al. 2008). In one such study, Allal et al. (2005) evaluated quality of life for 53 patients treated with twice-daily radiation (50 Gy in 40 fractions within 4 weeks), with or without biweekly gemcitabine, followed by abdominoperineal resection or low anterior resection. Quality of life was assessed with two self-rating questionnaires developed by the European Organisation for Research and Treatment of Cancer (EORTC): QLQ-C30 and QLQ-CR38. The questionnaires were completed once before and again 12 to 16 months after the radiation therapy. At 1 year after treatment, patients reported statistically significant improvements in their emotional state, perspective of the future, and global quality of life as well as a decrease in gastrointestinal symptoms relative to before treatment. However, sexual dysfunction scores were significantly higher, particularly in men, and a trend toward a lower body image score was observed after treatment. Contrary to expectations, quality of life, as defined in these instruments, was not improved in patients treated with sphincter-conserving approaches. If data were prospectively being collected in a database on all patients receiving standard conventional treatment and IMRT, then comparisons of tumor response, toxicity, and quality of life could be easily conducted in an unbiased fashion. This is but one an example of how prospectively collecting patient specific information and quality of life for all patients can be used to compare efficacy of different treatment modalities as opposed to conducting costly prospective clinical trials.

Quality-of-life questionnaires are becoming easier to administer as increasing numbers of patients have access to the Internet and to E-mail. Technologies to support online forms and communications are readily available and can be used to facilitate patient self-assessments during the long periods between follow-up visits (Basch et al. 2007; Trotti et al. 2007).

23.3 Tumor Assessment and Outcomes

Tumor control after therapy is typically monitored through imaging studies. One system for quantifying tumor control, the Response Evaluation Criteria in Solid Tumors (RECIST) (Eisenhauer et al. 2009) system, is a set of published rules that define when cancer improves ("response"), stays the same ("stable disease"), or worsens ("disease progression") during or after treatment. The RECIST system is often applied retrospectively by clinicians. Ideally, a disease response defined according to a RECIST criterion that is essentially based on tumor size before and after treatment should be determined by the radiologist at the time the scan is reviewed. This practice would improve the reliability and objectivity of assessments of tumor response.

Although sets of response criteria such as RECIST provide one means of quantitatively assessing tumor response, other outcomes including overall survival, progression-free survival, and local control are also important. Factors that can influence these outcomes include age, race, comorbid conditions, and

pathologic tumor- or treatment-related features such as surgical margins or lymph node status. Clinicians have to often "mine" patient and clinic charts to obtain this information. If these data were collected prospectively in a database, one could easily calculate univariate and multivariate analyses that are up-to-date and current based on patients recently treated. Because these data would be readily accessible, updating or repeating the analyses could be easily accomplished. One could subsequently analyze the data to determine how to best treat future patients with similar characteristics.

23.4 Data Collection in the Clinical Setting

Implementing medical IT in a clinical setting continues to be challenging. Clinicians tend to resist IT, perceiving it as a disruption in workflow that adds little value. Continued focus on the interaction between clinicians and the computer systems is required to further advance medical informatics. It is crucial that the computer system is implemented in such a way to assist clinicians, not to hinder or control them. As one example, physicians are typically mobile in the clinical setting, and computers have historically been designed for stationary desktops. Recent advances in IT systems provide the possibility of having a portable or fixed computer interface at any time and any location. These technologies must, however, be customized to meet the challenges of the clinical workflow and the patient–physician encounter. For example, radiation oncologists often need to review and sign portal images before a patient can begin radiation therapy. A handheld computer could be used to review data on that patient, enabling radiation oncologists to subsequently sign the films and assessments without having to stop and login to a desktop computer. Advances such as these can save time on many levels and lead to improved care and efficiency in the clinic.

An example clinical workflow in radiation oncology is illustrated in Figure 23.1. It has multiple stages, from simulation and treatment planning to recording of daily treatments and subsequent follow-up visits. During each stage of the process, significant amounts of image and treatment planning information are already collected electronically without affecting the workflow. Each patient–physician encounter is an opportunity to capture the results of therapy through assessment of toxicities, quality of life, laboratory tests, and survival. In addition, the Internet and Web enable more frequent updates through patient-reporting models that may include Web-based follow-up forms that can be sent periodically to the patient by e-mail.

Oncology information systems such as MOSAIQ (Elekta, Stockholm, Sweden), ARIA (Varian Medical Systems, Palo Alto, California), and Oncentra (Nucletron, Columbia, Maryland) have been developed to manage patient care in oncology and to store clinical information for permanent electronic records. Systems such as these allow users to specify the diagnosis [according to the *International Classification of Disease, Ninth*

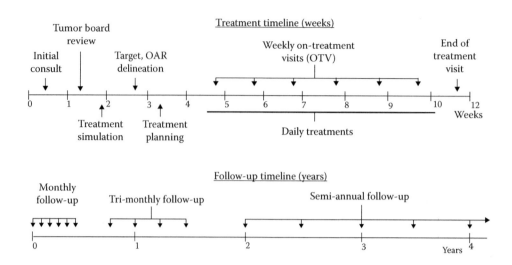

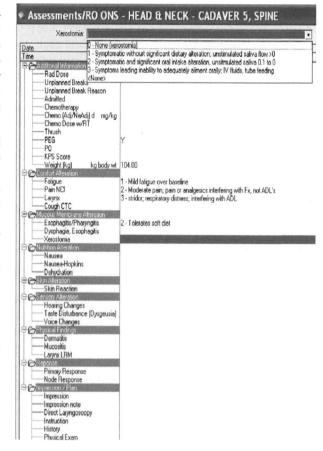

FIGURE 23.1   Timeline of a typical course of radiation therapy with periodic follow-up after treatment.

Edition (ICD-9)] and disease staging information; vital statistics and laboratory values; medications and chemotherapy; the radiation therapy prescriptions; and treatment plans and records of delivered treatments. These systems also have patient observation modules that enable the customization of forms and data items to allow the entry of structured observational data during the patient–physician encounter. These modules have been customized to support on-treatment assessments and follow-up visits and to use the NCI-CTCAE scoring model to quantify clinical observations. Use of these systems makes it both possible and feasible to integrate structured data collection into the daily clinical process.

Figure 23.2 shows an assessment entry form from the MOSAIQ system, which allows the user to enter the various assessments from selection lists or by using checkboxes, text, or single values. Figure 23.3 illustrates a typical view a physician would use to monitor patient assessments. With this system, toxicities can be numerically tracked over time and trends of concern identified. Warnings can also be posted by highlighting entries in yellow or red depending on low or high warning levels being breached.

The goal of data collection is to minimize the impact on the clinical workflow and to ensure that the patient–physician interaction is not impeded by the system of data collection. Although computerized systems have been deployed in clinical settings, the deployment has not always been a smooth transition.

The current weekly on-treatment visit begins with a nurse asking the patient a series of questions. During this conversation, the nurse must enter the assessments into the computer. Once the nurse has completed the initial assessment, the attending physician reviews that assessment and may ask additional questions to help refine the assessment and resolve any inconsistencies. The physician then continues with a physical examination to assess visible findings (e.g., mucositis and skin erythema). During this process, the findings must be entered into the computer. In some cases when both the nurse and physician are present, voice instruction from the examiner to another healthcare

FIGURE 23.2   Assessment entry screen in MOSAIQ, in which pull-down menus are used identifying the textual description and the grade level of the assessments. Items can be lists, single value numbers, or checkboxes depending on the type of assessment.

Assessments

Date	8/29	10/02	10/09	10/16	10/22	11/07	11/12	1/23	4/24	9/15
Additional Information										
Rad Dose		1400.00	2200.00	3400.00	4000.00	6400.00	7000.00			
Unplanned Breaks										
Unplanned Break Reason										
Admitted						Y				
Chemotherapy										
Chemo (Adj/NeAdj) d mg/kg										
Chemo Dose w/RT										
Thrush				Y	Y					
PEG				Y	Y	Y	Y		Y	
PO		Y	Y	Y	Y	Y	Y			
KPS Score										
Weight (kg) kg body wt	103.00		97.20	97.00				86.40	81.10	74.90
Comfort Alteration										
Fatigue		1	2	2	3	3	3			
Pain NCI		0	1	1	2	2	2		0	0
Larynx									0	0
Cough CTC			0	0	0	0	0		0	0
Mucous Membrane Alteration										
Esophagitis/Pharyngitis		1	2	2	3	3	3			
Dysphagia, Esophagitis		1	3	3	3	3	4			
Xerostomia		2	2	2	2	2	2		2	1
Nutrition Alteration										
Nausea		2	3	2	3	3	3	1	0	0
Nausea-Hopkins										
Dehydration										
Skin Alteration										
Skin Reaction			0	0	0	0	0			
Sensory Alteration										
Hearing Changes		0	0	0	0	1	1			
Taste Disturbance (Dysgeusia)		1	2	2	2	2	2		1	1
Voice Changes		0	0	0	0	0	0		0	0
Physical Findings										
Dermatitis			1	1	1	2	2			
Mucositis		2	2	2	3	3	3		0	0
Larynx LRM			0	0	0	0	0		0	0
Response										
Primary Response		NA	NA	NA	NA	NA	NA			
Node Response		NA	NA	NA	NA	NA	NA			
Impression / Plan										
Impression		0	0	0	0	0	0			
Impression note										
Direct Laryngoscopy										
Instruction			daily hydration			blood cultures	blood cultures			
History										
Physical Exam										
Plan										
Notes										
X-Ray Review Check										
Symptoms Management										
Oral Care										
Salt/Soda				Y	Y	Y	Y			
Oral Rinse										

FIGURE 23.3 MOSAIQ user interface for tracking weekly clinical assessments of a patient. Cells can be colored based on low and high alert levels for individual entries when items exceed desired warning tolerances.

professional operating the computer can be easily provided; however, most of the time, the computer entry in the exam room is more of a burden than the traditional clipboard with a form that can be carried around. Currently, a desktop computer is used. However, to be less intrusive, a portable computer/phone or notepad may be used in the future. Alternatively, voice-activated systems can be used, preferably with the results being briefly proofread before the examiner leaves the exam room.

Continued exploration of methods to improve the transparency of the IT system to clinicians will help improve the ability to monitor patients in real time. For example, repeat imaging is needed to confirm that patients are in the correct treatment position at each treatment session. These images are often not checked until the end of the day. If the patient was not in the correct treatment position, this could have a major impact on the accuracy of treatment. A small shift over the course of 25 to 30 treatments may have minimal impact; however, with hypo-fractionated radiation therapy (1–5 treatments), a small shift can have a major impact on tumor control and toxicity. Therefore, the ability of the radiation oncologist to review images in "real time" from any location can result in improved patient care. One option for this purpose is to use tablet computers or mobile computing with virtual sessions that enable clinicians to access their session from any display. In some instances, it may be possible to achieve self-documenting encounters.

23.5 Analysis of Assessments

Clinical assessments are the current means of assessing the positive and negative effects associated with radiation therapy. The primary use of these assessments is to monitor individuals throughout the course of treatment and to avoid severe early complications and/or treatment-related errors. The ability to quantify these assessments allows toxicities to be compared across large populations of patients, thus improving our understanding of the therapeutic outcome of various treatment techniques (such as comparing IMRT vs. conventional treatment for rectal cancer).

Clinical data are typically collected in one of two ways: retrospectively, by reviewing charts or electronic records, or prospectively, typically as part of a clinical trial. Retrospective data are fraught with bias and inconsistencies, which limit any interpretations based on those data. Clinical trials are costly, can also be biased by patient selection, and are typically designed to evaluate the effects of treatment on disease at only one anatomic site. Initiation of an electronic clinical assessment would allow prospective collection of data from all patients in a more cost-effective manner and could eliminate some of the biases outlined above.

The fundamental approach in most clinical trials in oncology is to separate (stratify) patients into subgroups based on their diagnosis, treatment technique, or demographics and to compare the outcomes and toxicities of treatments between subgroups. Properly structured data collection would allow clinical data to be analyzed in the same way, as is clinical trial data. A central but overlooked deficiency in current clinical practice is the lack of verification of the quality of practice in the community. The majority (97%) of cancer patients are treated in the community and not on clinical trials (Vickers 2008; Wassenaar et al. 2008). If those patients could be prospectively evaluated as part of the normal clinical structure, then those patients could be compared with patients on clinical trials. This approach could eliminate the need for "control" groups if one were able to otherwise standardize treatment delivery. Currently, community physicians often treat patients based on standard of care practice and have limited direct means of influencing or enhancing their practices. Successful implementation of electronic assessments may be a first step toward providing the infrastructure necessary to monitor the effectiveness of the clinical practice.

23.6 Future Possibilities to Improve Knowledge Gain Beyond Publication

Radiation therapy relies on the assessment of normal-tissue toxicity from patients enrolled in clinical trials to understand critical radiation dose limits when planning treatments. These data on normal-tissue toxicity are then communicated to clinicians through the publication of a trial's findings. Radiation dose delivered to normal structures during therapy is 3D in nature and is distributed at varying levels across the volume of the structure. Dose-volume histograms (DVHs) are used to analyze the 3D dose distribution as a plot of dose versus volume. In the analysis of toxicities, typically one point on the DVH is selected as the critical measure of dose. For example, to limit the occurrence of xerostomia (dry mouth), the radiation dose to at least half of the parotid gland volume is kept below 30 Gy. However, this simple rule represents a potential oversimplification of the complex dosimetry to meet the goals of disseminating the information via a journal publication.

The Quantitative Estimates of Normal Tissue Effects in the Clinic (QUANTEC) project (Jackson 2008; Marks 2008) seeks to document the normal tissue toxicity associated with radiation therapy by centralizing data found in the existing literature. The goal of the project is to publish a summary of knowledge acquired to date. Although much of the current knowledge of normal tissue toxicity has been published, information regarding the 3D dose distribution has been lost in the publication process. Preliminary evaluations by the QUANTEC project have determined that much of the data are sparse primarily because of

this loss of the original dosimetric data. If the 3D dose distributions for these studies had been saved in an analytical database, the knowledge that could be obtained through the QUANTEC project would be substantially greater. This example illustrates why textual publications are often limited as a means to transfer knowledge. The Internet style of Web-based portals into information of varied types has a much greater potential to transfer clinical knowledge.

The incidence, severity, and onset of toxicity due to radiation therapy are likely more complex than this traditional simple analysis; thus, a computer system that stores toxicity and 3D dose data will help us better understand the relationship between radiation dose and normal tissue toxicity. Although computers struggle with understanding textual data, they are quite capable of understanding complex multidimensional data that often cannot easily be translated to publication. For example, one might ask the database, "What is the risk of xerostomia given this 3D dose distribution?" and the computer could search the database, find patients with similar dose distributions (determined by multiple parameters), and then determine the estimated risk.

A future goal of medical informatics in radiation therapy is to accelerate the gain in knowledge that traditionally comes from clinical trials by making use of the full clinical data and having it accessible to the practicing clinician at any time. This acceleration is accomplished by providing a mechanism to present, in Web-based forms, processed knowledge in real time as new data become available to the system. For example, a plot of the incidence of rectal toxicity could be continuously updated. If that incidence began to rise, the system could alert clinicians. To reduce the time to alert, the time-to-event continual reassessment method or TITE-CRM (Braun 2006) could be used to statistically infer when a risk of toxicity is increased from a sparser set of data points that look at the time between treatment and the onset of toxicities. Such a system establishes a working model of evidence-based medicine for continual improvement in radiation therapy (Hazlehurst et al. 2005).

In the near future, we will see areas of significant advancement in the interfaces between physicians and computers, ranging from advanced user-interface development to simplify the physician experience to means of self-documenting patient–physician encounters in which technology is set up to "record" the encounter in a structured way. With improved data collection, we will also see advancements in data navigation and analysis tools that will enable physicians to learn from a broad set of past experience and to apply that accessible knowledge to improve the care of new patients.

References

Allal, A. S., Gervaz, P., Gertsch, P. et al. (2005). Assessment of quality of life in patients with rectal cancer treated by preoperative radiotherapy: A longitudinal prospective study. *Int J Radiat Oncol Biol Phys, 61*, 1129–1135.

Basch, E., Iasonos, A., Barz, A. et al. (2007). Long-term toxicity monitoring via electronic patient-reported outcomes in patients receiving chemotherapy. *J Clin Oncol, 25*, 5374–5380.

Braun, T. M. (2006). Generalizing the TITE-CRM to adapt for early- and late-onset toxicities. *Stat Med, 25*, 2071–2083.

caBIG Strategic Planning Workspace (2007). The Cancer Biomedical Informatics Grid (caBIG): Infrastructure and applications for a worldwide research community. *Stud Health Technol Inform, 129*, 330–334.

Chen, Y., Trotti, A., Coleman, C. N. et al. (2006). Adverse event reporting and developments in radiation biology after normal tissue injury: International Atomic Energy Agency Consultation. *Int J Radiat Oncol Biol Phys, 64*, 1442–1451.

Cimino, J. J., Hayamizu, T. F., Bodenreider, O. et al. (2009). The caBIG terminology review process. *J Biomed Inform, 42*, 571–580.

Davidson, S. E., Trotti, A., Ataman, O. U. et al. (2007). Improving the capture of adverse event data in clinical trials: The role of the international atomic energy agency. *Int J Radiat Oncol Biol Phys, 69*, 1218–1221.

DeHart, K. & Holbrook, J. (1992). Emergency department applications of digital dictation and natural language processing. *J Ambul Care Manage, 15*, 18–23.

Eisenhauer, E. A., Therasse, P., Bogaerts, J. et al. (2009). New response evaluation criteria in solid tumours: Revised RECIST guideline (version 1.1). *Eur J Cancer, 45*, 228–247.

Faria, S. L., Aslani, M., Tafazoli, F. S. et al. (2009). The challenge of scoring radiation-induced lung toxicity. *Clin Oncol (R Coll Radiol), 21*, 371–375.

Fenstermacher, D., Street, C., McSherry, T. et al. (2005). The cancer biomedical informatics grid (caBIG™). *Conf Proc IEEE Eng Med Biol Soc, 1*, 743–746.

Gervaz, P., Bucher, P., Konrad, B. et al. (2008). A prospective longitudinal evaluation of quality of life after abdominoperineal resection. *J Surg Oncol, 97*, 14–19.

Hazlehurst, B., Frost, H. R., Sittig, D. F. et al. (2005). MediClass: A system for detecting and classifying encounter-based clinical events in any electronic medical record. *J Am Med Inform Assoc, 12*, 517–529.

Jackson, A. (2008). Summarizing our knowledge of normal tissue tolerances: The progress and future directions of QUANTEC [abstract]. *Med Phys, 35*, 2863.

Marks, L. (2008). A clinician's view of QUANTEC [abstract]. *Med Phys, 35*, 2863–2864.

National Cancer Institute (2009). Common Terminology Criteria for Adverse Events, v4.0. Available through the Cancer Therapy Evaluation Program at http://ctep.cancer.gov/protocolDevelopment/electronic_applications/ctc.htm.

Palazzi, M., Tomatis, S., Orlandi, E. et al. (2008). Effects of treatment intensification on acute local toxicity during radiotherapy for head and neck cancer: Prospective observational study validating CTCAE, version 3.0, scoring system. *Int J Radiat Oncol Biol Phys, 70*, 330–337.

Trotti, A., Colevas, A. D., Setser, A. et al. (2003). CTCAE v3.0: Development of a comprehensive grading system for the adverse effects of cancer treatment. *Semin Radiat Oncol, 13*, 176–181.

Trotti, A., Colevas, A. D., Setser, A. et al. (2007). Patient-reported outcomes and the evolution of adverse event reporting in oncology. *J Clin Oncol, 25*, 5121–5127.

Velikova, G., Awad, N., Coles-Gale, R. et al. (2008). The clinical value of quality of life assessment in oncology practice—A qualitative study of patient and physician views. *Psychooncology, 17*, 690–698.

Vickers, A. J. (2008). Do we want more cancer patients on clinical trials and if so, what are the barriers to greater accrual. *Trials, 9*, 31.

Wassenaar, T. R., Walsh, M. C., Cleary, J. F. et al. (2008). Disparities in the clinical trial participation of adult cancer patients [abstract]. *J Clin Oncol, 26*, 9523.

Williams, J., Chen, Y., Rubin, P. et al. (2003). The biological basis of a comprehensive grading system for the adverse effects of cancer treatment. *Semin Radiat Oncol, 13*, 182–188.

VI

Outcomes Modeling and Quality Assurance Informatics

24

Outcomes Modeling

Issam El Naqa
McGill University

24.1 Introduction

Radiotherapy outcomes are determined by complex interactions among treatment, anatomical, and patient-related variables. A key component of radiation oncology research is to predict, at the time of treatment planning, or during the course of fractionated radiation treatment, the probability of tumor eradication and normal tissue risks for the type of treatment being considered for that particular patient (Torres-Roca & Stevens 2008). Outcomes in radiotherapy are usually characterized by tumor control probability (TCP) and the surrounding normal tissue complication probability (NTCP) (Webb 2001; Steel 2002). Traditionally, these outcomes are modeled using information about the dose distribution and the fractionation (Moissenko et al. 2005). However, it is recognized that radiation response may also be affected by multiple clinical prognostic factors (Marks 2002); more recently, inherited genetic variations have been suggested as playing an important role in radiosensitivity (West et al. 2007; Alsner et al. 2008). Therefore, recent approaches have utilized data-driven models incorporating advanced informatics tools in which dose-volume metrics are mixed with other patient or disease-based prognostic factors to improve outcomes prediction (El Naqa 2012). Accurate prediction of treatment outcomes would provide clinicians with better tools for informed decision-making about expected benefits versus anticipated risks.

In this chapter, we provide an overview of the current status of outcome modeling techniques for predicting tumor response and normal tissue toxicities for patients who receive radiation treatment with special focus on the emerging role of informatics approaches to improve outcomes modeling and response prediction. Then, we present examples of applying different informatics approaches for predicting tumor response and normal tissue toxicities in radiation oncology from our own experiences. Finally, we discuss the potentials and challenging obstacles to applying informatics strategies in radiotherapy outcomes modeling.

24.2 Background

24.2.1 Radiotherapy Outcomes Metrics

Recent years have witnessed tremendous technological advances in radiotherapy treatment planning, image guidance, and treatment delivery (Webb 2001; Bortfeld et al. 2006). Moreover, clinical trials examining treatment intensification in patients with locally advanced cancer have shown incremental improvements in local control and overall survival (Halperin et al. 2008). However, radiation-induced toxicities remain major dose-limiting factors (Bentzen et al. 2010; Jackson et al. 2010). Therefore, there is a need for studies directed toward predicting treatment benefit versus risk of failure. Clinically, such predictors would allow for more individualization of radiation treatment plans. In other words, physicians may prescribe a more or less intense radiation regimen for an individual based on model predictions of local control benefit and toxicity risk. Such an individualized regimen would aim toward an optimized radiation treatment response while keeping in mind that a more aggressive treatment with a promised improved tumor control will not translate into improved survival unless severe toxicities are accounted for and limited during treatment planning. Therefore, improved models for predicting both local control and normal tissue toxicity should be considered in the optimal treatment planning design process.

Radiotherapy outcomes are usually characterized by two metrics: the TCP and the NTCP of surrounding normal tissues (Webb 2001; Steel 2002). TCP/NTCP models could be used during the consultation period as a guide for ranking treatment options (Weinstein et al. 2001; Armstrong et al. 2005). Alternatively, once a decision has been reached, these models could be included in an objective function, and the optimization problem driving the actual patient's treatment plan can be formulated in terms relevant to maximizing tumor eradication benefit and minimizing complication risk (Brahme 1999; Moiseenko et al. 2004; Allen Li et al. 2012). Traditional models of TCP/NTCP models and their variations use information only about the dose distribution and fractionation. However, it is well known that radiotherapy outcomes may also be affected by multiple clinical and biological prognostic factors such as stage, volume, and tumor hypoxia (Fu et al. 1999; Choi et al. 2001) as depicted in Figure 24.1. Therefore, recent years have witnessed the emergence of data-driven models utilizing informatics techniques, in which dose-volume metrics are combined with other patient- or disease-based prognostic factors (Marks 2002; Bradley et al. 2004; Tucker et al. 2004; Blanco et al. 2005; Hope et al. 2005; El Naqa et al. 2006a; Deasy & El Naqa 2008; Bentzen et al. 2010; Jackson et al. 2010). However, before divulging into the details of outcomes modeling, it would be pedagogically necessary to provide a brief review of the basic relevant radiobiological principles.

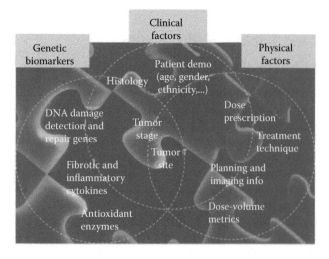

FIGURE 24.1 Radiotherapy treatment involves complex interaction of physical, biological, and clinical factors. The successful informatics approach should be able to resolve this interaction "puzzle" in the observed treatment outcome (e.g., local control or toxicity) for each individual patient. (From Spencer, S. et al., *Bioinformatics Methods for Learning Radiation-Induced Lung Inflammation from Heterogeneous Retrospective and Prospective Data*. Hindawi Publishing Corporation, 2009.)

24.2.2 Radiobiology of Radiotherapy Response

Classic radiobiology has been defined by "The Four R's" (cellular damage repair, cell-cycle redistribution, reoxygenation over a course of therapy, and cellular repopulation/division over a course of therapy) (Hall & Giaccia 2006). It is believed that radiation-induced cellular lethality is primarily caused by DNA damage in targeted cells. Two types of cell death have been linked to radiation: apoptosis and postmitotic cell death. However, tumor cell radiosensitivity is controlled via many factors (known and unknown) related to tumor DNA repair efficiency (e.g., homologous recombination or nonhomologous endjoining), cell cycle control, oxygen concentration, and the radiation dose rate (Hall & Giaccia 2006; Lehnert 2008; Joiner & Kogel 2009).

The seminal work of Fertil and Malaise has shown that the survival of cell lines given small doses of radiation in vitro correlates well with perceived ability to cure corresponding human tumors (Fertil & Malaise 1985). The preferred technique for deriving radiosensitivity data from biopsies was to allow plated cells to grow in vitro. These cells were then irradiated, typically to doses similar to the single fraction doses given in conventional radiotherapy (2 Gy), and then the rate of survival (SF2) was measured. This assay was found to be labor-intensive and not reproducible between laboratories. Other traditional markers of radiobiology were also noted, such as potential doubling time and the gamma factor, which represents the slope of the survival curve at 50% survival rate (Choi et al. 2001). These concepts comprise the basis for analytical (mechanistic) outcomes modeling techniques as discussed below.

24.2.3 Analytical Outcome Modeling

These are methods generally based on simplified biophysical understanding of irradiation effects mainly developed from in vitro assays as discussed above. Analytical, some of which are mechanistic and others are phenomenological, models have been applied widely for designing radiotherapy clinical trials and for modeling TCP and NTCP as discussed below.

24.2.3.1 TCP

Tumor control is strictly defined by the probability of the extinction of clonogenic tumor cells at the end of treatment (Munro & Gilbert 1961). Several radiobiological models have been proposed in the literature to model TCP. The linear-quadratic (LQ) model is the most frequently used model for including the effects of repair between treatment fractions. The LQ model is based on clonogenic cell survival curves and is parameterized by the radiosensitivity ratio (α/β). It is thought that it quantifies the effects of both unrepairable damage and repairable damage susceptible to misrepair after tumor sterilization by radiation (Hall & Giaccia 2006; Joiner & Kogel 2009):

$$SF = \exp(-(\alpha + \beta * d) * D + \ln 2 * t/T_{pot})), \qquad (24.1)$$

where d is the fraction size, D is the total delivered dose, t is the difference between the total treatment time (T) and the lag period before accelerated clonogen repopulation begins (T_K), and T_{pot} is the potential doubling time of the cells. The ratio $\ln 2/T_{pot}$ is referred to as the repopulation parameter. Several variations of this model have been proposed including a Poisson-based (Goitein 1987) and a birth-death model (Zaider & Minerbo 2000). Among the most commonly used LQ-based TCP models (Hall 1994) is

$$\text{TCP} = \exp(-N \exp(-(\alpha + \beta * d) * D + \ln 2 * t/T_{pot})). \quad (24.2)$$

A detailed review of analytical methods for TCP in radiation treatment has been recently published (Zaider & Hanin 2011).

24.2.3.2 NTCP

Radiation-induced injuries and toxicities involve a complex cascade of radiobiological processes that can begin within few hours after irradiation and progress over weeks, months, and years (Emami et al. 1991; Fajardo et al. 2001; Bentzen 2006). According to the onset time of these toxicities, they are clinically classified into early and late effects. Early effects are typically transient and manifest during treatment or within a few weeks of the completion of fractionated radiotherapy schedule. These effects include skin erythema, mucositis, esophagitis, nausea, and diarrhea. Late effects are typically expressed after a latent period that can extend from months to years. These effects include radiation-induced inflammation and fibrosis, atrophy, vascular degeneration, and neural damage (Shrieve & Loeffler 2011). The most commonly used NTCP model is the Lyman–Kutcher–Burman (LKB) model (Lyman 1985; Kutcher & Burman 1989), which could be written as

$$\text{NTCP}(D, D_{50}, m) = \frac{1}{\sqrt{2\pi}} \int_{-\infty}^{t} \exp(-u^2/2) du, \quad (24.3)$$

where $t = \dfrac{D - D_{50}}{m D_{50}}$, D_{50} is the position of the 50% probability dose point, and m is a parameter to control the slope of the dose response. Note that D_{50} is expressed as function of the partial organ volume (V):

$$D_{50}(V) = D_{50}(1)/V^n, \quad (24.4)$$

where $D_{50}(1)$ is D_{50} for the whole volume and n is a volume dependence parameter. Another commonly used NTCP model is the critical volume (CV) model (Niemierko & Goitein 1993; Stavrev et al. 2001), which is based on the idea that organs are composed of functional subunits (FSUs), which are arranged in series or parallel architectures:

$$\text{NTCP}(\bar{u}_d, \mu_{cr}, \sigma) = \frac{1}{\sqrt{2\pi}} \int_{-\infty}^{t} \exp(-u^2/2) du, \quad (24.5)$$

where $t = \dfrac{-\ln(-\ln \bar{\mu}_d) - \ln(-\ln \mu_{cr})}{\sigma}$, $\bar{\mu}_d$ is the mean relative damaged volume, μ_{cr} is the critical fraction of FSUs, and σ accounts for the interpatient variability.

24.2.3.3 Equivalent Uniform Dose

An alternative approach to the mechanistic TCP/NTCP mentioned above is using phenomenological models such as the equivalent uniform dose (EUD) or generalized EUD (gEUD) (Niemierko 1999):

$$\text{gEUD} = \left(\sum_i v_i D_i^a \right)^{1/a}, \quad (24.6)$$

where v_i is the fractional organ volume receiving a dose D_i and a is a tissue-specific parameter that describes the volume effect. To account for the effects of cold spots on TCP, the tumor is represented by a negative a (<-10). In the case of NTCP, the value of a depends on the FSU organization; for serial-organ complications, a large a (>10) is selected; for parallel-organs complications, $a \sim 1$.

24.2.4 Data-Driven Outcome Modeling

These are primarily phenomenological models and depend on parameters available from the collected clinical and dosimetric data (Deasy & El Naqa 2008). In the context of data-driven and multivariable modeling of outcomes, the observed treatment outcome (e.g., TCP or NTCP) is considered as the result of functional mapping of several dosimetric, clinical, or biological input variables (El Naqa et al. 2006a). Mathematically, this is expressed as $f(\mathbf{x}; \mathbf{w}^*): X \rightarrow Y$, where $x_i \in \mathbb{R}^N$ (an input variable vector of N dimensions) is composed of the input metrics (dose-volume metrics, patient disease-specific prognostic factors, or biological markers). The expression $y_i \in Y$ is the corresponding observed treatment outcome scalar. The variable $\mathbf{w}^*$ includes the optimal parameters of model $f(\cdot)$ obtained by optimizing a certain objective functional. Learning is defined in this context as estimating dependencies from data (Hastie et al. 2001). There are two common types of learning: supervised and unsupervised. Supervised learning is used when the endpoints of the treatments such as tumor control or toxicity grade are known; these endpoints are provided by experienced oncologists following Radiation Therapy Oncology Group (RTOG) or National Cancer Institute (NCI) criteria and it is the most commonly used learning method in outcomes modeling. Nevertheless, unsupervised methods such as principal component analysis (PCA) are also used to reduce dimensionality and to aid visualization of multivariate data and selection of learning method parameters as described later (Härdle & Simar 2003).

The selection of the functional form of the model $f(\cdot)$ is closely related to the prior knowledge of the problem. In mechanistic models, the shape of the functional form is selected based on the clinical or biological process at hand, however, in data-driven models; the objective is usually to find a functional form that fits the data (El Naqa et al. 2009a). In the following discussion, we will describe traditional regression methods and artificial intelligence techniques as two prominent informatics tools for outcomes modeling.

24.2.4.1 Logistic Regression Models

In radiation outcomes modeling, the response will usually follow an S-shaped curve. This suggests that models with sigmoidal shape are more appropriate to use (Marks 2002; Bradley et al. 2004, 2007; Tucker et al. 2004; Blanco et al. 2005; Hope et al. 2005; Huang et al. 2011, 2012). A commonly used sigmoidal form is the logistic model, which also has nice numerical stability properties. The logistic model is given by (Hosmer & Lemeshow 2000; Vittinghoff 2005):

$$f(\mathbf{x}_i) = \frac{e^{g(\mathbf{x}_i)}}{1 + e^{g(\mathbf{x}_i)}}, \quad i = 1, \ldots, n, \tag{24.7}$$

where n is the number of cases (patients) and $\mathbf{x}_i$ is a vector of the input variable values used to predict $f(\mathbf{x}_i)$ for outcome y_i of the i_{th} patient. $f(\cdot)$ is referred to as the logit transformation. The "x-axis" summation $g(\mathbf{x}_i)$ is given by

$$g(\mathbf{x}_i) = \beta_o + \sum_{j=1}^{s} \beta_j x_{ij}, i = 1, \ldots, n, j = 1, \ldots, s, \tag{24.8}$$

where s is the number of model variables and the β's are the set of model coefficients that are determined by maximizing the probability that the data gave rise to the observations (i.e., the likelihood function). Many commercially available software packages, such as SAS, SPSS, and Stata, provide estimates of the logistic regression model coefficients and their statistical significance. The results of this type of approach are not expressed in closed form as above, but instead the model parameters are chosen in a stepwise fashion to define the abscissa of a regression model as shown in Figure 24.2. However, it is the analyst's responsibility to test for interaction effects on the estimated response, which can potentially be corrected by adding cross terms to Equation 24.2. However, this transformation suffers from limited learning capacity. In such a model, it is the user's responsibility to determine whether interaction terms or higher-order terms should be added. A solution to ameliorate this problem is offered by applying artificial intelligence methods.

24.2.4.2 Artificial Intelligence Methods

Artificial intelligence techniques (e.g., neural networks and decision trees), which are able to emulate human intelligence by learning the surrounding environment from the given input data, have also been utilized because of their ability to detect nonlinear patterns in the data. In particular, neural networks were extensively investigated to model post-radiation treatment outcomes for cases of lung injury (Munley et al. 1999; Su et al. 2005) and biochemical failure and rectal bleeding in prostate cancer (Gulliford et al. 2004; Tomatis et al. 2012). However, these studies have mainly focused on using a single class of neural networks, i.e., feed-forward neural networks (Haykin 1999) with different types of activation functions.

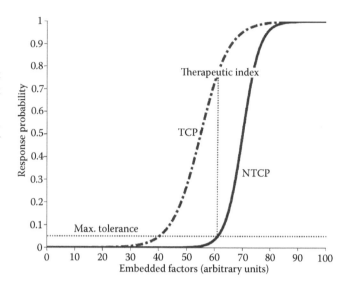

FIGURE 24.2 Sigmoidally shaped response curves (for TCP of NTCP) are constructed as a function of a linear weighting of various factors, for a given dose distribution, which may include multiple dose–volume metrics as well as clinical factors. The units of the x-axis may be thought of as "equivalent dose" units. (Reproduced from El Naqa, I. et al., *Int. J. Radiat. Oncol. Biol. Phys.*, 64, 1275, 2006.)

Feed-Forward Neural Networks (FFNN)

Neural networks are described as adaptive massively parallel-distributed computational models that consist of many nonlinear elements arranged in patterns similar to a simplistic biological neuron network. Typical neural network architecture is shown in Figure 24.3.

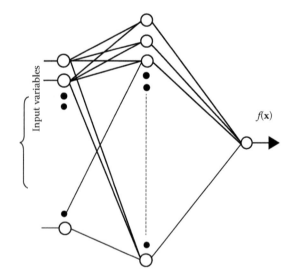

FIGURE 24.3 Neural network architecture consisting of an input layer, middle (hidden) layer(s), and an output layer. The synapses of the network consist of neurons that fire depending on their chosen activation functions.

Neural networks have been applied successfully to model many different types of complicated nonlinear processes, including many pattern recognition problems (Ripley 1996). A three-layer FFNN network would have the following model for the approximated functional:

$$f(\mathbf{x}) = \mathbf{y}^T \mathbf{w}^{(2)} + b^{(2)}, \tag{24.9}$$

where $\mathbf{y}$ is a vector, the elements of which are the output of the hidden neurons, i.e.,

$$y_i = s(\mathbf{x}^T, \mathbf{w}_i^{(1)} + b_i^{(1)}), \tag{24.10}$$

where $\mathbf{x}$ is the input vector and $\mathbf{w}^{(j)}$ and $b^{(j)}$ are the interconnect weight vector and the bias of layer j, respectively, $j = 1,2$. In the FFNN, the activation function $s(\cdot)$ is usually a sigmoid, but radial basis functions (RBF) were also used (Su et al. 2005). The FFNN could be trained in two ways: batch mode or sequential mode. In the batch mode, all the training examples are used at once; in sequential mode, the training examples are presented on a pattern basis, in an order that is randomized from one epoch (cycle) to another. The number of neurons is a user-defined parameter that determines the complexity of the network; the larger the number of neurons, the more complex the network would be. The number is determined during the training phase.

General Regression Neural Networks (GRNN)

The GRNN computes the best approximation model in the mean-squared sense. It is based on an estimate of the joint probability density of the input and the output obtained by the Parzen window method (Specht 1991). With training data, the output of the GRNN can be represented as:

$$f(\mathbf{x}) = \sum_{i=1}^{l'} y_i \exp\left(-\frac{\|\mathbf{x} - \mathbf{x}_i\|^2}{2\sigma^2}\right) \Bigg/ \sum_{i=1}^{l'} \exp\left(-\frac{\|\mathbf{x} - \mathbf{x}_i\|^2}{2\sigma^2}\right), \tag{24.11}$$

where y_i is the observed output for input variable vector $\mathbf{x}_i$ and $\sigma(> 0)$ defines empirically selected kernel width. Note that the implementation of GRNN does not require an optimization solver to obtain the weights, as in the case of FFNN. Also, the output is obtained a weighted sum of all the training samples, which could make it less efficient during running time. In our previous work, we demonstrated that GRNN can outperform traditional FFNN in radiotherapy outcomes prediction (El Naqa et al. 2005).

Kernel-Based Methods

Kernel-based methods and its most prominent member, support vector machines (SVMs), are universal constructive learning procedures based on the statistical learning theory (Vapnik 1998). For discrimination between patients who are at low risk versus patients who are at high risk of radiation therapy, the main idea of the kernel-based technique would be to separate these two classes with "hyperplanes" that maximizes the margin between them in the nonlinear feature space defined by implicit kernel mapping. The optimization problem is formulated as minimizing the following cost function:

$$L(\mathbf{w},\xi) = \frac{1}{2}\mathbf{w}^T\mathbf{w} + C\sum_{i=1}^{n}\xi_i, \tag{24.12}$$

subject to the constraint:

$$y_i(\mathbf{w}^T\Phi(\mathbf{x}_i) + b) \geq 1 - \zeta_i \quad i = 1,2,\dots,n$$
$$\zeta_i \geq 0 \text{ for all } i \quad,$$

where $\mathbf{w}$ is a weighting vector and $\Phi(\cdot)$ is a nonlinear mapping function. The ζ_i represents the tolerance error allowed for each sample being on the wrong side of the margin. Note that minimization of the first term in Equation 24.12 increases the separation (improves generalizabilty) between the two classes, whereas minimization of the second term (penalty term) improves fitting accuracy. The trade-off between complexity and fitting error is controlled by the regularization parameter C. However, such nonlinear formulation would suffer from the curse of dimensionality (i.e., the dimension of the problem becomes too large to solve) (Haykin 1999; Hastie et al. 2001). However, computational efficiency is achieved from solving the dual optimization problem instead of Equation 24.12, which is convex with a complexity that is dependent only on the number of samples (Vapnik 1998). The prediction function in this case is characterized only by a subset of the training data known as support vectors s_i:

$$f(\mathbf{x}) = \sum_{i=1}^{n_s} \alpha_i y_i K(\mathbf{s}_i, \mathbf{x}) + \alpha_0, \tag{24.13}$$

where n_s is the number of support vectors, α_i are the dual coefficients determined by quadratic programming, and $K(\cdot, \cdot)$ is the kernel function as discussed next.

Typically, used nonlinear kernels include

Polynomials: $K(\mathbf{x}, \mathbf{x}') = (\mathbf{x}^T\mathbf{x}' + c)^q$

Radial basis function (RBF): $K(\mathbf{x}, \mathbf{x}') = \exp\left(-\frac{1}{2\sigma^2}\|\mathbf{x} - \mathbf{x}'\|^2\right),$

$$\tag{24.14}$$

where c is a constant, q is the order of the polynomial, and σ is the width of the RBFs. The kernel-based approach is very flexible, which allows for constructing a neural network by using combination of sigmoidal kernels or choose a logistic regression equivalent kernel by replacing the hinge loss with a binomial deviance (Hastie et al. 2001).

24.2.4.3 Model Variable Selection

Any multivariate analysis often involves a large number of variables or features (Guyon & Elissee 2003). The main features that characterize the observations are usually unknown. Therefore, dimensionality reduction or subset selection aims to find the "significant" set of features. Finding the best subset of features is definitely challenging, especially in the case of nonlinear models. The objective is to reduce the model complexity, decrease the computational burden, and improve the generalizability on unseen data. The straightforward approach is to make an educated guess based on experience and domain knowledge and then to apply feature transformation (e.g., PCA) (Kennedy et al. 1998; Härdle & Simar 2003; Dawson et al. 2005), or sensitivity analysis by using organized search such as sequential forward selection, or sequential backward selection or combination of both (Kennedy et al. 1998). A recursive elimination technique that is based on machine learning has been also suggested (Guyon et al. 2002). In this technique, the data set is initialized to contain the whole set, train the predictor (e.g., SVM classifier) on the data, rank the features according to a certain criteria (e.g., ‖w‖), and keep iterating by eliminating the lowest ranked one. It should be noted that the specific definition of model order changes depending on the functional form. It could be identified by the number of parameters in logistic regression as in Equation 24.8 or by the number of neurons and layers in the case of neural networks (cf. Figure 24.3), etc. However, in any of these forms, the model order creates a balance between complexity (increased model order) and the model ability to generalize to unseen data. Finding this balance is referred to in statistical learning theory as the bias-variance dilemma (see Figure 24.4), in which an oversimple model is expected to underfit the data (large bias and small variance), whereas a too complex model is expected to overfit data (small bias and large variance) (Hastie et al. 2001). Hence, the objective is to achieve an optimal parsimonious model, that is, a model with the correct degree of complexity to fit the data and thus has a maximum ability to generalize to new, unseen, data sets.

Model Order Based on Information Theory

Information theory provides two intuitive measures of model order optimality such as: Akaike information criteria (AIC) and the Bayesian information criteria (BIC) (Burnham & Anderson 2002). AIC is an estimate of predictive power of a model, which includes both the maximum likelihood principle and a model complexity term that penalizes models with an increasing number of parameters (to avoid overfitting the data). BIC is derived from Bayesian theory, which results in a penalty term that increases linearly with the number of parameters.

Model Order Based on Resampling Methods

Resampling techniques are used for model selection and performance comparison purposes to provide statistically sound results when the available data set is limited (which almost always the case in radiotherapy). We use two types of fit-then-validate methods: cross-validation methods and bootstrap resampling techniques. Cross-validation (Kennedy et al. 1998) uses some of the data to train the model and some of the data to test the model validity. The type we most often use is the "leave-one-out" cross-validation (LOO-CV) procedure (also known as the "jackknife"). In each LOO-CV iteration, all the data are used for training/fitting except for one data point left out for testing, and this is repeated so that each data point is left out exactly once. The overall success of predicting the left-out data is a quantitative estimate of model performance on new data sets. Bootstrapping (Efron & Tibshirani 1993) is an inherently computationally intensive procedure but generates more realistic results. Typically, a bootstrap pseudo–data set is generated by making copies of original data points and randomly selected with a probability of inclusion of 63%. The bootstrap often works acceptably well even when data sets are small or unevenly distributed. To achieve valid results, this process must be repeated many times, typically several hundred or thousand times. Examples of applying these methods to outcomes modeling in radiotherapy could be found in our previous work (El Naqa et al. 2006a) and are discussed in detail by Deasy and El Naqa (2008).

24.2.4.4 Visualization of Data-Driven Models in Higher Dimensions

Prior to applying an informatics method, it is important to visualize the data distribution as a screening test. This requires projecting the data into a lower-dimensional space. As an example, one can choose the PCA approach due to its simplicity. In PCA analysis, the principal components (PC) of a data matrix **X** (with zero mean) are given by (Härdle & Simar 2003):

$$PC = U^T X = \sum V^T, \qquad (24.15)$$

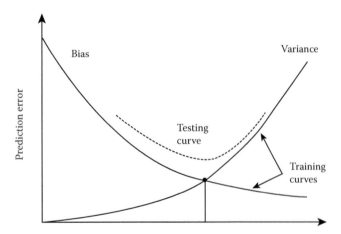

FIGURE 24.4 A common tradeoff in model predictive power between prediction bias (average error) and prediction variance (square error). As model complexity increases, the average prediction error (bias) tends to decrease, while the average square error tends to decrease. The point of optimal complexity tends to be near the point when average and square errors are of similar magnitude. (Reproduced from Deasy, J. O. & El Naqa, I., *Cancer Treat. Res.*, 139, 215, 2008. With permission.)

where $U\sum V^{T}$ is the singular value decomposition of $\mathbf{X}$. This is equivalent to transformation into a new coordinate system such that the greatest variance by any projection of the data would lie on the first coordinate (first PC), the second greatest variance on the second coordinate (second PC), and so on. For visualization purposes with the PCA, the heterogeneous variables are typically normalized using z-scoring (zero mean and unity variance). The term "Variance Explained," used in PCA plots (cf. Figure 24.5b), refers to the variance of the "data model" about the mean prognostic input factor values. The "data model" is

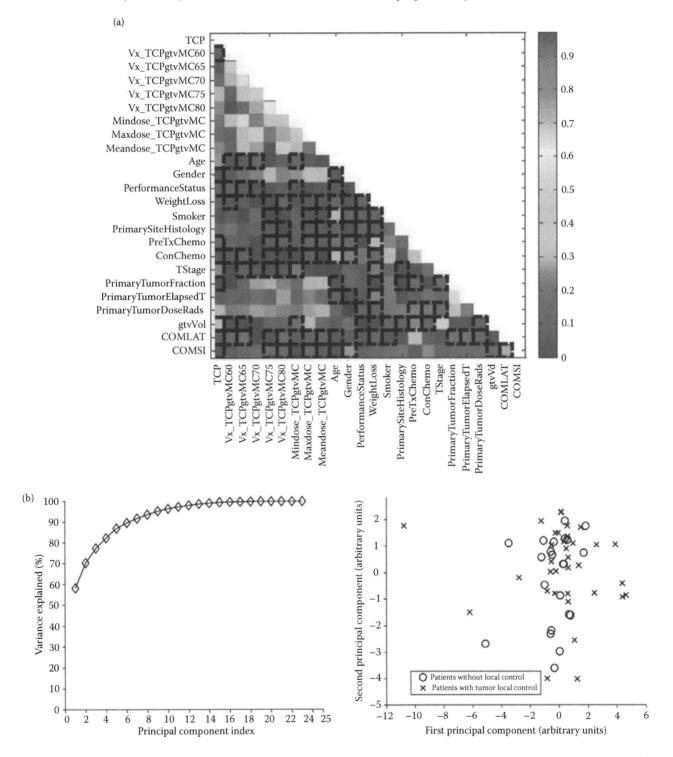

FIGURE 24.5 (a) Correlation matrix showing the candidate variables correlations with TCP and among the other candidate variables. (b) Visualization of higher-dimensional data by PCA. (left) The variation explanation versus PC index. (right) The data projection into the first two PC space. Note that the cases overlap.

formed as a linear combination of its PC. Thus, if the PC representation of the data "explains" the spread (variance) of the data about the full data mean, it would be expected that this PC representation capture enough information for modeling. Moreover, PCA analysis can provide an indication about class separability; however, it should be cautioned that PCA is an indicator and is not necessarily optimized for this purpose as supervised linear discriminant analysis, for instance (El Naqa et al. 2009a).

24.2.5 Evaluation and Validation Methods

24.2.5.1 Evaluation Metrics

To evaluate the performance of statistical classifiers, one can use the Matthews correlation coefficient (MCC) (Matthews 1975) as a performance evaluation metric for classification An MCC value of 1 would indicate perfect classification, a value of –1 would indicate anticlassification, and a value close to zero would mean no correlation. Alternatively, one can use the area under the receiver operating characteristics (AUROC) curve (Hanley & McNeil 1982), MCC and AUROC tends to be proportional, with ROC giving more pictorial representation of the performance. In case of regression, Spearman rank correlation (rs) has been widely applied which provides a simple robust estimator of trend (Sprent & Smeeton 2001).

24.2.5.2 Statistical Validation

Similar to model order selection, information theoretic methods (e.g., AIC or BIC) and resampling methods (e.g., cross-validation or bootstrapping) could be used for performance comparison purposes to provide statistically sound results when the available data set is limited (Efron & J. Tibshirani 1998; Kennedy et al. 1998). In the following examples, we will focus on resampling techniques because of their wide applicability.

24.3 Informatics for Radiotherapy Outcomes Modeling

In the following, we present different demonstrative examples from our work for applying informatics methods in radiotherapy TCP and NTCP outcomes modeling. Detailed information about these examples could be found in cited literature.

24.3.1 TCP Modeling

24.3.1.1 Data Set

A set of 56 patients diagnosed with non–small cell lung cancer (NSCLC) and have discrete primary lesions, complete dosimetric archives, and follow-up information for the endpoint of local control (22 locally failed cases) is used. The patients were treated with 3D conformal radiation therapy (3D-CRT) with a median prescription dose of 70 Gy (60–84 Gy). The dose distributions were corrected for heterogeneity using Monte Carlo simulations (Lindsay et al. 2007). The clinical data included age, gender,

performance status, weight loss, smoking, histology, neoadjuvant and concurrent chemotherapy, stage, number of fractions, tumor elapsed time, tumor volume, and prescription dose. Treatment planning data were de-archived and potential dose-volume prognostic metrics were extracted using Computational Environment for Radiotherapy Research (CERR) (Deasy et al. 2003). These metrics included Vx (percentage volume receiving at least x Gy), where x was varied from 60 to 80 Gy in steps of 5 Gy, mean dose, minimum and maximum doses, and center of mass (COM) location in the craniocaudal (COMSI) and lateral (COMLAT) directions. This resulted in a set of 23 candidate variables to model TCP. The modeling process using nonlinear statistical learning starts by applying PCA to visualize the data in 2D space and assess the separability of low-risk from high-risk patients. Nonseparable cases are modeled by nonlinear kernels. This step is preceded by a variable selection process and the generalizability of the model is evaluated using resampling techniques as discussed earlier and explained below (El Naqa et al. 2010).

24.3.1.2 Data Exploration

In Figure 24.5a, we show a correlation matrix representation of these variables with clinical TCP and cross-correlations among themselves using Spearman's rank correlation coefficient (rs). Note that many dose-volume histogram (DVH)–based dosimetric variables are highly cross-correlated, which complicate the analysis of such data. In Figure 24.5b, we summarize the PCA analysis of this data by projecting it into 2D space for visualization purposes. The plots show that two principle components are able to explain 70% of the data and reflects a relatively highly overlap between patients with and without local control, indicating potential benefit from using nonlinear kernel methods.

24.3.1.3 Logistic Regression Modeling Example

The multimetric model building using logistic regression is performed using a two-step procedure to estimate model order and parameters. In each step, a sequential forward selection strategy is used to build the model by selecting the next candidate variable from the available pool (23 variables in our case) based on increased significance using Wald's statistics (El Naqa et al. 2006a). In Figure 24.6a, we show the model order selection using the LOO-CV procedure. It is noticed that a model order of two parameters provides the best predictive power with rs = 0.4. In Figure 24.6b, we show the optimal model parameters' selection frequency on bootstrap samples (280 samples were generated in this case). A model consisting of gross tumor volume (GTV) ($\beta = -0.029$, $P = 0.006$) and GTV V75 ($\beta = +2.24$, $P = 0.016$) had the highest selection frequency (45% of the time). The model suggests that increase in tumor volume would lead to failure, as one would expect due to increase in the number of clonogens in larger tumor volumes. The V75 metric is related to dose coverage of the tumor, where it is noticed that patients who had less than 20% of their tumor covered by 75 Gy were at higher risk of failure. However, this approach does not account for possible interactions between these metrics nor does it account for higher order nonlinearities.

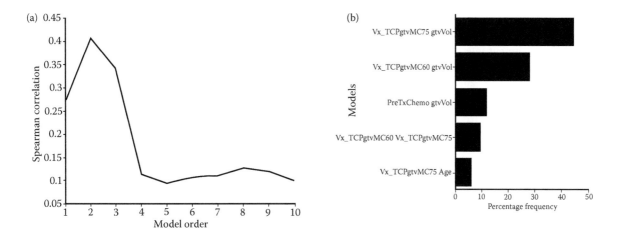

FIGURE 24.6 TCP model building using logistic regression. (a) Model order selection using LOO-CV. (b) Model parameters estimation by frequency selection on bootstrap samples.

24.3.1.4 Kernel-Based Modeling Example

To account for potential nonlinear interactions, we will apply kernel-based methods. Moreover, we will use the same variables selected by the logistic regression approach. We have demonstrated recently that such selection is more robust than other competitive techniques such as the recursive feature elimination (RFE) method used in microarray analysis. In this case, a vector of explored variables is generated by concatenation. The variables are normalized using the z-scoring approach to have a zero mean and unity variance (Kennedy et al. 1998). We experimented

with different kernel forms; best results are shown for the RBF in Figure 24.7a. The figure shows that the optimal kernel parameters are obtained with an RBF width σ = 2 and regularization parameter C = 10,000. This resulted in a predictive power on LOO-CV rs = 0.68, which represents 70% improvement over the logistic regression analysis results. This improvement could be further explained by examining Figure 24.7b, which shows how the RBF kernel tessellated the variable space nonlinearly into different regions of high and low risks of local failure. Four regions are shown in the figure representing high/low risks of local failure

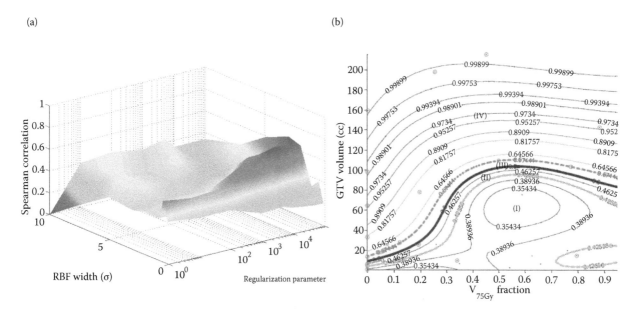

FIGURE 24.7 Kernel-based modeling of TCP in lung cancer using the GTV volume and V75 with SVM and a RBF kernel. Scatter plot of patient data (dots) being superimposed with failure cases represented with open circles. (a) Kernel parameter selection on LOO-CV with peak predictive power attained at σ = 2 and C = 10,000. (b) Plot of the kernel-based local failure (1-TCP) nonlinear prediction model with four different risk regions: (i) area of low-risk patients with high confidence prediction level, (ii) area of low-risk patients with lower confidence prediction level, (iii) area of high-risk patients with lower confidence prediction level, and (iv) area of high-risk patients with high confidence prediction level. Note that patients within the "margin" (cases ii and iii) represent intermediate-risk patients, which have border characteristics that could belong to either risk group.

with high/low confidence levels, respectively. Note that cases falling within the classification margin have low confidence prediction power and represent intermediate risk patients, that is, patients with "border-like" characteristics that could belong to either risk group (El Naqa et al. 2010).

24.3.1.5 Comparison with Other Known Models

For comparison purposes with mechanistic TCP models, we chose the Poisson-based TCP model and the cell kill EUD

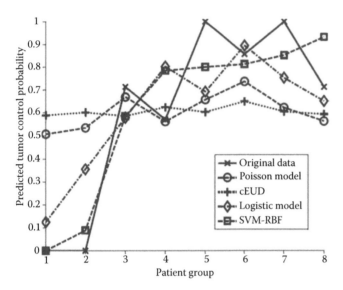

FIGURE 24.8 A TCP comparison plot of different models as a function of patients being binned into equal groups using the model with highest predictive power (SVM-RBF). The SVM-RBF is compared with Poisson-based TCP, cEUD, and best two-parameter logistic model. It is noted that prediction of low-risk (high control) patients is quite similar; however, the SVM-RBF provides a significant superior performance in predicting high-risk (low control) patients.

(cEUD) model. The Poisson-based TCP parameters for NSCLC were selected according to Willner et al. (2002), in which the sensitivity to dose per fraction (α/β = 10 Gy), dose for 50% control rate (D50 = 74.5 Gy), and the slope of the sigmoid-shaped dose-response at D50 (γ_{50} = 3.4). The resulting correlation of this model was rs = 0.33. Using D50 = 84.5 and γ_{50} = 1.5 (Martel et al. 1999; Mehta et al. 2001) yielded an rs = 0.33 also. For the cEUD model, we selected the survival fraction at 2 Gy (SF2 = 0.56) according to Brodin et al. (1991). The resulting correlation in this case was rs = 0.17. A summary plot of the different methods predictions as a function of binned patients into equal groups is shown in Figure 24.8. It is observed that the best performance was achieved by the nonlinear (SVM-RBF). This is particularly observed for predicting patients who are at high risk of local failure.

24.3.2 NTCP Modeling

24.3.2.1 Data Set

The data set consisted of NSCLC patients treated with 3D-CRT with median doses of approximately 70 Gy and 52 out of 219 patients were diagnosed with post-radiation pneumonitis (RP) (RTOG grade ≥3) (Hope et al. 2006). An independent data set from the RTOG-9311 after removing duplicate patients from our institution was used for evaluating generalizability to out-of-sample data (Bradley et al. 2005). These data sets contain clinical, dosimetric, and tumor location parameters. The clinical variables patient related information such as age, last follow-up date, status at follow-up, weight loss, gender, performance status, and smoking history. In addition to tumor characteristics such as tumor histology, GTV, and tumor stage, the dosimetric variables radiation prescription dose, mean lung dose (MLD), maximum dose (Gy), treatment time, fraction size (Gy), Vx, Dx, and the use of either sequential or concurrent chemotherapy. The high-dose location effect within the lung was analyzed using

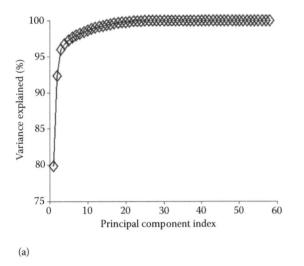

(a)

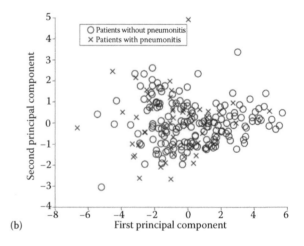

(b)

FIGURE 24.9 Visualization of the 58 variables related to RP by PCA. (a) Variation explanation versus PC index. This represents the variance of the data model about the mean prognostic input factor values. (b) Data projection into the first two components space. Note the high overlap in projection space, which suggests nonlinear kernel modeling to achieve better predictive power.

the center-of-mass (COM) of the GTV for each patient relative to lateral, anterior-to-posterior, and superior-to-inferior dimensions of the lung bounding box (Bradley et al. 2007).

24.3.2.2 Data Exploration

The PCA analysis is summarized in Figure 24.9. Notice that more than 99% of the variation in the input data was explained by the first two components. Additionally, the overlap between patients with and without RP is very high, suggesting that there is no linear classifier that can adequately separate these two classes. Similar overlap was also observed when other clinical and dosimetric variables were added emphasizing the nonlinear nature of this data (El Naqa et al. 2009a).

24.3.2.3 Logistic Regression Modeling Example

The results of the logistic regression model building are summarized in Figure 24.10. The best logistic model consisted of three parameters: D35, COMSI, and maximum dose (Hope et al. 2006). The model provided better fit to the data compared with classic models of V20 and MLD with rs = 0.22 using LOO-CV evaluation (Figure 24.10c).

24.3.2.4 Kernel-Based Modeling Examples

We will first consider dosimetric variables only to predict RP. Using the dosimetric variables Vx resulted in selecting polynomial kernels. The best performance was obtained with a polynomial kernel of order $p = 9$ and $C = 100$. It is noted that the

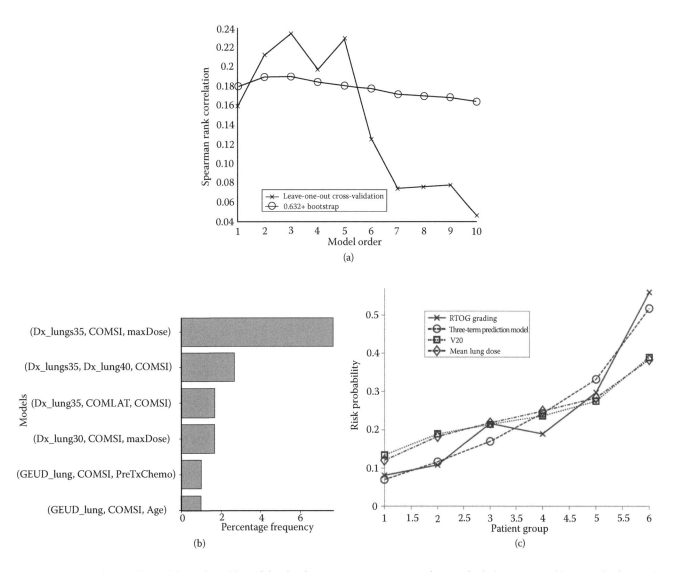

FIGURE 24.10 Multivariable modeling of RP. (a) Model order determination using resampling methods (LOO-CV and bootstrap) of RP with order of three and prediction power of 0.23 according to cross-validation. (b) Selection of best model ranking of RP using bootstrap. The best model consisted of D35, COMSI, and maximum dose. (c) Comparison of the three-term multivariable model with other existing metrics. Patients are binned according to predicted risk of pneumonitis by the model or parameter (equal patients per bin). Note that the relative risk classification between high-risk and low-risk groups is shallow (~3).

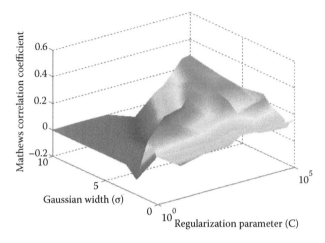
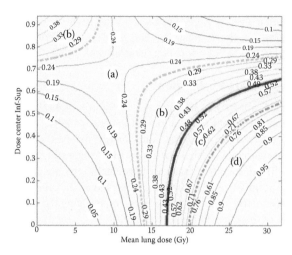

FIGURE 24.11 Kernel-based modeling in lung cancer using postradiation lung injury (pneumonitis). (left) The resulting SVM-RBF classifier on the LOO testing data. (right) The kernel-based pneumonitis nonlinear prediction model ($f(\mathbf{x})$) as a function of mean dose and dose center in the superior–inferior direction. The model is obtained using the WUSTL data (training set) and was evaluated on the independent RTOG data set (testing set). The plot shows four regions for risk prediction: (a) area of low-risk patients with high confidence prediction level, (b) area of low-risk patients with lower confidence prediction level, (c) area of high-risk patients with lower confidence prediction level, and (d) area of high-risk patients with lower confidence prediction level. Note patients within the "margin" (cases b and c) represent intermediate-risk patients, which have intertwined characteristics that could belong to either group.

higher the polynomial order, the better the prediction accuracy. However, a significant improved performance was noted by including nondosimetric variables when used combined data set with RTOG resulted in MLD and COMSI selection (Bradley et al. 2007). In Figure 24.11, we present the resulting RBF kernel-based RP nonlinear prediction model as a function f of MLD and COMSI with MCC = 0.36. There are four possible regions for prediction based on risk group and prediction confidence level: (1) region of low-risk patients with high confidence prediction level ($f \leq -1$), (2) region of low-risk patients with lower confidence prediction level ($-1 \leq f \leq 0$), (3) region of high-risk patients with lower confidence prediction level ($0 \leq f \leq 1$), and (4) region of high-risk patients with lower confidence prediction level ($f \geq 1$). These are translated into NTCP prediction probabilities using a sigmoidal function for illustration purposes. The lower confidence level group is patients whose characteristics lie within the margins for cases that are considered again "borderline" cases (El Naqa et al. 2009a).

24.3.2.5 Comparison with Other Known Models

For consistency with other reports, we will use Spearman's rank correlation. The SVM-RBF classifier for predicting RP risk achieved an rs of 0.37 (or MCC = 0.36) compared with V20 which yielded an rs = 0.18 (or MCC = 0.20) or with the best logistic regression model resulting in an rs = 0.22 using LOO-CV evaluation. These results thus provide a 60% improvement in prediction power. A comparison of risk prediction using proposed kernel-based approach versus conventional V20 and our previous three-term logistic regression model is shown in Figure 24.12, in which the patients are sorted in an ascending order and divided into eight equal groups. The distinctive ability of the kernel-based approach to fit both low-risk and high-risk groups

is demonstrated even in the case of imbalanced representation of events as in the RP case.

24.3.3 Software Tools for Outcomes Modeling

Many of the presented analytical/multimetric methods require dedicated software tools for implementation. As examples of such software tools in the literature are BIOPLAN and DREES. BIOPLAN uses several analytical models for evaluation of radiotherapy treatment plans (Sanchez-Nieto & Nahum 2000), while

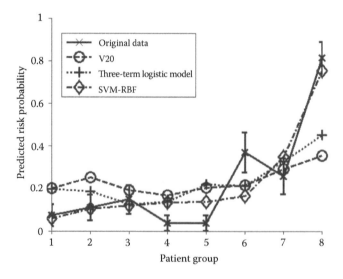

FIGURE 24.12 Risk plot comparison of different pneumonitis risk prediction models as a function of patients binned into equal groups. The SVM-RBF is compared with V20 and our previous best three-parameter logistic model.

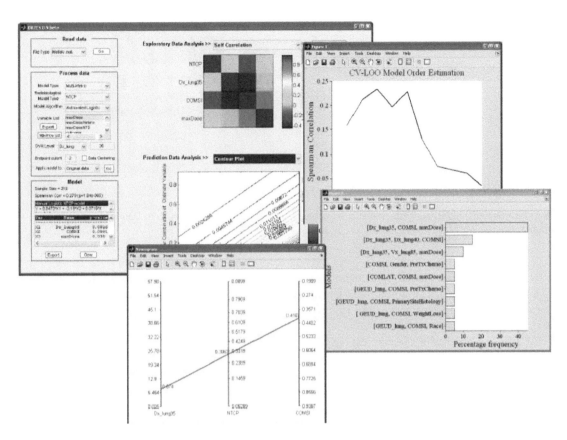

FIGURE 24.13 DREES allows for TCP/NTCP analytical and multivariate modeling of outcomes data. The example is for lung injury. The components shown here are main GUI, model order and parameter selection by resampling methods, and a nomogram of outcome as function of mean dose and location.

DREES is an open-source software package developed by our group for dose response modeling using analytical and multi-metric methods (El Naqa et al. 2006b) presented in Figure 24.13. It should be mentioned that several commercial treatment planning systems have currently incorporated different TCP/NTCP models, mainly analytical ones that could be used for plan ranking and biological optimization purposes. A discussion of these models and their quality assurance guidelines is provided in TG-166 (Allen Li et al. 2012).

24.4 Issues, Controversies, and Problems

24.4.1 Computational Modeling versus Statistics

There is a common mix-up between statistical analysis and outcomes modeling. The objective of statistical analysis is to use statistics to describe data and make inferences on the population for hypothesis testing purposes; for instance, variable x is significant while variable y is not in explaining the observed clinical endpoint (e.g., MLD for RP). In the case of computational modeling, the objective is to provide an adequate description of data and summarize its features for hypothesis generation as

summarized in Figure 24.14 (Berry & Linoff 2004). The arsenal of tools for outcomes modeling includes all available informatics approaches whether they were statistical or deterministic to enable proper simulation of events or prediction of outcomes.

24.4.2 Variable Selection Effects

Variable selection plays a crucial role in the performance of any prediction method (Guyon & Elissee 2003). It remains one of the main challenges in data-driven modeling, how to select the most relevant variables to include within the model. This is important clinically as well because it supports increased focus on potentially causative factors. In practice, however, the associations between the dosimetric, biological, clinical variables and the observed endpoints are generally unknown. Thus, the aim of the variable selection process is to find the best set of features that can improve the prediction power. In our previous work on multimetric modeling (El Naqa et al. 2006a), we investigated methods based on stepwise forward selection and resampling methods to extract significant variables and to demonstrate whether one variable set is robust versus a cohort of similarly performing models. Hence, future work will be dedicated to optimize the selection of the significant variables and interpret their clinical relevance. Moreover, based on our experiences,

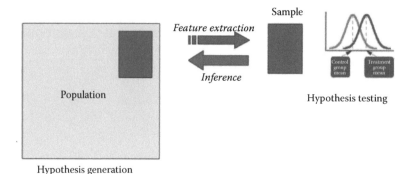

Hypothesis generation

FIGURE 24.14 Computational modeling versus statistics. (Adapted from Berry, M. J. A. & Linoff, G. *Data Mining Techniques: For Marketing, Sales, and Customer Relationship Management*, Wiley Pub., Indianapolis, IN, 2004.)

off-the-shelf techniques often fail to address the specificity of radiotherapy outcomes modeling. Using nonlinear sensitivity analysis of kernel methods (Rakotomamonjy 2003) in conjunction with resampling techniques may provide better opportunities for robust selection of relevant variables (El Naqa et al. 2009a).

24.4.3 The Need to Create Long-Lived Inter-Institutional Databases

Many of the hurdles toward creating and testing meaningful outcomes models in radiotherapy can at least partly be surmounted with the creation and evolution of long-lived, publicly available databases of treatment plans, imaging data, patient data, and resulting outcomes (Deasy & El Naqa 2008; Jackson et al. 2010). Whereas clinical trials are often designed to give the most unambiguous answer possible to a given single question, the need to model treatment response over a wide range of possible treatment and anatomy types means that it is desirable to collect a maximum range of dose distributions and relate those to treatment outcomes. This can be done by combining data sets from various sources because institutional or even interinstitutional databases related to a single clinical trial might have systematic trends that can hamper any type of extrapolation. Our experience combining multi-institutional data sets supports this view (Bradley et al. 2007). We share the belief that long-lived, publicly available archives of treatment plans and associated clinical and outcomes data represent the best tool for validating informatics outcomes models, which can be used across a wide range of treatment delivery techniques and patient characteristics.

24.5 Future Outcomes Modeling Research Directions

24.5.1 Beyond Dosimetric Variables

24.5.1.1 Imaging Features

Pretreatment or posttreatment information from anatomical or functional/molecular imaging could be used to monitor

and predict treatment outcomes in radiotherapy. For instance, changes in tumor volume on computed tomography (CT) have been used to predict radiotherapy response in NSCLC patients (Ramsey et al. 2006; Seibert et al. 2007). On the contrary, functional/molecular imaging, in particular positron emission tomography (PET) with fluorodeoxyglucose (FDG), has received special attention as a potential prognostic factor for predicting radiotherapy efficacy (Borst et al. 2005; Levine et al. 2006; Ben-Haim & Ell 2009; El Naqa et al. 2009b). For instance, high FDG-PET intensity has been shown to correlate with poor local control in lung cancer (Pieterman et al. 2000; Mac Manus et al. 2005; Yamamoto et al. 2006; Wong et al. 2007). In our previous work, new features based on image morphology, intensity, and texture/roughness can provide a more complete characterization of uptake heterogeneity (El Naqa et al. 2009b). Recently, we have shown that, in addition to PET features, CT-derived features (from the gross target volume) may also be used jointly to improve prediction of local tumor response (Vaidya et al. 2012).

24.5.1.2 Biological Markers

A biomarker is defined as "a characteristic that is objectively measured and evaluated as an indicator of normal biological processes, pathological processes, or pharmacological responses to a therapeutic intervention" (Biomarkers Definitions Working Group 2001). Biomarkers can be categorized based on the biochemical source of the marker into exogenous or endogenous.

Exogenous biomarkers are based on introducing a foreign substance into the patient's body such as those used in molecular imaging by PET as discussed above. Conversely, endogenous biomarkers can further be classified as (1) "expression biomarkers," measuring changes in gene expression or protein levels, or (2) "genetic biomarkers," based on variations, in tumors or normal tissues, in the underlying DNA genetic code. Measurements are typically based on tissue or fluid specimens, which are analyzed using molecular biology laboratory techniques (El Naqa et al. 2011). In our recent work, we have shown that integrating physical and biological variables can improve prediction of local control in NSCLC patients (Oh et al. 2011).

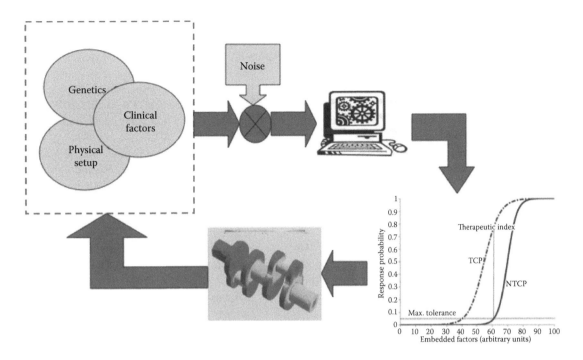

FIGURE 24.15 Informatics understanding of heterogeneous variables interactions as a feedback into the treatment planning system to improve patient's outcomes.

24.5.2 Systems Radiobiology

Systems biology has emerged as a new field to apply systematic study of complex interactions in biological systems (Alon 2007), but its application to radiobiology, despite its potential, has been unfortunately limited to date (Feinendegen et al. 2008; El Naqa 2012). Recently, Eschrich et al. presented a systems biology approach for identifying biomarkers related to radiosensitivity in different cancer cell lines using linear regression to correlate gene expression with SF2 measurements (Eschrich et al. 2009). However, such a linear regression model lacks the ability to account for higher-order interactions among the different genes and neglect the hierarchal relationships in signaling transduction of highly complex radiation response. It has been noted in the literature that modeling of molecular interactions could be represented using graphs of network connections as in power lines grids. In this case, radiobiological data can be represented as a graph (network) where the nodes represent genes or proteins and the edges may represent similarities or interactions between these nodes. We have utilized such approach based on Bayesian networks (BN) for modeling dosimetric RP relationships (Oh & El Naqa 2009) and most recently in predicting local control in lung cancer from biological and dosimetric data (Oh et al. 2011). In the more general realm of informatics, this systems approach could be represented as a part of a feedback treatment planning system as shown in Figure 24.15, in which informatics understanding of heterogeneous variables interactions could be used as an adaptive learning process to improve outcomes modeling and personalization of radiotherapy regimens.

24.6 Conclusions

Recent evolution in imaging and biotechnology has provided new opportunities for reshaping our understanding of radiotherapy response. However, the complexity of radiation-induced effects and the variability of tumor and normal tissue responses would render the utilization of advanced informatics and data mining methods as indispensible tools for better delineation of these complex interaction mechanisms and basically a cornerstone to "making data dream come true" (Nature Editorial 2004). Understanding the overlap of physical and biological factors in predicting complex radiotherapy response is necessary to make significant progress toward the goal of personalized treatment planning and improved quality of life for radiotherapy patients.

Acknowledgments

The author thanks Dr. Joseph O. Deasy from the Memorial Sloan Kettering Cancer Center and Dr. Jeffrey D. Bradley from the Washington University School of Medicine in St. Louis for their valuable insights and stimulating discussions. This work was partially supported by CIHR grant MOP-114910 and NSERC grant RGPIN 397711-11.

References

Allen Li, X., Alber, M., Deasy, J. O., Jackson, A., Ken Jee, K. W., Marks, L. B., Martel, M. K., Mayo, C., Moiseenko, V., Nahum, A. E. et al. (2012). The use and QA of biologically related models for treatment planning: Short report of the TG-166 of the therapy physics committee of the AAPM. *Med Phys, 39*, 1386–1409.

Alon, U. (2007). *An Introduction to Systems Biology: Design Principles of Biological Circuits*. Boca Raton, FL: Chapman & Hall/CRC.

Alsner, J., Andreassen, C. N., and Overgaard, J. (2008). Genetic markers for prediction of normal tissue toxicity after radiotherapy. *Semin Radiat Oncol, 18*, 126–135.

Armstrong, K., Weber, B., Ubel, P. A., Peters, N., Holmes, J., and Schwartz, J. S. (2005). Individualized survival curves improve satisfaction with cancer risk management decisions in women with BRCA1/2 mutations. *J Clin Oncol, 23*, 9319–9328.

Ben-Haim, S. & Ell, P. (2009). 18F-FDG PET and PET/CT in the evaluation of cancer treatment response. *J Nucl Med, 50*, 88–99.

Bentzen, S. M. (2006). Preventing or reducing late side effects of radiation therapy: Radiobiology meets molecular pathology. *Nat Rev Cancer, 6*, 702–713.

Bentzen, S. M., Constine, L. S., Deasy, J. O., Eisbruch, A., Jackson, A., Marks, L. B., Ten Haken, R. K., and Yorke, E. D. (2010). Quantitative Analyses of Normal Tissue Effects in the Clinic (QUANTEC): An introduction to the scientific issues. *Int J Radiat Oncol Biol Phys, 76*, S3–S9.

Berry, M. J. A. & Linoff, G. *Data Mining Techniques: For Marketing, Sales, and Customer Relationship Management*. Indianapolis, IN: Wiley Pub., 2004.

Biomarkers Definitions Working Group (2001). Biomarkers and surrogate endpoints: Preferred definitions and conceptual framework. *Clin Pharmacol Ther, 69*, 89–95.

Blanco, A. I., Chao, K. S., El Naqa, I., Franklin, G. E., Zakarian, K., Vicic, M., and Deasy, J. O. (2005). Dose-volume modeling of salivary function in patients with head-and-neck cancer receiving radiotherapy. *Int J Radiat Oncol Biol Phys, 62*, 1055–1069.

Borst, G. R., Belderbos, J. S., Boellaard, R., Comans, E. F., De Jaeger, K., Lammertsma, A. A., and Lebesque, J. V. (2005). Standardised FDG uptake: A prognostic factor for inoperable non-small cell lung cancer. *Eur J Cancer, 41*, 1533–1541.

Bortfeld, T., Schmidt-Ullrich, R., De Neve, W., and Wazer, D., Eds. *Image-Guided IMRT*. Berlin: Springer-Verlag, 2006.

Bradley, J., Deasy, J. O., Bentzen, S., and El-Naqa, I. (2004). Dosimetric correlates for acute esophagitis in patients treated with radiotherapy for lung carcinoma. *Int J Radiat Oncol Biol Phys, 58*, 1106–1113.

Bradley, J., Graham, M. V., Winter, K., Purdy, J. A., Komaki, R., Roa, W. H., Ryu, J. K., Bosch, W., and Emami, B. (2005). Toxicity and outcome results of RTOG 9311: A phase I-II dose-escalation study using three-dimensional conformal radiotherapy in patients with inoperable non-small-cell lung carcinoma. *Int J Radiat Oncol Biol Phys, 61*, 318–328.

Bradley, J. D., Hope, A., El Naqa, I., Apte, A., Lindsay, P. E., Bosch, W., Matthews, J., Sause, W., Graham, M. V., and Deasy, J. O. (2007). A nomogram to predict radiation pneumonitis, derived from a combined analysis of RTOG 9311 and institutional data. *Int J Radiat Oncol Biol Phys, 69*, 985–992.

Brahme, A. (1999). Optimized radiation therapy based on radiobiological objectives. *Semin Radiat Oncol, 9*, 35–47.

Brodin, O., Lennartsson, L., and Nilsson, S. (1991). Single-dose and fractionated irradiation of four human lung cancer cell lines in vitro. *Acta Oncol, 30*, 967–974.

Burnham, K. P. & Anderson, D. R. *Model Selection and Multimodal Inference: A Practical Information-Theoretic Approach*. New York: Springer, 2002.

Choi, N., Baumann, M., Flentjie, M., Kellokumpu-Lehtinen, P., Senan, S., Zamboglou, N., and Kosmidis, P. (2001). Predictive factors in radiotherapy for non-small cell lung cancer: Present status. *Lung Cancer, 31*, 43–56.

Dawson, L. A., Biersack, M., Lockwood, G., Eisbruch, A., Lawrence, T. S., and Ten Haken, R. K. (2005). Use of principal component analysis to evaluate the partial organ tolerance of normal tissues to radiation. *Int J Radiat Oncol Biol Phys, 62*, 829–837.

Deasy, J. O. & El Naqa, I. (2008). Image-based modeling of normal tissue complication probability for radiation therapy. *Cancer Treat Res, 139*, 215–256.

Deasy, J. O., Blanco, A. I., and Clark, V. H. (2003). CERR: A Computational Environment for Radiotherapy Research. *Med Phys, 30*, 979–985.

Efron, B. & Tibshirani, R. *An Introduction to the Bootstrap*. New York: Chapman & Hall, 1993.

Efron, B. & J. Tibshirani, R. *An Introduction to the Bootstrap*. Boca Raton: Chapman & Hall/CRC, 1998.

El Naqa, I. (2012). Machine learning methods for predicting tumor response in lung cancer. *Wiley Interdiscip Rev Data Mining Knowl Discov, 2*, 173–181.

El Naqa, I., Bradley, J., and Deasy, J. (2005). *Physical, Chemical, and Biological Targeting in Radiation Oncology*, Mehta, M. et al., Eds. Madison, WI: Medical Physics Publishing.

El Naqa, I., Bradley, J., Blanco, A. I., Lindsay, P. E., Vicic, M., Hope, A., and Deasy, J. O. (2006a). Multivariable modeling of radiotherapy outcomes, including dose-volume and clinical factors. *Int J Radiat Oncol Biol Phys, 64*, 1275–1286.

El Naqa, I., Suneja, G., Lindsay, P. E., Hope, A. J., Alaly, J. R., Vicic, M., Bradley, J. D., Apte, A., and Deasy, J. O. (2006b). Dose response explorer: An integrated open-source tool for exploring and modelling radiotherapy dose-volume outcome relationships. *Phys Med Biol, 51*, 5719–5735.

El Naqa, I., Bradley, J., Lindsay, P. E., Hope, A., and Deasy, J. O. (2009a). Predicting radiotherapy outcomes using statistical learning techniques. *Phys Med Biol, 54*, S9–S30.

El Naqa, I., Grigsby, P. W., Apte, A., Kidd, E., Donnelly, E., Khullar, D., Chaudhari, S., Yang, D., Schmitt, M., Laforest, R. et al. (2009b). Exploring feature-based approaches in PET images for predicting cancer treatment outcomes. *Pattern Recogn, 42*, 1162–1171.

El Naqa, I., Deasy, J., Mu, Y., Huang, E., Hope, A., Lindsay, P., Apte, A., Alaly, J., and Bradley, J. (2010). Datamining approaches for modeling tumor control probability. *Acta Oncol, 49*, 1363–1373.

El Naqa, I., Craft, J., Oh, J. and Deasy, J. *Adaptive Radiation Therapy*. Li, X. A., Ed. Boca Baton, FL: Taylor & Francis, 2011, 53–68.

Emami, B., Lyman, J., Brown, A., Coia, L., Goitein, M., Munzenrider, J. E., Shank, B., Solin, L. J., and Wesson, M. (1991). Tolerance of normal tissue to therapeutic irradiation. *Int J Radiat Oncol Biol Phys, 21*, 109–122.

Eschrich, S., Zhang, H., Zhao, H., Boulware, D., Lee, J.-H., Bloom, G., and Torres-Roca, J. F. (2009). Systems biology modeling of the radiation sensitivity network: A biomarker discovery platform. *Int J Radiat Oncol Biol Phys, 75*, 497–505.

Fajardo, L. F., Berthrong, M., and Anderson, R. E. *Radiation Pathology*. New York: Oxford University Press, 2001.

Feinendegen, L., Hahnfeldt, P., Schadt, E. E., Stumpf, M., and Voit, E. O. (2008). Systems biology and its potential role in radiobiology. *Radiat Environ Biophys, 47*, 5–23.

Fertil, B. & Malaise, E. (1985). Intrinsic radiosensitivity of human cell lines is correlated with radioresponsiveness of human tumors: Analysis of 101 published survival curves. *Int J Radiat Oncol Biol Phys, 11*, 1699–1707.

Fu, X. L., Zhu, X. Z., Shi, D. R., Xiu, L. Z., Wang, L. J., Zhao, S., Qian, H., Lu, H. F., Xiang, Y. B., and Jiang, G. L. (1999). Study of prognostic predictors for non-small cell lung cancer. *Lung Cancer, 23*, 143–152.

Goitein, M. *Evaluation of Treatment Planning for Particle Beam Radiotherapy*, Zink, S. Bethesda, MD: National Cancer Institute, 1987.

Gulliford, S. L., Webb, S., Rowbottom, C. G., Corne, D. W., and Dearnaley, D. P. (2004). Use of artificial neural networks to predict biological outcomes for patients receiving radical radiotherapy of the prostate. *Radiother Oncol, 71*, 3–12.

Guyon, I. & Elissee, A. (2003). An introduction to variable and feature selection. *J Mach Learn Res, 3*, 1157–1182.

Guyon, I., Weston, J., Barnhill, S., and Vapnik, V. (2002). Gene selection for cancer classification using support vector machines. *Mach Learn, 46*, 389–422.

Hall, E. J. *Radiobiology for the Radiologist*. Philadelphia: J.B. Lippincott, 1994.

Hall, E. J. & Giaccia, A. J. *Radiobiology for the Radiologist*. Philadelphia: Lippincott Williams & Wilkins, 2006.

Halperin, E. C., Perez, C. A., and Brady, L. W. *Perez and Brady's Principles and Practice of Radiation Oncology*. Philadelphia: Wolters Kluwer Health/Lippincott Williams & Wilkins, 2008.

Hanley, J. & McNeil, B. (1982). The meaning and use of the area under a receiver operating characteristic (ROC) curve. *Radiology, 143*, 29–36.

Härdle, W. & Simar, L. *Applied Multivariate Statistical Analysis*. Berlin; New York: Springer, 2003.

Hastie, T., Tibshirani, R., and Friedman, J. H. *The Elements of Statistical Learning: Data Mining, Inference, and Prediction: With 200 Full-Color Illustrations*. New York: Springer, 2001.

Haykin, S. *Neural Networks: A Comprehensive Foundation*. Prentice Hall, 1999.

Hope, A. J., Lindsay, P. E., El Naqa, I., Bradley, J. D., Vicic, M., and Deasy, J. O. Clinical, dosimetric, and location-related factors to predict local control in non-small cell lung cancer, in *ASTRO 47th Annual Meeting, Denver, CO*, 2005, S231.

Hope, A. J., Lindsay, P. E., El Naqa, I., Bradley, J. D., Alaly, J., Vicic, M., Purdy, J. A., and Deasy, J. O. (2006). Radiation pneumonitis risk based on clinical, dosimetric, and location related factors. *Int J Radiat Oncol Biol Phys, 65*, 112–124.

Hosmer, D. W. & Lemeshow, S. *Applied Logistic Regression*. John Wiley, 2000.

Huang, E. X., Hope, A. J., Lindsay, P. E., Trovo, M., El Naqa, I., Deasy, J. O., and Bradley, J. D. (2011). Heart irradiation as a risk factor for radiation pneumonitis. *Acta Oncol, 50*, 51–60.

Huang, E. X., Bradley, J. D., Naqa, I. E., Hope, A. J., Lindsay, P. E., Bosch, W. R., Matthews, J. W., Sause, W. T., Graham, M. V., and Deasy, J. O. (2012). Modeling the risk of radiation-induced acute esophagitis for combined Washington University and RTOG trial 93-11 lung cancer patients. *Int J Radiat Oncol Biol Phys, 82*, 1674–1679.

Jackson, A., Marks, L. B., Bentzen, S. M., Eisbruch, A., Yorke, E. D., Ten Haken, R. K., Constine, L. S., and Deasy, J. O. (2010). The lessons of QUANTEC: Recommendations for reporting and gathering data on dose-volume dependencies of treatment outcome. *Int J Radiat Oncol Biol Phys, 76*, S155–S160.

Joiner, M. & Kogel, A. v. d. *Basic Clinical Radiobiology*. London: Hodder Arnold, 2009.

Kennedy, R., Lee, Y., Van Roy, B., Reed, C. D., and Lippman, R. P. *Solving Data Mining Problems Through Pattern Recognition*. Prentice Hall, 1998.

Kutcher, G. J. & Burman, C. (1989). Calculation of complication probability factors for nonuniform normal tissue irradiation: The effective volume method. *Int J Radial Biol, 16*, 1623–1630.

Lehnert, S. *Biomolecular Action of Ionizing Radiation*. New York: Taylor & Francis, 2008.

Levine, E. A., Farmer, M. R., Clark, P., Mishra, G., Ho, C., Geisinger, K. R., Melin, S. A., Lovato, J., Oaks, T., and Blackstock, A. W. (2006). Predictive value of 18-fluoro-deoxy-glucose-positron emission tomography (18F-FDG-PET) in the identification of responders to chemoradiation therapy for the treatment of locally advanced esophageal cancer. *Ann Surg, 243*, 472–478.

Lindsay, P. E., El Naqa, I., Hope, A. J., Vicic, M., Cui, J., Bradley, J. D., and Deasy, J. O. (2007). Retrospective Monte Carlo dose calculations with limited beam weight information. *Med Phys, 34*, 334–346.

Lyman, J. T. (1985). Complication probability as assessed from dose-volume histogram. *Radiat Res Suppl, 8*, 13–19.

Mac Manus, M. P., Hicks, R. J., Matthews, J. P, Wirth, A., Rischin, D., and Ball, D. L. (2005). Metabolic (FDG-PET) response after radical radiotherapy/chemoradiotherapy for non-small cell lung cancer correlates with patterns of failure. *Lung Cancer, 49*, 95–108.

Marks, L. B. (2002). Dosimetric predictors of radiation-induced lung injury. *Int J Radiat Oncol Biol Phys, 54*, 313–316.

Martel, M. K., Ten Haken, R. K., Hazuka, M. B., Kessler, M. L., Strawderman, M., Turrisi, A. T., Lawrence, T. S., Fraass, B. A., and Lichter, A. S. (1999). Estimation of tumor control probability model parameters from 3-D dose distributions of non-small cell lung cancer patients. *Lung Cancer, 24,* 31–37.

Matthews, B. W. (1975). Comparison of the predicted and observed secondary structure of T4 phage lysozyme. *Biochim Biophys Acta, 405,* 442–451.

Mehta, M., Scrimger, R., Mackie, R., Paliwal, B., Chappell, R., and Fowler, J. (2001). A new approach to dose escalation in non-small-cell lung cancer. *Int J Radiat Oncol Biol Phys, 49,* 23–33.

Moiseenko, V., Kron, T., and Van Dyk, J. Biologically-based treatment plan optimization: A systematic comparison of NTCP models for tomotherapy treatment plans, in *Proceedings of the 14th International Conference on the Use of Computers in Radiation Therapy*, Seoul, Korea, 2004.

Moissenko, V., Deasy, J. O., and Van Dyk, J. *The Modern Technology of Radiation Oncology: A Compendium for Medical Physicists and Radiation Oncologists*. Van Dyk, J., Ed. Madison, WI: Medical Physics Publishing, 2005, 185–220.

Munley, M. T., Lo, J. Y., Sibley, G. S., Bentel, G. C., Anscher, M. S., and Marks, L. B. (1999). A neural network to predict symptomatic lung injury. *Phys Med Biol, 44,* 2241–2249.

Munro, T. R. & Gilbert, C. W. (1961). The relation between tumour lethal doses and the radiosensitivity of tumour cells. *Br J Radiol, 34,* 246–251.

Nature Editorial (2004). Making data dreams come true. *Nature, 428,* 239.

Niemierko, A. (1999). A generalized concept of equivalent uniform dose (EUD). *Med Phys, 26,* 1101.

Niemierko, A. & Goitein, M. (1993). Modeling of normal tissue response to radiation: The critical volume model. *Int J Radiat Oncol Biol Phys, 25,* 135–145.

Oh, J. H. & El Naqa, I. Bayesian network learning for detecting reliable interactions of dose-volume related parameters in radiation pneumonitis, in *International Conference on Machine Learning and Applications (ICMLA), Miami, FL*, 2009.

Oh, J. H., Craft, J., Al Lozi, R., Vaidya, M., Meng, Y., Deasy, J. O., Bradley, J. D., and El Naqa, I. (2011). A Bayesian network approach for modeling local failure in lung cancer. *Phys Med Biol, 56,* 1635–1651.

Pieterman, R. M., van Putten, J. W., Meuzelaar, J. J., Mooyaart, E. L., Vaalburg, W., Koeter, G. H., Fidler, V., Pruim, J., and Groen, H. J. (2000). Preoperative staging of non-small-cell lung cancer with positron-emission tomography. *N Engl J Med, 343,* 254–261.

Rakotomamonjy, A. (2003). Variable selection using SVM-based criteria. *J Mach Learn Res, 3,* 1357–1370.

Ramsey, C. R., Langen, K. M., Kupelian, P. A., Scaperoth, D. D., Meeks, S. L., Mahan, S. L., and Seibert, R. M. (2006). A technique for adaptive image-guided helical tomotherapy for lung cancer. *Int J Radiat Oncol Biol Phys, 64,* 1237–1244.

Ripley, B. *Pattern Recognition and Neural Networks*. Cambridge, 1996.

Sanchez-Nieto, B. & Nahum, A. E. (2000). Bioplan: Software for the biological evaluation of radiotherapy treatment plans. *Med Dosim, 25,* 71–76.

Seibert, R. M., Ramsey, C. R., Hines, J. W., Kupelian, P. A., Langen, K. M., Meeks, S. L., and Scaperoth, D. D. (2007). A model for predicting lung cancer response to therapy. *Int J Radiat Oncol Biol Phys, 67,* 601–609.

Shrieve, D. C. & Loeffler, J. S. *Human Radiation Injury*. Philadelphia: Wolters Kluwer Health/Lippincott Williams & Wilkins, 2011.

Specht, D. F. (1991). A general regression neural network. *IEEE Trans Neural Network, 2,* 568–576.

Spencer, S., Bonnin, D. A., Deasy, J. O., Bradley, J. D., and El Naqa, I. *Bioinformatics Methods for Learning Radiation-Induced Lung Inflammation from Heterogeneous Retrospective and Prospective Data*. Hindawi Publishing Corporation, 2009.

Sprent, P. and Smeeton, N. C. *Applied Nonparametric Statistical Methods*. Boca Raton: Chapman & Hall/CRC, 2001.

Stavrev, P., Stavreva, N., Niemierko, A., and Goitein, M. (2001). Generalization of a model of tissue response to radiation based on the idea of functional subunits and binomial statistics. *Phys Med Biol, 46,* 1501–1518.

Steel, G. G. *Basic Clinical Radiobiology*. London; New York: Arnold; Oxford University Press, 2002.

Su, M., Miften, M., Whiddon, C., Sun, X., Light, K., and Marks, L. (2005). An artificial neural network for predicting the incidence of radiation pneumonitis. *Med Phys, 32,* 318–325.

Tomatis, S., Rancati, T., Fiorino, C., Vavassori, V., Fellin, G., Cagna, E., Mauro, F. A., Girelli, G., Monti, A., Baccolini, M. et al. (2012). Late rectal bleeding after 3D-CRT for prostate cancer: Development of a neural-network-based predictive model. *Phys Med Biol, 57,* 1399.

Torres-Roca, J. F. & Stevens, C. W. (2008). Predicting response to clinical radiotherapy: Past, present, and future directions. *Cancer Control, 15,* 151–156.

Tucker, S. L., Cheung, R., Dong, L., Liu, H. H., Thames, H. D., Huang, E. H., Kuban, D., and Mohan, R. (2004). Dose-volume response analyses of late rectal bleeding after radiotherapy for prostate cancer. *Int J Radiat Oncol Biol Phys, 59,* 353–365.

Vaidya, M., Creach, K. M., Frye, J., Dehdashti, F., Bradley, J. D., and El Naqa, I. (2012). Combined PET/CT image characteristics for radiotherapy tumor response in lung cancer. *Radiother Oncol, 102,* 239–245.

Vapnik, V. *Statistical Learning Theory*. New York: Wiley, 1998.

Vittinghoff, E. *Regression Methods in Biostatistics: Linear, Logistic, Survival, and Repeated Measures Models*. New York: Springer, 2005.

Webb, S. *The Physics of Three-Dimensional Radiation Therapy: Conformal Radiotherapy, Radiosurgery, and Treatment Planning*. Bristol, UK; Philadelphia: Institute of Physics Pub., 2001.

Weinstein, M. C., Toy, E. L., Sandberg, E. A., Neumann, P. J., Evans, J. S., Kuntz, K. M., Graham, J. D. and Hammitt, J. K. (2001). Modeling for health care and other policy decisions: Uses, roles, and validity. *Value Health, 4,* 348–361.

West, C. M. L., Elliott, R. M., and Burnet, N. G. (2007). The genomics revolution and radiotherapy. *Clin Oncol, 19,* 470–480.

Willner, J., Baier, K., Caragiani, E., Tschammler, A., and Flentje, M. (2002). Dose, volume, and tumor control prediction in primary radiotherapy of non-small-cell lung cancer. *Int J Radiat Oncol Biol Phys, 52,* 382–389.

Wong, C. Y., Schmidt, J., Bong, J. S., Chundru, S., Kestin, L., Yan, D., Grills, I., Gaskill, M., Cheng, V., Martinez, A. A., and Fink-Bennett, D. (2007). Correlating metabolic and anatomic responses of primary lung cancers to radiotherapy by combined F-18 FDG PET-CT imaging. *Radiat Oncol, 2,* 18.

Yamamoto, Y., Nishiyama, Y., Monden, T., Sasakawa, Y., Ohkawa, M., Gotoh, M., Kameyama, K., and Haba, R. (2006). Correlation of FDG-PET findings with histopathology in the assessment of response to induction chemoradiotherapy in non-small cell lung cancer. *Eur J Nucl Med Mol Imaging, 33,* 140–147.

Zaider, M. & Hanin, L. (2011). Tumor control probability in radiation treatment. *Med Phys, 38,* 574–583.

Zaider, M. & Minerbo, G. N. (2000). Tumour control probability: A formulation applicable to any temporal protocol of dose delivery. *Phys Med Biol, 45,* 279–293.

Quality Assurance in Informatics

Sasa Mutic
Washington University
School of Medicine

Scott Brame
Radialogica LLC

25.1 Introduction

To paraphrase Andrew Carnegie, there is nothing more important to business than quality. In modern radiation therapy (RT), quality assurance (QA) is ubiquitous; every organization devotes a significant amount of effort to QA, numerous papers are published on the topic each month, and various advisory or regulatory bodies have issued QA guidelines. Two terms that are used less frequently are quality control (QC) and quality improvement (QI). Formal distinctions between QA, QC, and QI are discussed in the next section; for the present purposes, it is sufficient to understand that there is more to the pursuit of quality than QA.

In an age when RT treatment planning, treatment delivery, measurements, record keeping, and various other activities are increasingly computerized or computer-controlled, opportunities abound to reevaluate the approach to quality in RT. The expanded use of computers for managing or performing RT tasks has created a foundation for the development of an informatics infrastructure with the potential to revolutionize how the quality and safety of RT operations are quantified and managed. Existing processes can be adapted to more intelligently reflect the current state of computerized clinical workflow (both in terms of realizing promise and avoiding danger), and new, more powerful approaches and techniques can be pursued that were previously impractical in typical clinical settings owing to resource limitations. These new opportunities will facilitate changes in all aspects of quality-related efforts, from individual patient treatments and treatment planning to improvements in hardware and software operation to improvements in the operations of individual RT facilities and in radiation oncology as a field. Individual patient treatments will benefit, processes can be standardized between facilities of various sizes, locations, resources, and experience, and RT as a whole can become more consistent, affordable, and effective.

As is often the case, there is a gap between what is possible and what is realized, and numerous obstacles must be overcome if the full promise of informatics for QM in RT is to be realized. A few examples of these obstacles include the limited availability of informatics tools and infrastructures, the relatively slow development of recommendations by professional societies, perceptions that the existing processes are already adequate, and, when applicable, even changes in regulatory requirements. The purpose of this chapter is to provide a broad perspective on RT quality, discuss how the proliferation of computer-based process management and paperless departments creates new opportunities, and challenges, for managing quality, and highlight some of areas that are, or soon will be, benefiting from new and emerging informatics tools.

25.1.1 QC, QA, and QI

Successful implementation of QC, QA, and QI programs requires definition and understanding of these terms and cultivation of an organizational culture that places the utmost importance on quality-related efforts. Defining QC, QA, and QI is a difficult task that is complicated by the widespread variability in the descriptions and usage of the terms (Hoyle 2007). An in-depth analysis of these terms is beyond the scope of this text, and a first principles overview of these concepts should be adequate to understand the importance of informatics for the QM effort in RT. QA can generally be described as efforts to ensure that processes and operation of an organization are designed to consistently produce deliverables that satisfy predetermined minimum levels of quality during the lifetime of the organization. Therefore, QA efforts begin before any products or service are produced or delivered. QC, on the contrary, is a planned inspection of products or services designed to verify whether the aforementioned minimum levels of quality are met throughout the production or service delivery process. For the purposes of this chapter, QC is defined as activities that focus on the testing of individual deliverables while preventing patient harm at each major step in the RT process, whereas QA activities are focused on improving and stabilizing these processes and proactively minimizing issues that lead to errors and suboptimal deliverables. With these definitions, it should be clear that much of what is referred to as QA in RT can be considered as QC. This is neither surprising nor unique to RT: the interdependence of QA and QC often leads practitioners to think of them as one and the same. This lack of differentiation also tends to be more prevalent in institutions where QA and QC are the responsibilities of the same personnel, which is often the case in RT. The price of this intermingling is that the precepts that justify the differentiation in the formal quality literature are often misunderstood, resulting in suboptimal implementation of QA and QC programs. On a final note, neither QA nor QC stops with products or services: both should be applied to personnel as well. If those who produce deliverables are poorly trained, do not understand processes, or cannot interpret instructions, the quality of other dependent deliverables will suffer. Regardless of the vernacular, QA and QC expectations in RT should be defined by clinicians based on their understanding of requirements to safely and effectively deliver radiation to patients in such as way as to maximize the likelihood of a cure while minimizing the chance of complications.

QI is the mechanism by which the information and findings of QA and QC activities are acted upon. It is a continuous process that identifies problems, determines solutions or improvements, and monitors and evaluates the effects of these solutions and improvements. The aim of QI activities is to determine where the quality of deliverables stands and how to improve this quality and to implement these improvements in the operation and processes to monitor and evaluate their effects. QI activities are present in all RT facilities and are typically subsumed as part of routine clinical management and QA/QC efforts.

Collectively, QA, QC, and QI are part of an organization's quality management (QM) program. The QM program is typically embodied in a document, or set of documents, that lay out the overall process for managing quality. In RT, the level of formality and complexity in QM program documentation is not as important as in other disciplines (e.g., nuclear power or medical device industries); what is important is that an organization have, and communicate, formalized mechanisms for the improvement of its operations and evaluation of the effects of these improvements.

25.1.2 QC in Radiation Oncology

The sine qua non of quality in RT is patient safety. Given the potential for catastrophic errors that affect dozens of patients, the design and implementation of safety-related processes in RT are inconsistently approached by regulatory bodies. Several states and the U.S. Nuclear Regulatory Commission have substantial regulations for approach to QA and QC in RT. In many jurisdictions, guidance and support for QA/QC requirements in significant components of RT are left to professional societies or various national and international organizations. The QA and QC for the field are often determined by the work of early adopters of new technologies. From a technical and procedural perspective, RT is a rapidly developing specialty with significant new technologies being constantly introduced. A few technologies and processes that have been either introduced or have gone from a few users to widespread acceptance in recent years are the various forms of intensity-modulated RT (IMRT), ring-based linear accelerators (e.g., Hi-Art system, Tomotherapy, Inc., Madison, Wisconsin), linear accelerators mounted on a robotic arm (e.g., CyberKnife system, Accuray, Inc., Sunnyvale, California), planar kV x-ray and cone-beam computed tomography (CT) imaging capabilities on conventional linear accelerators, deformable image registration, extracranial stereotactic RT, various in-room localization and tracking devices, and proton beam RT. On many levels, an asynchrony is present between the implementation of new technologies and procedures and the ability of various regulatory bodies and other organizations to provide timely guidance for the development of QM processes for these technologies. Thus, early adaptors often develop QA and QC procedures that are then distributed to other practitioners and eventually endorsed by regulatory bodies and various other organizations. Once the formal regulations and recommendations are developed, they tend to be followed without routine evaluation of their efficiency and efficacy. The reason for this is that these recommendations tend to be prescriptive in nature. One of the shortcomings of these regulations and recommendations is the implicit position that everything that can be checked should be checked. This approach does not establish a framework for optimal distribution of resources for QM activities to maximize the safety and quality of patient care.

The American Association of Physicists in Medicine recognized these problems with the current approach to QM in RT and formed a task group (TG-100) to develop a new paradigm. This task group has identified that a systematic understanding of the likelihood and clinical impact of possible errors over the course of an RT treatment is needed to manage QM activities within the limitations of locally available resources such that maximum benefit is conferred to the safety and quality of patient care. This task group is developing a framework for designing QM activities, and hence allocating resources, based on estimates of clinical outcome, risk assessment, and failure modes. Toward this goal, the task group has adopted a Failure Modes and Effects Analysis (FMEA) tool (http://www.fmeainfo centre.com/) for the design of QM programs in RT. In FMEA, failure modes are prioritized based on the severity of their consequences, likelihood of failure, and likelihood of failure detection. Ideally, FMEA begins at the earliest part of development of a clinical service or technology implementation and continues throughout the life of the service or technology. FMEA is an analytical tool requiring failure rate–related data (e.g., likelihood of failure and likelihood of detection) and identification of potential failure modes. To be most effective, failure data should be measured whenever possible and failure modes documented in detail. This is where QC and FMEA intersect and where results obtained during QC can be used to govern the design of the FMEA process.

When this chapter was written, the TG-100 report had not been published, but it is clear that actual QC data will have a significant role in the development and definition of RT QM activities. The report is expected to require that QC data be systematically collected and organized so it can be used for development of QM efforts. Collection and organization of such data rely strongly on informatics systems in RT. These systems must introduce minimal impediments in routine clinical processes while gathering performance data at each major step in the RT process. Development of such informatics systems in RT is possible with an understanding of clinical operations and of the data needed to design meaningful changes in the existing processes. As discussed in later sections, one of the paradigm shifts presaged by this work is that processes, not their deliverables, will increasingly become a deliverable upon which QC is conducted.

25.1.3 Paperless Environment

The design of QC processes and systematic collection of performance and failure data in RT can potentially be revolutionized as more clinics become paperless in numerous aspects of their processes and activities. A paperless environment in RT initially implied the elimination of a paper-based daily dose delivery record. The concept of a paperless environment has since been expanded to include machines and other equipment, QA, chart checks, plan evaluations, daily patient localization reviews, many aspects of communication, patient visits, and numerous other activities. What is important to point out is that implementation of a paperless environment has two distinct stages. The first of these stages is simply a replacement of paper forms with electronic forms, and the second is a change of processes and operations based on the existence of an electronic environment. In the first stage, an organization is often focused only on taking paper-based forms and processes and implementing them in an electronic environment. This is not only a formidable and costly task but also one that offers a limited return on investment, as many of these processes tend to be cumbersome and difficult to follow. The first stage is often poorly accepted by staff because of the additional work required, inefficiencies, and cost. The second stage of the paperless environment acknowledges the recognition by an organization that reorganizing paper-based processes to embrace electronic and paperless operations offers numerous opportunities for improved clinical operations. In other words, Stage 1 is "going paperless for the sake of going paperless," and Stage 2 is "going paperless for the sake of more efficient, safer, automated, smarter, more effective, and better service processes." Operating in Stage 2 leads to the realization that an electronic world and informatics can be used to redefine how QC is performed and organized and how QC results are used to drive future services and clinical operations.

In RT, a paperless environment allows many aspects of QC efforts to become automated, more efficient, and more effective in identifying failures and problems; it further allows collection and application of past performance data to improve QC efforts as well as the overall efficacy of the operation. Examples of the use of electronic tools and informatics in RT QC are provided in subsequent sections of this chapter. It should be noted that these are only examples of what can be done and that an understanding of the full potential of informatics for improvement of quality in RT can be realized only as more clinics become paperless and collectively start gathering, exchanging, and analyzing data and developing processes that take advantage of these data.

A few words of caution are in order: the promise of new opportunities must be balanced with an awareness of the new dangers that get introduced with each new technology. Some of the greatest perils of today's paperless environment are incompatibilities between various clinical systems, inadequate design of many systems with limited ability to support various clinical workflows, and, perhaps the largest problem, the propensity of clinical staff to make data entry errors. Modern record and verify (R&V) systems are far from being robust; although their use generally can reduce errors (Schwarz et al. 2001), there is certainly opportunity for the introduction of errors as well (Patton et al. 2003). Use of such systems has been demonstrated to create a whole new set of different types of errors, inaccuracies, and challenges. It must be understood that, although information systems can improve many aspects of treatment planning and delivery, the technology is not perfect and it alone does not solve quality issues in RT.

25.2 Opportunities for Informatics in Radiation Oncology QC

Unquestionably, the quality of patient care and patient and staff safety can be improved in RT through increased gathering and analysis of performance data for various clinical systems, processes, and individuals. These data, when properly organized, analyzed, and integrated in daily workflows, can be used to identify performance issues as they arise and anticipate problems before they occur. To implement this infrastructure, a concept of RT QA Information Systems (QA-IS) needs to be developed and broadly accepted. Development of such systems is necessary, first of all, for the support of paperless QA activities within individual RT departments. More importantly, these systems are needed for the systematic collection, analysis, and benchmarking of clinical performance data. These data can then be used for improvement of QA and QC activities of individual patient treatments, various clinical areas, individual RT departments, and the practice of radiation oncology as a field. QA-IS should contain the following components:

- Error, near-miss, and corrective action data
- Individual equipment performance data and trends
- Individual process performance data
- Clinical area performance data
- Individual staff performance data

As clinics become paperless in the areas of treatment planning, treatment delivery, and treatment management, there is an increasing need for QA and QC activities to become paperless as well. Although paperless practices are currently not widespread, individual clinics and commercial vendors are pursuing this goal. One such example is the Mallinckrodt Institute of Radiology (MIR) Radiation Oncology Department, where all patient treatment operations as well as QA/QC activities are virtually paperless. Because commercial systems have not matured to the point of providing comprehensive and customizable solutions for paperless RT environments, the current MIR implementation for patient treatment as well as QA/QC operations is a combination of commercial and homegrown paperless solutions. The concept implemented by the department is to use commercial tools whenever possible and to develop custom solutions whenever necessary. For example, daily QA for linear accelerators is managed by a commercial product with a centralized server that connects all accelerators at the main site and a satellite; all other QA activities for linear accelerators, simulators, high-dose rate (HDR) machines, and Gamma Knife (among others) are managed by a centralized homegrown QA-IS. Patient-specific QA is managed, performed, and tracked through a combination of commercial and homegrown systems. The management and performance of QC activities is also facilitated by a combination of commercial and homegrown systems, whereas tracking and analysis of failures detected during QC activities is managed through a homegrown QC system. These systems have significantly improved the efficiency of numerous clinical activities and have translated to improved quality of patient treatments.

Clearly, this implementation relies heavily on homegrown tools that require significant resources for development and support and may not necessarily be practical at most other centers. To fully realize the potential of informatics in radiation oncology, standardized commercial systems that support QA/QC activities and the QA-IS concept will have to become commonly available and accepted by the field. The examples and concepts provided in this chapter should support motivation for the development of such systems.

25.2.1 QC for Best Practices

Perhaps the greatest opportunity for informatics in RT QC is the standardization of best practices. To facilitate this, QA-IS for individual facilities must be able to pass performance information to centralized databases for analysis, benchmarking, and, hopefully, for the standardization of processes across participating facilities. Figure 25.1 shows a diagram for collecting QC data in individual facilities through a QA-IS and exchange of those data with one or more centralized databases.

To date, significant efforts have been made to standardize best practices through cooperative group multi-institutional clinical trials. Unfortunately, many of these efforts have lacked validation of treatment delivery based on the organized collection and standardization of QC data that minimize process variations and reduce the uncertainty of subsequent analyses. Through the efforts of the Radiologic Physics Center at The University of Texas M. D. Anderson Cancer Center, it is understood that at

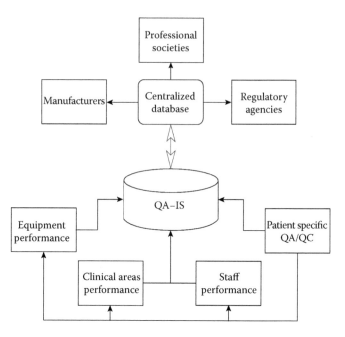

FIGURE 25.1 Schematic diagram of a QA-IS.

least some components of treatment delivery quality vary significantly among different institutions. For example, data from the Radiologic Physics Center showed that approximately 30% of institutions participating in credentialing for an IMRT delivery-related trial failed to deliver the dose indicated in their treatment plans to within 7% dose agreement or 4 mm distance to agreement to an anthropomorphic phantom used for credentialing trial participation (Ibbott et al. 2008). (Interestingly, most institutions have adopted 3% or 3 mm as their internal tolerances.) Although it can be argued that a portion of these observed failures is attributable to the complexity of the credentialing process and, perhaps, phantom design, some proportion of institutions undoubtedly had, or may still have, inaccuracies inherent in their treatment planning, delivery equipment, and corresponding processes. In that same report, Ibbott et al. (2008) noted that the process of irradiating phantoms as part of credentialing has identified the following errors: (1) incorrect output factors and percentage depth dose data entered by institutions; (2) inadequate modeling of the penumbra at multileaf collimator leaf ends; (3) incorrect application of QA calculations or measurements; (4) inadequate QA of multileaf collimators; (5) incorrect patient positioning, including couch indexing errors with serial tomography system; and (6) errors in treatment-planning software.

Pawlicki and Mundt (2007) also question the current approach to QC in RT and offer as an example the requirement to perform a measurement for every IMRT case before treatment. They note that whether or not this actually improves quality is still a point of debate but that independent anthropomorphic-based IMRT plan verification performed by an outside group can bring user errors to light and thus lead to improvements in quality. Understanding and benchmarking process variations between

RT institutions is key to an individual institution's ability to drive toward best practices and to the field's collective pursuit of standardized and equitable best practices. As Pawlicki and Mundt (2007) discuss, high-quality treatments require both minimizing process variation and moving the average closer to the optimum value. For RT as a field, it is required to understand what this optimum value is and what the acceptable limits of variation are.

The establishment of centralized healthcare databases for the collection of QI data has been a significant topic of interest since the Institute of Medicine published two reports, *To Err is Human* (IOM 2000) and *Crossing the Quality Chasm* (IOM 2001). These reports demonstrated the need to improve the quality of healthcare delivery as a whole and the disparity between the ability to delivery high-quality care and the quality of care that most Americans actually receive. In a report titled *Building a Better Delivery System*, the National Academy of Engineering and Institute of Medicine (2005) acknowledged the problems presented in the earlier two reports and provided a framework for a systems approach to healthcare delivery and a plan to "transform the U.S. healthcare sector from an underperforming conglomerate of independent entities (individual practitioners, small group practices, clinics, hospitals, pharmacies, community health centers, etc.) into a high-performance 'system' in which participating units recognize their interdependence and the implications and repercussions of their actions on the system as a whole."

One traditional obstacle for the exchange of QC statistics, errors, and failures among institutions has been concerns regarding litigation risk. The Patient Safety and Quality Improvement Act of 2005 created a mechanism for the establishment of "patient safety organizations" to collect and analyze confidential information reported by healthcare institutions. The Act provides federal legal privilege and confidentiality protections for information that is collected or developed by patient safety organizations for the purposes of improving patient safety. The Act significantly limits the use of this information in criminal, civil, and administrative proceedings and includes provisions for monetary penalties for confidentiality violations. The Act also calls for the establishment of a network of patient safety databases to provide an interactive, evidence-based management resource for healthcare providers, patient safety organizations, and other entities. The hope is that, by using common formats (definitions, data elements, and communication standards) and promoting interoperability among reporting systems, the network of patient safety databases will facilitate the analysis of national and regional statistics, including trends and patterns of events related to patient safety. The principles of this Act provide means for the establishment of centralized databases for QC/QA in RT and for the collection and analysis of data for the radiation oncology field as a whole. This is a lofty goal for the use of informatics in RT QC, and the form in which it will eventually materialize is not yet clear. However, if individual RT facilities are to learn collectively and be able to benchmark their performances and identify best practices, such systems must be established. Semblances of such systems already exist. One such system is the Radiologic Physics Center; another is the Radiation Oncology Safety Information System (http://www.clin.radfys.lu.se/default.asp), a voluntary Web-based safety information database for RT designed for individual institutions to report clinical incidents and corrective actions.

25.2.2 QC of RT Processes

As noted above, the aim of QC is to verify the performance of a system to either demonstrate and document correct functioning or to process defects and verify and monitor their correction. For much of the history of RT, the focus of quality-oriented tasks has been on deliverables produced throughout the treatment planning process, with the goals of ensuring that the plan generated for a patient is consistent with current best practices, that what is produced in the treatment planning process is indeed delivered to the patient, and that catastrophic errors are avoided or minimized. However, as is true in all facets of healthcare, radiation oncology is continually expanding its view of quality, pursuing not only avoidance of harm but also provision of the full potential of today's technology and science to as many patients as possible in as cost-effective a manner as possible. For an organization to meet this challenge, it must continually assess its weakness and gaps and then identify and implement processes and corrective actions that redress vulnerabilities. The logical extension of historical QA/QC practices in RT is to begin to see processes themselves as deliverables whose quality should be monitored and actively managed. As mentioned above, TG-100 presages this evolution, as does the salience of topics like comparative effectiveness or pay-for-performance. In this section, several techniques for quantifying and controlling the quality of processes are discussed as well as several informatics tools that are currently being used for QC of the treatment planning and machine QA process and will, in the future, allow QC of the scheduling process.

To support this new focus, RT practitioners have increasingly begun to look for answers in the tools and techniques that have been developed over many decades for the management of industrial processes. An in-depth explanation of these techniques or this general area of investigation is beyond the scope this chapter, but a brief overview of their philosophies and tools will help to frame opportunities for informatics-based QC of RT processes. To limit the scope of the introduction, two of the more common techniques of process management in use today, "Lean" and "Six Sigma" (Pyzdek 2003; Carroll 2008), are discussed below.

As implied by the name, "Lean" process management is concerned with efficiency and waste in processes. One of its core precepts is that slow processes tend to be overly expensive and of low quality. Slow-moving inventory must be moved, counted, stored, moved again, recounted, and so on. Every handoff creates an opportunity for an error or inefficiency to be introduced. Moreover, when a problem arises, the progress of the other inventory in the supply chain is in jeopardy, as is the quality of the final products. The simplest way to accommodate slow-moving inventory is with excess capacity, be it staff or other resources. In the process improvement literature, this excess capacity is referred to as the "hidden factory" and is something to be minimized (George 2002). Aside from the direct expenses associated with maintaining excess capacity, there are many indirect costs, process complexity being an obvious and potentially damaging one. Lean process management aims at finding the points in a process that do not directly add value and either engineering these points out of the process or reducing the time that a product spends therein.

"Six Sigma" process management, on the contrary, is concerned with placing processes under statistical control. Unlike the Lean approach, which indirectly influences quality by optimizing the predictability and efficiency of a process, the Six Sigma approach is focused more directly on the quality of process deliverables. The name Six Sigma comes from the goal of reducing defects to the 6σ level (i.e., 3.4 defects per million opportunities). Defects almost always reference variation; as noted earlier, good processes are notable for the average performance as well as limited variability about the mean. It is also worth mentioning that Six Sigma concerns itself as much with an organization's culture as it does with the data collection and analytical framework need for statistical process control. The key elements of a Six Sigma program are a management system committed to top performance, a measure to define the capability of any process, and goals for improvement.

Accepting that the quality of a process itself is an important deliverable, one must ask how approaches such as Lean or Six Sigma can be leveraged for use in RT. From a Lean point of view, for example, the treatment planning process can be seen to have many hand-offs and is punctuated by long periods where the inventory (individual patient plans) sit idle, waiting for the next team member to pick them up. Clearly, there are opportunities to improve this workflow and create a more robust process that is characterized by more consistent "inventory" flow. On the contrary, Six Sigma techniques force administrators and clinicians to engage in discussions related to what constitutes a "defect," with the goals of helping the organization work as a whole toward minimizing near-misses and process inefficiencies that are too often harbingers of actual treatment errors.

Regardless of the particular approach, the mechanics of most process improvement approaches can be grouped into the following five components (commonly referred to as Define Measure Analyze Improve Control (DMAIC) (De Feo 2005):

- Define: Identify opportunities for improvement, goals, and value to organization.
- Measure: Collect data on the problem.
- Analyze: Use the data to separate symptoms from problems and probe the real depths of the process improvement opportunities.
- Improve: Systematically strive to reduce defects and improve process velocity.
- Control: Once the problem has been solved, put measures into place to ensure that the problem is either controlled or prevented from recurring.

For the RT treatment planning process, one can apply QC to a variety of intermediate work products at different stages of the planning process; examples include instructions given, or settings used, for simulation, registrations between multimodality images, contours, prescriptions, and beam angle. However, from an industrial process perspective, it is clear that the length of, and variability in, the time required to perform many of the individual tasks should also be subjected to QC techniques, as should the frequency of deviations or mistakes that require corrective actions. The length of time required to generate, for example, a treatment plan for head-and-neck cancer matters, not just for the purposes of scheduling patient start dates but also because of the tumor growth that is often observed between the time of simulation and start of treatment (Seel & Foroudi 2002). Moreover, it is well known that seemingly benign process inefficiencies or defects can lead to sentinel events (Reason 1997). The process itself matters, not just its output.

While few would disagree with the potential benefits of QC for RT processes or, conversely, most would agree with the definition of the problem and its potential value to the organization and the field, the reality has been that, until recently, neither the tools nor the imperative have existed that would allow RT practitioners to move beyond the definition phase of the DMAIC process. Fortunately, the transition to paperless environments, the proliferation of Web-based technologies, and the increasing focus on the quality and cost of healthcare are all combining to make QC of RT processes both possible and more common.

A logical first step in measuring process performance is process mapping. A variety of techniques and variants of process maps have been described (George 2002). Two recent examples in the RT literature were reported by Ford et al. (2009), and Mutic et al. (2010) and Oddiraju et al. (2009); the latter is shown in Figure 25.2. One of the benefits of process mapping is that it allows the user to identify points in the process where interesting data can be collected. Another is that process mapping begs the question of what sorts of data should be collected.

If process efficiency is accepted as being strongly correlated with process quality, then the distinction made between errors and near-misses or inefficiencies may be an artificial one. Consider the frequency with which plans submitted for physician approval are rejected. From one point of view, plan rejection is validation of the process; something substandard was caught before it affected a patient. However, from a Six Sigma point of view, something that could have led to substandard treatment existed in the system; this is a defect, a near-miss, and effort should be applied to reducing or eliminating the occurrence of substandard plans. Likewise, from a Lean point of view, the rejection of a plan adds stress and expense to the system; depending on the workload, other patients may be affected or the patient start date may have to be moved. In other words, under a Lean scheme, no additional value is added in a replan, so the frequency of replanning should be minimized and the planning/replanning process should be optimized so that the turnaround time is minimized.

At the MIR, several systems have been developed and implemented to support the measurement and analysis phases of

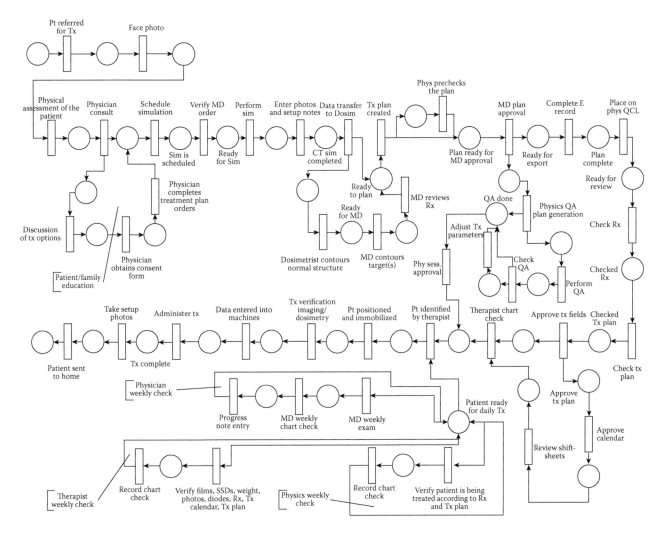

FIGURE 25.2 Treatment planning process map with Petri nets.

DMAIC. The first is a Web-based event reporting system (Oddiraju et al. 2009 and Mutic et al. 2010). The taxonomy of the events is informed by the process map and reinforced through periodic training. Linking the process map to the event taxonomy allows identification of points in the processes that either maintain a large number of events or show significant variability. Figure 25.3 shows how event frequencies are mapped onto a portion of the process map shown in Figure 25.2. The analysis, improvement, and control components of the process improvement program are effected by a management team populated with members from each division, ensuring that a breadth of expertise informs root cause analyses and that all participants in the process are invested in its improvement.

The second system is a Web-based program for managing and collecting data on simulation, treatment planning, and QA processes. Figure 25.4 illustrates a computer screenshot from this system, commonly referred to as an "electronic whiteboard," other examples of which have been described elsewhere (Vaarkamp et al. 2009). This system, and derivatives of it, is used to manage everything except the actual patient treatment, for which the existing R&V system is judged sufficient. On the surface, the board functions as any whiteboard: it can be viewed in a central location, assignments and current task states are updated as the process evolves, and so on. However, as others have noted (Vaarkamp et al. 2009), the tool improves throughput and has the advantage that all of the time data are stored and available to support process analysis—either to identify problems or to assess the impact of corrective actions.

There are several keys to leveraging these tools, and the data they provide, in support of the analysis, improvement, and control phases of DMAIC:

- Consistent terminology driven by the process map and event taxonomy.
- Ubiquitous access—Both the reporting and managing/collecting systems are Web-based and available throughout the department.
- Ease of data collection—When possible, data should be collected automatically; when manual reporting or entry is required, it should take no more than a couple of minutes.
- Use of standard code bases and tool sets for the underlying databases and analysis tools.

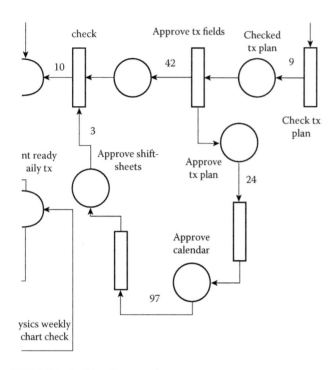

FIGURE 25.3 Mapping event frequency onto process map.

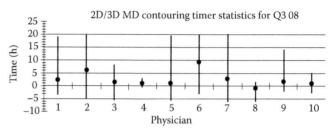

FIGURE 25.5 Actual versus allotted physician contouring times.

Data and inference exist to support decisions, meaning these tools and the data they collect are nothing more than a means to an end. Next, consider three examples of how these tools and the QC they enable can be used to affect the quality of RT processes and, ultimately, the quality of patient treatments. The first example shows how QC can affect the treatment planning process, the second the machine QA process, and the last the scheduling process.

Figure 25.5 shows agreement between the time allotted for the physician to contour the patient CT data set versus that actually required (collected via the electronic whiteboard). As expected, some physicians are consistently on time, some are consistently early, and some are consistently late. Intuitively, one would

expect that those that are consistently late would introduce stress into the treatment planning process. The effects of this stress can be examined by inspection of the event reporting data with an eye toward establishing correlations between timeliness and number and severity of near-misses and events. With this type of data, the MIR system can make an informed assessment of how severe the effects of the inefficiency are, propose solutions (e.g., new tools, more staff, different processes), implement them, and then monitor their effect. This example of how QC of the process (versus a more traditional focus on intermediate deliverables, which would have only verified the quality of the contours produced) can drive change and improve the quality of treatments and reduce errors.

The second example deals with machine QA. Historically, machine QA has focused on detecting faults. At the MIR, the combination of the electronic whiteboard and a paperless QA allows the physics staff to proactively anticipate equipment problems and, thanks to mature literature in the field of predictive maintenance, assess the cost associated with different maintenance schemes. Mills et al. (2008) give a nice overview of some techniques that can be used for proactive monitoring of equipment performance, all of which are made more practical because of the informatics infrastructure previously discussed. These authors describe how intelligent systems can be applied for QC of RT machines and immediate, short-term, and long-term monitoring. The authors also note the need in RT to move

Therapy #	Last	First	Site	Tomo	Seq #	Att. MD	Res. MD	Dos	Sim Date	Fusion	MD Appr of Sim/Ctr/Fus/Blks (if app)	Dos Ctr	Plan Comp	Prechck	MD Plan Appr	Chart to Phys	Pre port date	Start Date	Notes	Dos Done
				No		DBM		MBW		N/a	Yes	Yes	Yes	No	No	No				No
				Yes	IZ		DP	TSJ		N/a	Yes	Yes	Yes	Yes	No	No				No
				No		DBM		RJB	Complete	Yes	Yes	Yes	Yes	Yes	No				No
				No		JDB	KJB	MBW	Complete	Yes	Yes	Yes	Yes	Yes	Yes				No
				No		JMM		MBW	Complete	Yes	Yes	Yes	Yes	Yes	No				No
				Yes		DBM		LAW		N/a	Yes	Yes	Yes	Yes	No	No				No
				No		JMM		RJB		N/a	Yes	Yes	Yes	Yes	No	No				No
				No		WLT	RMT	MBW		N/a	Yes	Yes	Yes	Yes	No	No				No
				No		RJM		RJB		N/a	Yes	Yes	Yes	Yes	Yes	No				No
				No		WLT	RMT	LAW		N/a	Yes	Yes	No	No	No	No				No
				No		WLT	RMT	SHE		N/a	Yes	Yes	No	No	No	No				No
				Yes		PWG	JKS	TSJ		N/a	Yes	Yes	No	No	No	No				No
				No		WLT	RMT	MSR	Complete	Yes	Yes	No	No	No	No				No
				No		PJP	DJM	MSR		N/a	Yes	Yes	No	No	No	No				No
				No		JMM		MSR	Complete	Yes	Yes	No	No	No	No				No
				No		DBM		MBW	Complete	Yes	Yes	No	No	No	No				No
				Yes		JRS	DAM	TSJ	Complete	Yes	Yes	No	No	No	No				No
				No		JMM		MSR		N/a	Yes	Yes	No	No	No	No				No
				No		PWG	JKS		Complete	No	Yes	No	No	No	No				No
				No		JDB	KJB	KWB	Complete	Yes	Yes	No	No	No	No				No

Add Record   Physics checklist   Physics sort

FIGURE 25.4 Screenshot of the MIR electronic whiteboard.

away from QC being just a fault detection system to where QC is used as an indicator of potential failures. In this concept, QC results would be used to determine when intervention should be performed and proactively avoid failures.

The third example deals with a topic that has been the subject of increasing focus in the literature, namely, patient scheduling (Song et al. 2007; Mishra et al. 2008; Petrovic & Leite-Rocha 2008). The patient schedule is relevant with respect to patient and employee satisfaction, treatment efficiency, and perhaps, as noted earlier, treatment efficacy. The difficulty in developing optimal schedules for RT has been noted by several authors (Petrovic et al. 2006) and involves the heterogeneity of the patient population and treatment techniques and the need for accommodating machine uptimes, among other factors. In this example, only the scheduling of patient start dates is considered—a difficult problem, but much simpler than managing an adaptive schedule for the entire patient treatment process.

At present, two models exist for scheduling treatment start dates. The first is based on setting the start data between simulation and treatment start based on a fixed interval that typically depends on treatment type, meaning 3D treatments get one interval, IMRT gets another, and so on. With this approach, patients receive their start dates around the time of simulation. The second approach sets the start date later in the process, typically notifying the patient a few days before the treatment is to begin. Although neither approach is "wrong," both would benefit from more quantitative data on actual process flows and sophistication in how the data are used. It is straightforward to envision how data from the first example of this section could be used to produce robust start dates that reflect the dual pursuits of efficiency and safety/minimal process stress. Interest in the RT scheduling process is only going to increase as financial pressures grow and the sophistication and diversity of treatment techniques continue to evolve. This reality can be anticipated and accommodated by making the schedule itself a deliverable and applying QC measures to it using informatics tools such as those that have been developed and deployed at the MIR.

These examples are brief, but it should be clear that RT processes themselves can and should be the subject of routine QC. Thanks to the emergence of new informatics tools, this is rapidly becoming possible. In all cases, the proliferation of data-driven processes reinforces a foundation for the sort of field-wide benchmarking and data exchange needed to revolutionize the QC of best practices.

25.2.3 QC of Plan Quality

The evaluation of treatment plan quality has grown in complexity as treatment planning and delivery capabilities have become more sophisticated. This is especially true for IMRT planning, where plan quality for the same CT data set and planning goals can vary significantly among different planning systems, planners, or institutions (Bohsung et al. 2005; Williams et al. 2007). Too often, plan quality is judged based on its individual merits

with respect to target coverage, dose conformality, critical structure sparing, plan complexity, and numerous other merits. Evaluating a single plan individually, without consideration of similar plans that have been completed previously, relies on the reviewer's ability to discern if a particular plan is as good as possible, and, if the plan does not meet certain minimal quality requirements, that the discrepancy is due to the difficulty of the case rather than to a suboptimal or erroneous creation of the treatment plan. For IMRT planning, even if all plan objectives are met, the likelihood is still high that the calculated dose distributions are not as good as those that can be practically achieved. This inability to consistently create plans of high- or highest-quality results largely from the complexity of the state-of-the-art system, the immaturity of plan optimization algorithms and their implementation, and the varying experience of treatment planners.

A comparison of individual treatment plans against a cohort of high-quality treatment plans with similar features offers a possibility for the identification of substandard or erroneous plans and reduced variability in plan quality. Azmandian et al. (2007) described an electronic plan checker that relies on data mining and machine learning techniques to electronically check treatment plans and identify outliers and potentially erroneous treatment plans. Similar applications that look for inconsistent plan parameters have been described by others as well (Furhang et al. 2009; Siochi et al. 2009). These applications mainly concentrate on automatically identifying specific treatment plan parameter outliers against a cohort of acceptable standards. Items compared in these applications include a magnitude of monitor units, linear accelerator settings, beam energy selection, and other similar parameters. A step further is an application where dose distributions and target and critical doses are compared against those of similar plans that have been previously verified to be of sufficient quality. In the extended utility, the software would compare dose conformality, dose volume histograms, the magnitude and location of hot spots, and other similar parameters. It is not assumed that these applications would be a replacement for manual plan quality verification and review. It is likely that a plan would first be processed through such a utility and any potential concerns would be identified by the software and later analyzed by a human reviewer. Such a process would increase the likelihood of identifying possible planning errors and substandard dose distribution and could result in reduced plan quality variability and an overall increase in treatment quality.

One problem with benchmarking plan quality is the continuing evolution of treatment planning systems and the introduction of new treatment planning and delivery techniques. However, the past few years have brought stabilization in the understanding of IMRT delivery capabilities and, as a result, the capability to perform interinstitutional and intrainstitutional benchmarking of IMRT plan quality is expanding. Logically, one would expect that benchmarking and standardization of plan quality would start within an individual institution and then expand to support interinstitutional comparisons.

25.2.4 QC of Plan Consistency

A large majority of patient treatments in modern RT are delivered using computer-controlled treatment delivery processes as defined by Fraass (2008). The field has made significant progress moving from manual entry of treatment data and manual collection of treatment verification information to computer-controlled processes. Although state-of-the-art R&V systems still do not interface with many delivery devices (e.g., HDR machines, specialized radiosurgery devices, and some newer external beam delivery devices), most modern linear accelerators are capable of computer-controlled treatment delivery. Not only is the input, control, verification, and recording of patient treatment information computerized but so is the collection and processing of daily patient positioning verification data and in vivo dosimetry data. As Fraass (2008) has noted, computer-controlled delivery and verification processes require a change in conventional QC activities for verification of patient treatments and offer opportunities for improvement of these processes. Specifically, computerized treatment delivery and verification allows:

- Automation of many related QC processes
- Computer-aided QC activities
- Increased QC frequency
- Use of statistical processing tools for trend analysis and proactive implementation of corrective actions

- Improved distribution of QC results and notification of out of tolerance items
- Benchmarking
- Improved evaluation of QA programs

Some of these items require systematic data collection and analysis for QC management and necessitate the use of some form of QA-IS. Such systems are still not widely available, but several applications have been published (Patel & Kirby 2007; Fraass 2008; Yan 2008; Furhang et al. 2009; Gerard et al. 2009; Siochi et al. 2009) that have components of the above listed features. Furhang et al. (2009) described a process for automating initial physics chart-checking. In their implementation, the process is divided into two components, intraplan and interplan reviews. In the intraplan review, R&V system parameters are checked against an approved treatment plan through a computer interface that facilitates automation of this process. The interplan review involves comparison of current plan parameters using statistical process control against similar plans that were previously processed. Figure 25.6 shows a report page from a system described by Furhang et al. (2009). Pawlicki et al. (2005) described use of statistical process control for RT QA. In this implementation, similar cases are categorized according to diagnosis, anatomic site, laterality, delivery technique, and fractionation scheme. Various field parameters (Linac settings, treatment aids, patient positioning, and others) are then compared. Each

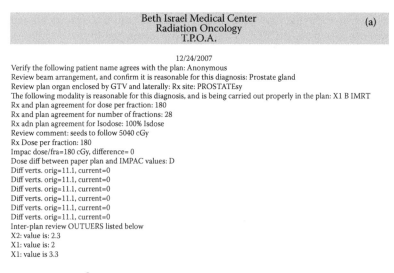

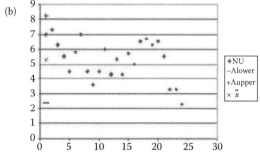

FIGURE 25.6 Report page. (From Furhang, E. E. et al., *J. Appl. Clin. Med. Phys., 10*(1), 129–135.)

tracked plan parameter is automatically compared against its upper and lower thresholds, and flags are created when a parameter exceeds its upper or lower threshold. Manual comparisons such as these are practically impossible and, more importantly, in this application physicist time is concentrated on investigating potential problems rather than on pure comparison of values for various plan parameters.

Siochi et al. (2009) also describe a similar process for plan checks in paperless RT clinics. In this application, the authors concentrate mainly on the automation of clinical processes, but their approach demonstrates the important concepts for use of informatics in RT QC. These authors describe a clinically functioning electronic RT plan QA system that is used for QC of plan quality assessment, treatment plan parameter export, and data integrity verification between the treatment planning system and the R&V system. The system is first used to extract plan data from the planning system. During this step, dosimetrists can add additional information. The authors noted that one of the shortcomings of modern planning systems is that they do not contain all of the relevant information about patient treatment; for example, prescriptions are incomplete, there is no information about gating, and detailed information on use of treatment aids is missing. This import process allows dosimetrists to supplement treatment planning system data with additional treatment information. The software compares individual plans during export against clinical practice standards and provides a warning when certain plan parameter inconsistencies are discovered. The software further verifies that all the fields in the R&V system match those of the plan in the treatment planning system and verifies a logical consistency of some other manually entered parameters (beam energy, fraction does, etc.) in the R&V system and the treatment plan. The authors report that the system has discovered several errors in their processes, and although it has not reduced the actual amount of time required to perform a chart check, the automation has allowed the practice to check much more data for individual patients. Some of the errors in their processes were not discovered because of the relatively early stage in the use and design of the system. From their experience, the authors recommend that vendors of treatment planning system and R&V databases should adopt standards that define the minimum required data sets and locations where they should be located for the purpose of electronic data interchange and integrity checks.

As the above two example demonstrate, a paperless environment and computer-controlled treatment delivery enable improved verification and quality of patient treatment plans and reduced variability between planned and delivered radiation doses. Clearly, these systems will evolve with wider clinical acceptance, as planning processes in RT become more automated and as treatment planning practices change with the development of adaptive RT. Although the form of eventual implementation of informatics-based QC processes for treatment delivery are less important in the context of this text, it is much more important to note that systems that can detect inconsistencies in individual patient treatments with respect to planned delivery or statistics from a cohort of similar patients can offer improvement in the quality of patient treatments and efficiency and scope of QC activities. Verification of daily patient localization, image-guided delivery, and corresponding issues (van Herk et al. 2000; Lam et al. 2005) are beyond the scope of this chapter, but the concepts presented here for verification of patient treatment information can be applied to this area as well.

25.3 Conclusions

QC and QA are critical components in modern RT; how these practices are defined and performed must be continually assessed and improved. Paperless clinical environments, computer-controlled treatment delivery, increased use of measurement and QA systems that are connected to a database, and maturity of treatment planning techniques all present significant opportunities for QA/QC improvements. The development of workflow management tools, event reporting systems, electronic QA systems, and QA-IS has the potential to facilitate improved quality of RT services, benchmarking and standardization across multiple practices, and improved efficiency of treatment planning and delivery operations and QC. The ability to automate many QC processes and to develop systems that would enable trend detection and use of statistical processes for QA management could drastically change the RT approach to the design of QA systems. This is demonstrated by formation of the American Association of Physicists in Medicine Task Group (TG-100), which has been charged with addressing the design of QA processes in RT.

Although many technologies in RT have been rapidly accepted and adopted, informatics-based QC processes are emerging more slowly. The pace of development can be improved with better standardization of communication processes among different clinical systems and better guidance and encouragement from regulatory and professional bodies.

References

Azmandian, F. et al. (2007). Towards development of an error checker for radiotherapy treatment plans: a preliminary study. *Phys. Med. Biol., 52*, 6511–6524.

Bohsung, J. et al. (2005). IMRT treatment planning—A comparative inter-system and inter-center planning exercise of the QUASIMODO group. *Radiother. Oncol., 76*, 354–361.

Carroll, B. J. *Lean Performance ERP Project Management: Implementing the Virtual Lean Enterprise.* Boca Raton, FL: Auerbach Publications, 2008.

De Feo, J. A. *JURAN Institute's Six Sigma Breakthrough and Beyond—Quality Performance Breakthrough Methods.* New York, McGraw-Hill, 2005.

Ford, E. C. et al. (2009). Evaluation of safety in radiation oncology setting using failure mode and effects analysis. *Int. J. Radiat. Oncol. Biol. Phys., 74*(3), 852–858.

Fraass, B. A. (2008). QA issues for computer-controlled treatment delivery: This is not your old R/V system and more! *Int. J. Radiat. Oncol. Biol. Phys., 71*(1), S98–S102.

Furhang, E. E. et al. (2009). Automating the initial physics chart-checking process. *J. Appl. Clin. Med. Phys., 10*(1), 129–135.

George, M. *Lean Six Sigma*. New York: McGraw-Hill, 2002.

Gerard, K. et al. (2009). A comprehensive analysis of IMRT dose delivery process using statistical process control (SPC). *Med. Phys., 36*(4), 1275–1285.

Hoyle, D., *Quality Management Essentials*. Oxford, UK: Elsevier Ltd., 2007.

Ibbott, G. S. et al. (2008). Challenges in credentialing institutions and participants in advanced technology multi-institutional clinical trials. *Int. J. Radiat. Oncol. Biol. Phys., 71*(S1), S71–S75.

Institute of Medicine (IOM). *To Err is Human: Building a Safer Health System*. Kohn, L. T. et al., Eds. Washington, DC: National Academy Press, 2000.

Institute of Medicine (IOM). *Crossing the Quality Chasm: A New Health System for the 21st Century*. Washington, DC, National Academy Press, 2001.

Lam, K. L. et al. (2005). An application of Bayesian statistical methods to adaptive radiotherapy. *Phys. Med. Biol., 50*, 3849–2858.

Mills, J. A. et al. Control methods and intelligent systems in the quality control of megavoltage treatment machines, in *Intelligent and Adaptive Systems in Medicine*. Haas, O. C. L. & Burnham, K. J., Eds. New York: Taylor & Francis, 2008.

Mishra, N. et al. A novel case based reasoning approach to radiotherapy planning, in *18th Triennial Conference of the International Federation of Operational Research Societies, Sandton, South Africa*, 2008.

Mutic, S. et al. (2010). Design and mapping of complex healthcare processes using IDEF0: A radiotherapy example. *Int. J. Coll. Enter. 1*, 316–331.

National Academy of Engineering and Institute of Medicine. *Building a Better Delivery System*. Reid, P. P. et al., Eds. Washington, DC, National Academies Press, 2005.

Oddiraju, S. et al. (2009). Modeling and error analysis of the clinical process in radiation therapy. *Med. Phys., 36*, 2769.

Patel, I. & Kirby, M. C. (2007). Design and implementation of an electronic data recording and processing system for physics quality control checks in external beam radiotherapy. *Br. J. Radiol., 80*, 126–131.

Patton, G. A. et al. (2003). Facilitation of radiotherapeutic error by computerized record and verify systems. *Int. J. Radiat. Oncol. Biol. Phys., 56*(1), 50–57.

Pawlicki, T. et al. (2005). "Statistical process control for radiotherapy quality assurance," *Med. Phys. 32*, 2777–2786

Pawlicki, T. & Mundt, A. J. (2007). Quality in radiation oncology. *Med. Phys., 34*(5), 1529–1533.

Petrovic, S. & Leite-Rocha, P. Constructive approaches to radiotherapy scheduling, in *World Congress on Engineering and Computer Science, San Francisco*, 2008.

Petrovic, S. et al. Algorithms for radiotherapy treatment booking, in *25th Workshop of the UK Planning and Scheduling Special Interest Group, Nottingham, UK*, 2006.

Pyzdek, T. *The Six Sigma Handbook: The Complete Guide for Green Belts, Black Belts, and Managers at All Levels*. New York: McGraw-Hill, 2003.

Reason, J. *Managing the Risks of Organizational Accidents*. Hampshire, England: Ashgate Publishing Ltd., 1997.

Schwarz, M. et al. (2001). Implementation of an integrated record and verify system for data and images in radiotherapy. *Tumori, 87*(1), 36–41.

Seel, M. & Foroudi, F. (2002). Waiting for radiation therapy: Does is matter? *Australas. Radiol., 46*(5), 275–279.

Siochi, R. A. et al. (2009). Radiation therapy plan checks in a paperless clinic. *J. Appl. Clin. Med. Phys., 10*(1), 43–62.

Song, X. et al. A case-based reasoning approach to dose planning in radiotherapy, in *The Seventh International Conference on Case-Based Reasoning, Belfast, Northern Ireland*, 2007.

Vaarkamp, J. et al. (2009). Managing workflow in treatment planning using standard spreadsheet software. *J. Radiother. Pract., 7*, 213–221.

van Herk, M. et al. (2000). The probability of correct target dosage: dose-population histograms for deriving treatment margins in radiotherapy. *Int. J. Radiat. Oncol. Biol. Phys., 47*, 1121–1135.

Williams, M. J. et al. (2007). Multicentre quality assurance of intensity-modulated radiation therapy plans: A precursor to clinical trials. *Australas. Radiol., 51*, 472–479.

Yan, D. (2008). Developing quality assurance process for image-guided adaptive radiation therapy. *Int. J. Radiat. Oncol. Biol. Phys., 71*(1), S28–S32.

26

Quality Assurance and the Informatics Environment

Collin D. Brack
University of Texas Medical Branch

Ivan L. Kessel
University of Texas Medical Branch

This chapter explores aspects of quality within medical informatics environments including the differentiators between Quality Assurance, Quality Control, Quality Improvement, and use cases in the context of Radiation Oncology and the ROIS. We evaluate how quality controlled data affect data workflows and provide specific examples describing a typical ROIS data QA process.

26.1 Background

In 1999, the Institute of Medicine's (IOM) published the report "To Err Is Human: Building a Safer Health System" (Kohn et al. 1999). The report detailed the prevalence of medical errors in the healthcare industry in the United States and suggested measures that should be taken to prevent them. They estimated that between 44,000 and 98,000 Americans were dying in hospitals each year as a result of medical errors. Even at the lower estimate of 44,000, deaths in hospitals due to medical errors exceeded the annual deaths attributable at the time to motor vehicle accidents (43,458), breast cancer (42,297), or AIDS (16,516). The recommendations of the report included the need for a greater emphasis by healthcare professionals to evaluate their current approaches and to create new systems to reduce the incidence of medical errors. This resulted in the adoption of more regimented Quality Assessment and Quality Management initiatives focused around incident reporting and reactive to medical errors and when they occurred. The objective was to understand why an incident occurred as well as measures to ensure such an error was less likely to occur in the future. Technology is central to these "quality initiatives," thanks to the advances in centralized data collection in the form of a hospital-wide electronic medical records (EMR) and specialty systems such as the

Radiation Oncology Information System (ROIS). At a high level, the quality process involves: recognizing and documenting the error, investigating the root cause of the incident, documenting the root cause, and discussing the findings at a formalized Quality Assurance (QA) meeting in which therapists, dosimetrists, physicists, nurses, physicians, and administrators review existing quality-based policies and guidelines to make operational changes to avoid repetition of the relevant error. There is, however, a move toward proactive risk prevention strategies to avoid errors, and it is within this context that informatics can have a critical role.

26.2 Classifying Quality: QA, Control, and Improvement

There are several terms that we use that describe various aspects of quality.

Quality Assurance (QA) refers to the planned and systematic activities that are implemented to ensure that the quality requirements for a service (Radiation Therapy) will be fulfilled. It involves the systematic measurement, comparison with a standard, monitoring of processes, and an associated feedback loop designed to ensure error prevention. This can be contrasted with quality control, which is focused on process outputs. Two guiding principles included in QA are "Fit for purpose," the principle that the management plan should be suitable and appropriate for the intended diagnosis, stage, etc., and "Right first time," which is the principle that mistakes should be prevented rather than requiring that they be detected and corrected. QA includes management of the quality of the entire process of oncologic management, including initial assessment, special investigations, and appropriate staging, that all prognostic information that could influence the plan of management is taken into account and documented, and that the radiation therapy plan conforms to established guidelines for dosimetry, both to the target volumes and the adjacent normal tissues.

Quality Control (QC) is a process involving ongoing review of the quality of radiation therapy planning and delivery. This approach places an emphasis on controls, assigning responsibility for different parts of the assessment, and planning and delivery of treatment-specific information. QC presupposes that measurable performance and integrity criteria have been defined to ensure accuracy of records. Part of this process requires ensuring that the appropriate team members have the necessary qualifications, competence, knowledge, skills, and experience. It also requires standardized naming conventions of the morphology (histology) and topography (anatomic site) of a diagnosis, such as the *International Classification of Diseases, Tenth Edition (ICD-10)* (World Health Organization 1992), to ensure that outcomes can be linked to the appropriate diagnosis, prescription, and anatomic site.

Quality Improvement (QI) involves the formalized analysis of performance and systematic efforts to improve it, by the prospective and retrospective evaluation of the entire radiation therapy process. The goal of QI is to design and implement improvements based on predefined quality benchmarks. It avoids attributing blame but rather creates an environment or system in which the risk of errors is minimized. There are many methods for achieving QI, including data quality reporting (exception reporting) and process improvement in which critical business processes, clinical processes, and data transfer process are captured and optimized around a quality-based metric.

Performance Improvement (PI) involves quantifying the output of a particular process or procedure and then exploring ways of modifying the process or procedure to improve the output, increase efficiency, or increase the efficacy of the process or procedure.

26.3 ROIS QA

We first differentiate between the aspects of quality because these quality descriptors are used broadly within the context of medicine to describe "clinical quality measures" and it is important for the Radiation Oncology informatics environment to be an extension of "clinical quality" and thus borrow from the same descriptors, definitions, and goals. The clinical definition of QA was described above as including "systematic measurement, comparison with a standard, monitoring of processes and an associated feedback loop designed to ensure error prevention." QA within the context of the ROIS first begins with Data Integrity—the primary tenant in data management that must be in place to trust that the underlying data are accurate. The second data management concept addresses the "comparison with a standard" aspect of clinical QA. The ROIS must codify or store a subset of the clinic data based on well-defined standards. These standards have been defined by industry or specialty standards bodies and are more important than the ROIS vendor's

underlying data schema. The concept of "process monitoring" is another data management concept in which a ROIS-derived clinical outcome measure is generated. The outcome measure can then be monitored manually via a custom ROIS report or the outcome measure report can run in an automated fashion in which upper and lower boundaries trigger an alert (E-mail, system message, etc). The "feedback loop" described in clinical QA is achieved within the ROIS by taking corrective action against bad or null (empty) data which can contribute to erroneous clinical data and can jeopardize the stability of the system itself. For example, the inability of the ROIS to calculate an outcome measure due to missing data is just as important as an alert designed to report against an outcomes measure that is out of tolerance.

26.4 Data Integrity

Data integrity is an attribute of data management that refers to the ability to maintain the exact contents of the information contained within the data over the lifecycle of data storage, transmission, and transformation. The Digital Imaging and Communications in Medicine (DICOM) standard (National Electrical Manufacturers Association 2001) defines data integrity as "the property that data has not been altered or destroyed in an unauthorized manner." For example, data archival systems must be able to store information with full data integrity to have the ability to retrieve the exact information. The term "lossless" is another qualitative data descriptor in which the exact source data are maintained with full data integrity upon the compression and un-compression of the source data, as in the common method of "zipping data," which uses the lossless compression algorithm Lempel–Ziv–Welch (LZW) compression (Welch 1984). Controlling for data integrity is especially important in the transmission of large data sets over networks subject to latency issues, transmission loss, and packet corruption. Consider the data integrity issues in the following example: A rural imaging clinic in Africa is attempting to send medical imaging data overseas to a radiologist in Canada. The medical imaging data set is a 1024 MB compressed binary "zip file" that contains 500 medical images. The rural imaging clinic is connected to the public Internet via satellite, subject to high latency and packet loss. The data set in this scenario must first be transferred with 100% transmission integrity, before it can be uncompressed (lossless) and viewed in its original form. Transmission integrity in this example could be accomplished by employing a data transfer method that incorporates "transfer resuming," which is supported by a number of secure file transfer protocol (SFTP) clients such as WinSCP (Přikryl 2012) and PuTTY (Tatham 2012).

When transferring mission critical binary data, regardless of size, a cyclic redundancy check (CRC) can be employed to ensure full data integrity. A CRC is a method of detecting for data errors in which the data contents are processed through a predefined calculation before data transmission and then recalculated upon data receipt to ensure the integrity of the transferred data. This calculated CRC check value must be transferred with the original data in order for the calculation to be performed a second time so that the CRC values can be compared and validated.

26.5 Data Integrity in Radiation Oncology

Within the context of the ROIS, data integrity measures must be employed to transfer treatment parameter data over electronic networks and between systems. For example, consider the transfer of "treatment files" with treatment parameter data from a Treatment Planning System (TPS) to a treatment delivery system, record & verify (R&V) system, or linear accelerator console. Treatment planning data can be encoded in a vendor-specific data format or in the medical imaging standard format, DICOM, in which the treatment plan data would be transferred and encoded as a "DICOM-RT Plan" file. Systems capable of creating, sending, and receiving DICOM data over electronic networks benefit from a base level of data integrity that is inherent in Transmission Control Protocol (TCP) networks as described by the Checksum features in the TCP segment structure (RFC 793: TCP 1981). However, TCP-based data integrity may not provide adequate error control for mission critical data transfer, especially in the presence of the transmission issues described earlier and large binary data sets, such as DICOM-RT files.

The DICOM standard does not have CRC-based error checking at the individual DICOM element or DICOM-RT file; the standard itself incorporates a level of data integrity in conjunction with the security layer known as the Transport Layer Security (TLS) layer, which is described in the *DICOM—Part 15: Security Profiles* publication (National Electrical Manufacturers Association 2001). However, both sending and receiving systems must support the TLS feature set in order for this additional layer of error correction to occur. The TLS layer would be an ideal solution for transferring data between institutions as it incorporates a security layer as well as an error checking layer; however, for interdepartment data transfer, for example from Radiology to Radiation Oncology within the same hospital, the TLS layer may not be active because the networks have been designed to only transfer data internally and have been optimized for speed (the security introduction of a layer of encryption slows data transfer speed).

The presence of any controlled data standard is by definition a welcome error-checking tool. The DICOM-RT standard provides this level of data integrity in the sense that the information within the digital data set is consistent, complete, and faithful to the DICOM standard. The majority of data integrity efforts should be built around the generation, transmission, and receipt of nonstandard radiotherapy data, for example, when transferring TPS data encoded in a proprietary format (Vendor A) to a treatment delivery system or R&V system (Vendor B). Data interoperability between vendors, absent of industry standards, must include a robust layer of error checking, data completeness checks (truncating), and sophisticated accuracy analysis to ensure tolerances. The DICOM-RT standard was designed to remove these burdens from the vendor's software development life cycle and thus minimize the likelihood of an adverse event related to data integrity.

26.6 Controlled Data Entry and QA Reporting

Data quality can be achieved by controlling data before they are entered into the ROIS (at the point of entry). Once the data have been entered, the primary QA mechanism becomes the data audit, whereby the logical integrity of the data is audited. The data audit can take the shape of a QA report highlighting critical data fields that are null or have been identified as outside normal controls. Data fields critical to patient care are typically programmatically controlled within the system architecture and at the database level. For example, some systems have been designed to not allow the entry of a known lethal value into a prescription data field. However, a Cancer Center or Hospital may have an internal data quality program that calls for even tighter data controls than those available at the system level. The first step is to discover the preprogrammed data value tolerances built into the system. Database field properties describing value tolerances should be available from the systems vendor in the form of a data dictionary or similar data schema documentation. Once the system controls are understood, additional internal controls from QA/QI programs can be codified in reporting software capable of querying the system and returning data fields that fall out of acceptable tolerances. Actively maintaining and updating these data limits in a separate system of record can be accomplished more easily within a database. A more robust "QA System" can take the form of a data mart or data warehouse consisting of a database containing these internal QA data limits that can then be referenced when reporting against the actual clinical data fields. The tolerances and limits that describe a custom QA measure should "live" (codified or stored) outside the QA report itself. This is a data management concept that will prevent stale reporting data in situations where the quality measure is changed in one QA report but not another. The goal is to have a central repository of QA measures, which can be modified in one location, and the changes propagate to all QA reports that rely on the measures. Industry standard reporting systems capable of centrally managing your own set of quality measures include Microsoft's SQL Server Reporting Services (SSRS) and Crystal Reports from Business Objects. Another broader category of database reporting, Business Intelligence (BI), is based on analysis and decision-making yet does not necessarily incorporate concepts of data warehousing. BI tools tend to be "read only" analytical views of data; therefore, when selecting a BI reporting platform, it is important to decide where to "write" the QA measures used to derive the QA report itself. Gartner publishes an annual "Magic Quadrant for Business Intelligence Platforms Report," which lists the main software vendors active in the BI software sector. The report is accessible online at http://www.microstrategy.com/company/gartnerquadrant.asp.

26.7 Clinical Case for Informatics-Based QA

Although an informatics system requires its own QA measures, once that is established, it is possible to use informatics tools to perform QA on the clinical processes within radiotherapy. This can remove or limit the effects of human error within a process. The importance of this has been highlighted by the attention of the media to radiotherapy errors.

In a series of articles in the *New York Times*, published in January 2010, several cases of patients suffering from major complications after receiving excessive doses of radiation therapy were described. These were due to errors in the programming of instructions for treatment delivery to the linear accelerators, failures in QA and calibration, or due to therapists ignoring warnings of machine error on their monitors. According to these reports, these errors lead to one man being treated for tongue cancer receiving excessive dose to his brain stem and a woman being treated for breast cancer receiving three times the prescribed radiation dose due to a field flattening filter not being activated. Some patients received excess radiation due to human error, some by machine error, but in most cases the medical team failed to detect the errors, despite warnings displayed on the treatment consoles. In all these cases, the clinical consequences of the error were avoidable. There were several opportunities for the error to be detected and the erroneous treatment to be aborted: if the plan had been verified on a phantom, the error in delivery would have been easily apparent; if the therapist had reacted to the error message on the screen, treatment could have been aborted and the error corrected. In addition, once the machine detected the problem, the default should have been not to allow treatment. None of these apparently occurred in these cases, at significant cost in morbidity and even mortality for the patients involved.

We are avoiding more adverse events through QA measures. One example of a systematic QI initiative has been illustrated by a simple experiment using a Surgical Checklist prior to operative procedures. The checklist consisted of 19 basic items, including the verification that the correct procedure is being performed on the appropriate patient, venous access is available as needed, and that all members of the team have been introduced and are in communication from before the procedure commences. The checklist also asks if prophylactic antibiotics have been given where indicated, appropriate measures taken if excessive blood loss anticipated, and after the procedure, the correct procedure and complications, if any, are documented, and that instrument and sponge counts are accurate. These simple steps, in a multinational study, were shown to reduce the perioperative mortality rate from 1.5% before the checklist was introduced to 0.8% afterward ($P = 0.003$). Inpatient complications were also reduced from 11.0% of patients at baseline to 7.0% after the introduction of the checklist ($P < 0.001$) (Haynes et al. 2009).

In the field of Radiation Oncology, there are multiple opportunities for achieving similar, or even greater, improvements in morbidity and mortality, especially considering the toxicity of cancer treatments. A simple checklist developed for use in Radiation Oncology could ensure that the correct patient and anatomic site is being treated, port films or image guidance have been verified, correct monitor units are set, concurrent systemic therapy (where indicated) is being given at the appropriate time, blood counts (where indicated) are adequate, and that the patient does not have any adverse symptoms or signs (such as a severe mucositis or skin reaction or diarrhea). Any red flag should result in a time-out to assess whether corrective action or alteration of the treatment plan or schedule needs to occur.

There are tremendous opportunities for improvements in the quality and safety of medical practice using information technology (Bates and Gawande 2003) by providing access to information, improving communication between members of the healthcare team, decision support, and monitoring. Safety can also be enhanced by requiring certain information before a treatment can proceed. Radiation Oncology practice is in a good position to benefit from such a process. There is a vast amount of information that is currently collected in the radiotherapy process, with more sophisticated digital imaging, treatment planning, information management (R&V) systems, and integrated image guidance systems. Informatics is helping to improve the efficiency and safety of advanced radiotherapy treatments. Multimodality images are frequently used to develop optimal treatment plans. We rely on robust networks to ensure the accurate transmission of digital images from the Picture Archiving and Communication System (PACS) to the treatment planning computer and treatment plan information between the treatment planning computer and linear accelerator. This presents many opportunities for the utilization of informatics tools to automate the QA of the process.

An example of such a system designed to verify treatment parameters and identify any data corruption was described by Kim (2007). When treating a patient, upon execution of the "open patient" instruction by the therapist at the treatment console, the program sampled the instructions regarding treatment parameters such as monitor units, leaf positions, and gantry positions and compared these with the previous fraction. Any variation could generate a warning that the instructions had been corrupted.

A further example of the use of informatics tools in improving radiation oncology is with portal imaging. Historically, portal images were obtained on film and compared with simulation films to verify patient setup positioning and field location. Many linear accelerators today have onboard digital portal imaging that permits the acquisition of digital images of the patient setup. These can be compared with digitally reconstructed images generated from the planning computed tomography (CT) data set, both visually and by using automated comparison programs. More sophisticated image guidance systems, based on cone beam CT or on the detection and localization of fiducial seeds implanted into the tumor, can further improve the accuracy of treatment. This can reduce errors of setup and compensate for internal motion. This is especially important in prostate intensity-modulated radiation therapy (IMRT), where small margins can be used around the target if localization is precise. Information can be obtained regarding the position of the target, either from cone beam CT or orthogonal films displaying the location of the fiducial seed, can be analyzed, correlated with plan data, and generate precise adjustments in couch position prior to treatment. There is potential for further refinement of this process, with visualization during treatment, to account for intrafraction movement (i.e., while the patient is on the linear accelerator receiving a fraction of treatment). This would require software to rapidly acquire the position data, to infer the required adjustments, and to transmit those instructions to the linear accelerator controls, for adjustment of couch or field position.

The IT infrastructure of the radiation oncology department, particularly with the trend toward paperless medical records, lends itself to the introduction of QI protocols within the system to reduce errors and improve safety. The clinical data that are entered could be crosschecked against other entries in the EMR to ensure consistency and accuracy. QA methods can be utilized not only to ensure accuracy in billing but also to improve clinical outcomes by verifying that diagnostic workup is appropriate and treatment planning and delivery standards are maintained. For example, the use of interactive screens for data input during clinical encounters provides a layer of data quality and can ensure that the appropriate questions are not missed during the patient interview and that significant findings are not missed. If a patient were to complain of chest pain, the screen could be populated with questions to help differentiate between cardiac and noncardiac causes of that pain, which may lead to differing paths of investigation and treatment. Once the clinical encounter is completed, a checklist specific to the oncologic diagnosis (as defined by the *ICD-10* topography and morphology codes and by the stage) can be generated to ensure that the appropriate special investigations have been performed, prognostic information has been extracted from those investigations, and staging has been performed. Treatment recommendations can then be input and crosschecked with published guidelines.

The utility of this program can only be as good as the quality and accuracy of the data entered. Errors can occur if questions are omitted or if incorrect data are entered. For example, the pathologic diagnosis could be confirmed by interrogating the pathology report. The treatment plan can be interrogated to ensure that it matches all the objectives. The plan that is delivered to the R&V system and to the treatment console should be verified to ensure that data are not corrupted in the transfer. Finally, the machine parameters can be verified to ensure that they correspond to the parameters suggested by the treatment plan. Finally, with all the data collected, there is an opportunity for data mining to identify potential sources of errors, especially near misses (where an error is averted by human intervention), to help develop improved protocols that would obviate the need of that corrective action in the future.

Data mining is a key component in the use of informatics to improve radiation therapy, using a process known as Business Intelligence, defined by Luhn as "the ability to apprehend the interrelationships of presented facts in such a way as to guide action toward a desired goal" (Luhn 1958).

The concept of data mining is used for knowledge discovery by looking for patterns in the data. It does not involve posing specific questions (such as the incidence of a disease in a given population), but rather uses a technique asking more generic questions, with multiple queries, and using algorithms to analyze results (and frequently to generate further questions). The challenge in extracting meaningful information from data mining is that the patterns may be hidden in massive amounts of data. We therefore need to have or develop the appropriate algorithms to uncover the associations. Before we can analyze the data, it needs to be prepared (e.g., selection of attributes and classifiers). Variables can be classified using predictive labeling (e.g., "high risk," "intermediate risk," and "low risk") and then a process of regression analysis (e.g., using neural nets) can used to generate a numerical prediction. An alternate method is to cluster data sets and develop rules to account for the associations observed. One attribute of the data is then characterized as the classifier (e.g., level of risk). A training set of these previously classified patients is analyzed based on a variety of other attributes, and algorithms are used to find the combination of attributes most likely to yield a given classification.

An example of the application of data mining is described in a paper by Razavi et al. (2008). They used not only a decision tree induction to classify cases, but also a root cause analysis of data acquired by data mining and by manual inspection of patient records, to assess noncompliance with postmastectomy radiotherapy guidelines in a group of patients treated for breast cancer. They demonstrated that data mining could be used as a medical audit to highlight noncompliance patterns. This can be valuable for ensuring future adherence to appropriate standards of care and also, using feedback from data mining, to make modifications and improve the design of those guidelines.

The major challenge in this approach is ensuring the accuracy of the contours of target volumes and avoidance structures. There are several software programs based on contour libraries that can be used to assist in this process, but currently none of these replace the expert eye of the experienced clinician. Manual contouring, especially for head-and-neck cancer IMRT, is labor intensive and time consuming and could be automated using informatics techniques of autosegmentation with comparison with reference atlases. The potential for this approach has been demonstrated by Stapleford et al. (2010), who were able to achieve a time savings of 68% to 87% using atlas-based segmentation for head-and-neck cancer from a single best-matched atlas subject. Their technique was to create a 20 subject head-and-neck cancer atlas containing targets and normal structures. Each atlas-subject contained lymph level targets (Levels I–VI), manually defined according to Radiation Therapy Oncology Group (RTOG)/European Organization for Research and Treatment of Cancer (EORTC) guidelines, and normal structures. They then compared with the contours generated using up to five different atlases to manually generated contours. They then calculated an average Dice Similarity Coefficient (DSC) for seven key structures, including the neck levels, mandible, left and right parotid, larynx, spinal cord, and brainstem. They found their best correlation across all the contours was achieved when five atlases were used (average DSC of 0.762 ± 0.093). If four atlases were used, the average DSC was 0.761 ± 0.099, the average DSC was 0.748 ± 0.11 for three atlases, and the single best atlas achieved a DSC of 0.710 ± 0.134. The process took 47 ± 7 seconds to segment using a single atlas.

A further example of potential utility of informatics techniques is in the evaluation of treatment plans. Within the treatment planning paradigm, QA is easier to design, as most parameters can be assessed automatically to ensure that they conform to the treatment objective to deliver an appropriate dose of radiation therapy to the targets while limiting the dose to adjacent critical structures to within the predefined dose tolerances of those organs or tissues. Further, the appropriateness of the objectives entered can easily be defined and verified based on established guidelines such as the National Comprehensive Cancer Network (NCCN) for treatment objectives and the Quantitative Analysis of Normal Tissue Effects in the Clinic (QUANTEC) (Bentzen et al. 2010) for normal tissue tolerance. Automated tools can be used to analyze whether the plans meet all the objectives regarding target coverage with a prescribed dose as well as the doses delivered to adjacent organs at risk. Currently, plan evaluation is largely conducted by personal evaluation, but it would be beneficial to have analytic tools to evaluate the plan and to highlight sources of concern.

A demonstration project to detect errors in radiation therapy has been described by Azmandian et al. (2007). They propose using machine learning and data mining techniques to help detect major human errors in radiotherapy treatment plans as a complement to systems of manual inspection. One such technique they suggest is using clustering algorithms for outlier detection. The data from a large number of patient treatment plans are clustered based on the treatment parameters. Then, while checking a new treatment plan, the parameters of the plan will be tested to establish whether or not they belong to one of the existing clusters. If the plan does not appear to fit in with an established cluster, they designated that plan as an "outlier" and identified that plan as requiring further attention of the human chart checkers. They verified this approach, analyzing 1000 prostate cancer four field box plans, using the K-means clustering algorithm to generate the learning set clusters, which they tested by analyzing a further 650 plans. They identified eight distinct clusters. They were able to demonstrate a very high sensitivity (77%–100% depending on the parameter and error level used) and specificity (90%) in detection of errors using this technique. This preliminary work demonstrates the promise for development of automatic outlier detection software.

Radiation Oncology departments collect large volumes of information in their EMRs and R&V systems. They have, however, been slow to adopt tools that may help analyze and improve their operations. This large database could be used to improve safety and quality at every stage. Through a process of data mining, text mining, and online analytic processing, and comparing the information to benchmarks, process deviation events could be identified. These could be analyzed to try and find association

rules identifying those interactions that are more prone to error. This could not only result in improvements in quality and safety but could also be used to reduce costs and as an aid to research, especially comparative effectiveness research.

The process of radiation therapy starts with planning, which relies on the importation of images from the CT scanner, and often also requires the importation and fusion of diagnostic images [CT, positron emission tomography (PET), or magnetic resonance imaging (MRI) images] to the simulation images. These diagnostic images may have been obtained with the patient lying in a different position than the simulation/treatment position. The process of fusion requires 3D geometric alignment that may require translation, rotation, and even deformation of the diagnostic image to try and approximate the simulation position and thus allow registration of the two image sets. Once this has been achieved, the targets and avoidance structures need to be contoured, using the definitions of target volumes proposed by the International Commission on Radiation Units and Measurements (ICRU) (ICRU Report 62 1999). They define the following volumes: gross tumor volume (GTV), clinical target volume (CTV), and planning target volume (PTV). The GTV is the part of the tumor that is visible on 3D imaging. The volume delineated is dependent on the imaging modality utilized and the quality of data acquisition. The most relevant volume is the CTV that includes the GTV as well as estimates of subclinical or microscopic spread, which are not visible with current imaging technology, and requires knowledge and inferences related to patterns of spread. The process of contouring the GTV can be assisted by automated programs that outline structures based on density interfaces (such as the difference between the Hounsfield number of lung and soft tissue). The process is not exact and still requires a subjective assessment. This results in uncertainty in the accuracy of the delineation of those structures and targets, and a large potential source of error in the treatment planning process, as a large interobserver variation in target definition, has been demonstrated (Hong et al. 2004; Villeirs et al. 2005; Njeh 2008; Moore 2009). The process can be improved by using a library of shapes of normal structures, and atlases of nodal beds, for reference comparison when outlining the target, organ-at-risk, or draining uninvolved nodes. Informatics tools can be used to quantify the extent of this uncertainty, by anatomic site and by number and type of imaging modalities used. Quantitative metric derivation has been described to assess the accuracy of an individual contouring, by comparing the entered contour to an expert atlas, or to assess consistency based on repeated contouring by the same individual (Moore et al. 2012). Once the contours have been entered, uncertainties need to be accounted for in designing PTVs for the targets and organs-at-risk. There are explicit guidelines for applying geometric margins around targets and organs-at-risk to account for setup error and internal organ motion.

This whole process of correlation of outcomes with variables depends on the quality of the data entered in the database. While we are capturing a wide array of data for each patient that we treat, including laboratory results, images, treatment plans, these are stored in disparate databases and therefore are not routinely analyzed in an integrated manner. Integrating all these data into a single EMR could facilitate this analysis and improve the quality and safety of patient management. In addition, we tend to discard potentially useful data (such as discarded plans) that would help us understand the decision-making process and factors that may influence the process, such as experience. Information that is not captured systematically cannot improve patient care.

References

Azmandian, F., Kaeli, D., Dy, J. G. et al. (2007). Towards the development of an error checker for radiotherapy treatment plans: A preliminary study. *Physics in Medicine and Biology, 52*, 6511.

Bates, D. W. & Gawande A. A. (2003). Improving safety with information technology. *New England Journal Medicine, 348*(25), 2526–2534.

Bentzen, S. M., Constine, L. S., Deasy, J. O. et al. (2010). Quantitative analyses of normal tissue effects in the clinic (QUANTEC): An introduction to the scientific issues. *International Journal of Radiation Oncology Biology Physics, 76*, S3–S9.

Haynes, A. B., Weiser, T. G., Berry, W. R. et al. (2009). A surgical safety checklist to reduce morbidity and mortality in a global population. *New England Journal of Medicine, 360*, 491–499.

Hong, T., Tome, W., Chappell, R. et al. (2004). Variations in target delineation for head and neck IMRT: An international multi-institutional study. *International Journal of Radiation Oncology Biology Physics, 60*, S157–S158.

International Commission on Radiation Units. *Prescribing, Recording, and Reporting Electron Beam Therapy.* Oxford University Press, 2004.

Kim, D. Y. (2007). Just-in-time quality assurance auditing of radiotherapy control informatics. *International Journal of Radiation Oncology Biology Physics, 69*(3), S566.

Kohn, L. T., Corrigan, J. M., and Donaldson, M. S. *To Err Is Human. Building a Safer Health System.* Washington: National Academy Press, 1999.

Luhn, H. P. (1958). A business intelligence system. *IBM Journal of Research and Development, 2*, 314–319.

Moore, K. (2009). SU-FF-I-87: DTA-based metrics for the evaluation of autosegmentation algorithms in clinical radiotherapy workflow. *Medical Physics, 36*, 2454–2454.

Moore, K. L., Brame, R. S., Low, D. A. et al. (2012). Quantitative metrics for assessing plan quality. *Seminars in Radiation Oncology, 22*(1), 62–69.

National Electrical Manufacturers Association. *Digital Imaging and Communications in Medicine (DICOM)—Part 15: Security Profiles.* Virginia: National Electrical Manufacturers Association, 2001.

Njeh, C. (2008). Tumor delineation: The weakest link in the search for accuracy in radiotherapy. *Journal of Medical Physics/Association of Medical Physicists of India, 33*, 136.

Postel, J. (2003). RFC 793: Transmission control protocol, September 1981. Status: Standard.

Přikryl, M. (2012). WinSCP. http://WinSCP.net. Accessed May 7, 2013.

Razavi, A. R., Gill, H., Ahlfeldt, H. et al. (2008). Non-compliance with a postmastectomy radiotherapy guideline: Decision tree and cause analysis. *BMC Medical Informatics and Decision Making, 8*.

Stapleford, L. J., Lawson, J. D., Perkins, C. et al. (2010). Evaluation of automatic atlas-based lymph node segmentation for head-and-neck cancer. *International Journal of Radiation Oncology Biology Physics, 77*, 959–966.

Tatham, S. (2012). PuTTY. http://bitflop.com/document/80.

Villeirs, G. M., Van Vaerenbergh, K., Vakaet, L. et al. (2005). Interobserver delineation variation using CT versus combined CT+MRI in intensity-modulated radiotherapy for prostate cancer. *Strahlentherapie und Onkologie, 181*, 424–430.

Welch, T. (1984). Technique for high-performance data compression. *Computer, 17*, 8–19.

World Health Organization, *International Statistical Classification of Diseases and Related Health Problems, Tenth Revision.* World Health Organization, 1992.

Index

Page numbers f and t indicate figures and tables, respectively.